INTERNATIONAL CHRONOSTRATIGRAPHIC CHART

www.stratigraphy.org — **International Commission on Stratigraphy** — **v 2013/01**

IUGS

Eonothem / Eon	Erathem / Era	System / Period	Series / Epoch	Stage / Age	numerical age (Ma)
					present
Phanerozoic	Cenozoic	Quaternary	Holocene		0.0117
			Pleistocene	Upper	0.126
				Middle	0.781
				Calabrian	1.806
				Gelasian	2.588
		Neogene	Pliocene	Piacenzian	3.600
				Zanclean	5.333
			Miocene	Messinian	7.246
				Tortonian	11.62
				Serravallian	13.82
				Langhian	15.97
				Burdigalian	20.44
				Aquitanian	23.03
		Paleogene	Oligocene	Chattian	28.1
				Rupelian	33.9
			Eocene	Priabonian	38.0
				Bartonian	41.3
				Lutetian	47.8
				Ypresian	56.0
			Paleocene	Thanetian	59.2
				Selandian	61.6
				Danian	66.0
	Mesozoic	Cretaceous	Upper	Maastrichtian	72.1 ±0.2
				Campanian	83.6 ±0.2
				Santonian	86.3 ±0.5
				Coniacian	89.8 ±0.3
				Turonian	93.9
				Cenomanian	100.5
			Lower	Albian	~ 113.0
				Aptian	~ 125.0
				Barremian	~ 129.4
				Hauterivian	~ 132.9
				Valanginian	~ 139.8
				Berriasian	~ 145.0

Eonothem / Eon	Erathem / Era	System / Period	Series / Epoch	Stage / Age	numerical age (Ma)
					~ 145.0
Phanerozoic	Mesozoic	Jurassic	Upper	Tithonian	152.1 ±0.9
				Kimmeridgian	157.3 ±1.0
				Oxfordian	163.5 ±1.0
			Middle	Callovian	166.1 ±1.2
				Bathonian	168.3 ±1.3
				Bajocian	170.3 ±1.4
				Aalenian	174.1 ±1.0
			Lower	Toarcian	182.7 ±0.7
				Pliensbachian	190.8 ±1.0
				Sinemurian	199.3 ±0.3
				Hettangian	201.3 ±0.2
		Triassic	Upper	Rhaetian	~ 208.5
				Norian	~ 227
				Carnian	~ 237
			Middle	Ladinian	~ 242
				Anisian	247.2
			Lower	Olenekian	251.2
				Induan	252.17 ±0.06
	Paleozoic	Permian	Lopingian	Changhsingian	254.14 ±0.07
				Wuchiapingian	259.8 ±0.4
			Guadalupian	Capitanian	265.1 ±0.4
				Wordian	268.8 ±0.5
				Roadian	272.3 ±0.5
			Cisuralian	Kungurian	283.5 ±0.6
				Artinskian	290.1 ±0.26
				Sakmarian	295.0 ±0.18
				Asselian	298.9 ±0.15
		Carboniferous — Pennsylvanian	Upper	Gzhelian	303.7 ±0.1
				Kasimovian	307.0 ±0.1
			Middle	Moscovian	315.2 ±0.2
			Lower	Bashkirian	323.2 ±0.4
		Carboniferous — Mississippian	Upper	Serpukhovian	330.9 ±0.2
			Middle	Visean	346.7 ±0.4
			Lower	Tournaisian	358.9 ±0.4

Eonothem / Eon	Erathem / Era	System / Period	Series / Epoch	Stage / Age	numerical age (Ma)
					358.9 ±0.4
Phanerozoic	Paleozoic	Devonian	Upper	Famennian	372.2 ±1.6
				Frasnian	382.7 ±1.6
			Middle	Givetian	387.7 ±0.8
				Eifelian	393.3 ±1.2
			Lower	Emsian	407.6 ±2.6
				Pragian	410.8 ±2.8
				Lochkovian	419.2 ±3.2
		Silurian	Pridoli		423.0 ±2.3
			Ludlow	Ludfordian	425.6 ±0.9
				Gorstian	427.4 ±0.5
			Wenlock	Homerian	430.5 ±0.7
				Sheinwoodian	433.4 ±0.8
			Llandovery	Telychian	438.5 ±1.1
				Aeronian	440.8 ±1.2
				Rhuddanian	443.4 ±1.5
		Ordovician	Upper	Hirnantian	445.2 ±1.4
				Katian	453.0 ±0.7
				Sandbian	458.4 ±0.9
			Middle	Darriwilian	467.3 ±1.1
				Dapingian	470.0 ±1.4
			Lower	Floian	477.7 ±1.4
				Tremadocian	485.4 ±1.9
		Cambrian	Furongian	Stage 10	~ 489.5
				Jiangshanian	~ 494
				Paibian	~ 497
			Series 3	Guzhangian	~ 500.5
				Drumian	~ 504.5
				Stage 5	~ 509
			Series 2	Stage 4	~ 514
				Stage 3	~ 521
			Terreneuvian	Stage 2	~ 529
				Fortunian	541.0 ±1.0

Eonothem / Eon		Erathem / Era	System / Period	numerical age (Ma)
				541.0 ±1.0
Precambrian	Proterozoic	Neoproterozoic	Ediacaran	~ 635
			Cryogenian	850
			Tonian	1000
		Mesoproterozoic	Stenian	1200
			Ectasian	1400
			Calymmian	1600
		Paleoproterozoic	Statherian	1800
			Orosirian	2050
			Rhyacian	2300
			Siderian	2500
	Archean	Neoarchean		2800
		Mesoarchean		3200
		Paleoarchean		3600
		Eoarchean		4000
	Hadean			~ 4600

Units of all ranks are in the process of being defined by Global Boundary Stratotype Section and Points (GSSP) for their lower boundaries, including those of the Archean and Proterozoic, long defined by Global Standard Stratigraphic Ages (GSSA). Charts and detailed information on ratified GSSPs are available at the website http://www.stratigraphy.org. The URL to this chart is found below.

Numerical ages are subject to revision and do not define units in the Phanerozoic and the Ediacaran; only GSSPs do. For boundaries in the Phanerozoic without ratified GSSPs or without constrained numerical ages, an approximate numerical age (~) is provided.

Numerical ages for all systems except Permian, Triassic, Cretaceous and Precambrian are taken from 'A Geologic Time Scale 2012' by Gradstein et al. (2012); those for the Permian, Triassic and Cretaceous were provided by the relevant ICS subcommissions.

Coloring follows the Commission for the Geological Map of the World. http://www.ccgm.org

Chart drafted by K.M. Cohen, S. Finney, P.L. Gibbard

CCGM CGMW

http://www.stratigraphy.org/ICSchart/ChronostratChart2013-01.pdf

This book is dedicated to Klaus Vogel, the father of modern small Earth modelling studies.

On the Origin of Continents and Oceans:

A Paradigm Shift in Understanding

Dr James Maxlow

And some rin up hill and down dale,
Knapping the chunky stanes to pieces wi' hammers,
Like sae many road makers run daft.
They say it is to see how the warld was made.

Sir Walter Scott, St. Ronan's Well. 1824

Terrella Press 2014

First published in Australia 2014 by Terrella Press, Perth, Western Australia.

ISBN 978-0-9925652-0-6

Photography by Anita and James Maxlow.

Graphics by Anita Maxlow of Terrella Press.

Permission was granted by the Commission de la Carte Geologique du Monda, Paris to use and digitise the *Geological Map of the World at 1:25,000,000 M scale (1990)*, published by the Commission for the Geological Map of the World and UNESCO © (1990).

Digitising of the Geological Map of the World (1990) © was originally carried out by Simon Brown of Geoviz International.

Front cover and portrait photograph by Norman Bailey.

Inside front cover: International Chronostratigraphic Chart (International Commission on Stratigraphy, 2013).

Inside back cover: 24-gore equal area sinusoidal bedrock geological map projections based on the *Geological Map of the World* © (1990). Enlarge these to make your own present-day geological globe of the world.

About the Author

James Maxlow was born in Middlesbrough, England in May 1949, exactly 23 years after the birth of his good friend and fellow expansionist Klaus Vogel. James' passion for geology was inherited from a family history of "ironstone workers" supplying iron ores mined from the Eston Ironstone Mine to the foundries and steel rolling mills of Middlesbrough during the 1800s to mid-1900s.

James immigrated to Australia with his parents in 1953, where he grew up in Melbourne. He studied Civil Engineering at the then Swinburne College, but soon became disillusioned with engineering and redirected himself to a degree in geology at the Royal Melbourne Institute of Technology, graduating in 1971. It was in Melbourne where he later met and married his lovely wife Anita and during their work and travels throughout Australia they had three children, Jason, Karena, and Jarred.

James' interest in the expanding Earth theory stems from working in the Pilbara region of Western Australia during the late 1970s where he first read the book *"The Expanding Earth"* written by Professor Samuel Warren Carey. The Pilbara region is a huge, Precambrian domal structure, several hundreds of kilometres across. It occurred to James that this relatively undisturbed domal structure may have been a fragment of a much smaller radius ancient Earth.

James gained his Master of Science in geology in 1995, followed by a Doctorate of Philosophy in 2002 at Curtin University of Technology, Perth, Western Australia, including a letter of commendation from the university Chancellor for thought provoking research into Expansion Tectonics.

During his academic years James met and communicated with many wonderful "expansionists" from around the world. Most notable of which was the late Professor Samuel Warren Carey from Tasmania, the father of modern Earth Expansion, Jan Koziar from Poland who was the first scientist to measure and calculate the ancient Earth radius, and Klaus Vogel from Germany, the father of modern Expanding Earth modelling studies. It was during James' academic studies that Professor Carey recognized the potential of his research into Expansion Tectonics. Carey then kindly *"passed on his Expanding Earth baton"* to James, an honour that James deeply cherishes.

Contents

Preface

"The main cause of intellectual stagnation [is] the herd instinct which causes many scientists to stay with mainstream ideas and concepts rather than risk alienating their peers by investigating controversial new ideas. The peer review system thus tends to maintain the intellectual status quo by rejecting new ideas." Gold, 2009

The title of this book is an intentional play on words, capturing the essence of famous publications by Charles Darwin and Alfred Wegener. Charles Darwin, with his book *On the Origin of Species* is famous for his painstaking research and contribution towards natural selection and the evolution of species. Alfred Wegener, with his book *The Origins of Continents and Oceans* is less well known but, with his work on Continental Drift, is now credited for giving rise to the modern theory of Plate Tectonics.

The significance of both of these scientists is that, rather than simply following mainstream science prevailing during their times, they were both prepared to go into the world and gather new factual evidence and observations prior to making their own now famous contributions to their respective fields of science. Whether this was intentional or simply opportunistic is uncertain, however, in both cases their worldly observations were shown to contradict what mainstream science existing at the time was insisting. Fortunately, even though both were initially ridiculed and rejected, they chose to ignore prevailing opinions and peer-pressures and instead focused on continuing their methodical research. They then chose to present and publish their own factual evidence and observations as they saw and understood it.

As the introductory caption above suggests, this book is about looking beyond conventional or mainstream opinions and concepts about the origin of the continents and oceans. In this book I will be focusing on and using freely available modern scientific information and evidence gathered about the Earth, but the intent is to look at this evidence from a new and somewhat controversial tectonic perspective. This world has enough tectonic theories the reader may say, so why introduce or even contemplate others, in particular theories that challenge the fundamental assumptions and acceptances of today's mainstream Plate Tectonic theory. To me, the answer to this statement is quite simple:

There is an unrecognized fundamental flaw in Plate Tectonic thinking whereby the size of the Earth is assumed, without query or question, to be constant throughout time.

As will be rigorously demonstrated throughout this book, the available modern scientific evidence shows that this fundamental Plate Tectonic assumption is now outdated and the evidence is not consistent with, nor does it continue to substantiate conventional Plate Tectonic theory. This modern evidence is instead being unconsciously forced into a stereotypical Plate Tectonic mold simply because there has been no viable alternative theory available to convincingly challenge the stranglehold that Plate Tectonics has on science today.

My first recollection of anything to do with *tectonics*—a field of study within geology—was during my late teens. There was something on television about how the world was shrinking and in my wisdom I proceeded to enlighten the rest of my family by saying that it was because the Earth was cooling and getting smaller. Hence, not only was the world getting smaller but wrinkling of the crust was making the mountains grow taller. My father quickly chastised me and said no, it was because, with the advent of modern travel and communications, the world was perceived to be getting smaller, not actually getting smaller. Sufficient to say though, this book is definitely not about proposing a cooling or shrinking Earth, and certainly not about modern travel or communications.

My late teens also coincided with the late 1960s when I was doing an undergraduate course in geology. At that time there was great debate and open discussion on a wide range of competing theories about the origin of continents and oceans, along with a multitude of others that have long since been abandoned—including the shrinking Earth theory. To this day I cannot recollect studying Plate Tectonics during my undergraduate years, only basic mention of land bridges connecting the continents, probably some mention of Continental Drift, and maybe brief mention of other, now largely historical theories. Being the late 1960s, Plate Tectonics was only just being established as a theorem so I can only assume my professors at the time did not subscribe to this new theory so did not teach it.

As most people know and appreciate, Plate Tectonics is now firmly established in science today and is currently touted as having revolutionized modern scientific thinking. Plate Tectonic theory is generally considered to be ranked with the best of revolutionary theories, including Charles Darwin's Theory of Evolution and Einstein's Theory of Relativity. It is also widely accepted throughout most countries and is taught almost religiously in modern school curriculum, even to the point where every pre-school child, it seems, is subject to the basics of the theory by way of children's books and animated videos so prevalent in the popular media.

When fostering and promoting new theories, such as when Plate Tectonics was first proposed during the 1960s for example, it should be realized that certain assumptions often have to be made in order to overcome any limitations to the concepts and ideas that were not fully understood at the time the theory was first promoted. Over time, true science insists that we must then continually challenge these assumptions until such time as the original concepts and ideas are indeed fully understood and can be confidently accepted within the discipline as truth. If time shows that these basic assumptions are incorrect, true science also insists that the particular theory or discipline being promoted must either be modified or rejected in favour of a theory that better suites the available evidence.

Unfortunately, some theories do slip through this net. The Earth-centric theory of the perceived universe lasted for some 1,500 years up to the middle ages and is a good historical example. The primary reason why this theory persisted for so long was because of pious religious insistence that man—and hence the Earth—must be located at the centre of the known universe and therefore everything must revolve around a stationary Earth. What this lesson should have shown is that because of unforeseen, blinkered, or simply unrecognized flaws in the basic assumptions and insistences, many theories can be perpetuated well beyond their usefulness has expired.

It is also an unfortunate trait of humankind that, as well as making fundamental assumptions, our concepts, ideas, and thought processes are often unknowingly led and misguided by established and generally unrecognized memes. A *meme* is defined as *an idea, behavior, or style that spreads from person to person within a culture*. As such, a meme acts as a mechanism for promoting and maintaining cultural ideas, symbols, or practices, which can be transmitted from one person to another through writing, speech, gestures, rituals, or other imitable means.

What I mean by this is that most, if not all of us have pre-conceived ideas and notions embedded in our minds that defy thinking about or challenging. These are generally accepted as basic, fundamental truths. In most cases we are not even aware or conscious of the memes we use. As an example, essentially all humans unknowingly accept the meme that the size of the Earth has always been constant, or near constant since its creation. At no point does science challenge this meme, it is simply accepted as a truth. All assumptions made towards promoting new theories about the Earth are therefore unknowingly influenced by this underlying meme. The popular concept of a *paradigm* is now frequently used as a synonym for a meme, with a *paradigm shift* referring, in essence, to a change in which memes are still recognized, promoted, or used.

A serious challenger to Plate Tectonic theory during the 1950s and 1960s was the Expanding Earth theory, promoted by a number of researchers but most notably by the late Professor Samuel Warren Carey, Emeritus Professor of Geology at the University of Tasmania. The fundamental difference between Plate Tectonics and Earth Expansion was, and still is, that Plate Tectonics assumes—based on the underlying constant Earth size meme—that the radius of the Earth has remained constant or near constant throughout time. In contrast, the Expanding Earth theory challenges this meme by removing the constant Earth radius assumption and allows Earth radius to change with time.

While this understanding is generally not common knowledge, what should be realized is when these two theories were first developed there was no real evidence to properly support, quantify, or promote either theory. Regardless of this, for those who still have some knowledge of the mid- to late-1900s, Plate Tectonics is now seen to have majestically thwarted the Expanding Earth theory, to the point where very little is known about the latter theory today. Most people are now oblivious to the fact that this alternative theory still exists, or that the Earth can indeed potentially change in size.

As may or may not be appreciated, challenging Plate Tectonics in any shape, form, or format, is an almost impossible task. As noted by Gold in 2009, "*The peer review system tends to maintain the intellectual status quo by rejecting new ideas.*" What this simply means is that there are enormous amounts of research funds that are currently available for Plate Tectonic research from all over the world. The last thing these researchers would want is for their hard earned efforts to be shown to be wrong, and in particular for their academic or even international reputations to be tainted. In order to retain their hard earned reputations, as well as control the outcomes of any outside research, efforts to publish new ideas in so called respected publications must be *peer reviewed*. Because this peer review process is dominated by advocates of Plate Tectonic theory, rejection of alternative ideas or outcomes is essentially guaranteed because of the ingrained scientific consensus of opinion.

Scientific consensus of opinion is a *collective judgment, position, or opinion of a community of scientists in a particular field of study*. This consensus implies that there is general agreement amongst the various scientists, though not necessarily unanimity. It should be appreciated that scientific consensus is not in itself a scientific argument, and is not part of the established scientific method, however, consensus of opinion is generally based on a collection of both scientific arguments and scientific methods adopted by the scientific community.

As is also generally not fully appreciated, the modelling of modern scientific evidence using Plate Tectonic theory is essentially restricted to about the past 500 million years or so of past geological history. Because of the limitations imposed by the basic assumptions of conventional plate theory, modelling further back in time can only be speculated on prior to that time. I preempt myself here, but in this book this same modern evidence, collected primarily in support of Plate Tectonic theory, will be used to recreate the entire 4,000 million years of Earth's known geological history. Not cartoon schematics of fictitious past continental assemblages, but accurate reconstructions of precise continental locations. Accurate enough for anyone to know precisely where their particular project, mineral discovery, climate change indicator, pet project, or fossil find was at any moment in history. In addition, and more importantly, this modelling will then be used as a basis to investigate how all of this evidence relates globally in both time and space.

Investigating this modern global evidence on anything but a constant radius Earth has never been done before. Because it has never been done before, scientists have, in essence, been deprived of a valuable scientific basis to thoroughly test or independently evaluate the merits of this modern evidence. The reason why this investigation and modelling has never been done before is simply because our culture has been stereotyped into firmly believing that Earth radius has always been the same size as it is today—remember the meme. Because of this, Plate Tectonics then insists that continental and seafloor crustal development is a random, non-predictive, and somewhat catastrophic process.

So, imagine a tectonic process whereby the formation and break-up of the continents, as well as the sympathetic opening of each of the oceans is progressive, predictive, and evolutionary. All ancient magnetic north and south poles are precisely located, and the established poles and equator coincide fully with observed climate zonation and plant and animal species development. Plant and animal species evolution is then intimately related to continental development and changes to climate zonation, and global extinction events coincide with naturally occurring changes in coastal distributions, sea-levels, and climate.

Plate Tectonics cannot do this, even for evidence reflecting the more recent times. So why does science continue to support this theory? Because, as I will reiterate time and again, the established peer review system applied to Plate Tectonic research and publication does not allow for, nor does it encourage the acknowledgement or investigation of alternative considerations, in particular if these alternative considerations have been rejected in the past. Why? Here is an example.

Popular, albeit misinformed, opinion decrees that the concept of an Earth increasing its radius with time has been adequately tested, and hence adequately rejected. The main reason for not adopting an Expanding Earth concept in the past was because of the lack of a suitable mechanism to explain an increase in Earth radius. As will be shown in the following chapters, even though there is now a considerable amount of empirical evidence to quantify an increase in Earth radius, historically, it is still not fully understood why this is occurring, hence the continued rejection. This is how science appears to work; the concept is tested, and if it does not stand up to the peer review system, it is rejected—seemingly forever.

Unfortunately, science is littered with theories and concepts that were once considered to be adequately tested and adequately rejected only to be later resurrected as a revolutionary new theorem. Continental Drift, for instance, was chastised and rejected for many years only to be resurrected as Plate Tectonics. This acceptance of Plate Tectonics came about because new geophysical evidence became available and static ideas prevailing at the time about the formation of the Earth and movement of the continents were very much relaxed.

In practice, the theory of Earth Expansion has indeed been adequately rejected to the satisfaction of a scientific community constrained by a constant radius Earth pre-conception. However, I can assure the reader, like Continental Drift was many years ago, the concept of an Earth increasing its radius with time has been quietly waiting while scientists from around the world spend millions of dollars on research funded projects accumulating a vast array of modern scientific evidence. Modern technology and space exploration has also been advancing in leaps and bounds, keeping pace with our insatiable appetite for acquiring, manipulating, and displaying this evidence. Science has now reached a stage where this evidence and technology is freely available to test any theory we desire.

It should also be appreciated that the concept of an Earth increasing its radius with time was rejected during the 1950s to 1970s, well before the advent of this modern global data gathering frenzy, and certainly well before modern computer and space technology. The point I am again reiterating here is that, since the time the theory of Earth Expansion was first introduced, the theory has never been scientifically tested at all using modern scientific evidence. So it is a mystery why science must be continually compelled to reject this theory out of hand, without first applying modern evidence and modern principles to test the theory in order to provide alternatives to the origin of our continents and oceans? If the theory truly has been adequately tested, then yes, by all means reject it. Unfortunately, the theory has not been tested at all so the primary intent here is to do something about applying the newly available modern evidence and modern principles to quantify Earth Expansion.

The historical terms Earth Expansion, Expanding Earth, and similarly Growing Earth refer to changes to the shape and size of the Earth, as well as the motions of the various crusts on an Earth whose surface area increases with time in sympathy with an increase in Earth radius. These terms are synonymous and, to avoid confusion, I have introduced and will continue to use throughout this book the all-embracing new term ***Expansion Tectonics*** to encompass each of these now largely unpopular terms. The term Expansion Tectonics was first introduced in 1995 within my Master of Science research thesis as Global Expansion Tectonics.

It is my intent to structure this book in two main parts. The first part will focus on presenting and displaying scientific evidence and observations on purpose-built spherical scale models of the Earth. These models will be used to reassemble the Earth's crustal plates back in time, and from these the origins of not only the continents and oceans but also the origins of the more ancient supercontinents and primitive seas will be discussed. The content of this section of the book is designed to appeal to a wide audience of interested persons, and as such its technical content will be kept to a minimum. It is appreciated there will be a large number of unanswered questions emerging from this first part. Most of these questions will initially be suppressed in order to focus on presenting unbiased empirical observations from the modelling studies, rather than potentially clouding, rejecting, or prejudging these observations through the influence of pre-conceived conventional theory.

The second part of this book will focus on introducing new scientific evidence from a wide range of more specialized fields of Earth science in order to more specifically answer the questions raised in the first section. The opportunity will also be taken to speculate on the implications raised by this new shift in thinking in an attempt to further promote original thought and to create new opportunities for on-going research.

While both parts of this book are inter-related, the technical content of the second part will be lifted somewhat to reflect the increase in technical input. However, this will still definitely appeal to all persons with an innate interest in the natural sciences, in particular to persons with prior exposure to the various Earth sciences.

Much of this book is based on research originally carried out as part of the requirement for the award of the Doctor of Philosophy of Curtin University of Technology, Perth, Western Australia, completed in 2001, with some extracts from my Master of Science award of Curtin University of Technology, Perth, Western Australia, completed in 1995. This book is also an extension of my earlier book, *Terra Non Firma Earth*, first published in 2005.

Acknowledgements

In 1993 Samuel Warren Carey, Emeritus Professor of Geology at the University of Tasmania, wrote to me and said: *"Thank you for sending me your draft manuscript* [an early draft of my Master of Science Global Expansion Tectonics thesis submitted in 1995]. *I am most impressed. I now believe that you are the one to pick up the baton, now I am bowing out. I took it from Alex du Toit...With warm greetings, and my welcome to you as the one whom I must pass my baton"*. Professor Carey has since passed away in 2002, and in essence this book is a culmination of my subsequent research carried out in order to fulfil this esteemed honour that Carey has bestowed on me.

In addition to the memory of Professor Carey, this book is dedicated to a number of close friends and valued researchers that I have been associated with for many years now. In particular Klaus Vogel from East Germany and Jan Koziar from Poland who have been researching Earth Expansion since at least the 1970s.

The author (left) with Klaus Vogel (centre) and Jan Koziar (right) attending the 37th Interdisciplinary Workshop; *The Earth Expansion Evidence: A challenge for Geology, Geophysics and Astronomy*, held at the Ettore Majorana Foundation and Centre for Scientific Culture, Erice, Sicily, 4-9 October, 2011.

Klaus Vogel is renowned for having constructed the first modern small Earth models based on seafloor mapping, and Jan Koziar was the first to mathematically derive a formula for calculating ancient Earth radius as well as crustal modelling and related studies from seafloor mapping (see his website at: www.wrocgeolab.pl). Both are now retired and this book quantifies their groundbreaking research into Earth Expansion, which will hopefully bring our common cause to fruition.

Acknowledgements and thanks must especially go to my wife Anita Maxlow of Terrella Press for her support, encouragement, and assistance in editing, and for her involvement in graphical display and publication of this book. Sincere thanks must also go to Philippe Bouysse of the Commission de la Carte Geologique du Monde, Paris, for originally granting permission to use and digitize the Geological Map of the World at 1:25,000,000 M scale (1990), published by the Commission for the Geological Map of the World and UNESCO©. Simon Brown of Geoviz International, Perth, Western Australia, who originally digitized the Geological Map of the World which was then used as a base map in my original research. Professor Cliff Ollier, John B. Eichler, Jan Koziar, and Stephen Hurrell, amongst many others, who have provided invaluable assistance with editing and stimulating discussion on mountains, causal mechanism, crustal processes, and dinosaurs respectively, as well as ongoing encouragement towards completion of the book.

Dr James Maxlow

An Important Preliminary Consideration

"Science, does not - in the proper sense - discover new facts or regularities in nature, but rather offers some new ways of seeing and understanding the physical world." Zarebski 2009

Throughout this book the most challenging question the reader will need to confront and seriously consider is: ***has the past Earth radius really remained constant or has it been changing over time?*** The importance of this question in understanding and fully appreciating the origin of continents and oceans cannot be overstated. In general, this question has rarely been considered and, more importantly, has rarely if ever been taken seriously by most scientists. In addition, this question also leads to further related questions, such as: is Earth mass changing with time, is the volume of ocean waters constant, is the dynamics of the Earths motion around the solar system constant, and so on? Conventional science does not need to consider any of these questions simply because the constancy of these questions is assumed to be true despite a lack of evidence or proof to suggest otherwise.

What most people do not comprehend is that conventional Plate Tectonic theory is governed by the most fundamental of assumptions which insists that Earth radius, and hence its mass, has remained constant, or near constant, since the Earth was first formed. In other words, Plate Tectonics can just as easily be called the non-Expanding Earth theory. This assumption is never discussed, rarely if ever investigated, and never challenged in modern science today. It is unconsciously accepted as a blind truth, a truth that seemingly does not need to be proven. By constraining Earth radius to a constant value the unforeseen outcome of this assumption is that all subsequent assumptions, models, research, and theses based on evidence gathered about the Earth are also constrained to a constant Earth model. If this most fundamental assumption about Earth radius is shown to be wrong, then it stands to reason that all subsequent models and theories based on this constant Earth radius assumption will also be wrong and therefore Plate Tectonics must be wrong.

Years ago an old Professor of mine once said to me, *"James, in order to convince me that the Earth is expanding you must provide definitive proof"*. What he meant by this statement was that, by scientific convention, in order to convince him that the Earth is expanding I must be able to either measure the increase in Earth radius or at least observe it.

There are basically two ways that an ancient Earth radius can be measured. And by ancient Earth radius I am talking about the radius of Earth at any moment in time before the present-day. These techniques include palaeomagnetic studies—the study of Earth's remnant magnetism preserved in rocks, as will be discussed in Chapter 13—and measuring the surface areas of seafloor volcanic rocks, as will be introduced in Chapter 1. Others would argue that modern space geodetic studies—the use of a network of radio telescopes, satellites, and ground-based receiver and transmitter stations to routinely measure the dimensions of the Earth, as will be discussed in Chapter 17—also constitutes a third measurement technique.

Both palaeomagnetic and space geodetic measurement techniques are routinely used in support of Plate Tectonic studies, such as in determining plate assemblage and plate motions on a constant radius Earth. The outcomes of these studies are then promoted in modern science as confirmatory evidence in support of a constant radius Plate Tectonic Earth model. In contrast, the method of measuring seafloor surface areas to determine an ancient Earth radius was pioneered by Jan Koziar during the early 1980s and this technique will be introduced in Chapter 3. This technique is unique to Expansion Tectonics, was only available for study since the 1980s, and has no relevance to Plate Tectonics.

Palaeomagnetics is used in Plate Tectonic theory to supply evidence about movement of the various continental crusts, continental growth, mountain formation, apparent-polar-wander, and prior to 1975, for measurement of the Earth's ancient radius. It was palaeomagnetics that provided the first clear geophysical evidence for Continental Drift during the 1950s and its use is held in high esteem in Plate Tectonic studies. Even though early Earth radius measurements were deemed inconclusive by palaeomagneticians, it was concluded from the studies of McElhinny and Brock in 1975 that "*...within the limits of confidence, theses of exponential Earth expansion, or even moderate expansion of the Earth are contradicted by the palaeomagnetic evidence.*" This then led McElhinny and Brock to further conclude, "*...there has been no significant change in the ancient radius of the Earth with time.*"

It will be shown in Chapter 13 that, while McElhinny and Brock went to great lengths to present quality data and sound methodology, prior to 1975 there was, and in fact still is, very little agreement as to what a potential ancient Earth radius may or may not have been. What McElhinny and Brock failed to fully appreciate was the significance of their ancient magnetic pole locations, as determined from conventional palaeomagnetic formulae, and the application of these formulae towards determining an ancient Earth radius. In contrast, when McElhinny and Brock's ancient magnetic pole data are used to determine an Earth radius based on an Expansion Tectonic Earth model, it will be shown in Chapter 13 that the ancient radius calculates exactly the same as the ancient Earth radius established from measurements of seafloor surface area data—as outlined in Chapter 3. This, in effect, contradicts the conclusions of McElhinny and Brock and instead represents definitive palaeomagnetic proof in support of an Expansion Tectonic Earth.

The application of advanced space geodetic techniques to studies of the Earth has now progressed to the point where estimation of present-day radius and plate movement is quoted to sub-millimetre accuracy. Results of Shen and others in 2011 now show that "*...both geodetic and gravimetric observations support the conclusions that the Earth is expanding at a rate of 0.2 millimetres per year in recent decades.*" While this is encouraging, the result falls well short of the expected current rate of 22 millimetres per year increase in Earth radius as measured and presented in Chapter 3 of this book using published seafloor surface area data.

From the discussion on space geodetics in Chapter 17, it will be shown that in order for Shen and others to calculate Earth radius and come to this conclusion, 60 percent of the raw satellite observational data is eliminated before calculating a rate of change in Earth radius. In other words, what they do is remove all data that might otherwise indicate an increase in Earth radius simply because this data does not fit a pre-conceived constant radius Plate Tectonic model of the Earth. This, in effect, smooth's out and constrains the raw data before making their calculation and conclusion. In reality though, space geodetics measures the mean radius of the present-day Earth, not an ancient Earth radius. This is the primary reason why I do not include space geodetics as a means of measuring an ancient Earth radius. Its primary function is to measure the present-day radius of the Earth in order to then measure the present-day movement of the various crustal plates.

The pioneering seafloor surface area work of Koziar during the 1980s, and to a lesser extent similar work by Blinov, was in direct response to preliminary completion of geophysical mapping of the seafloor crusts throughout all of the oceans. A more updated version of this seafloor mapping was published by the Commission for the Bedrock Geological Map of the World and UNESCO in 1990—as will be introduced in Chapter 1—which forms the basis of quantification and modeling studies presented throughout this book. The primary outcome of this seafloor mapping is that new seafloor volcanic lava is shown to be continually intruded along the full length of all mid-ocean-ridge spreading zones within each ocean. This intrusion of lava causes all of the oceans to increase their surface areas over time, and, as a consequence, all continental crusts are shown to be moving away from each other. Comprehension of this phenomenon is the primary aim of this book, and the outcome of this discussion rests squarely on convincing the reader whether or not Earth radius is constant or is in fact increasing over time.

In this book it will be shown that, on an Expansion Tectonic Earth, by moving back in time the surface area of each of the oceans must be progressively reduced and each of the oceans must be progressively closed in strict accordance with the published seafloor geologic mapping data. This seafloor mapping data will be shown to provide a unique means of measuring ancient Earth radius and from this, a method of establishing a formula for calculating the ancient Earth radius at any moment back or forward in time.

By progressively closing off each of the oceans it will be shown that there is no need to invoke the presence of hypothetical ancient oceans, nor is there a need to arbitrarily fragment any existing continental crusts, or invent prior seafloor crusts, as insisted in Plate Tectonic theory. At approximately 250 million years in the past, it will be shown in Chapter 5 that, by progressively closing each of the modern oceans, all of the modern continents assemble together to form a single unique ancient Pangaea supercontinent that encompasses the entire ancient Earth surface on a reduced radius Expansion Tectonic Earth. Additionally, by utilizing the continental mapping data shown on the published Geological Map of the World, it will also be shown that it is possible to extend assemblage of the Earth's crusts further back in time to the beginning of Earth history. This is something that has never been done before and something that Plate Tectonics is unable to accomplish.

In this book, answering the fundamental question regarding a causal mechanism for Earth expansion will be intentionally left until last, regardless of when this most fundamental of questions first arises. This will be done to allow all empirical evidence to be presented in a logical manner, without biasing any outcomes of this evidence. It is considered unfortunate that most humans instinctively want to know, or at least comprehend a cause well before acknowledging any form of physical evidence. This also touches on our innate and instinctive hesitation towards change, preferring instead to consider that it is better to err towards what science already knows and is comfortable with rather than consider something new or different. These hesitations and concerns remain equally true today since historically, in the evolution of knowledge, empirical phenomena have often been recognized long before their cause or reason has been fully comprehended—such as during the introduction of Continental Drift and even Plate Tectonics.

Since the theory of Earth Expansion was first proposed towards the end of the eighteenth century a vast amount of new scientific evidence has become available, in particular over the past few decades, to seriously challenge conventional science and offer a plausible causal mechanism for Expansion Tectonics. The causal mechanism proposed in this book, while speculative, is based on a proposal put forward by Eichler in 2011. In promoting this causal mechanism Eichler first posed the question *"Does plasma from the Sun cause the Earth to expand?"* and by presenting arguments based on known physical phenomena, he suggests that this might indeed be the case.

1 Controversial New Ideas

"Scientists still do not appear to understand sufficiently that all earth sciences must contribute evidence toward unveiling the state of our planet in earlier times, and that the truth of the matter can only be reached by combing all this evidence. . . It is only by combing the information furnished by all the earth sciences that we can hope to determine 'truth' here, that is to say, to find the picture that sets out all the known facts in the best arrangement and that therefore has the highest degree of probability. Further, we have to be prepared always for the possibility that each new discovery, no matter what science furnishes it, may modify the conclusions we draw." Alfred Wegener. *The Origin of Continents and Oceans* (1915)

It is hard for most of us to imagine that at one time our ancestors used to think the Earth was flat and we could actually fall off an imagined edge if one strayed too far. This belief that the Earth was flat was justifiable at the time and was no doubt the result of a limited understanding of the Earth when compared to our own modern understanding. This belief may have also been a result of fertile imagination, which may in turn have been governed by religious and socially acceptable beliefs of the time and, of course, a strong maritime fear of the unknown.

The idea of a flat Earth depicts the Earth as a flat, plate-like body and reference to this belief is found in mankind's oldest writings. In early Mesopotamian thought the world was portrayed as a flat disk floating on an ocean. This was a common belief for many hundreds of years, surviving in some instances even to this day, until the Classical Greeks began to discuss the Earth's shape during the 4th century BCE and demonstrated that the Earth is in fact a sphere.

By the early middle ages, long sea fairing voyages of discovery had become common place. By then, it quickly became apparent that the Earth did not have an edge to fall off and because of this observation was indeed confirmed to be a sphere. As early as 1596 map makers of the time had also made a significant observation, that the opposite coasts of the Atlantic Ocean have similar shapes and the west coast of Africa and east coast of South America seemed to have once fitted together. As can be appreciated this was a profound observation for their time!

With this well-founded knowledge, the emerging science of geology enabled a number of scientists to begin to speculate on the origins and subsequent geological history of the continents and oceans of the Earth. During the late 19th and early 20th centuries geologists assumed that the

Earth's major features, such as the continents and oceans, were fixed. Other geologic features, such as sedimentary basins and mountains, were explained by vertical crustal movement described in what was called the *geosynclinal theory*—which is still a very good, albeit discarded theory. Generally, at that time, these phenomena were placed in the context of a contracting Earth theory. Contractions of the Earth radius were considered to have been due to heat loss, which was based on the assumption that the Earth was originally molten and had cooled over the course of a relatively short period of geological time.

These rather primitive ideas were justifiable at that time, however they have long since been rejected in modern science. Since then, science has had an abundance of new concepts and theories, including Continental Drift and Earth Expansion, followed by Plate Tectonics to add to our ever changing perspective of the Earth.

By 1915 Alfred Wegener, a PhD university professor, was making serious arguments for the idea of Continental Drift in the first edition of his book, *Die entstehung der kontinente und ozeane* [*The Origin of Continents and Oceans*]. In his book, as did mapmakers before him, he once again noted how the shape of the east coast of South America and the west coast of Africa looked as if they were once joined. When Wegener first presented his arguments for the idea of Continental Drift he became the first to gather significant fossil and geological evidence to support his simple observations for the break-up and subsequent movement of the continents through time. From these beginnings, Wegener went further to suggest that the present continents once formed a single land mass—later to be called *Pangaea*. This land mass was inferred by Wegener to have subsequently broken up and drifted apart, "*...thus releasing the continents from the Earth's mantle.*" Wegener likened this too "*...icebergs of low density granite floating on a sea of denser basalt*".

At that time his ideas were not taken seriously by most geologists. They rightly pointed out that there was no apparent mechanism for Continental Drift. Without detailed evidence, or a force sufficient to drive the movement, the theory was discounted: *...the Earth might have a solid crust and mantle and a liquid core, but there seemed to be no way that portions of the crust could move around the surface of the Earth.* Unfortunately Wegener could not explain the force that drove Continental Drift, and vindication for his efforts did not come until well after his untimely death. Other responses were less than sympathetic, including: "*Utter damned rot*" (the then President of the American Philosophical Society). "*If we are to believe this hypothesis we must forget everything we learned in the last seventy years and start all over again*"

(Thomas Chamberlin). "*Anyone who valued his reputation or scientific sanity would never dare support such a theory*" (a British geologist).

Although initially rejected for many years, when Wegener introduced his theory of Continental Drift he did in fact set in motion a completely new train of thinking and speculation about the origin of our continents and oceans. As Wegener correctly promoted, the fit of the Americas against Africa and Europe was real and had to be explained. Time has, of course, since shown that it was only the mechanism behind Continental Drift that was difficult to explain, not the actual fit of the continents. Since then, with changing ideas about the Earth and also a name change, Wegener's theory of Continental Drift is now credited with having given rise to the modern theory of Plate Tectonics. Most people have now come to accept this Plate Tectonic theory without question and without prior concern for this considerable amount of initial and still relevant rejection.

Notwithstanding the opposition, the concept of Continental Drift eventually gained support and a lively debate started between proponents of the theory. Proponents were referred to as *drifters* or *mobilists*, and opponents to the theory were called *fixists*. During the 1920s to 1940s the former reached important milestones, proposing that convection currents might have driven the crustal movements and that spreading may have occurred below the sea within the seafloor crust. Concepts close to these elements are now incorporated in Plate Tectonics, as initially proposed by various geophysicists and geologists such as Vening-Meinesz, Holmes, and Umbgrove.

All this evidence, including new evidence emerging from both the seafloor and the continental margins, made it clear during the early 1960s that Continental Drift was indeed feasible. From these early beginnings the theory of Plate Tectonics, which was defined in a series of papers published between 1965 and 1967, was born. This new theory of Plate Tectonics is now considered to have revolutionized the Earth sciences and is portrayed as explaining a diverse range of geological phenomena, with supporting implications from other studies such as ancient geography and ancient biogeography.

1.1 Conventional Wisdom

The theory of Plate Tectonics has been extensively promoted in science to explain a diverse range of observed global geological phenomena and this theory is now widely accepted by both scientists and the general public. The theory is considered by most scientists to adequately link all geologic phenomena, from the age and composition of ocean floors, to the rise of the mountains, as well as the past distributions of plant and animal species.

In 1992 Lowman, in assessing the strengths and weakness of Plate Tectonic theory, considered that the theory is portrayed as a useful conceptual framework for teaching geology. It is fundamentally simple and is easy to understand. But, he further considered that the strengths of this theory are far outweighed by its inherent weaknesses. These weaknesses include its deceptive simplicity, its limited scope, and its possible fundamental incorrectness. Lowman suggested that these weaknesses are, in fact, *"...slowing geologic progress by inducing narrow, rigid viewpoints, which discourage alternative research by convincing students of science that the main global tectonic problems are now solved."*

In discussing alternatives to Plate Tectonics, Saull in 1986 commented that the inherent non-uniqueness of Global Tectonic theory raises a logical possibility that the Plate Tectonic model may be false. He considered this to be disturbing, as the model is very widely accepted and fundamental geological conclusions flow from its application. If Plate Tectonic theory were false, Saull concluded that it would nevertheless be difficult to refute for the following reasons:

- *"Although experts question virtually every widely accepted explanation of the Plate Tectonic model, the processes supposed to be responsible are rooted in regions so poorly known that it would be hard to prove or disprove any geophysical model of them."* Moreover, popular acceptance of Plate Tectonics was considered by Saull *"...not to be based on knowing the mechanism that supposedly moves the plates."*

- *"The hard core of belief in Plate Tectonics is protected from direct assault by auxiliary hypotheses that are still being generated (e.g. polar wandering, hot spot evolution, and accretionary complexes)."*

- *"The Plate Tectonic theory is so widely believed to be correct that it would be difficult to publish alternative interpretations* [such as Earth expansion.] *Lacking well-known alternatives, the dominant Plate Tectonic theory would still not be rejected because, in science, it would seem that error is better than confusion."*

In science and philosophy, the auxiliary hypotheses that Saull mentions are also referred to as *ad hoc hypotheses*. These are hypotheses added to a theory in order to save it from being falsified. Ad hoc hypothesizing is a way of compensating for anomalies not anticipated by the theory in its unmodified form, such as anomalies related to increasing surface area on a static radius Earth model. While scientists are often sceptical of ad hoc hypotheses that rely on frequent unsupported adjustments to sustain the primary theory, there is no limit to the number of hypotheses that they could add. Thus, the particular theory runs the risk of becoming more and more complex, but without viable alternatives it is never falsified, even at a cost to the theory's credibility and predictive power.

Despite these inherent weaknesses, Plate Tectonics is today considered the main tectonic theory in Earth Sciences. It is therefore appropriate to dedicate some space in this introduction to explain what the theory is about. Tectonics—from the Late Latin *tectonicus*, pertaining to building—is a scientific term that describes the large scale motions of the Earth's crust. Tectonics, and in particular the generic term *Global Tectonics*, is now generally considered synonymous with Plate Tectonics. As will be discussed in this book, I take this opportunity to point out that, by its very definition Global Tectonics is, however, just as equally pertinent to all tectonic theories. As such, it definitely should not be confused with, nor its relevance constrained to just Plate Tectonic theory.

In Plate Tectonic theory—as well as Expansion Tectonic theory—the Earth's crust is now known to be broken into a series of seven or eight major and many minor plates, made up of both continental and seafloor crusts (Figure 1.1). These crustal plates move in relation to one another at one of three types of plate boundaries called: *convergent* or *collisional* boundaries, where plates are said to collide resulting in the formation of mountains; *divergent* boundaries, where ocean crusts break apart and new volcanic crust is erupted along seafloor spreading centres; and *conservative* or *transform-fault* boundaries, where plates are faulted relative to each other. Earthquakes, volcanic activity, mountain-building, and seafloor trench formation are also said to occur along each of these plate boundaries, with relative movement of the plates typically varying from between 0 to 100 millimetres annually.

What must be appreciated from this modern description is that each tectonic plate, as defined by Plate Tectonic usage, is composed of two types of crust: thick continental crusts surrounded by relatively thin seafloor crusts. This could be seen as analogous to a piece of meat sitting in the middle of a dinner plate. This modern description is totally different to Wegener's, and also Continental Drift's original concept whereby the continents where likened to "*...icebergs of low density granite floating on a sea of denser basalt*". In order to now have any form of plate movement the entire plate, as defined by the heavy black outlines in Figure 1.1, must move relative to adjoining plates, not just the continent alone. What this also means is that before a continent can interact or collide with another continent, much of the surrounding seafloor crust must be completely disposed of by the ad hoc hypothesis called *subduction.*

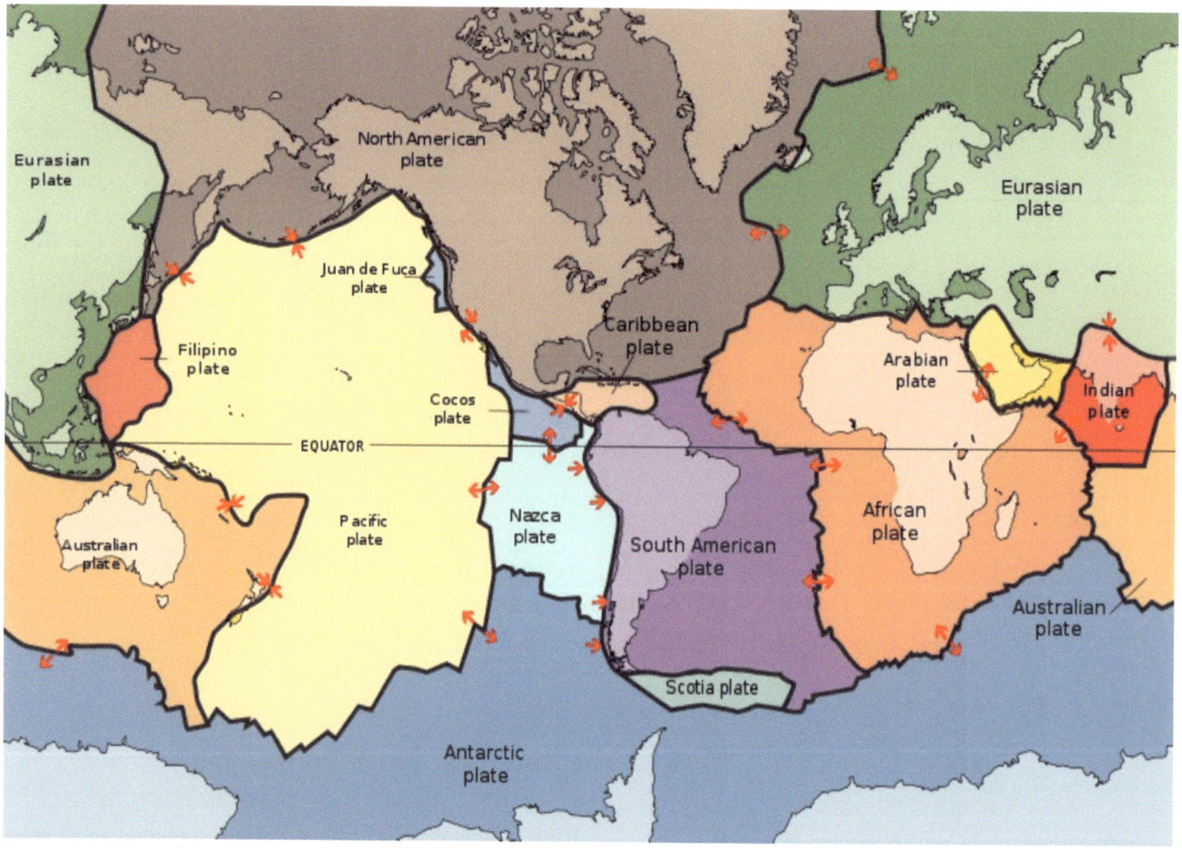

Figure 1.1 Map of the Earth showing the distribution of tectonic plates comprising continental and seafloor crustal rocks. (http://en.wikipedia.org/wiki/File:Plates_tect2_en.svg)

In order to dispose of this seafloor crust, Plate Tectonic theory insists that an equal amount of surface area of the plates must disappear into the mantle along convergent boundaries by subduction (Figure 1.2). This disappearance of seafloor crust must, of course, be more or less in equilibrium with generation of new seafloor crust being formed along the divergent margins by seafloor spreading. This process is also referred to as the *conveyor belt principle*—another ad hoc hypothesis. In this way it is assumed, but never acknowledged as a basic premise, that the total surface area of the Earth remains constant throughout time and hence it is also assumed that the radius of the Earth must remain constant—a circular argument.

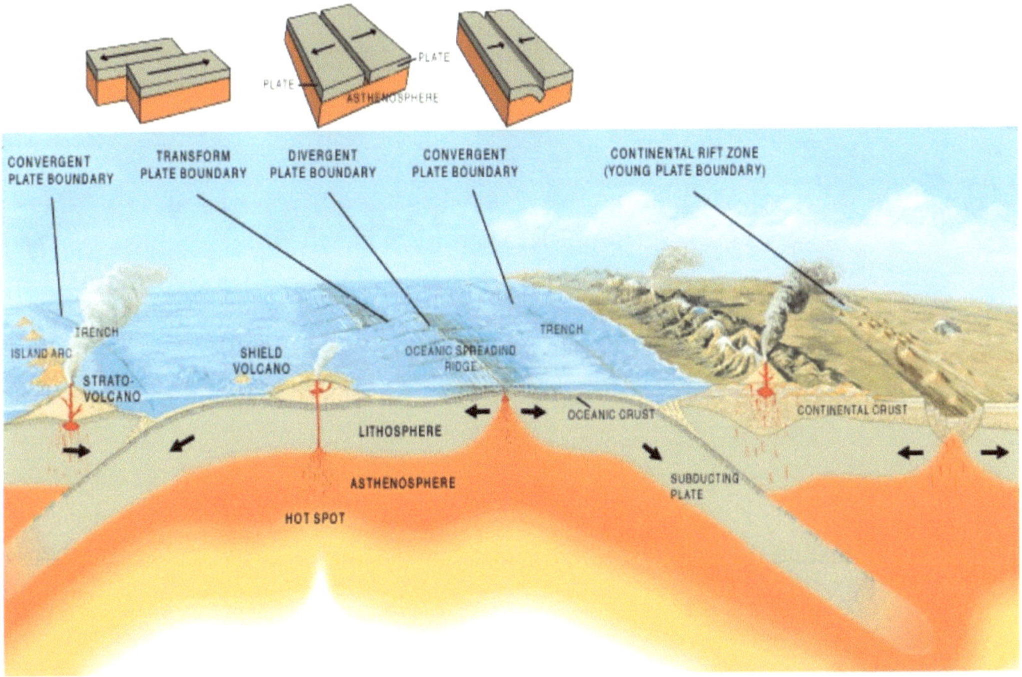

Figure 1.2 Idealised schematic cross section of the Earth showing the Plate Tectonic model for generation of new seafloor crust along mid-ocean spreading centres and disposal of excess crust within subduction zones. (http://en.wikipedia.org/wiki/File:Tectonic_plate_boundaries.png)

Stepping back in time to the 1960s, it was initially recognised that the Earth's crust was expanding along centrally located seafloor ridges within each of the oceans. Early researchers, such as Hess and Dietz, reasoned, like Holmes and others had also done before them, that because of this expansion a similar area of crust must then be shrinking somewhere else. Hess suggested that new seafloor crust continuously spreads away from the seafloor ridges in a conveyor belt-like motion. Using the mobilistic concepts developed before him, Hess concluded that many millions of years later the seafloor crust eventually descends along the continental margins where seafloor *trenches*—very deep, narrow canyons located along the margins of some continents—are formed, for example around the margins of the Pacific Ocean.

The proposal Hess suggested was that mantle convection currents were the driving force behind this process, which primarily uses the mechanism of spreading along the ridges to drive the currents. Hess therefore concluded that the Atlantic Ocean was expanding while the Pacific Ocean was shrinking. He further suggested that, as old seafloor crust is *consumed* in the trenches around the Pacific Ocean, new magma rises and erupts along the spreading ridges to form new crust. In effect, the ocean basins were said to be perpetually being *recycled*, with the creation of new crust and the destruction of old seafloor crust occurring simultaneously.

The majority of Earth scientists now agree on the observation and deduction that the crustal plates move one with respect to the other. From this they consider that all that is left to do is debate and find agreements on how and when the plates move. But, what is rarely considered is that there is still a major question remaining regarding what the motor behind this plate movement is—referred to as the *geodynamic mechanism*. Scientific opinion still diverges here and a large number of different theories have been proposed.

Generally, it is considered by most scientists that tectonic plates are able to move because of the relative density of the seafloor crust—which is the outer, rigid part of the Earth's seafloor—and the relative weakness of the underlying lower, low strength, low rigidity part of the Earth's crust. Dissipation of heat from the underlying mantle was suggested to be the original source of energy driving plate movement, which resulted in convection or large scale upwelling and doming. As a consequence, in the current view, the excess density of the seafloor crust is said to cause the crust to sink within subduction zones whereby a powerful source of plate motion is then said to be generated. When new crust forms at the mid-ocean spreading ridges, this seafloor crust is considered to be initially less dense than the underlying lower crust, but becomes denser with age as it progressively cools and thickens. The greater density of old crust relative to the underlying crust then allows it to sink into the deep mantle at subduction zones. The weakness of the lower crust is then considered to allow the tectonic plates to move towards a subduction zone.

Unfortunately, the lack of explanation as to how the seafloor crust could progressively thicken and become denser with age raises some very pertinent questions as to how this could possibly happen. In particular, the seafloor crust is, in reality, simply cooled and quenched volcanic rock which originates from the upper mantle. Any convective process would simply return mantle rocks back into the mantle where they came from, so why should there be a requirement for thickening or density differences?

Although subduction is now believed by plate tectonists to be the strongest force driving plate motion, it is also acknowledged by many researchers that it cannot be the only force since there are a number of plates, such as the North American Plate, that are moving, yet are nowhere being subducted. The same is true for the enormous European and Asian Plate, and especially the Antarctican Plate. So, even though Plate Tectonics is currently considered the main tectonic theory in the Earth Sciences, the mechanisms for plate motion and conservation of surface areas are still a matter of intensive research and debate among many Earth scientists.

According to Cwojdzinski, in Maxlow 2005, "*Nobody appears to appreciate that...* [Plate Tectonics] *...contain*[s] *an assumption, a hypothesis, not a proven fact, and that based on this assumption the whole gigantic edifice of plate tectonics is built. The present generation of geologists does not reach to the classics of plate tectonics. They are satisfied by the knowledge delivered by authority and the soothing consciousness of a full domination of professed paradigm.*" Cwojdzinski further considered that "*The real scientific discussion, in reality, must therefore concentrate on proof based on facts, proofs which in fact directly show an expansion of the Earth.*"

1.2 It doesn't have to be Like That

Somewhat like the way Wegner's original arguments for Continental Drift were initially heckled and rejected, the concept of an Earth increasing its radius also continues to this day to be unfairly rejected, based on very similar rather emotive argument. In all fairness though, during the pre-1960s the gathering of global-scale data and evidence was simply not able to be done, nor was the available evidence extensive enough for early researchers into Earth Expansion to argue their case with *sound geological evidence*. Nobody was privy to the vast amount of global geological data available now, so scientists could only provide cursory, albeit emotive comment on the merits of any alternative theories. Since then, there has been a frenzied increase in global-scale geological, geophysical, space-based, and geographical research and data gathering. This data gathering has now vastly increased the amount of evidence available to thoroughly evaluate and test any global tectonic theory that science desires.

It should also be noted that this new global geological, geophysical, space-based, and geographical evidence, while routinely collected and used in Plate Tectonic studies, has never been seriously looked at other than from a constant radius Earth perspective. More importantly, this new information has never been applied to the alternative theories, such as Earth Expansion, which unfortunately leaves science without an acceptable alternative geological viewpoint or basis for constructive scientific argument.

Many theories have come and many more have gone throughout the ages, in particular after the science of geology was first conceived during the 1700s. The Flat Earth theory, while popular in ancient times, is now largely historical but it serves as a useful starting point in understanding the progression of knowledge about the Earth through history. This Flat Earth theory stems from the limited knowledge of the size and shape of the Earth in ancient times and also from the limited number of scientists, or philosophers, at the time capable of gathering enough information to make meaningful sense of the knowledge or observations available to them.

The suggestion that continents have not always been at their present positions was introduced as early as 1596 by the Dutch map maker Abraham Ortelius. Ortelius suggested, based on the symmetric outlines of the Atlantic coastlines, that the Americas, Europe, Asia, and Africa were once joined and have since drifted apart "*...by earthquakes and floods*", creating the modern Atlantic Ocean. For evidence he wrote: "*The vestiges of the rupture reveal themselves, if someone brings forward a map of the world and considers carefully the coasts of the three continents.*" As previously mentioned, in 1915 Alfred Wegener also noted how the east coast of South America and the west coast of Africa looked as if they were once joined. Wegener went further to suggest that the present continents once formed a single, what is now referred to as the "*Pangaean land mass*"–which is the ancient supercontinent called Pangaea–that subsequently broke up and drifted apart.

So, did Wegener get it all wrong? No, he didn't get it wrong. In fact, he is justifiably credited with being one of the first scientists to apply correct geological observation to support and substantiate his claims. But, what he and others subsequently didn't do was to go far enough. What he and others failed to recognise was that, as well as fitting the South American and African coastlines together to come up with a reasonably approximate fit-together, the remaining Indian, Pacific, and Southern Ocean coastlines can just as easily be fitted together, with similar fossil and geological evidence to support these observations.

In researching the concept of Continental Drift during the 1950s, Professor Carey made a scale model of the present-day Earth to investigate the potential fit of the continents during closure of each of the oceans. In addition to the Atlantic Ocean, his investigation was extended to also consider fitting the various continents together within the Indian and Pacific Oceans. It is important to mention that Carey made comment that the trans-

Atlantic fit was not as good a fit as Wegener and others had claimed. His comments and conclusions from his research are reproduced in full as follows:

"At an early stage in my investigations I went to some pains to ensure that I compared and transferred shapes and sizes of the continental blocks accurately. I have spent tedious years plotting large oblique stereographic projections about diverse centres not only for Africa and South America but for every piece of the Earth's surface. I combined this with spherical tracings from the globe, working on a spherical table. The reward for this zeal for accuracy was frustration. Again and again over the years I have assembled Pangaea but could never attain a whole Pangaea. I could make satisfactory sketches like Wegener's classic assembly, but I could never put it all together on the globe, or a rigorous projection. I could reconstruct satisfactorily any sector I might choose but never the whole. If I started from the assembly of South America...by the time I reached Indonesia there was a yawning gulf to Australia, although I felt sure from the oroclines that Indonesia and Australia belonged together...If I started from Australia and Indonesia I had no hope of closing the Arctic Sphenochasm [where the split occurred]..., which I was convinced was basically correct...I was painfully aware that there was a crucial link missing from the global synthesis. I was tempted to abandon the quantitative assembly and resort to sketches which would show every block related as I inferred they should be, even though I knew I could not bring them together that way with the rigour I sought."

"But in the end the rigorous approach has paid off. For it has revealed a discrepancy which had not been apparent. It was not my method that was at fault, but my assumption that the earth of Pangaea was the same size as the earth of today. The assembly of Pangaea is not possible on the earth of the present radius, but on a smaller globe, a globe such as is demanded by the orocline analysis, these difficulties vanish."

Unfortunately, with the subsequent promotion of Plate Tectonics, these very important physical observations and conclusions of Carey continue to be neglected and totally ignored to this day.

During the 20th century there were also a number of additional independent thinkers who considered opening of the oceans could be attributed to an increase in Earth radius. In 1859, Alfred W. Drayson published his book *The Earth We Inhabit: Its past, present, and probable future* in which he speculated that the Earth had undergone an expansion over time. In 1888, Yarkovski was the first to postulate a growth of the Earth mass. Similarly, Roberto Mantovani in 1889, and again in 1909, published a theory of "*...earth expansion and continental drift.*" In this theory Mantovani considered that a closed continent covered the entire surface on a smaller radius Earth.

He suggested that "*...thermal expansion led to volcanic activity, which broke the land mass into smaller continents.*" These continents then drifted away from each other because of further expansion at the "*rip-zones,*" where the oceans currently lie. This was followed later by the pioneering work and publications of Lindemann in 1927, small Earth modelling by Ott Christoph Hilgenberg during the 1930s, S. Warren Carey during the 1950s to late 1990s, Jan Koziar during the 1980s, and small Earth modelling by Klaus Vogel during the 1980s and 1990s.

Hilgenberg, stimulated by the pioneering work of Wegener on Continental Drift, has been attributed as being the first model maker to fit all of the present-day land masses together to completely enclose a small papier-mâché globe and in 1933 his work was published in his classical work *Vom Wachsenden Erdball [About the growing Earth]*. On each of his globes (Figure 1.3) all oceans were progressively eliminated and the remaining continental crusts eventually enclosed the entire Earth on a globe at about 60 percent of the diameter of the present Earth.

As can probably be seen in Figure 1.3 the main limitation to accepting Hilgenberg's globes was the purely visual fit-together of continents across each of the oceans? In 1990 Vogel commented that, "*...though fairly exact in several regions, the totality of Hilgenberg's result was not entirely convincing because, especially in the Indian, Pacific, and Arctic Ocean regions, numerous gaps and overlaps appeared between continental fragments.*" Hilgenberg also gave no explanation for the creation of the oceans, nor an explanation as to what happens to the present-day volume of seawater on a much reduced radius Earth.

More specifically, Hilgenberg's reconstruction across the Atlantic was considered to be convincing, however difficulties were encountered in the Indian Ocean due to a greater dispersion of continents and an uncertain initial position of India and Madagascar. The Pacific region was the most difficult to reconstruct, as workers to follow also found. Unlike the Atlantic and Indian Oceans, where the borders of these oceans retained their shapes, the Pacific borders were considered by Hilgenberg to have opened much earlier and hence the shape of these borders remained tectonically active throughout the continental dispersal times.

To explain the expansion process, Hilgenberg postulated that the mass of the Earth, as well as its volume "*...waxed with time.*" Because of this stance and the several problems inherent in his reconstructions, Hilgenberg's ideas on Earth expansion were largely ignored and he has since received scant recognition for his efforts.

Figure 1.3 Reproductions of Hilgenberg's Expansion Tectonic Earth models, attributed to being the first small Earth models constructed. Size of the small globe to the left is approximately 60 percent of the present Earth shown on the right.

In 1983 Vogel published a comprehensive set of scaled small Earth models, or "*terrella*"—meaning small Earth—as he referred to them, at various diameters, including a unique representation of a 55 percent reassembled globe inside a transparent plastic sphere of the present-day Earth (Figure 1.4). Each of Vogel's models is unique in that his work coincided with the first publication of seafloor mapping—more on this subject later. For the first time, this mapping enabled Vogel to accurately constrain both continental and seafloor crustal assemblages back in time, without having to resort to arbitrary fragmentation of the continents or to visual fitting-together of the various crustal plates.

This is a very important point to take note of because all previous model makers, while having enough foresight and courage to remove seafloor crusts, were faced with the less than envious task of having to visually fit-together the remaining continental crusts. Their visual fitting was done without the benefit of being able to accurately constrain or reposition the remaining crustal fragments.

Also, the prevailing geological viewpoint during the 20th century, and to a lesser extent still remaining today, was that each of the continental crusts acted as rigid bodies. In other words, the crusts were considered to be solid, immovable rock incapable of any form of distortion other than fragmentation due to faulting and earthquake activity. The early small Earth model makers were then faced with the additional fixation that continental crusts were brittle and hence needed to be fragmented like an egg shell in order to be reassembled on a smaller size globe when moving back in time.

Figure 1.4 Vogel's 1983 *"terrella"* models at various stages of expansion commencing on the far left with a continental reconstruction, without continental shelves, at 40 percent of the present Earth radius. A 55 percent radius model is also shown within a transparent sphere of the present day Earth at the right, demonstrating a radial motion of Earth expansion.

Vogel was an engineer, so he was not as fixated on geological protocol as most geologists at the time were. Because of this, he was able to construct stunning models with a high degree of precision. Each of his models demonstrated a greater dispersion of the southern continents, compared to those in the northern hemisphere, and he also noted a marked westward movement of all the northern continents relative to the southern continents. His models demonstrated that, in general, the continents tended to move out radially from their ancient positions to reach their modern positions. Vogel commented that this is a *"...odd coincidence for any theory except that of expansion of the Earth."*

From his extensive modelling studies, Vogel published a number of articles and gave a comprehensive outline of the fit of continental fragments under the headings of the ancient supercontinents called *Gondwana* and *Laurasia*. On Vogel's models, these ancient supercontinents represent an assemblage of the ancient continental crusts, which agree in principal to conventional tectonic theory, however, on Vogel's models the continental fragments are more tightly assembled on a reduced radius Earth. Vogel considered that development of the oceans commenced during the early Mesozoic Era—starting around 200 million years ago—and break-up and dislocation of the continental fragments was considered to be due to a widening of the oceans, centred along the mid-oceanic crustal spreading zones. Vogel went on further to consider the two hemispheres of the Earth as complementary counterparts, with no need for consideration of additional ancient oceans, or to arbitrary break-up or fragment the continents, as is required in Plate Tectonic theory.

Vogel concluded from his modelling studies that:

- At a reduced Earth radius of between 55 to 60 percent of the present radius, the continental outlines can be neatly fitted together to form a closed crust.

- The positions of the different continents with respect to each other remain generally constant, with their separation caused by a radial expansion of the Earth.

- The cause of the movements of continents has resulted from an accelerating increase in Earth radius with time, in accordance with seafloor spreading.

Vogel also made comment that "*...an accordance of these three phenomena cannot be accidental,*" but must be due to "*...processes operating from within the interior of the Earth resulting in Earth expansion.*" In addition to these observations, Vogel realized that it was theoretically possible for the continents, without the continental shelves, to fit together on an even smaller Earth globe calculated to be approximately 40 percent of the size of the present Earth. He based this observation on his comment that "*...the continental shelves must have formed only after the brittle upper crust had broken into pieces.*" This simple observation therefore alludes to the further potential that expansion of the Earth has been operating throughout earlier Earth history and has been active long before the time that modelling of the seafloor crusts suggests.

What should be noted here is that these researchers all showed, as well as other small Earth model makers at the time, that, like Wegener and others had suggested for the Atlantic Ocean, if each of the oceans were removed and the remaining continents were physically fitted together they would neatly envelope the Earth with continental crust on a small Earth globe some 50 to 55 percent of its present size (Figures 1.3 and 1.4). This coincidence led both Hilgenberg and Vogel, and similarly Carey from his early Continental Drift studies and Koziar from his extensive mathematical and crustal modelling, to come to similar conclusions that *terrestrial expansion has brought about the splitting and gradual dispersal of continents as they moved radially outwards during geological time*.

From this brief review of historical Earth expansion and small Earth modelling it must be noted that, although there is extensive literature on the subject of Earth expansion, the small Earth models briefly outlined above, as well as other lesser published models, represent essentially the sum-total from which Earth expansion has been judged in the past. These models were developed and the majority conceived prior to or during the early stages of investigation into seafloor spreading and prior to the complete and accurate geological mapping and age dating of the oceans. Most of these early reconstructions of continents on small Earth models suffered from a lack of precise cartographic methods, as well as quantitative constraint of both ancient Earth radius and crustal assemblage. Only Vogel was fortunate enough to have been able to use published seafloor mapping to constrain crustal assemblage on his models. All of these models have, however, since been totally ignored in mainstream science and tectonics.

Throughout the subsequent literature small Earth models, and hence Earth expansion, continue to be judged by the scientific community to be speculative and inconclusive. It is considered that one of the main reasons for this judgment is because, with early crustal assemblages based primarily on a visual fit-together of opposing continental margins, the small Earth models often gave rise to a wide variation of crustal fits, in particular the Pacific Ocean region. Similarly, a conclusive, quantifiable, *motor and mechanism* for Earth expansion was not able to be given.

The small Earth models of Hilgenberg and Vogel indicate, however, that an ancient Pangaean crustal assemblage on a small Earth globe, at between 50 to 60 percent of the present Earth radius, can produce a tight and coherent fit of all continents. Similarly, as will be shown in this book, remnant mountain belts from the various continents match consistently, geological boundaries are maintained and the ancient geographical and biological boundaries match precisely.

1.3 Important Contributions to Tectonic Theory

As is well known, planet Earth is broadly made up of a core, a layered mantle, an outer continental crust, seafloor crust, the ocean waters, and an atmosphere (Figure 1.5). The continental crust typically varies from 30 to 50 kilometres thick, which is less than 1 percent of the present Earth radius. Similarly, the seafloor crust in turn is about 5 to 10 kilometres thick, which is only about 0.2 percent of the present Earth radius. These thicknesses are vaguely similar to the proportions of the skin of an apple or the shell and membrane of an egg.

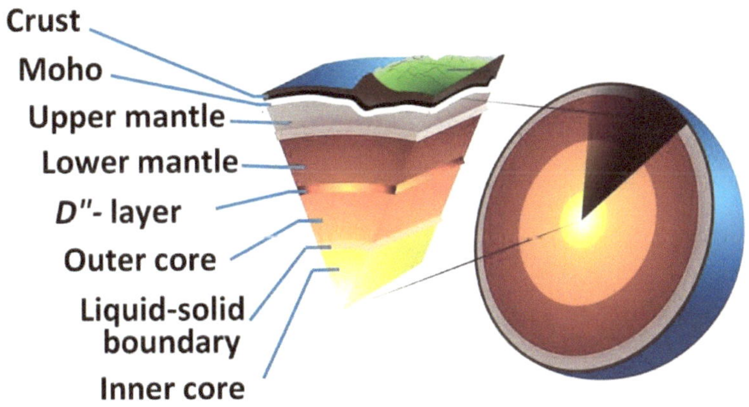

Figure 1.5 Schematic models of the Earth's layered structure showing the crust, upper mantle, lower mantle, outer and inner core. (http://en.wikipedia.org/wiki/Crust_(geology))

In geology, the crust is traditionally seen as the outermost solid shell of the Earth which is, in turn, also considered to be chemically distinct from the underlying mantle–in other words the rocks comprising the crusts are traditionally thought to be different to those of the underlying mantle. These continental and seafloor crustal rocks are often referred to as the *lithosphere*. The boundary between the base of the continental and seafloor crusts and the mantle is conventionally placed at the *Mohorovičić discontinuity*–shown in Figure 1.5 as the Moho–a boundary defined by a contrast in seismic velocity. The underlying relatively rigid uppermost part of the mantle is, in turn, referred to as the *asthenosphere*.

With the advent of modern geophysical and oceanographic studies it is now known that the seafloor crusts are, in fact, completely separate and distinct from the continental crusts. This will be elaborated on further at a later stage, sufficient to say that seafloor crusts are now known to comprise almost entirely of intruded volcanic lava which, in effect, represents cooled and solidified upper mantle rocks.

The seafloor crust can therefore just as easily be considered as exposed mantle rocks. In contrast, the continental crusts broadly consist of granite, as well as vast areas of sedimentary and metamorphic rocks originally derived from more ancient granite or volcanic rocks. In this context, the continental crustal rocks can be vaguely likened to the fragmented remains of more silica-rich rocks covering more extensive iron-rich upper mantle rocks.

This revelation in our knowledge about the seafloor crust is in stark contrast to past, pre-Plate Tectonic, thinking where seafloor crusts were originally thought to represent drowned and *oceanised* continental rocks covered by seawater. The outlines of the continents were then simply considered to be the result of periodic uplift of the crusts to form continents, followed by periodic down-warping and drowning of the crusts to form seas or oceans. The positions of all continents were then considered to be fixed and inter-continental land connections enabling species migration were the result of periodic land-bridges joining the various continents—which have now long since mysteriously disappeared.

At the human scale, the continental and seafloor crusts are also known to be very complex and are made up of a wide range of rock types formed at various times during the past 4,000 million years of recorded Earth's history—the age of the oldest known rocks on Earth. This relatively insignificant continental crustal and volcanic seafloor crustal skin is all that is available to measure and observe the unfolding geological history preserved on our Earth; all else is mere speculation or hypothetical guesswork.

1.3.1 Bathymetric Seafloor Mapping

Knowledge about the seafloor crusts has mainly come about as a result of geophysical surveys completed since the 1940s. Because of the rise of strategic submarine defence needs during the post-war years, knowledge of bathymetry and other oceanographic parameters became critical. As a consequence, the view we have of the seafloor today primarily stems from U.S. Navy supported marine geologists and geophysicists from government agencies and oceanographic institutions. Beginning in the late 1940's, ships from Columbia University's Lamont Geological Observatory, the Woods Hole Oceanographic Institution, and the Scripps Institution of Oceanography fanned out over the world's oceans collecting bathymetrical and geophysical data along with other oceanographic information. Once completed, for the first time scientists from all disciplines had a better understanding of what the previously unavailable 70 percent of Earth's surface really looked like.

Bathymetry is the study and mapping of the topography of the seafloor. It involves obtaining measurements of the depth of the ocean and is equivalent to mapping the surface topography of the continents. The name bathymetry comes from Greek *bathus* meaning deep and *metron* to measure. Originally, bathymetry involved the measurement of ocean depths through depth sounding. Early techniques used pre-measured heavy rope or cable lowered over a ship's side. In the twentieth century, sounding lines were entirely replaced by sonar systems. Sonar, invented during the Second World War, measures distances to the seafloor by emitting a short pulse of high-frequency sound and measuring the interval of time elapsing until a return echo is heard. Ships with sonar units attached to their hulls have since crisscrossed the world's oceans systematically measuring the ocean depths. Data collected has then made possible the publication of complete bathymetric maps of the contours and features of all of the world's oceans (Figure 1.6).

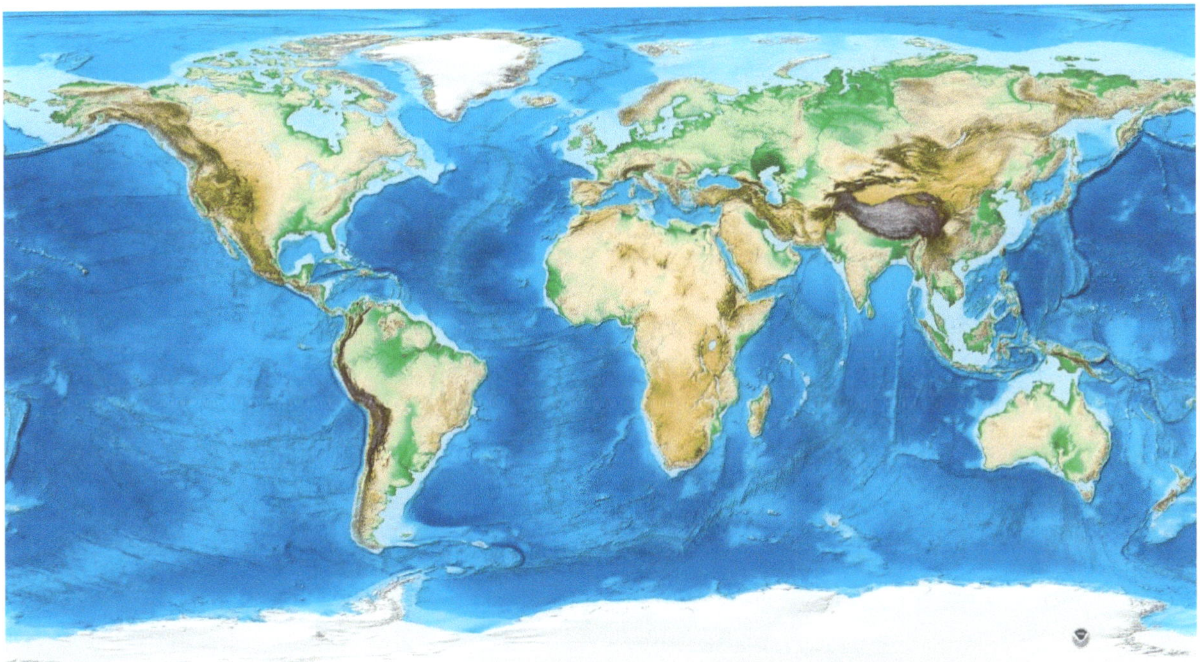

Figure 1.6 Continental topography and seafloor bathymetry map, compiled by the National Geophysical Data Centre: http://www.worldmapsonline.com/global_topography_bathymetry_wall_map_borders_labels.htm.

Within less than a quarter of a century, the vision of a featureless and static seafloor was shattered and, instead, the detailed seafloor bathymetry has revealed a far more complex terrane. The regional and world views of seafloor bathymetry produced by these investigators were instrumental in helping form the concepts of seafloor spreading, Continental Drift, and the theory of Plate Tectonics as we know them today. The advent of multi-beam sonar, along with access to the accurate global positioning system, satellite altimetry remote sensing technology, and computer-aided analysis and interpolation has since led to further advancements in bathymetric mapping.

By definition an ocean, from Ancient Greek *Okeanos*, is a body of saline water that composes a large part of the Earth's hydrosphere. In the context of Earth, it refers to one or all of the major divisions of the world oceans. These are, in descending order of area, the Pacific, Atlantic, Indian, Southern, and Arctic Oceans. The word sea is often used interchangeably with ocean but, strictly speaking, a sea is a body of saline water partly or fully enclosed by land. The oceans begin at the shore of each of the exposed continents and islands, the irregular boundary where the surface of a continent descends first to sea-level and then beneath it. If the depth of the ocean is measured along a line drawn straight out from a continental shore, the following sequence of seafloor features are typically seen.

For many kilometres out from the shore the ocean is only a few hundred metres deep and deepens quite slowly with slopes falling, on average, around 1.7 metres per kilometre. This flat, wide margin is found around every continent and is known as the continental shelf. The average width of a continental shelf is around 70 kilometres. The continental shelf ends at a sudden drop-off called the shelf break. Beyond the shelf break, the slope of the seafloor becomes much steeper, typically around 70 metres per kilometre. This steep embankment is called the continental slope and is generally grooved by submarine canyons and gullies. The continental slope is, on average, about 16 kilometres wide and descends to a depth of about 2.4 kilometres. There it ends at the abyssal plain as the slope moderates to a mere degree or two from horizontal. This gradual zone, which may be several hundred kilometres wide, is called the continental rise and is composed of fine-grained continental sediments washed down the many submarine canyons that notch the continental slope.

The abyssal plain, which is the deepest and most level part of the ocean, is often dotted with thousands of small extinct volcanoes called abyssal hills. The abyssal plains of the Atlantic Ocean appear smooth because its abyssal hills are buried under a thick blanket of continental sediment, but in the Pacific Ocean basin, which is ringed by trenches that trap sediments before they can spread over the seafloor, tens of thousands of unburied abyssal hills have been observed. Abyssal hills more than one kilometre high are called *seamounts*; whereas seamounts with flat tops are called *guyots* or tablemounts. Guyots are simply drowned volcanic islands.

Beyond the abyssal plain, which may be several hundred to thousands of kilometres wide, the seafloor begins to ascend again with a gentle slope. This area represents the flank of the mid-ocean-ridge, a long, undersea mountain chain that extends down the middle of the ocean. The Mid-Atlantic Ridge, for example, snakes down the middle of the Atlantic Ocean from the North Pole to Antarctica. As the ocean floor climbs slowly toward the centre of the mid-ocean-ridge its thin sediment blanket gets thinner and its surface more irregular. Here, the seafloor is also marked by thousand-kilometre long cracks called fracture zones that lie at right angles across the mid-ocean-ridge.

Along the centre of the mid-ocean-ridge is a rift valley, a deep V-shaped notch in the seafloor. From this valley, new seafloor crust is constantly being extruded from the Earth's mantle. Sheets of fresh, thick volcanic lava emerge from the mantle along the sides of the rift valley and are quickly quenched by the sea water. These growing sheets of seafloor crust have the effect of forcing the distant continents farther apart, a process called sea-floor spreading.

Beyond the rift valley, the depth sequence described above is observed in reverse: a gently declining slope of ridges and fractures, an abyssal plain, a continental rise, a steep continental slope, a somewhat flat continental shelf, and finally dry continental land again. This idyllic ocean setting is typical of the Atlantic Ocean. Elsewhere, however, such neat symmetry is not always found.

1.3.2 Magnetic Seafloor Mapping

During the 1950s scientists, using sensitive instruments called magnetometers adapted from airborne devices developed during World War II to detect submarines, also began to recognize strange magnetic patterns across the seafloor. This finding, though unexpected, was not entirely surprising because it was known that basalt–the iron-rich volcanic rock making up the seafloor crust–contains a strongly magnetic mineral called magnetite, which can locally distort compass readings. More importantly, because the presence of magnetite gives the basalt measurable magnetic properties, these newly discovered magnetic seafloor patterns provided an important means to study the distribution of intruded volcanic rocks throughout each of the oceans.

As more and more of the seafloors were mapped during the 1950s and 1960s, these magnetic patterns turned out to be not random or isolated occurrences but instead revealed a predictable zebra-stripe like pattern. These stripes were in turn found to be symmetrical about the centrally located mid-ocean-ridges (Figure 1.7). From this mapping, alternating stripes of magnetised basalt rock were shown to be laid out in parallel rows on either side of the mid-ocean-ridge, where one stripe showed a normal magnetic polarity and the adjoining stripe showed a reversed polarity. While the magnetic stripes were initially not linked to inversions of the Earth's magnetic field, this connection was later explained by Vine and Mathews in 1963. The overall magnetic pattern, as defined by these alternating bands of normally and reversely polarized rock, then became known as *magnetic striping*.

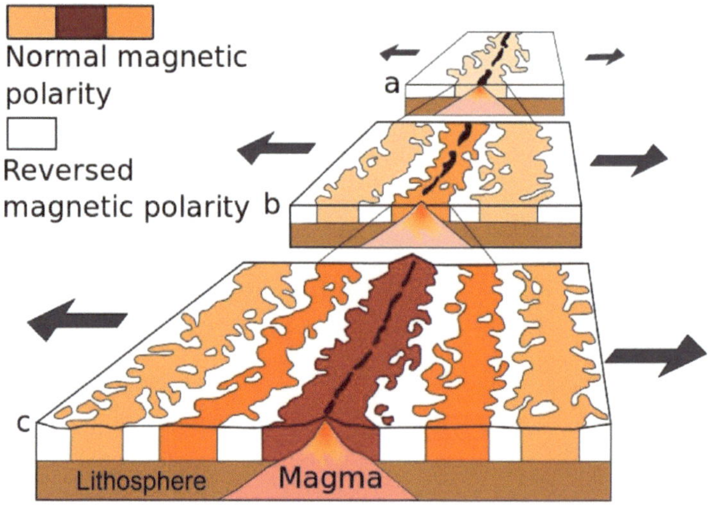

Figure 1.7 Symmetrical magnetic striping across a small part of the North Atlantic Ocean centred over the mid-ocean-ridge showing a progressive opening and increase in surface area of the seafloor crusts.
(http://en.wikipedia.org/wiki/File:Oceanic.Stripe.Magnetic.Anomalies.Scheme.svg)

The discovery of this symmetrical magnetic striping pattern suggested a close relationship between the mid-ocean-ridges, bathymetry, and the formation of the stripes. Seafloor spreading, as it later became known as, was first recognised by Carey in 1958 and again by Heezen in 1960. In 1961 scientists—most notably the American geologist Harry Hess—began to theorize that the mid-ocean-ridges mark structurally weak zones, where the seafloor was considered as being *"ripped apart"* lengthwise along the crest of the mid-ocean-ridges (Figure 1.8). From this, it was suggested that new volcanic lava from deep within the Earth must rise through these structurally weak zones to eventually erupt along the crest of the ridges and cool to form new basaltic seafloor crust.

It was further appreciated from dating the ages of the various newly formed seafloor crustal rocks that this process has operated over many millions of years. Subsequent mapping has since shown that this process is continuing to form new seafloor crust along the entire 65,000 kilometre-long system of centrally located mid-ocean-ridges, now known to be present throughout each of the oceans.

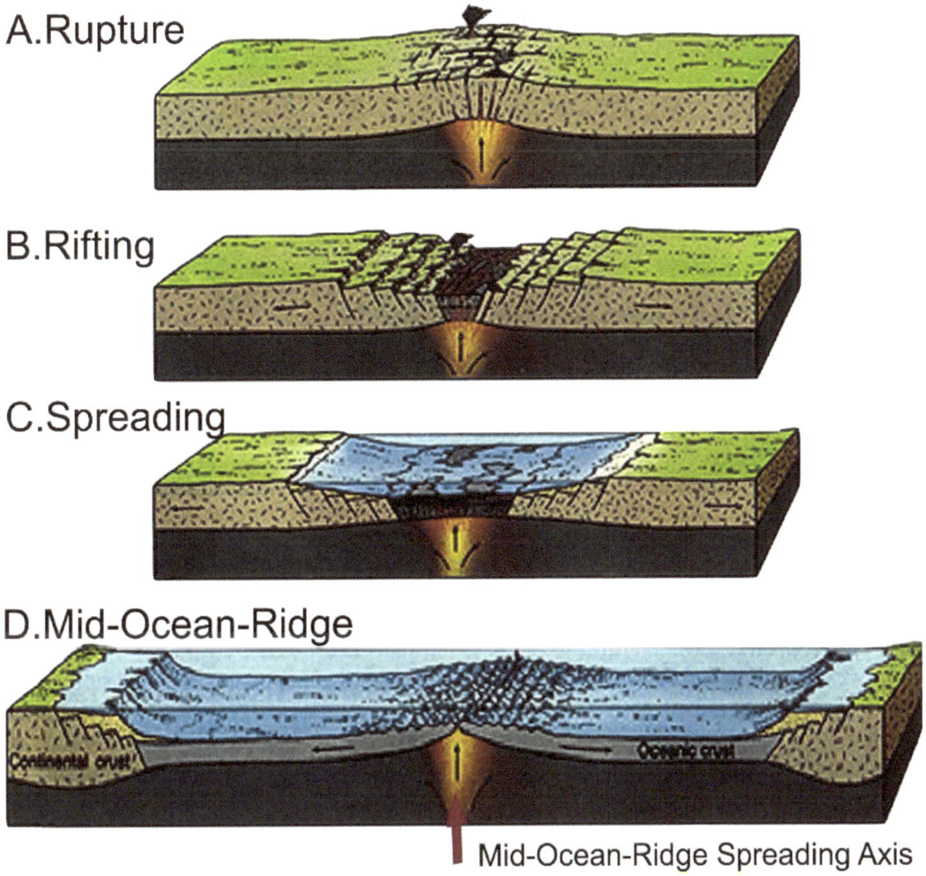

Figure 1.8 Schematic cross-sections showing structurally weak seafloor crustal zones opening along the crest of a mid-ocean-ridge to form a new ocean.

This seafloor spreading hypothesis was based primarily on the magnetic mapping evidence. It was also supported by several additional lines of evidence available at the time including evidence from age dating and bathymetric surveys. At or near the crest of the mid-ocean-ridges, the seafloor crustal rocks were shown to be very young and these rocks become progressively older when moving away from the ridge crests. The youngest rocks at the ridge crests always have present-day normal magnetic polarity. Moving away from the ridge crests the stripes of rock parallel to the ridges were shown to have alternated in magnetic polarity from normal to reverse to normal and so on. This suggested that the Earth's magnetic field has reversed many times throughout its history. By explaining both the zebra-like magnetic striping and the construction of the mid-ocean-ridge system the seafloor spreading hypothesis quickly gained converts. Furthermore, this seafloor crustal mapping is now universally appreciated to be a *natural tape recording* of both the history of the reversals in the Earth's magnetic field and opening of each of the oceans.

A profound consequence of this observation of seafloor spreading is that new crust is being continually intruded along the full length of the seafloor spreading ridges. It is interesting to note that this observation was initially—and still is—considered to support the theory of Earth Expansion, where new crust was formed at the mid-ocean-ridges as a consequence of an increase in Earth radius. History now shows that subsequent work has favoured the Plate Tectonic theory, where excess crust generated at the mid-ocean-ridge spreading centres is presumed to eventually disappear along seafloor trenches located along the margins of some continents where *subduction* of the seafloor crustal rocks is inferred.

Subsequent oceanographic work by the Commission for the Geological Map of the World and UNESCO during the 1980s has since led to the publication of the *Geological Map of the World* in 1990 (Figure 1.9). Later versions of this map are now available but is not used here because the editors have since changed the colours to a more subdued pallet. A legend for each of the colours depicted in this map is shown in Figure 1.10. Similarly, an *Age of Oceanic Lithosphere* map was published in collaboration with the National Geophysical Data Centre (NGDC) and the National Oceanic and Atmospheric Administration (NOAA) in 2008 (Figure 1.11).

In traditional geological usage the term *bedrock* refers to *the solid rock underlying unconsolidated surface materials*, such as soil or alluvium. In stratigraphy—a separate field of geology—bedrock is also regarded as *the native consolidated rock underlying the surface of a terrestrial planet*. In the context of geological mapping, however, this basic and somewhat historical definition is further extended to consider that; *a solid* [bedrock] *geologic map of an area will usually show the distribution of differing rock types, that is, rock that would be exposed at the surface if all soil or other superficial deposits were removed.*

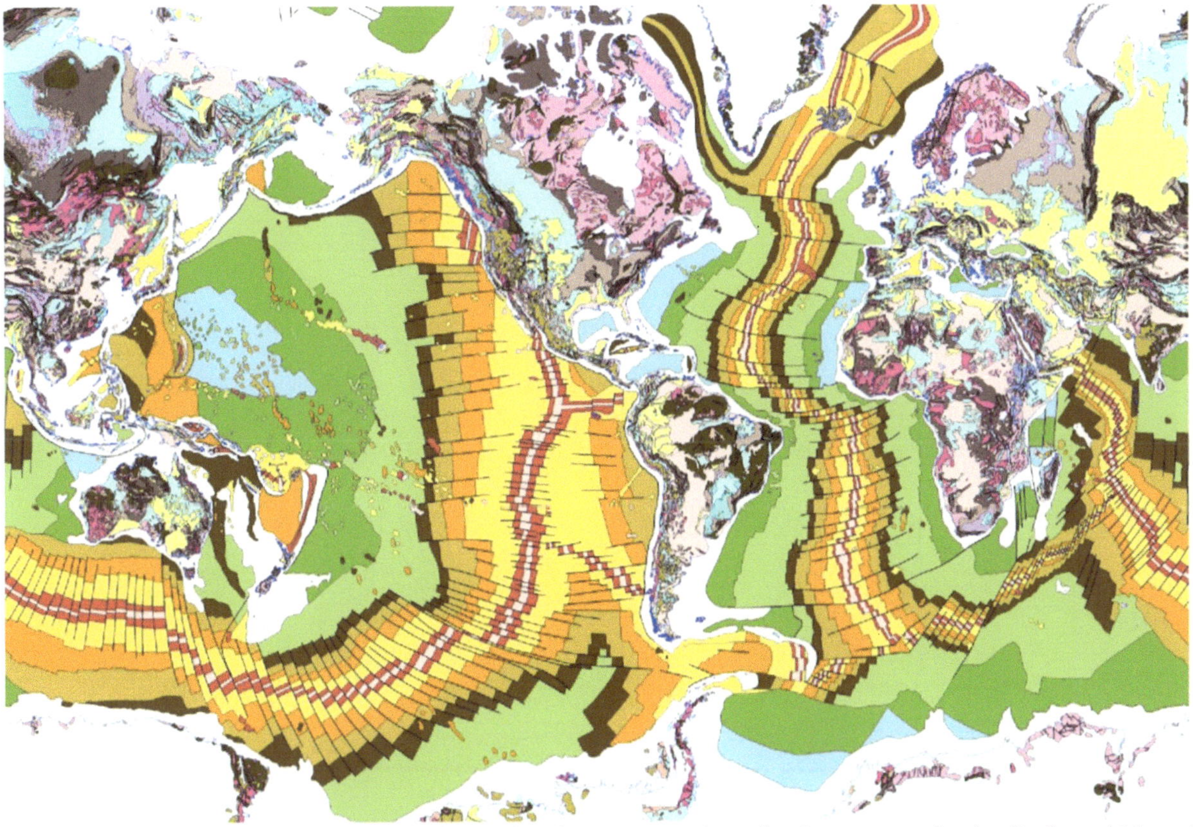

Figure 1.9 Geological Map of the World (digitized with permission from the Commission for the Geological Map of the World and UNESCO, 1990).

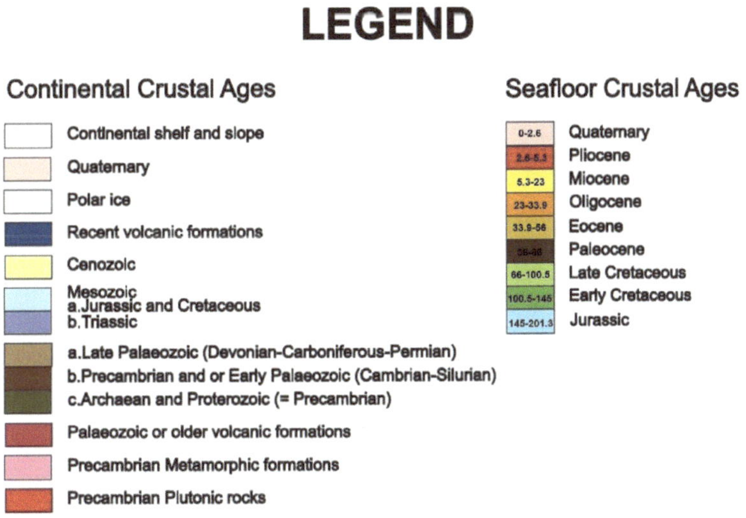

Figure 1.10 Geological timescale legend showing the various colours of the continental and seafloor crustal ages as shown in Figure 1.9. Seafloor crustal ages are in millions of years before the present-day.

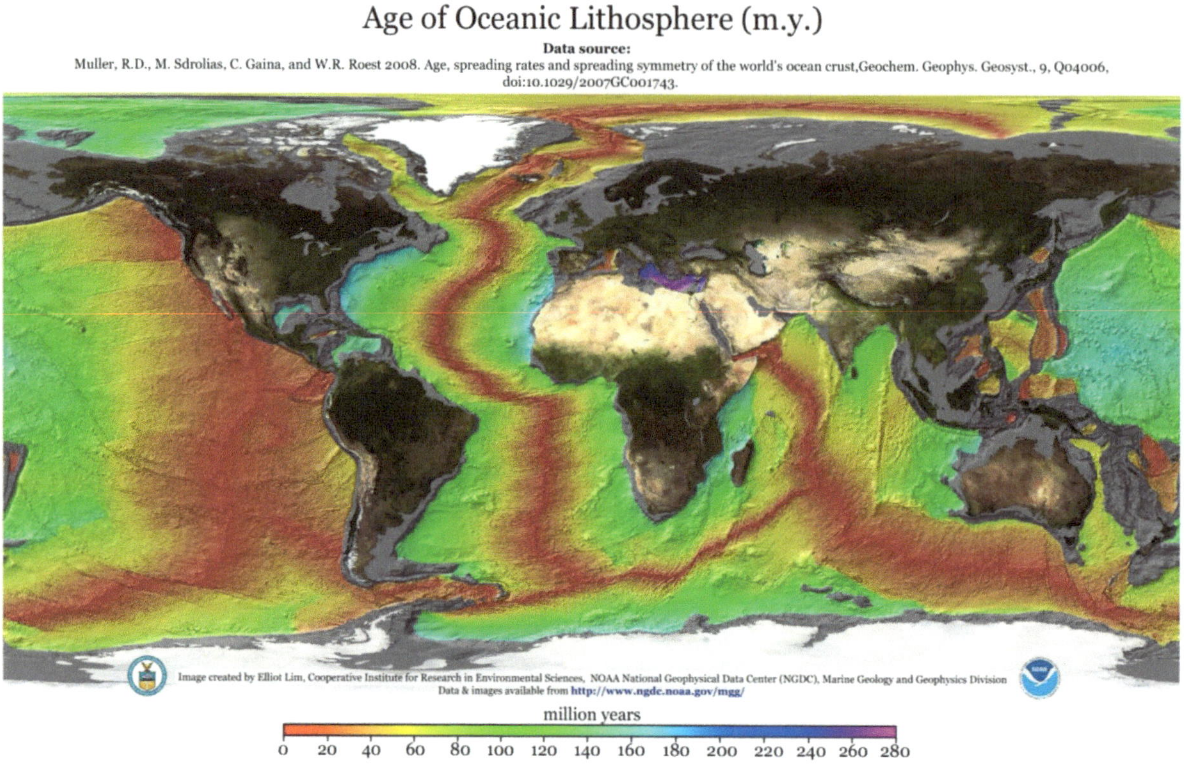

Figure 1.11 Age of Oceanic Lithosphere map (NOAA & NGDC, 2008).

What this simply means is that the bedrock geology depicted in the Geological Map of the World in Figure 1.9 shows the distribution of the various rocks occurring beneath the layer of surface soils. This mapping also includes bedrock located below other unconsolidated surface rocks, such as alluvial gravels in stream deposits, glacial debris, beach sands, recent volcanic eruptions, seafloor sediments, and so on. Removal of this surface material during mapping is necessary in order to give a truer depiction of the underlying distribution of more ancient rocks.

In each of the published maps shown in both Figures 1.9 and 1.11, the geophysical data gathering process introduced above has been extended to cover all of the oceans. In both maps the magnetic striping evidence has also been taken a step further. By dating the ages of samples of rock collected from the seafloor crust at regular intervals across the bottom of each of the oceans, and by comparing these ages with the magnetic striping, the seafloor crust in these maps is then depicted as intervals of geological time. The coloured striping shown in Figure 1.9 then represents age data displayed as set intervals of geological time, while in Figure 1.11 the age data is displayed as a continuous rainbow time-spectrum of colour.

What this magnetic striping evidence means is that the yellow stripes shown in Figure 1.9, for instance, located between the younger red stripes and the older orange stripes, represent volcanic rock—the volcanic lava called basalt—that was progressively intruded along the ancient mid-ocean-ridge spreading centres during the Miocene Epoch. The Miocene Epoch is the interval of time extending from 6 to 23 million years ago. During that time, the younger red and pink rocks respectively did not exist. It is easy to imagine that the two adjoining yellow Miocene stripes must have then been joined together and remained assembled along their common mid-ocean-ridges throughout this interval of time.

A good analogy for the seafloor stripes shown in Figure 1.9 is growth rings on a tree. For most trees, a new growth ring is added around the outside perimeter of a trunk or branch for each year of growth. By moving back in time and removing each growth ring in turn you can then measure the diameter of the ancient tree and by counting the remaining rings you also know how old the tree was. Similarly, the volcanic seafloor stripes shown in Figure 1.9 can be visualised as the growth pattern of each plate, where each stripe represents millions to tens of millions of year's crustal growth. By moving back in time and removing each growth stripe in turn the remaining surface area of each plate can be measured and an estimate of the ancient Earth radius can be made. This basic understanding is fundamental to further discussions on the origin of the oceans on an Expansion Tectonic Earth. It is also the fundamental process used here in the assemblage of crustal plates back in time to the Triassic Period—to about 200 million years ago.

At this stage there are a number of very important considerations relating to the bedrock geological mapping shown in Figure 1.9 that must be fully appreciated. As noted previously, this seafloor mapping represents a natural tape recording of the growth history of the reversals in the Earth's magnetic field and similarly represents the factual distribution and intrusive growth history of the volcanic rocks that make up the seafloor crusts. This preserved history must therefore be strictly adhered to during any theoretical crustal modelling study as well as during assemblage of the various crustal plates and continents back in time.

It should also be appreciated that none, or very little of this magnetic striping, age dating, or bathymetric evidence was available when both Expansion Tectonic and Plate Tectonic theories were first proposed. The global distribution of this magnetic striping, age dating, and bathymetry was, in fact, completed later in order to assist with, and substantiate assemblage of the various crustal plates and continents on a Plate Tectonic Earth model.

The bedrock geology map shown in Figure 1.9 will be discussed in more detail later. There are, however, a number of important observations that can be appreciated from an initial examination of this map. These observations include:

- The pattern of colours representing the ages of the crustal rocks shown in Figure 1.9 confirm that the seafloor rocks are vastly different from the continental crustal rocks. Similarly, the continental rocks, in general, are shown to be more ancient than the seafloor rocks.

- The intrusion of volcanic lava along each of the mid-ocean-ridge spreading centres represents quenched and cooled mantle rock; not oceanised continental rocks.

- The coloured striping shown in Figure 1.9 confirms that all of the oceans contain a mid-ocean-ridge—currently centred beneath the present day pink coloured stripes—and each ocean is increasing its surface area with time.

- This increase in surface area is symmetrical about the mid-ocean-ridges within each ocean. The maximum age of exposed seafloor volcanic crust, located along the continental margins, is early-Jurassic in age—about 170 million years old—which are shown as areas of pale blue striping.

- If it were possible to move back in time, each of the coloured stripes shown in Figure 1.9 must be progressively removed in turn. The corresponding edges of each coloured stripe must then be moved closer together. That is, the erupted volcanic rocks within each young coloured stripe must be progressively returned to the mantle where they originally came from.

- When moving back in time, each of the continents must move closer together in strict accordance with the coloured striping evidence. This phenomenon can then be used to accurately constrain the location of the various crustal plates during modelling of the ancient continents and oceans back in time.

- By measuring the surface area of each coloured stripe in turn, this information can also be used to investigate the change in seafloor surface area with time, and from this to investigate the change in Earth radius with time.

In summary, in Plate Tectonic theory the radius of the Earth is assumed to remain constant with time. As the ocean basins widen, new volcanic lava intrudes along the mid-ocean-ridge spreading centres, allowing new seafloor crust to solidify and form. To maintain a constant radius Earth, an equal amount of pre-existing crust must be disposed of elsewhere and returned to the mantle by a theorized process called subduction. This subduction process forms the basis of Plate Tectonic theory and, as a consequence, is essential in maintaining a static radius Earth premise.

Alternatively, for an Expansion Tectonic Earth the very same volcanic lava intruded along the mid-ocean-ridge spreading centres again widens the oceans and increases the surface area of seafloor crust. For an Expansion Tectonic Earth, this increase in surface area of all seafloors is a direct result of an increase in Earth radius and is cumulative with time. There is therefore no requirement for any net disposal of excess crust by subduction processes, nor is there a need to consider the existence of pre-existing crusts in order to maintain a static surface area.

What this simply means is that, for an Expansion Tectonic Earth, prior to about 250 million years ago—to be discussed in later chapters—none of the modern oceans existed. Instead, at that time all continental crust was assembled and united to form a single supercontinent called Pangaea enclosing the entire surface on a smaller radius Earth. As will also be elaborated on in later chapters, instead of the presence of modern oceans, a network of relatively shallow continental seas then covered low-lying parts of the ancient Pangaea supercontinent. At that time all of the relatively young seafloor volcanic crust, as well as much of the ocean water and atmosphere, were retained within the mantle where they originally came from.

While arguments exist for and against both theories, it is emphasized that small Earth model makers have consistently demonstrated, from a purely empirical perspective, that the continental crustal fragments making up both the ancient supercontinents and modern continents can indeed be fitted together precisely on a smaller radius Earth model. This assemblage process is somewhat like a spherical jigsaw puzzle, where continental crusts are assembled on a smaller radius Earth to form a single Pangaea supercontinent.

At this stage a fundamental question arises that must be seriously considered and further investigated:

Is this empirical phenomenon fact, or is it mere coincidence?

2 Preliminary Geological Considerations

"A Geologist is a scientist who understands the language of geology. Geology–from the Greek gê, earth and logos, study–is the science comprising the study of solid Earth, the rocks of which it is composed, and the processes by which they evolve. Geology gives insight into the history of the Earth, as it provides the primary evidence for tectonics, the evolutionary history of life, and past climates."

Before commencing our quest to consider the origin of continents and oceans on an Expansion Tectonic Earth I must first introduce a few basic geological considerations. This is necessary in order to help understand what it means when I talk about geologic time, the ancient oceans and seas, and the ancient supercontinents, especially as promoted in conventional geology and Plate Tectonic studies.

The geologic time scale plays a fundamental role in the geological sciences of today and this will need to be continually referred to when discussing the various intervals of geologic time within this book. The names of some of the ancient oceans and supercontinents are relatively well known in the modern literature, but these will also need a brief introduction before I can talk about the same oceans and supercontinents on an Expansion Tectonic Earth.

This introductory overview will simply introduce the commonly used terms and names used in modern geology, but I won't be delving too deep as I will be returning to the ancient seas and continents in later chapters once crustal modelling studies have been completed. It must be appreciated though, that in all conventional tectonic reconstructions of the continents and oceans there is a requirement for inclusion of often very large and generally hypothetical ancient oceans in order to maintain a constant radius Earth. Similarly, there is a requirement for relatively discrete fragments of prior supercontinents to collide and assemble together during a succession of complex but random *dispersion-amalgamation-dispersal mega-cycles*. These then form the basis for formation of conventional supercontinents, which are not applicable to what will be presented in this book.

Throughout this book I will also use the terms *modern oceans* and *modern continents*. I do this in order to distinguish between the present-day and ancient oceans, and similarly the present-day continents and ancient supercontinents. Sounds straight forward but this is necessary because, on an

Expansion Tectonic Earth, the origins of modern continents and modern oceans each formed together and have only existed since about 250 million years ago. It will be shown that the modern continents on an Expansion Tectonic Earth formed during break-up of the ancient Pangaea supercontinent and the modern oceans simply occupy the spaces that opened between these modern continents.

As mentioned earlier, the rest of this book is primarily about a new way of looking at the modern global geological, geophysical, and geographical evidence available today. In effect, I will be letting planet Earth tell its own story without imposing prior assumptions on the evidence or forcing the evidence to fit a preconceived theory. As such, the presently available conventional views and reconstructions of each of the ancient supercontinents and oceans described below cannot be reconciled on an Expansion Tectonic Earth. The primary crustal evidence used to reconstruct these ancient supercontinents and oceans is, however, real and factual. As the reader progresses further into this book it will become more and more evident when I say that these same oceans and supercontinents can be far better reconstructed and explained on Expansion Tectonic small Earth models.

2.1 Geologic Time Scale

The basis of this book, and in fact the basis of the science of geology, is the *geologic time scale*. The geologic time scale is defined as a system of time measurements that relates *stratigraphy*—the succession of rock strata—to time. It is used by all Earth scientists from around the world to define the timing and relationships between various geologic *events* that have occurred at different times throughout Earth's history. When talking about geologic time though, it relates to intervals of time lasting many millions, to tens or even hundreds of millions of years. Geologic time represents an interval of time between various identified geologic events—for example between two extinction events. These events have long been studied and recognised in geology from all countries of the world. With the advent of *radiometric age dating*—the use of radiometric decay of various elements within minerals to physically measure time—science is now able to measure the actual ages of various rocks or events. These ages are then used to provide further absolute numerical ages to the more commonly used intervals of geologic time.

When talking about the age of rocks, it should be appreciated that, strictly speaking, scientists are actually referring to the age of the various elements contained within minerals. These minerals in turn are what the rocks are made of and for igneous and volcanic rocks the age of the elements represents the moment in time when the various minerals first crystallised and the rocks first cooled and solidified. However, complications do arise with overprinting of ages when the rocks themselves have undergone periods of renewed heating and recrystallization, such as for the various metamorphic rocks. For igneous rocks that haven't been subjected to metamorphic changes, such as granites and volcanic rocks, the age of the dated minerals represent the age of the rocks when they were first formed. In contrast, for sedimentary rocks, where the minerals making up the rocks may have been derived from weathering and decomposition of a wide variety of igneous, metamorphic, or other sedimentary rocks, age dates are highly variable and generally inconclusive. In science, the measured dates of sedimentary rocks are only used to give broad indications of when the sediments were deposited or when they were metamorphosed, not for absolute ages of the rocks themselves.

The principles underlying the geological time scale were first developed by Nicholas Steno in the late 17th century. Steno argued that all rock layers—now called *strata*—must have been laid down in succession and that each rock layer represents an interval of time. He also formulated *"the law of superposition,"* which simply means that any given rock layer is probably older than those above it and younger than those below it.

While Steno's principles were simple and easy to understand, applying them to real rocks proved more complex and, over the course of the 18th century, geologists realized that:

- Sequences of rock strata were often eroded, folded, tilted, or even overturned after deposition and subsequent folding had occurred.

- Rock strata laid down at the same time but in different areas could have entirely different appearances and characteristics.

- The rock strata of any given area may represent only a small part of the Earth's long tectonic history.

During the early 1800s, when William Smith and Sir Charles Lyell first began recognizing that rock strata represent successive intervals of time, they had no means of measuring or even estimating absolute time scales or rock ages. The various rates of change in time used in their estimations were highly variable because of the limitations in both their dating technology and in their understanding of the amount of time available. While religion proposed dates of around six to seven thousand years for the age of the Earth based on Biblical records, early geologists of the time were instead suggesting millions of years for geologic periods, with some even suggesting a virtually infinite age for the Earth. In general though, geologists of the time had to resort to constructing the geologic time scale based on the relative positions of different rock strata and fossils. They then had to estimate the ages of the time scales based on studying rates of different kinds of weathering, erosion, deposition of sediment, and rock formation.

The identification of sedimentary strata by the fossils they contained was pioneered by William Smith, Georges Cuvier, Jean d'Omalius d'Halloy, and Alexandre Brogniart in the early nineteenth century. This observation enabled geologists to divide Earth history more precisely, but they still could not provide absolute ages. It also enabled strata to be correlated over large distances and sometimes even across continental boundaries. The principle behind this correlation was that, if two strata contained the same fossils then chances were they had been laid down at the same time and under similar conditions.

Naming the various intervals of geologic time was dominated by British geologists and the names chosen for the periods of Earth history reflect that dominance. The *Cambrian*—the classical name for Wales—and the *Ordovician* and *Silurian* were named after ancient Welsh tribes and these periods were defined using rock sequences first discovered from Wales. The *Devonian* was named for the English county of Devon and the name *Carboniferous* was simply an adaptation of the term *coal measures*, the old British geologists' term for the same set of strata. The *Permian* was named after Perm in Russia and was defined using strata in that region by Scottish geologist Roderick Murchison. Other periods were also defined by geologists from other countries. The *Triassic* was named by a German geologist Friedrich Von Alberti from three distinct sedimentary rock layers that are found throughout Germany and Northwest Europe. The *Jurassic* was named by the French geologist Alexandre Brogniart for the extensive marine limestone exposures of the Jura Mountains. The *Cretaceous*—from Latin *creta* meaning chalk—was defined by Belgian geologist Jean d'Omalius d'Halloy using strata in the Paris basin and named for the extensive beds of chalk.

Radioactivity in various naturally occurring minerals was first discovered in 1896. During the first half of the twentieth century, the development of its geological applications through radiometric age dating of rocks and minerals was first pioneered by British geologists, such as Arthur Holmes. This development allowed for more precise absolute dating of rocks and the first geologic time scale was eventually published by Holmes in 1913. From this, Holmes was able to further the newly created discipline of *geochronology*—the science of age dating of rocks—and published his world renowned book *The Age of the Earth* in which he estimated the Earth's age to be at least 1.6 billion years old.

Subsequent radiometric dating of rocks from all over the world now show that the oldest rocks found on Earth are more like 4 billion years old. Dating of extra-terrestrial rocks, such as meteors and Moon rocks, is thought to potentially extend the age of the Earth to around 4.54 billion years old. Throughout this book though I will limit reference to the age of the Earth to the age of the oldest known terrestrial rocks themselves and will adopt 4 billion years as the oldest known age of the Earth.

The modern chart of geologic time scales shown in Figure 2.1 was published by the International Commission on Stratigraphy. There are many more charts published by various organisations and each country generally has its own often more detailed charts representing their specific rock strata. Any of these time scales can be used, but I have adopted the chart shown in Figure 2.1 as it is the official, internationally recognised time scale. For the small Earth models that will be presented here I have also adopted the same colour scheme shown in this chart, which is in turn the same colour scheme used for the various intervals of geologic time shown in the Geological Map of the World.

This stratigraphic chart, in effect, summarizes the major geologic events that have influenced the history of the Earth since its formation. Recorded *Biblical Time*, out of interest, is represented by the Holocene Epoch in the upper left corner. The chart is arranged with the most recent geologic time periods at the top left hand corner and the most ancient at the bottom right hand corner—called the Hadean Eon. In this book, the time scales for the first three to four columns—the eon, era, period, and epoch times—in each table will be primarily used. Occasional reference will be made to the fifth column as required. The age symbol *Ma* simply means *millions of years before the present-day*. Each of the remaining columns is far too detailed for what will be presented here and won't be used.

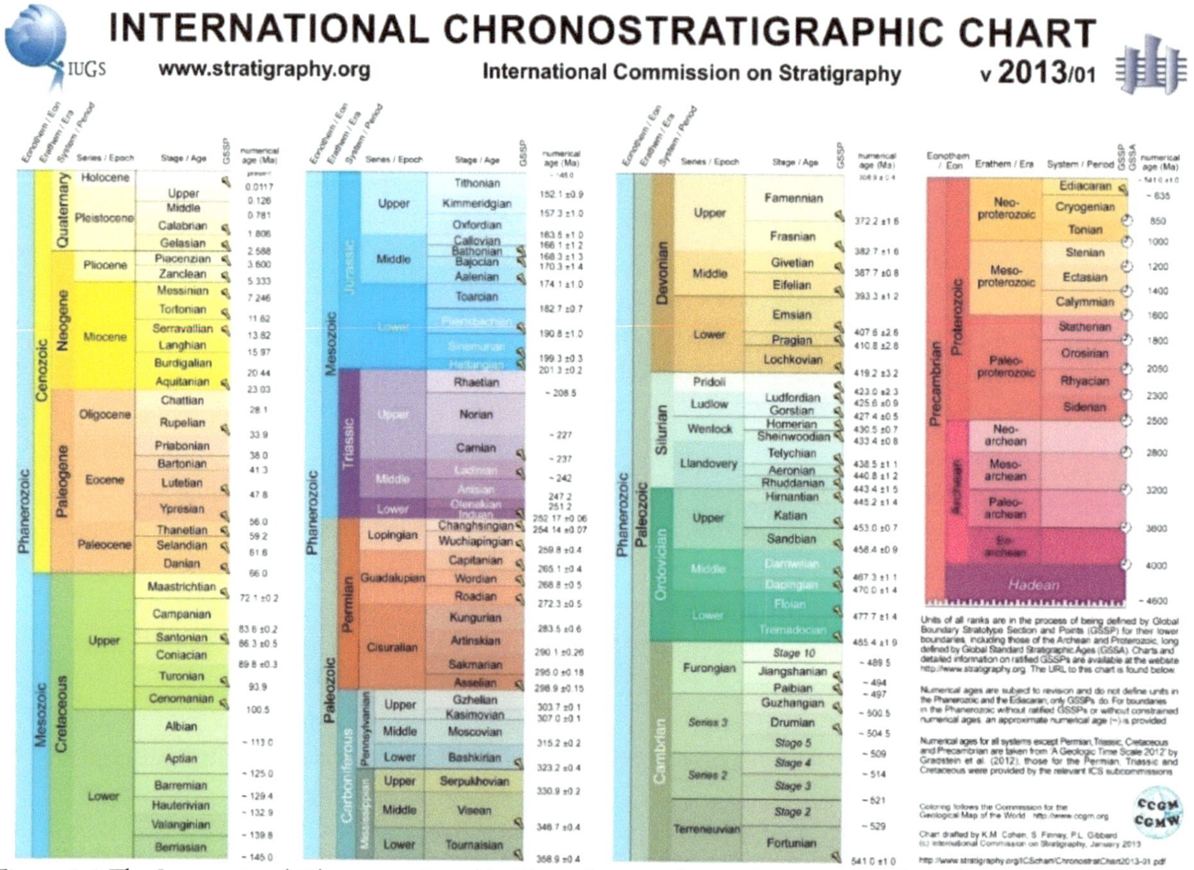

Figure 2.1 The International Chronostratigraphic Chart showing the subdivision of geological time into the various eons, eras, periods, epochs and ages, along with their measured absolute ages (International Commission on Stratigraphy, 2013).

The geologic time scale shown in this chart is organized into intervals of geologic time which accord with the various identified geologic events. The largest defined unit of time is the *supereon*, which is in turn subdivided into various *eons*. Eons are further divided into *eras*, which are then divided into *periods*, *epochs*, and *ages*. It should be appreciated that this table is a non-linear representation of time where the spacing of each time interval is graphical only, not actual lengths of time.

In general, the Precambrian Supereon is by far the longest interval of time, representing the first, approximately 3.4 billion years of Earth history. This supereon is subdivided further into the Archaean and Proterozoic Eons. This is followed by the Phanerozoic Eon, which is subdivided in turn into three eras: the Paleozoic, Mesozoic, and Cenozoic.

The Paleozoic Era started at the beginning of the Cambrian Period, which is defined by the first introduction and rapid evolution of primitive marine animals. The rest of the Paleozoic Era saw the first land plants evolve during the Ordovician Period, the first vascular plants during the Silurian Period, the first insects during the Devonian Period, and the first reptiles during the Carboniferous Period. The Mesozoic Era started with the Triassic Period when the first dinosaurs appeared and about 10 million years later the first mammals appeared. The Jurassic Period saw the first birds and the Cretaceous Period saw the first flowering plants. The Mesozoic Era ended as a mass extinction of many plant and animal species. The following Cenozoic Era is noted as the era of mammals and their dominance continues to the present-day.

2.2 Ancient Supercontinents

In Plate Tectonic studies the ancient supercontinents represent the random break-up and reassemblage of fragmented remains of more ancient supercontinental crusts. The modern continents in turn represent the fragmented remains of the former Pangaea supercontinent. Geographers studying the assemblage of the ancient landmasses use the term *supercontinent* to refer to *a single landmass consisting of fragments of all or several of the modern continents*. In conventional tectonic studies the earliest identified supercontinent is called *Vaalbara*. This supercontinent is said to have formed from the assemblage of very ancient continents and became a supercontinent by about 3.1 billion years ago. Vaalbara is then said to have broken up by about 2.8 billion years ago. The supercontinent *Kenorland* formed around 2.7 billion years ago and then broke up again sometime after 2.5 billion years ago. The remnant fragments of this supercontinent gave rise to *Laurentia*, *Baltica*, *Australia*, and *Kalahari*. The supercontinent *Columbia*, sometimes called *Nuna*, formed around 2.0 to 1.8 billion years ago and broke up between about 1.5 to 1.3 billion years ago.

The supercontinent *Rodinia* formed about 1.1 billion years ago and broke up roughly 750 million years ago. One of the fragments of Rodinia included large parts of the continents now located in the southern hemisphere. Plate Tectonics then brings the fragments of Rodinia back together into a different configuration during the late Paleozoic Era—about 300 million years ago—forming the better known supercontinent *Pangaea*. Following this assemblage, Pangaea is said to have subsequently broken up again to form the northern and southern supercontinents called *Laurasia* and

Gondwana about 200 million years ago, as well as lesser continents called *Laurentia* and *Baltica* plus a multitude of even smaller remnant fragments.

The following very brief introductory descriptions will only deal with the better known conventional supercontinents. Many of the more ancient supercontinental fragments are purely local names or assemblages which, on an Expansion tectonic Earth model, often simply represent islands within a network of continental seas. I also emphasise that the following descriptions and figures are Plate Tectonic descriptors and figures only and bear no, to very little, resemblance to Expansion Tectonic small Earth models that will shortly be presented here. These small Earth models will be further elaborated on in later chapters when I come to discuss the distribution of the ancient supercontinents and seas on an Expansion Tectonic Earth.

2.2.1 Pangaea

The name *Pangaea*–also spelt Pangæa or Pangea–is derived from the Ancient Greek words *pan* meaning *entire* and *gaia* meaning *Earth*. The name was adopted during a 1927 symposium discussing Alfred Wegener's theory of continental drift. In his book *The Origin of Continents and Oceans*, first published in 1915, Wegener postulated that all the modern continents had at one time formed a single supercontinent, which he called the *"Urkontinent,"* before later breaking up and drifting to their present locations.

In Plate Tectonic studies Pangaea (Figure 2.2) was a supercontinent that existed during the late Paleozoic and early Mesozoic times. This first formed about 300 million years ago and began to break up and rift apart around 200 million years ago before the component continents were separated into their current present-day configurations. A single ancient ocean called the *Panthalassa Ocean* is inferred to have surrounded Pangaea and at that time covered the rest of the world.

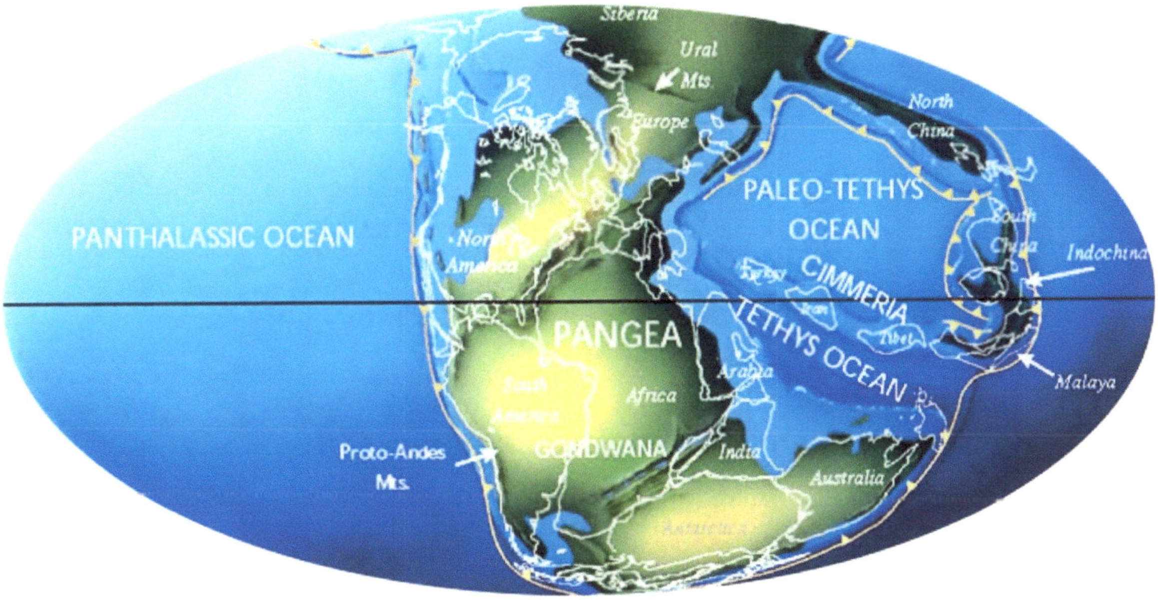

Pangaea: Early-Triassic 237 Ma

Figure 2.2 Plate Tectonic Pangaea supercontinent crustal assemblages (after Scotese, 2001).

2.2.2 Gondwana

The ancient supercontinent of *Gondwana* preceded Pangaea and was first named by the Austrian scientist Eduard Suess after the Gondwana region of central northern India. This name is derived from Sanskrit *gondavana* meaning *forest of the Gonds*, from which the Gondwanan Permian and Triassic sedimentary rock sequences were also described.

In conventional tectonic studies, Gondwana (Figure 2.3), originally called *Gondwanaland*, was the southernmost of two supercontinents–the other being *Laurasia*–that later became parts of the Pangaea supercontinent. Gondwana is said to have existed from approximately 510 to 180 million years ago and is believed to have *sutured*–meaning the joining together of fragments of various continents–between about 570 and 510 million years ago, thus joining East Gondwana to West Gondwana. These fragments separated from Laurasia from around 200 to 180 million years ago during the break-up of Pangaea, which then drifted further south after the break-up.

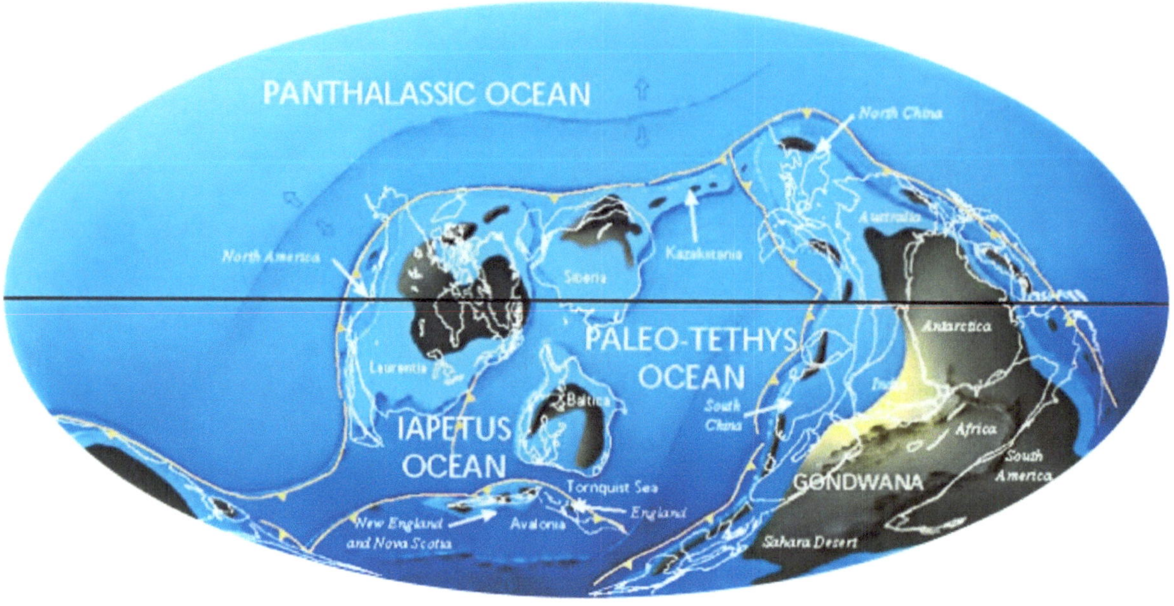

Gondwana: Mid-Ordovician 458Ma

Figure 2.3 Plate Tectonic Gondwana supercontinent crustal assemblages (after Scotese, 2001).

The Gondwana supercontinent included most of the landmasses present in today's southern hemisphere, including Antarctica, South America, Africa, Madagascar, and the Australian continent, as well as the Arabian Peninsula and the Indian subcontinent, which have since moved into the Northern Hemisphere.

2.2.3 Rodinia

In geology, *Rodinia*—from the Russian word *rodit* meaning *to give birth*—is the name of an ancient supercontinent that preceded Gondwana and is said to have contained most, if not all, of Earth's landmass (Figure 2.4). According to conventional tectonic studies Rodinia existed between 1.1 billion and 750 million years ago, during the late-Proterozoic times. It is said to have formed at about 1.0 billion years ago by the accretion and collision of fragments produced by the break-up of an even older supercontinent called *Columbia*, which was said to have assembled during global-scale *collisional* events around 2.0 to 1.8 billion years ago.

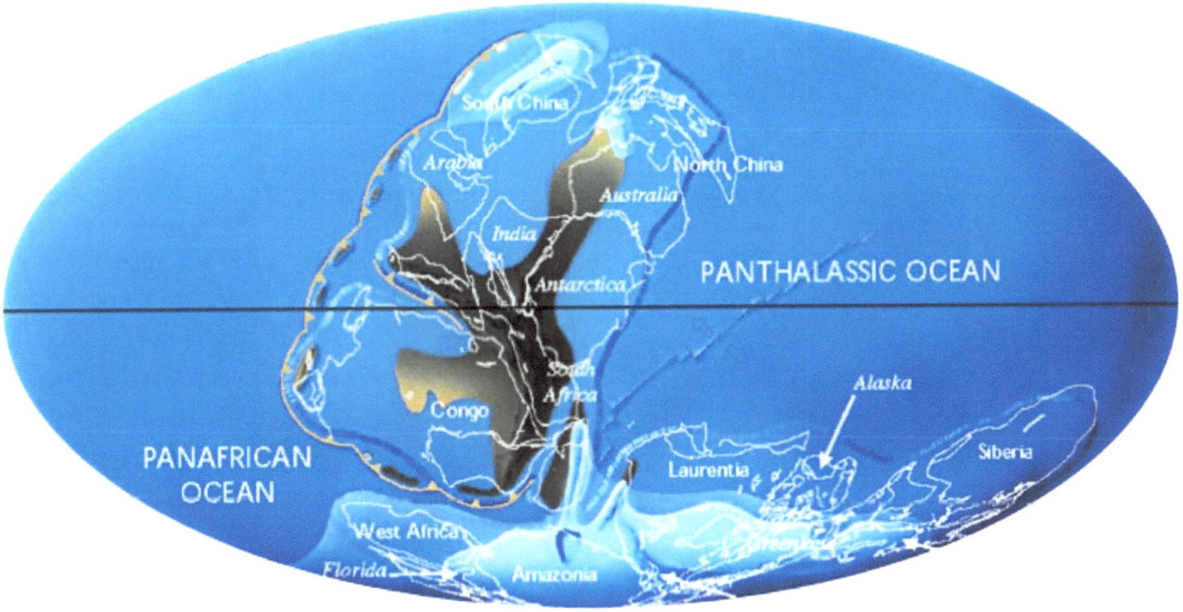

Rodinia: Late-Proterozoic 650 Ma

Figure 2.4 Plate Tectonic Rodinia supercontinent crustal assemblages (after Scotese, 2001).

In conventional tectonic studies Rodinia broke up during the late-Proterozoic and its continental fragments were re-assembled to form Gondwana and later as Pangaea some 300 to 250 million years ago. In contrast to Pangaea and Gondwana, little is known about the exact configuration or history of Rodinia. Cooling of the global climate around 700 million years ago and the rapid evolution of primitive life during the subsequent Cambrian times are often thought to have been triggered by the break-up of Rodinia.

2.2.4 Laurasia

In Plate Tectonic studies, *Laurasia* was the northernmost of two supercontinents–the other being Gondwana–that formed part of the Pangaea supercontinent from approximately 510 to 200 million years ago. It is said to have separated from Gondwana about 200 to 180 million years ago during the break-up of Pangaea and drifted further north after the break-up.

The name combines the names of the supercontinent *Laurentia*, the name given to the ancient North American continent, and Eurasia–the combination of the words Europe and Asia. As suggested by the name,

Laurasia included most of the landmasses that make up today's continents of the northern hemisphere, chiefly Laurentia, Baltica, Siberia, Kazakhstania, and the North China and East China cratons.

2.2.5 Laurentia

In the past, *Laurentia* was the ancient North American continent which still forms a large area of the continental landmass. It now forms the ancient geological core of the North American continent. In conventional tectonic studies, in its past history Laurentia remained a separate continent, as it is now in the form of North America, although originally it also included the areas of Greenland and the north-western part of Scotland. During other times, it is said that Laurentia may have also formed part of larger continents and supercontinents. The supercontinent is named after the Canadian Laurentian Shield which is in turn named after the St. Lawrence River.

2.2.6 Baltica

The supercontinent *Baltica* is a name applied by geologists to a late-Proterozoic to early-Palaeozoic continent that now includes the East European landmasses of north-western Europe and Asia. In conventional tectonic studies, Baltica was formed as an entity no earlier than 1.8 billion years ago. Before that time, the three fragments of ancient continents that now comprise the East European landmass originated from different places on the ancient Earth, which are then said to have drifted and sutured together to form Baltica.

2.3 Ancient Seas and Oceans

As with the ancient supercontinents, the ancient seas and oceans, as described in Plate Tectonic studies, cannot be reconciled in their entirety with Expansion Tectonics. As will be shown in this book, when moving back in time to around 250 million years ago, on an Expansion Tectonic Earth the surface area and radius of the Earth was much reduced. All of the modern continents were then assembled together as a single Pangaea supercontinent covering the entire surface of the ancient Earth. At that time there were no modern oceans, only a network of interconnected continental seas, and I emphasise the use of the term seas as distinct from oceans which will be reserved here for the modern oceans. It should also be appreciated that evidence for the existence of these ancient seas is now preserved in the rocks exposed on each of the present-day continents. So, when I say that conventional ancient seas and oceans cannot be reconciled with Expansion Tectonics, the factual geological and geographical evidence from each of these continents is definitely reconciled but the interpretation of this evidence from a conventional tectonic perspective cannot.

Once again, the following descriptions are simply brief introductions to the names themselves and in no way forms a complete coverage of all of the ancient oceans or seas referred to in conventional tectonic studies. The intent here is to simply introduce these more commonly used names in order to familiarise with what conventional tectonic studies are suggesting, before discussing both the ancient supercontinents and seas on Expansion Tectonic small Earth models.

2.3.1 Panthalassa Ocean

On all Plate Tectonic assemblages, as the ancient Atlantic and Indian Oceans are closed off there is a requirement to then open the Pacific Ocean in order to maintain a static radius Earth premise. This ancient ocean is called *Panthalassa*, which is a Greek word meaning *all sea*, also sometimes known as the *Panthalassic Ocean*. In conventional tectonic studies this ancient ocean was a vast area that surrounded the supercontinent Pangaea during the late-Paleozoic and the early-Mesozoic times, as shown in Figures 2.2 to 2.4 above. It included the precursor to the Pacific Ocean to the west and north, which in turn merged with the ancient *Tethys Ocean* to the southeast. Following the break-up of Pangaea and subsequent closing of the Tethys Ocean, Panthalassa was then reduced in total area to form the Pacific Ocean as each of the Atlantic, Arctic, and Indian Oceans opened in turn. The Panthalassa Ocean is also often called the *Palaeo-Pacific*—meaning old Pacific—because the present-day Pacific Ocean is said to have developed from it during the Mesozoic to the present-day.

Conventional tectonic studies also insist that, during Paleozoic and Mesozoic times, small fragments of pre-existing continents were drifting across the Panthalassa Ocean until they collided with and accreted against surrounding continental margins. These fragments are then said to be currently preserved as rocks along the North American and Asian margins.

2.3.2 Tethys Ocean

In conventional tectonic assemblages the ancient *Tethys Ocean* existed between the supercontinents of Gondwana and Laurasia during much of the Mesozoic Era. This ocean continued until opening of the Indian and Atlantic Oceans commenced during the Cretaceous Period. This ocean is, in effect, depicted as being located between the fragmented remains of the present European and Asian continent and its existence is necessary in order to close off the North Atlantic Ocean. As conventional theory has been modified over the years, scientists have also extended the Tethys name to refer to similar, more ancient oceans that may have preceded it.

The ancient Tethys Ocean is said to have existed from Silurian through to Jurassic times. Before then a more ancient Tethys Ocean existed prior to the Cambrian which extended into the Devonian Period. This ancient ocean was located between Baltica and Laurentia to the north and Gondwana to the south. The ocean is shown in Figures 2.2 and 2.3 above and labelled as the *Palaeo-Tethys Ocean.* The western part of the Tethys Ocean is also referred to as the Tethys Sea, the Western Tethys Ocean, or Alpine Tethys Ocean. The present-day Black Sea, Caspian Sea, and Aral Sea are in turn thought to be its remains. In conventional tectonic studies the western Tethys Ocean is inferred to have covered many small crustal plates, as well as Cretaceous *island-arcs* and smaller continents. The eastern part of the Tethys Ocean is likewise sometimes referred to as the Eastern Tethys.

During the Oligocene Period, large parts of central and Eastern Europe are said to have been covered by a northern branch of the Tethys Ocean. This branch was separated from the Tethys by the formation of the Alps, Carpathians, Dinarides, Taurus, and Elburz mountains before it gradually disappeared during the late Miocene Period and became an isolated inland sea.

2.3.3 Iapetus Ocean

The ancient *Iapetus Ocean* existed during the late Proterozoic and Paleozoic times between about 600 and 400 million years ago. The Iapetus Ocean was situated in the southern hemisphere, located between the ancient continents of Laurentia, Baltica, and *Avalonia* as shown in Figure 2.3. This ocean then disappeared when the three continents joined to form yet another large supercontinental landmass called *Laurussia.*

In conventional tectonic studies, because the Iapetus Ocean was positioned between ancient continental landmasses that would later roughly form the opposite shores of the Atlantic Ocean, it was considered to be a precursor of the modern Atlantic Ocean. The Iapetus Ocean was named after the *titan Iapetus*, who in Greek mythology was the father of Atlas after whom the Atlantic Ocean was named.

3 Measuring Earth Radius

"The oceanic crust of any epoch, originating as a result of spreading, is fully fixed in the geological record, and continues to increase the surface area." Blinov, 1983

The assumption that the radius of the Earth has remained constant, or essentially constant throughout Earth history was regarded by Carey as a *meme*—a notion or idea that has remained both untested and uncontested for so long that it is now unconsciously considered a truth. Because it is considered a truth, science is mistakenly led to believe there is no need to test it regardless of whether or not it has ever been tested at all. In fact, this meme has become so deeply ingrained into our culture and for so long now that people do not realize that it needs to be tested. Hence, when carrying out any form of research about the Earth, all assumptions based on a constant Earth size are therefore unconsciously perpetuating a potentially false premise.

In modern usage this conundrum is also referred to as a *paradigm*, which is defined in the Oxford English Dictionary as *"...a pattern or model, an exemplar."* The historian of science Thomas Kuhn gave a paradigm its contemporary meaning when he adopted the word to refer to *"...the set of practices that define a scientific discipline at any particular period of time."* In his book *The Structure of Scientific Revolutions* Kuhn defines a scientific paradigm as *"...universally recognized scientific achievements that, for a time, provide model problems and solutions for a community of researchers."* Based on this Kuhn saw the sciences as going through alternating periods of normal science, where an existing model of reality dominates a protracted period of puzzle-solving leading to an irresolvable crisis situation, followed by revolution where the model of reality itself undergoes sudden drastic change.

For members or researchers of a particular discipline, such as Plate Tectonics today, the established paradigm—referring to the assumption that the radius of the Earth is constant—is universally considered so convincing that it renders even the possibility of alternative theories unconvincing and counter-intuitive. In such a situation Kuhn considered the paradigm to be opaque, thus appearing to be a direct view of perceived reality itself, which obscures the possibility that there might be other alternative possibilities hidden behind it. The universal conviction that the current Plate Tectonic paradigm is true then disqualifies any evidence that might undermine the paradigm itself, which in turn leads to a build-up of un-reconciled anomalies in the fundamental discipline.

The eventual revolutionary overthrow of an incumbent paradigm and its replacement by a new one is called a *paradigm shift*, which can be likened to the change that occurs when our visual interpretation of an ambiguous image flips-over from one state to another. Kuhn further wrote that "*Successive transition from one paradigm to another via revolution is the usual developmental pattern of mature science.*" This paradigm shift tends to be most dramatic in sciences that appear to be stable and mature, such as Plate Tectonic theory. In most cases the greatest barrier to a paradigm shift, however, is the problem of *paradigm paralysis*—the inability or refusal to see beyond the current models of thinking.

Moving back to the problem of measuring Earth radius, if I were to ask what is the radius of the Earth now most people would understandably respond by saying they don't know. Many people could make a reasonably educated guess and maybe a handful could tell me exactly what the current Earth radius is. The reader may well ask what does this have to do with the origin of continents and oceans. My response would be, if Earth radius has remained relatively constant since the beginning of geological time—that moment in time when minerals were cool or stable enough to be preserved as rocks—the radius of the Earth has no real relevance to the origin of continents and oceans. If this were the case, then current tectonic theory is correct and science does not need to bother with measuring Earth radius, we can simply leave that to the geodesists.

If I were to then ask what the radius of the Earth was at say one million years in the past, most, if not all, people should say they do not know; human capabilities were not advanced enough one million years ago so how can we possibly know what the radius was? In reality, most people would probably err towards saying that the ancient Earth radius was the same as it is today. If prompted as to how they know this, I would envisage that many of the answers would range from being a guess, to reciting religious beliefs, through to consideration of the Kant-Laplace theory of the creation of the Solar System, or to quotations from the Big Bang theory.

All things considered though, the answer to my question cannot be answered by mainstream science because of the ingrained belief that Earth radius is constant and therefore the question doesn't need to be asked let alone answered. Regardless of this, true science, however, still insists that:

Proof of any theory comes through direct observation or by direct measurement.

What this simply means is that if scientists cannot physically measure the ancient Earth radius, or provide unequivocal evidence to support conclusions about the ancient Earth radius, all answers to my question are mere speculation. In other words, the answers are not factual if they cannot support their conclusions with hard factual evidence. Speculation based on a speculative premise is not a precursor to fact. If conclusive evidence cannot be provided to say the ancient Earth radius was, or was not, the same as today then current tectonic theory is potentially incorrect and hence an understanding of Earth radius has enormous implications towards understanding the origin of the continents and oceans.

In this chapter I will delve a bit further into understanding what the Earth's radius is and what it has been in the past, so it is important to understand a little about what Earth radius actually means, as well as about the history behind determining an accurate Earth radius. While this may seem rather mundane, it is extremely important when I later discuss whether Earth radius has remained constant throughout time, or has, as I will show, been changing with time. Understandably this is not something that immediately springs to mind when we think about modern geology or global tectonics, even though it most definitely should. The primary intent of this exercise will become increasingly evident in later chapters. Once established, I will then use this ancient Earth radius information to construct accurately scaled *small Earth models* in order to then investigate the distribution of modern geological, geophysical, and geographical evidence.

Ever since the dawning of human history humankind has always been interested in the Earth, after all it is where we all live, gather food, seek shelter, and have successfully evolved. During these very early ancestral times this interest was, of course, limited to the immediate vicinity of a particular person's tribal hunting grounds, their village, home, or place of residence. The fact that we live on a near spherical planet would not have been apparent at that time or of much interest to most people struggling to merely survive. As humanity developed and the sciences became increasingly popular, and in particular as long distance voyages of trade or discovery became more commonplace, so did our interest in understanding and mapping the size, shape, and composition of the Earth.

While the term radius is normally a characteristic of perfect spheres, the term, when applied to the Earth, more generally means the distance from the centre of the Earth to a point on the surface. It can also mean the radius of a sphere whose curvature matches the curvature of an idealized model of the Earth at any given point. Similarly, it can also mean some kind of average of such distances to then give a mean value for Earth radius.

Regardless of which idealized model is used to define the Earth radius, all modern determinations of Earth radius fall between a polar minimum radius of about 6,357 kilometres and an equatorial maximum of about 6,378 kilometres with a mean radius of 6371 kilometres—although this figure may vary depending on which model of the Earth is used. In other words, the Earth is not perfectly round. These varying measurements for Earth radius now tell us that the Earth in fact deviates from a perfect sphere by about a third of a percent, which is still sufficiently close to treat it as a sphere in most contexts, including the small Earth modelling studies I will shortly introduce here. From this understanding, it is important to appreciate that local variations in modelling Earth radius effectively prevent the definition of a single, precise Earth radius and scientists therefore generally adopt an idealized mean Earth radius model.

Ever since Eratosthenes made the first estimate of Earth radius in the third century BCE, many Earth radius models have been devised to describe the shape of the Earth. Historically, these models were based on regional topography, in particular topography local to where the philosopher lived or worked which then gave the best reference spherical model for the area under study. As modern satellite remote sensing, especially the Global Positioning System, rose in importance during the late twentieth century true global models have now been developed which best approximate the mean radius of the Earth as a whole.

Early ideas about the shape and size of the Earth include the early Greeks who, in their speculation and theorizing, ranged from considering a flat disc, as advocated by Homer, to a spherical body as postulated by Pythagoras—an idea supported one hundred years later by Aristotle. In contrast, Anaximenes, another early Greek philosopher, also strongly believed that the shape of the Earth was rectangular. While there were many contrasting ideas, most scientists and philosophers of the time favoured a spherical Earth model. This consensus was based on the knowledge that lunar eclipses were seen as circular shadows, which could only be caused by a spherical Earth, and that the star Polaris is seen lower in the Mediterranean sky as one travels south. The mathematician Pythagoras also considered that the most perfect figure imaginable was a sphere. From this, he reasoned that the *"gods"* would only create a perfect figure for the Earth and therefore the Earth must be spherical in shape—circular reasoning maybe.

The spherical shape of the Earth was most widely supported during the Greek Era and efforts to determine the Earth's size soon followed. Plato determined the circumference of the Earth to be 400,000 stadia—somewhere between 62,800 and 74,000 kilometres depending on the size of the individual *stadion* used to convert the measurement, while Archimedes estimated it to be 300,000 stadia or 55,500 kilometres using the Hellenic stadia. Scholars now generally take the length of a stadion to be about 185 metres. It is now known that the present day circumference of the Earth is 40,075.16 kilometres, so, in reality, Plato's figure was actually a guess and Archimedes' was a more conservative approximation.

In Egypt, the Greek scholar and philosopher Eratosthenes (276–195 BCE) is said to have made the first scientific estimation of the circumference of the Earth—from which Earth radius can be derived. He had heard that on the longest day of the summer solstice the midday sun shone to the bottom of a well in the town of Syene (Aswan). At the same time he observed that the sun was not directly overhead at Alexandria. Instead, it cast a shadow with the difference in vertical angle equal to 1/50th of a circle (7°12'). Legend also has it that in order to complete his calculation he had someone walk from Alexandria to Syene to measure the distance between the two towns, which turned out to be 5,000 stadia or, at the usual Hellenic 185 metres per stadion, about 925 kilometres.

From these observations, Eratosthenes concluded that the linear distance between Alexandria and Syene was $1/50^{th}$ of the circumference of the Earth, which he then calculated to be 50×5,000 = 250,000 stadia. It is uncertain what his stadion equals in today's units but most current specialists in antiquities accept that it was the regular Hellenic stadion and hence the circumference of the Earth determined by Eratosthenes was 46,250 kilometres.

Another legendary influential Greek scholar Posidonius (135–51 BCE) also made an ancient measurement of the size of the Earth. He noted that the star Canopus, hidden from view in most parts of Greece, just grazed the horizon at Rhodes. Posidonius measured the angular elevation of Canopus at Alexandria and determined that the angle was 1/48th of a circle. He assumed the distance from Alexandria to Rhodes to be 5,000 stadia and from this he computed the Earth's circumference to be 240,000 stadia, or 44,400 kilometres. Later, either he or a follower may have altered the base distance to agree with Eratosthenes's Alexandria-to-Rhodes figure of 3,750 stadia and Posidonius' final circumference was determined to be 180,000 stadia which equals 33,300 kilometres.

The Indian mathematician Aryabhata (476-550 ADE) is considered a pioneer of mathematical astronomy. He described the Earth as being spherical and that it rotates on an axis. He explained that the apparent motion of heavenly bodies is only an illusion based on the following simile, "*Just as a passenger in a boat moving downstream sees the stationary* [objects on the river banks] *as traversing upstream, so does an observer on Earth see the fixed stars as moving towards the west at exactly the same speed.*" Aryabhata estimated the circumference of Earth to an accuracy of 1 percent of the present known value, which was remarkable for his time. Aryabhata also went on to describe the radius of planets in terms of the Earth-Sun distance as essentially their periods of rotation around the Sun. He also gave the correct explanation for lunar and solar eclipses and considered that the Moon shines by reflecting sunlight.

The Muslim scholars, who also held to the spherical Earth theory, used this knowledge to calculate the distance and direction from any given point on the Earth to Mecca. This determined the *Qibla*, or Muslim direction of prayer. Muslim mathematicians also developed spherical trigonometry which formed the basis of these calculations. Around 830 ADE, Caliph al-Ma'mun commissioned a group of astronomers to measure the distance from Tadmur (Palmyra) to al-Raqqah in modern Syria. They found the cities were separated by one degree of latitude and thus calculated the Earth's circumference to be 40,248 kilometres, which is very close to the current modern value of Earth circumference. Similarly, Abu Rayhan Biruni introduced techniques to measure the size of the Earth and distances using triangulation. He found the radius of the Earth to be 6,339.6 kilometres, compared to today's mean radius of 6,371 kilometres, a value he determined with such precision that was not obtained in the West until the 16th century.

Revising the figures attributed to Posidonius, another Greek philosopher determined the Earth's circumference to be 28,968 kilometres and this last figure was promulgated by Ptolemy through use of his world maps. These maps of Ptolemy strongly influenced the cartographers of the middle Ages and it is probable that Christopher Columbus, by using these maps, was mistakenly led to believe that Asia was only five to six thousand kilometres west of Europe. It was not until the 16th century that this concept of the Earth's size was revised to accord with reality. During that same period, the Flemish cartographer Mercator made successive changes in the size of the

Mediterranean Sea and all of Europe. This had the effect of increasing the size of the Earth, again to a more realistic size.

The next great advance in determining the size of the Earth came with the invention of the telescope and theodolite, as well as the development of trigonometry and logarithm tables. In Europe, Jean Picard performed the first modern meridian arc measurement in the years 1699 to 1670. He measured a base line with the aid of wooden rods, used a telescope to measure his angles and computed using logarithms. Jacques Cassini later continued Picard's arc northward to Dunkirk and southward to the Spanish border. Cassini divided the measured arc into two parts, one northward from Paris and another southward. When he computed the length of a degree from both parts he found that the length of one degree in the northern part was shorter than that in the southern part. This result, if correct, showed that the Earth was not a sphere, as previously thought, but an oblong ellipsoid which had an extended elliptical shape. This also contradicted the earlier computations made by Isaac Newton and Christiaan Huygens, who predicted the Earth to be an oblate or pear shaped spheroid with a flattening ratio of 1:230.

Further measurements by the French Academy of Sciences during the 1700s conclusively showed, however, that the Earth was indeed an oblate pear shape with a flattening ratio of 1:210. This oblate approximation to the true figure of the Earth also showed that the Earth had an oblong ellipsoid of revolution around the Sun.

Following on from these discoveries and measurements, in South America Pierre Bouguer noticed, as did George Everest in the 19th century Great Trigonometric Survey of India, that the astronomical vertical angle tended to be pulled in the direction of large mountain ranges due to the gravitational attraction of these huge piles of rock. As this vertical angle should be everywhere perpendicular to the idealized surface of mean sea-level, or the *geoid* as it is formally known, this demonstrated that the figure of the Earth has an even more irregular ellipsoid shape. From this knowledge the study of the *undulation of the geoid* became the next great undertaking in the science of studying the size and shape of the Earth.

In the late 19th century, the *Zentralbüro für die Internationale Erdmessung* (Central Bureau for International Geodesy) was established by Austria-Hungary and Germany. One of the most important goals of the Bureau was the derivation of an international ellipsoid shape and a gravity formula, which had to be optimal not only for Europe but also for the whole world. This Zentralbüro was an early predecessor of the International

Association of Geodesy (IAG) and the International Union of Geodesy and Geophysics (IUGG), founded in 1919.

Most of the relevant theories of geodesy were later derived by the German geodesist Friedrich Robert Helmert in his famous book *Die mathematischen und physikalischen Theorieen der höheren Geodäsie, Einleitung und 1 Teil* (1880) (Mathematical and Physical Theories of Higher Geodesy, Vol. 1). Helmert also derived the first global ellipsoid in 1906 with an accuracy of 100 metres (0.002 percent of the Earth's radius). The US geodesist Hayford also derived a global ellipsoid in about 1910 based on intercontinental measurements, with an accuracy of 200 metres, which was adopted by the IUGG as the "*International Ellipsoid 1924.*"

With the advent of modern satellite and computer technologies during the 20th century, our present knowledge of the size and shape of the Earth is, by comparison, at a very advanced state of accuracy. The International Earth Rotation Service (IERS) (2003) now quotes the equatorial Earth radius as 6,378,136.6 metres and similarly quotes the polar radius as 6,356,751.9 metres.

The point I am alluding to in this extended discussion about measuring an ancient Earth radius is that, when anyone talks about determining the size or shape of the Earth they are actually talking about determining the size or shape of the Earth at that moment in time when the particular person was making the determination. At no time do any of these past or present scientists talk about determining the size or radius of an ancient Earth, it always relates to determining the present-day radius of the Earth at that moment in time. Why? Simply because it is assumed that Earth radius has always been constant. Therefore, if, for instance, scientists tell us that modern satellite technology can now measure the radius of the Earth to sub-centimetre accuracy, they are actually saying that they can now measure the present-day radius of the Earth to sub-centimetre accuracy, not the past ancient Earth radius.

I dwell on this point because, when science discusses, talks about, or researches any form of global tectonic theory they are actually dealing with extended periods of geological time. In other words they are using timeframes that are many tens to hundreds, or even thousands of millions of years in the past. Scientists therefore have no justification whatsoever to simply assume the radius of the Earth now is the same, or even very similar to the ancient Earth radius without first making an attempt to prove conclusively one way or the other.

At the beginning of human history the size and shape of the present-day Earth was unknown. It was a struggle for the early scientists or philosophers to calculate the circumference of the Earth to the nearest hundreds of kilometres and hence mathematically derive an Earth radius, let alone to accuracies suitable for considering any changes in the Earth's size. From this introductory discussion I will now state categorically that without fully appreciating or comprehending that the Earth's radius could possibly be changing with time it was, and still is, not possible for science to measure the past ancient Earth radius.

How can I say measurement is not possible when I will later show the reader how to calculate the past mean radius of the Earth and will present a formula that will enable us to determine the ancient Earth radius at any moment in time? Simple! It is an unfortunate human trait that many and in fact, I may even be as bold as to say, most humans tend to condition themselves to fail to recognize evidence for something they cannot comprehend—again remember the constant Earth radius meme. Therefore, throughout history very few people have seriously considered that Earth radius may have, or may still be changing with time simply because they have failed to recognize or even acknowledge the evidence available to test or contradict their preconceived assumptions.

Contrary to all of this though we can in fact measure the ancient Earth's radius at any moment in time, we just need to recognize what it is that we should measure. As will be shown the key to measuring the ancient Earth radius is written and preserved in the Earth's *rock-record*, that is, in the rocks themselves. The evidence for determining ancient Earth radii for the various geological time periods and epochs has only been available since the late-1980s, well after Plate Tectonics was first conceived and popularized. This evidence is shown preserved in the *Geological Map of the World* as previously introduced in Figure 1.7. The reader therefore needs to fully comprehend and appreciate what it is that we are actually looking at in this map before we can utilize this information to physically measure ancient Earth radius.

3.1 Geological Map of the World

As everyone knows the Earth is not flat, in particular not the way the Geological Map of the World is depicted in conventional Mercator projection previously shown in Figure 1.7; it is a sphere. The only reason why most people use flat maps is because of the ease and convenience in being able to display or fold a flat map as against carrying around a spherical globe. Of course, maps work extremely well at scales that are typically dealt with on a day to day basis, such as a street directory, an electronic navigational aid, or topographical map, but not when discussing or measuring the dimensions of the Earth. So, from now on I will be displaying and discussing the bedrock geological map in spherical format (Figure 3.1). The reason for this is because the information displayed on a spherical globe is realistic, true to scale, and distortion free, which then enables us to confidently view, measure, and model the data as required.

Figure 3.1 Present-day bedrock geological map of the world shown in spherical Earth format, based on the Geological Map of the World, 1990.

While appearing complicated, once again it should be appreciated that the various views of this globe show the continental and seafloor crustal geology of the Earth. The coloured geology shown in Figure 3.1 represents rocks that were formed during common intervals of geological time. These intervals of geological time have been studied and documented for many hundreds of years and form the basis of modern geology. The age relationships shown in this figure were originally established from the distribution of fossil plant and animal species in relation to major recognizable geologic events. These established age relationships are now further substantiated by modern age dating techniques.

The colours shown on this geological globe of the Earth simply mean that each colour represents rocks that have been deposited, intruded, or extruded during the same interval of geological time. The colours do not represent specific rock types, although at the map scale used there is a broad correlation between the most ancient continental rocks shown as red and pink colours, belts of deformed and folded rocks shown as khaki colour, younger sedimentary rocks shown as browns, blues and yellow colours, and the seafloor crusts shown as various coloured stripes.

The colours depicting the continental crustal geology shown in Figure 3.1 represent rocks that were formed during the five major intervals of geological time, including the most ancient Archaean Eon—beginning around 4,000 million years ago—followed by the Proterozoic Eon and then the Palaeozoic, Mesozoic, and Cenozoic Eras. Similarly, the colours depicting the seafloor crustal geology represent rocks that were formed during the geological time periods and epochs, ranging from the Jurassic—beginning around 200 million years ago—through to present-day times. These periods and epochs also represent subdivisions of the Mesozoic and Cenozoic Eras. For the actual age relationships of these geological times refer to the legend previously shown in Figure 1.9 as well as the International Chronostratigraphic Chart shown in Figure 2.1.

While geological mapping of the continents commenced many centuries ago, geological and geophysical mapping of the oceans first began in the 1950s and complete coverage was essentially finished during the late 1980s. This completed seafloor mapping, as shown in the Geological Map of the World (Figure 1.9 and Figure 3.1), now provides a unique insight into what the seafloor crust is made of, when it was formed, and the age distribution of the various seafloor volcanic rocks that are present. It is emphasised that this seafloor mapping was not available to early researchers into global tectonics and it was originally commissioned to provide an important tool towards constraining reconstructions of past Plate Tectonic crustal assemblages.

As can be seen in Figure 3.1, each ocean contains a mid-ocean-ridge and these ridges are approximately centrally located within each ocean coinciding with the pink Pleistocene stripes. These observations are further substantiated by seafloor bathymetric surveys which show that the ridges also coincide with the distribution of an extensive, centrally located, network of submarine mountain ranges. As is known from modern tectonic studies these mid-ocean-ridges subdivide the entire Earth's crust into very large, convex, plate-like crustal fragments. Each plate includes both continental and seafloor crust, although some plates have seafloor crust only, and the plates are generally centred on and around each of the continents.

The magnetic patterns and crustal age dating recognised and collected during the various seafloor mapping campaigns were historically interpreted as evidence for seafloor growth and spreading. As previously mentioned, the seafloor stripes shown in Figure 3.1 can be visualised as the preserved historical growth pattern of the Earth, where each stripe represents millions to tens of millions of year's crustal growth. This was an enormous revelation to the Earth sciences at that time. The scientists were finally confronted with physical evidence to show that the continents are in fact mobile and can actually move around the surface of the Earth. How this occurred and what caused the continents to move was not apparent to the early researchers. This observation was then subsequently adopted as the primary basis for Plate Tectonic theory.

From the early researchers' interpretations, basaltic lava was shown to have been intruded within fissures located along the entire length of the mid-ocean spreading ridges. These fissures were considered to have formed as the central spreading zones were *torn apart* and were simply viewed as huge cracks in the seafloor crust which formed as each of the oceans enlarged. Any new cracks were then further intruded with more volcanic lava as the fissures continued to open. What should be appreciated is that this volcanic lava making up the seafloor crust originates from deep within the Earth's mantle and as such the seafloor crusts actually represent quenched and exposed mantle volcanic rocks, not continental crust. These volcanic rocks are therefore very different to the observed continental crustal rocks, which are predominantly made up of sediments ultimately derived from erosion of much older eroded granite, as well as metamorphic, and lesser volcanic rocks.

In Figure 3.1 it can also be seen that the coloured patterns of seafloor crustal ages are symmetrical about the mid-ocean-ridges and their ages become progressively older when moving away from the ridges. This aging occurs because new crust is continually being added along the full length of the mid-ocean-ridge spreading centres and each oceanic plate enlarges with time to accommodate for the new seafloor volcanic lava being intruded. The oldest seafloor crust in each ocean is now known to be early-Jurassic in age—shown as pale blue coloured areas—and these rocks are mainly located along the present continental margins. Further to this, it should be apparent from the distribution of coloured stripes that the continents must in fact all be moving away from each other in order to accommodate for this growth pattern of new seafloor volcanic crust. As a consequence of this observation, the surface area of all of the oceans must then be increasing and each of the continents must be moving away from each other.

As previously introduced in Section 1.3 the seafloor mapping shown in this bedrock geological map represents a *natural tape recording* of the growth history of the reversals in the Earth's magnetic field. These magnetic reversals, in turn, are used to show the factual distribution and intrusive history of the volcanic rocks that now make up the seafloor crusts. By moving back in time and rewinding the tape recording it is very important to appreciate that each of the coloured stripes shown in Figure 3.1 must be progressively removed and the intruded volcanic lava within each stripe must be returned to the mantle, where it originally came from. The corresponding edges of each coloured stripe must then be moved closer together and, similarly, each of the continents must move closer together in strict accordance with the coloured striping evidence—regardless of which tectonic theory the reader adheres to.

In Plate Tectonic theory, the fundamental premise is that the radius of the Earth has remained essentially constant with time. As such, as new volcanic lava is intruded along the mid-ocean-ridge spreading centres the seafloor widens further and preserves this lava as new seafloor crust. To maintain an assumed constant radius Earth, an equal amount of new or presumed pre-existing seafloor or continental crust must be disposed of elsewhere and returned to the mantle by a theorised process called subduction. This subduction process forms the most important basis of Plate Tectonic theory and is essential for maintaining a static radius Earth premise.

Alternatively, for an Earth undergoing an increase in radius and surface area over time, the very same volcanic lava intruded along the mid-ocean-ridge spreading centres is again preserved and added to the surface area of seafloor crusts. For an Expansion Tectonic Earth, this increase in surface area of each of the seafloors is a direct result of an increase in Earth radius and there is no requirement for any net disposal of excess crust by subduction processes. There is also no requirement to assume, or even consider, the presence of pre-existing seafloor or continental crusts that are currently not present on the present-day Earth surface.

To account for this increase in surface area on an Expansion Tectonic Earth various researchers in the past have speculated that an increase in Earth radius may have varied considerably, ranging from a partial expansion process, where some subduction may have occurred, to full expansion. For full expansion, no pre-existing crust or subduction is required to account for the observed geological and geophysical features present on the present-day seafloor, whereas partial expansion was originally proffered to appease scientists and account for subduction-related phenomena.

Other suggestions have also included a possible pulsating Earth process. Here, expansion was said to have caused crustal stretching and later contractions were said to have given rise to crustal uplift to form mountains. In this process it was argued that, in order to have mountains there must have been compression and in order to have mid-ocean-rifting there must have expansion. A pulsating expansion-contraction process was then said to have given rise to the various mountain-building episodes which were in turn accompanied by opening of each of the modern oceans. While hypothetically feasible, this concept is now contrary to modern global mapping and tectonic observation. There is no evidence for this cyclical process shown in the seafloor spreading patterns. This pulsating Earth process also conflicts with current understanding of mechanisms for mountain building.

From this brief introduction to the *Geological Map of the World*, it is important to fully appreciate that by moving back in time we are compelled to restore the seafloor back to its original configuration along each of the mid-ocean-ridge axes. There is no other physically valid or proper method available to explain this unique seafloor mapping phenomenon, as shown throughout all of the oceans. Similarly, each of the continents must be moved closer together and at the same time the surface area of each of the oceans must be reduced. The seafloor mapping shown in Figure 3.1 is therefore unique in that it shows precisely where the edges of each plate, and hence mid-ocean-ridges, were at any moment back in time. This phenomenon then provides an important mechanism to reassemble and constrain the positions of adjoining crustal plates with a high degree of precision, for any time period ranging from the present-day back to around 200 million years in the past.

3.2 Seafloor Crustal Geology

At this stage there is a need to query and ask some fundamental questions about the validity of the seafloor mapping data shown in Figures 1.9 and 3.1. Does this seafloor mapping evidence really define the plate motion history of the Earth? Did the early researchers interpret this evidence correctly? Does the age dating really reflect the age distribution of the various seafloor crustal rocks? In addition, are there other alternative interpretations for this evidence that should be considered?

The gathering and mapping of magnetic data over large areas of the Earth's surface is now a basic and well-established geophysical tool at our disposal, not only for the seafloors but also for the continental crusts. This mapping is fundamental to land-based mineral search and a valuable aid to geological mapping of large areas of land surface, typically covered by sands, soil, and vegetation. Whatever rocks are present, albeit volcanic rocks with strong magnetic signatures or sedimentary rocks that may have a very weak to no signature, the remnant magnetic signatures preserved in these rocks can be detected by the various geophysical instruments. These signatures, in turn, are a reflection of the type of rocks lying beneath the Earth's surface. With continental rocks there is an added advantage that any magnetic features detected can be further tested by mapping and studying the rocks from surface outcrops, or by drilling the various magnetic features if buried. From this, geologists can physically identify and map the various rock types and then extend this knowledge over large areas just as the seafloor researchers did with the seafloor magnetic data.

Virtually every land-based exploration geologist or geophysicist uses magnetic data at some stage in his or her career, whether for mineral search or for mapping the distribution of the various rocks. The basic principles behind collecting and interpreting this magnetic data are now well established and has proved to be extremely valuable in mapping large areas of buried land surface. Therefore, extending this basic geophysical tool to mapping the seafloors is considered valid and does not contravene any geophysical or geological principles behind the methodology.

The fundamental difference between data collected from continental rocks and data collected from the seafloor is the differing complexity of the magnetic patterns identified. Magnetic patterns from continental rocks are typically very complex and convoluted which highlight the presence of a broad range of rock types ranging from sediments, to granite intrusions, to metamorphic and volcanic rocks. In addition, continental rocks have a long and complex history of folding, faulting, metamorphism, and so on as well as multiple periods of weathering and erosion. This contrasts strongly with the magnetic patterns observed from the seafloors, where patterns have a regular and characteristic zebra-stripe-like pattern as well as well-defined crosscutting patterns of faulting and crustal dislocation.

Volcanic lava intruded along sub-vertical fissures within the continental rocks is called a *dyke;* or a *sill* if it is sub-horizontal. These dykes and sills are relatively common in the ancient continental rocks, albeit on a much smaller scale than in the seafloor crusts. Unlike the continental dyke rocks which, if they reach the surface, can then extrude onto the surface as volcanic lava flows, the seafloor volcanic lava is intruded along fissures within the mid-ocean-ridges and is rapidly quenched and cooled by seawater once it reaches the seafloor. As the mid-ocean-ridges continue to open and widen, new volcanic lava is further intruded and continues to accumulate over time. All of these intruded rocks are then preserved in-situ.

By drilling into the seabed and recovering drill core rock samples the scientists are then able to identify and date the rocks, either by dating the various fossils recovered from the thin layer of sediment covering the seafloor or by directly dating minerals preserved within the volcanic rocks. Just as is done for the continental rocks, this dating evidence can then be used to constrain the age of the magnetic striping detected during magnetic mapping and to also extend the observations along the full length of the various patterns detected. Again, this is a basic and fundamental mapping tool that scientists have at their disposal and similarly does not contravene basic principles or methodology.

Also, the modern age dating techniques used to date the various minerals contained within the rocks are now at a very advanced stage of sophistication and precision. There are a wide range of dating techniques available to the technician, which are also supported by more traditional dating of fossils. Again, these age dating techniques do not contravene the basic principles or methodology behind seafloor crustal dating and mapping.

Does the seafloor magnetic evidence really define the plate motion history of the Earth? The magnetic instruments used to detect the variations in magnetism preserved in the seafloor rocks are actually seeing through the relatively thin layers of sedimentary sludge covering the seafloor. Because the volcanic lava has such a strong magnetic response the instruments are identifying patterns from the volcanic bedrock itself, with only minor, if any, influence from the overlying sedimentary sludge. The extensive amount of deep sea drilling throughout all of the oceans now shows that this bedrock is predominantly volcanic in origin and the age dating shows that the youngest volcanic rocks occur along the mid-ocean-ridges. These volcanic rocks are also shown to increase in age away from the ridges. In addition, modern satellite-based geodetic measurements—for example GPS—independently show that most continents are moving away from each other—to be discussed further in a later chapter—and hence the surface areas of the oceans are increasing. If the continents are moving away from each other then it is plausible to accept that the magnetic patterns do indeed reflect the plate motion and growth history of the seafloors as predicted.

In order to fully appreciate the implications of what is recorded in the Geological Map of the World, I will now digress somewhat and talk through a couple of fictitious scenarios, with implications for the various tectonic options available. This is primarily to put us in a frame of mind to enable us to visualize and appreciate what it is that we are actually seeing and what the seafloor spreading patterns shown in the geological map are telling us. This is necessary in order to remain open minded enough to focus on the new perspectives into the origins of continents and oceans that will be discussed later.

Firstly, let us visualize a fictitious spinning globe filled with a semi-molten material that erupts at the surface. This semi-molten material has enough viscosity to circulate below the surface as convection cells. The basic laws of fluid dynamics enable us to predict and model these convection cells circulating beneath the surface which, because the sphere is spinning, will generate characteristic circulation cells that obey the conventional *Coriolis Effect*. This coriolis circulation pattern is more familiar as surface weather patterns, whereby atmospheric circulation patterns in the northern hemisphere rotate clockwise along the equatorial belt and counterclockwise in the southern hemisphere due to the rotation of the Earth. Obviously, if there were objects capable of floating on the surface of this fictitious globe, it would be expected that they would move in a predictable circular motion according to these same coriolis circulation patterns. As they circulate, these objects may also periodically collide with other objects then separate, rotate, and continue along their merry way.

If the semi-molten material filling this sphere erupts at the surface we can safely presume that it will be associated with large rising *thermal plumes* and these will erupt at various *hot-spot* locations similar to large shield volcanoes or other volcanic complexes familiar to us on Earth. Eruptions within spreading zones located along the trailing side of the floating fragments would also be expected. This fictitious globe is also covered by a fluid which is cool enough to quench the erupted material before it spreads too far. It would be anticipated that any magnetic patterns that are preserved when the erupted material cools and solidifies would form patterns in sympathy with the motion of the floating objects, as well as the exterior shape of the floating objects. In order to preserve the radius of this globe, the erupted material must also be viscous enough to be absorbed back into the circulating convection pattern as required. As the size and surface area of the globe is already known we do not need to measure these phenomena and we can therefore ignore this need from any future considerations of the erupted material or patterns of the floating objects.

In the second analogy, let us imagine another fictitious spinning globe filled with a semi-molten material. In contrast to the first example, the semi-molten material here slowly increases in volume prior to erupting at surface. Because the globe is also spinning the semi-molten material may also circulate according to the same laws of fluid dynamics as required. This globe is again covered with a cool liquid at surface.

As the semi-molten filling slowly increases in volume, the already solidified surface material is subject to on-going fragmentation, stretching, and distortion, which, due to its inherent strength, initially retains its physical integrity but gradually increases in surface area. Over time, as the filling continues to increase in volume, this solidified surface material eventually reaches the limit of its ability to stretch and it then starts to crack, rupture, and tear apart along linear rip-lines. The semi-molten filling is then able to erupt along the full length of these rip-lines, filling and healing the tears, but is quenched and cooled by the surface liquid before spreading too far. On-going increases in volume will repeatedly tear open the surface material, this time focusing along the pre-existing lines of weakness where previous semi-molten filling has already erupted. If we were to measure the surface area of the newly erupted material and compare this with the surface area of the original globe, this information could be used to calculate the change in surface area and from this to determine the change in radius of the globe.

In other words, by using the natural tape recording of the history of the increase in surface area of the globe we can directly measure the radius of the globe at any moment back in time. This facility then becomes a very useful tool for determining a past radius of the globe, as well as a means to accurately predict the movement history of the fragmented surface material and opening along the rip-lines.

Back to the real Earth, while the first scenario is hypothetically feasible, when this scenario is extrapolated to the size of the Earth it must be remembered that the seafloor volcanic crustal rocks on Earth are seven to ten kilometres thick and once cooled are solid rock. The mistake that Wegener made when he introduced his Continental Drift theory was that he likened the continents to *"...icebergs of low density granite floating on a sea of denser basalt"*. What should now be appreciated is that the seafloor crust is not a fluid and as Wegener's opponents rightly pointed out it is very difficult to imagine a floating object, such as a continent, ploughing its way through a sea of solid rock without leaving any evidence of its passage. A more likely scenario would have been likening it to an icebreaker ship attempting to plough its way through a sheet of floating sea-ice, the evidence for its passage is then, of course, in the ship's wake.

It must also be appreciated that, in order to retain a constant radius Earth, plate consumption must occur along the leading edge of the crustal plate—referred to as subduction in Plate Tectonic usage—and plate creation—called seafloor spreading—must occur along the trailing edge. This process should, in theory, leave behind a characteristic, albeit complex, random pattern or tape recording of its plate motion history which should of course be reversible when moving back in time. The question that must be considered is: are the symmetric globally connected patterns shown in the Geological Map of the World truly representative of a random plate motion or are they telling a different story?

In contrast to conventional wisdom, a cursory inspection of the crustal rocks in the Geological Map of the World (Figure 3.1) shows that the global distribution of the coloured seafloor crustal striping, for all oceans, occurs as long interconnected linear growth patterns which are symmetrical about the mid-ocean-ridge spreading zones. Similarly, the oldest rocks in each of the oceans are identical and the maximum age of exposed seafloor volcanic rocks, located along the continental margins, is early-Jurassic—the pale blue areas shown in Figure 1.9 and 3.1. What the bedrock geological mapping is not showing is a pattern that should be expected from circulating mantle convection cells. Nor does it show crosscutting seafloor crustal relationships that should be expected as the plates randomly migrate across the surface and collide, nor remnant fragments of pre-existing seafloor or continental crusts as expected from Plate Tectonic theory.

This is a very significant and major concern, a concern that the early researchers during the 1950s and 1960s were unable to appreciate or consider. They did not consider this because they only had basic seafloor data from a very small area of the North Atlantic Ocean, which they then extrapolated to the rest of the Earth. If they did have the full complement of mapping data that is now available then they may have reconsidered disposing vast amounts of excess crust within the Pacific Ocean, an area that also has its own very substantial mid-ocean-spreading zone.

Regardless of the simplicity of the analogies used above, not only does the magnetic mapping shown in Figure 3.1 show interconnected north-south orientated mid-ocean-ridge spreading centres for each of the major oceans, which alone may support an east-west *conveyor belt* mantle circulation pattern, it also shows a substantial east-west orientated Southern Ocean mid-ocean-spreading pattern. This Southern Ocean mid-ocean-ridge pattern not only does not have an associated subduction zone it is also completely contrary to the conventional theorized conveyor belt mantle circulation pattern.

These fictitious scenarios are introduced here to justify the next step in our quest to measure an ancient Earth radius. In order to measure an ancient radius of the Earth we must now take a bold and presumptuous step and remove the assumption that Earth radius is constant from our minds. In other words, we are going to allow the seafloor crustal pattern depicted in the bedrock geological mapping to represent seafloor crusts that have been intruded and preserved without subduction of pre-existing crusts. A bold step, but no more so than continuing to assume that Earth radius has been static without first attempting to at least measure and substantiate this new surface area evidence that has only become available since the late 1980s.

3.3 Measuring Ancient Earth Radius

From discussions so far it should be evident that there are basically two types of crusts covering the Earth's surface; relatively thick continental crusts and relatively thin seafloor crusts. Each of these crusts has different physical characteristics and each also has completely different age characteristics—continental crusts are predominantly relatively older and seafloor crusts are predominantly younger. From previous discussions, it was shown that new seafloor crusts are being added continuously along the entire length of the mid-ocean-ridge spreading centres and in doing so progressively increase the surface areas of each of the oceans over time.

Historically, determining an ancient radius of the Earth has been one of the most fundamental requirements of any Expansion Tectonic Earth thesis. Prior to completion of the bedrock geological mapping of all the oceans during the late 1980s attempts were limited to using unconstrained small Earth models, as well as measurements from remnant magnetism preserved in rocks, early measurement of areas of seafloor spreading, mantle plume separation, and astronomical observations. However, in all cases, because of the limitations in the quality and quantity of data available at the time, history has shown that none of these techniques were sufficiently accurate to determine an ancient Earth radius, at least convincingly enough to sway conventional thinking.

Consideration of available literature suggests there was also no unified opinion amongst most authors as to whether the Earth may be expanding, contracting, pulsating, rhythmical, or static over time. Depending on which method was used to determine an ancient Earth radius, reference is made in the literature of a potential large expansion of the Earth equated with a rate of increase in Earth radius of between 0.4 to 2 centimetres per year. There are also descriptions of gross radial expansion reaching several metres per year and another widely held opinion that allowed for small to negligible radial increases in Earth radius of up to 1 millimetre per year.

Koziar in 1980, and later Blinov in 1983 were the first scientists to measure seafloor surface areas from modern seafloor mapping in order to determine an ancient Earth radius. Koziar was also the first scientist to then express these measurements as a mathematical formula. His measurements included seafloor and continental crusts from the whole of the Phanerozoic Eon—extending back in time to about 550 million years ago—and its oldest calculated point was based on the total surface area of remaining Precambrian crusts. From this formula Koziar was able to calculate the present annual increase in Earth radius as 2.6 centimetres per year.

With the completion of the Geological Map of the World it is now a relatively easy process to determine ancient Earth radii using measured surface areas of seafloor crusts. The added advantage of this map is that the seafloor crusts now have known ages, which are in turn used to accurately constrain the moment in time when these rocks were first formed. Once the areas of each of the coloured seafloor crustal stripes are known, the cumulative areas of the seafloor crust—that is, the total area of crusts younger than the time we want to calculate a radius for—is then simply taken away from the present-day Earth surface area to mathematically derive an ancient Earth radius. This presumes, of course, that any increase in surface area is confined to the seafloor crusts which is not strictly correct but is a reasonable approximation. As will be shown later, increases in surface area of continental crustal rocks are primarily the result of crustal stretching which is basically due to on-going changes to surface curvature.

It should also be appreciated that the sum of the time periods shown by the coloured seafloor bedrock mapping represents only about 5 percent of the total known age of Earth history. Whenever Earth expansion is mentioned in the published literature it is this limited timeframe that is typically associated with the concept of Earth expansion. It is rarely, if ever, appreciated that an increase in Earth radius may in fact have also occurred before this time to include the entire pre-Triassic history of the Earth—times prior to about 200 million years ago.

To determine ancient Earth radii for each of the time intervals shown by the coloured seafloor spreading stripes in Figure 3.1 I will initially focus on the seafloor crusts, extending in time from the present-day back to about 200 million years in the past. This 200 million year time-frame covers the age of the oldest known seafloor crusts plus continental shelf sediments deposited around the margins of the continents. By calculating Earth radii from the surface area measurements, scaled small Earth models of the ancient crustal plate assemblages will then be constructed for each of the coloured time periods back to the early-Jurassic Period. These constructed models will in turn be used as a basis to further investigate the origin of the modern continents and oceans.

This modeling will also enable any assumptions about an increasing Earth radius scenario to be tested. If these assumptions are correct and an increasing Earth radius concept is shown to be feasible, we will then be in a better position to take the more challenging step towards modelling the remaining pre-Triassic aged continental crusts. As will be outlined, changes to pre-Triassic Earth radius will be further investigated in later chapters and, depending on the outcomes of this investigation, small Earth modelling will be continued back to the beginning of geological time—around 4,000 million years ago. These models will then be used to investigate the origins of the ancient supercontinents.

The term *geological crustal budget* is a term I use to account for the surface areas of each of the coloured crustal rocks shown in Figure 3.1. What is meant by this term is that if we create something, in this case an area of new volcanic seafloor crust added along the mid-ocean-ridge spreading axes, we must be able to account for it. We must also be able to balance the overall global geological crustal budget for each of the time-periods involved—in this case the total ancient surface area of the Earth at any moment in time.

As mentioned, the coloured patterns of seafloor crustal rocks shown in Figure 3.1 represent volcanic lava that has been intruded and preserved over a set interval of time along the entire length of the mid-ocean-ridge spreading axes. These spreading axes are shown to be centrally located in each of the oceans. If the surface areas of each of the seafloor crustal stripes are measured, then this information can be used to investigate the variation in the Earth's surface area with time. From this, it is a relatively straightforward process to mathematically derive and investigate any potential changes in Earth radius with time.

These may still seem to be ambitious statements, however, in all global tectonic concepts the generation of new seafloor crusts during the past 200 million years is universally accepted by scientists to have been continuously created along the mid-oceanic-ridge seafloor spreading axes. There is no evidence for seafloor crusts older than 200 million years remaining within the modern oceans, only relatively minor occurrences of buried seafloor volcanic rocks located within some continental sedimentary basins. Only in Plate Tectonic studies is there a need to invoke pre-200 million year old seafloor crusts in order to maintain a constant Earth surface area.

To measure the surface areas of each of the coloured seafloor stripes all projection distortion must be eliminated from the bedrock map and, strictly speaking, this information must be displayed in spherical format, that is, on a spherical globe. The method adopted here requires the existing Geological Map of the World (Figure 1.9) to be displayed as a *24-gore sinusoidal projection map* (Figure 3.2). A sinusoidal projection format is unique in that it gives undistorted, true-to-scale, geological information from anywhere within the map area enabling the information to be both measured and confidently modeled on spherical globes.

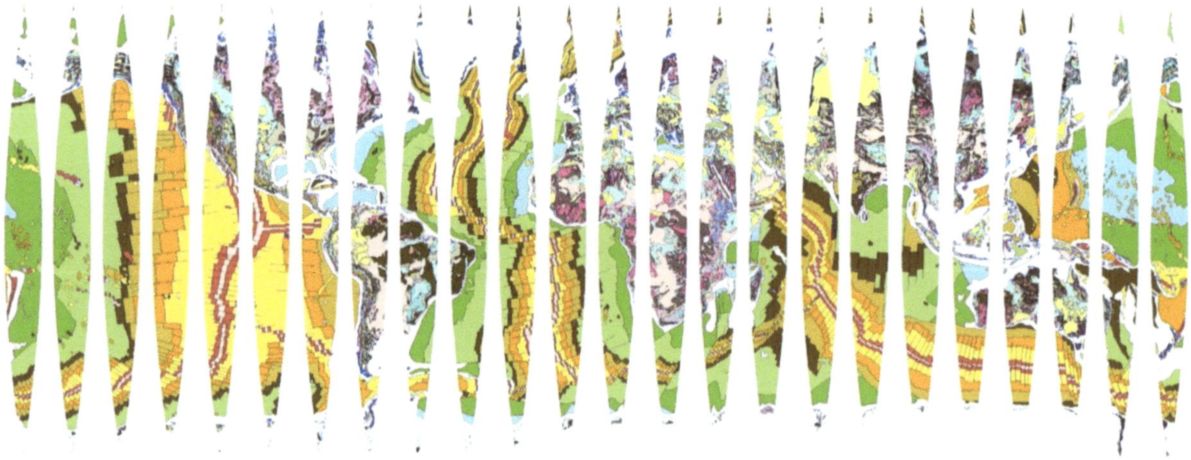

Figure 3.2 24-gore sinusoidal map projection of the Geological Map of the World. This projection enables the geological map to be displayed in distortion-free spherical format and forms the primary base-map for both surface area measurement and small Earth model constructions.

The term *gore* simply means that each curved stripe, which in this figure represents fifteen degrees of longitude at the equator, tapers to a width of zero degrees longitude at each pole. In Figure 3.2, twenty four gores have been used to represent the original map area. The unique quality of this type of map projection is that it can then be cut and pasted directly onto a globe during model construction, as shown in Figure 3.1. This sinusoidal map will then form the basis for further detailed small Earth modelling, to be introduced later.

By using this sinusoidal map, the outline of each coloured seafloor stripe can be digitized in turn. The surface areas of successive intervals can then be measured, and an ancient Earth radius derived for each time period shown. The raw data from this exercise is shown in Table 3.1, and a graphical plot of the cumulative surface areas against time is shown in Figure 3.3.

Age		Surface Area			Ancient Radius
Chron	Years ($x10^6$)	dS ($x10^7$ km^2)	SdS ($x10^7$ km^2)	$S_a=S_0$-SdS ($x10^7$ km^2)	R_a (km)
C0	0	0	0	51	6370.8
C2	-1.9	0.5342	0.5342	50.4658	6337.15
C3a	-5.9	1.3328	1.867	49.33	6265.43
C6b	-23	4.9213	6.7883	44.2117	5931.49
C15	-37.7	4.1624	10.9507	40.0493	5645.37
C25	-59.2	4.1649	15.1156	35.8844	5343.77
C29	-66.2	1.0462	16.1618	34.8382	5265.3
C34	-84	4.7956	20.9574	30.0426	4889.49
M0	-118.7	5.6758	26.6332	24.3668	4403.46
M17	-143.8	1.9348	28.568	22.432	4225.02
M38	-205	1.9386	30.5066	20.4934	4038.31

Table 3.1 Empirical surface area data derived from mapping by Larson et al. (1985) and CMGW & UNESCO (1990). Areas were originally digitized using a CAD based Graphical Design System software package and ancient Earth radius calculated from this data. Arbitrarily assigned errors are ±5% (from Maxlow, 1995).

The chart shown in Figure 3.3 is a means of graphically displaying the distribution and accountability of the measured empirical surface areas of all Earth's continental and seafloor crusts. This chart covers approximately 200 million years of Earth history, extending from the present-day back to the beginning of the Jurassic Period. In this figure, each of the measured seafloor stripes is simply added together when moving forward in time from the early-Jurassic to the present-day. In other words, the area of new crust intruded along the mid-ocean-spreading-ridges is considered to be cumulative over time and is progressively added to the area of previous seafloor crust for each successive time interval shown.

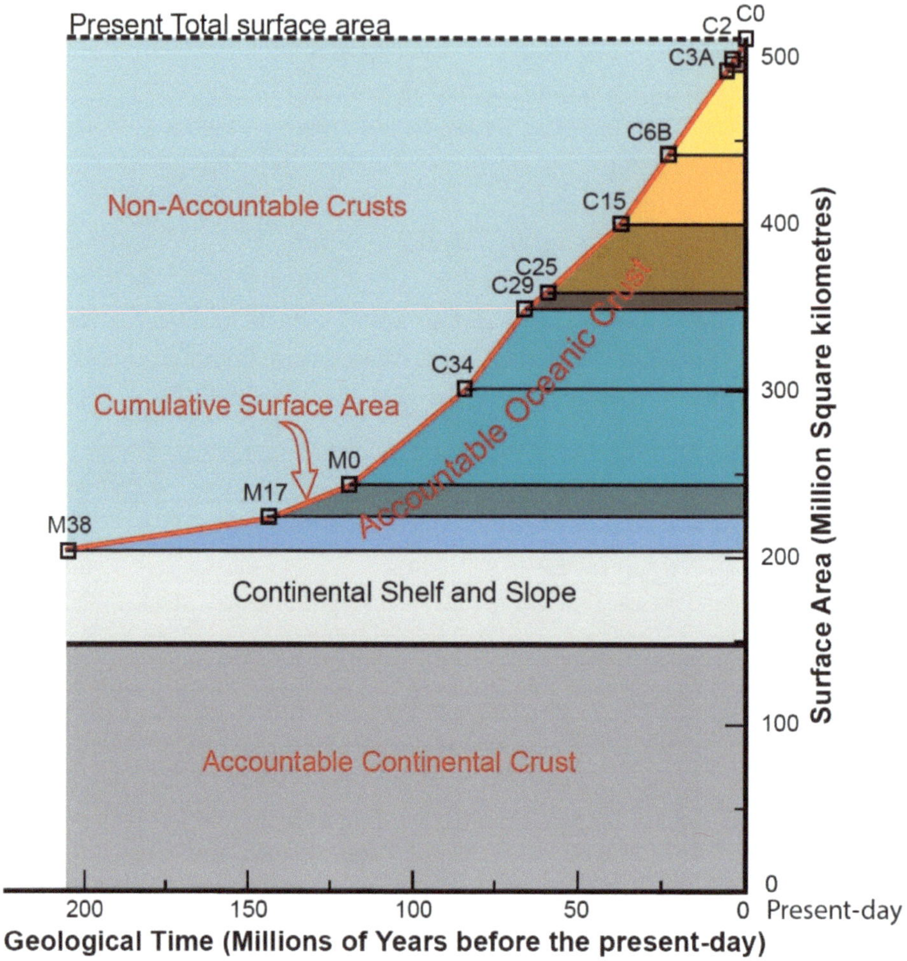

Figure 3.3 Cumulative surface areas of both continental and seafloor crustal rocks as measured from the Geological Map of the World sinusoidal base-map (Figure 3.2). The time scale represents millions of years before the present.

In this chart the cumulative surface areas of each of the crusts can be compared to the present-day total surface area, shown as a dashed line at the top of the chart. Here, areas of both accountable-oceanic and accountable-continental crusts are shown below the red surface area curve. In this context, accountable-crust represents all continental and oceanic crusts currently existing on the Earth's surface today. The cumulative surface area of each of these accountable-crusts must then equal the present surface area of the Earth at time zero—which is the present-day.

Also shown on this chart is a pale blue area located to the upper left, referred to as non-accountable crusts. This area is located above the measured red cumulative surface area curve and below the dashed total surface area line at the top of the chart. This non-accountable crust represents the additional surface area of inferred crust that must have been present in order to maintain a constant surface area Plate Tectonic Earth. It is important to note that no record of this non-accountable crust exists on Earth today and there is only very limited evidence to suggest that it ever did.

If a constant surface area Plate Tectonic Earth is to be maintained, this non-accountable crust represents the additional surface area of continental and seafloor crusts that must have previously existed since the early-Jurassic Period. This area also represents the total area of crusts that must have been removed by the inferred subduction process. A subtle outcome of this charting is that an explanation for the accelerating increase in non-accountable crust shown in this chart, and by inference an accelerating increase in subduction, must also be explained. What I mean by this is that the rate of increase in surface areas shown in this chart is actually increasing with time and hence an explanation and a mechanism for this phenomenon must be given by advocates of Plate Tectonic theory.

In contrast, accountable-crust on an Expansion Tectonic Earth is a reflection of the actual variation in Earth radius with time. On an Expansion Tectonic Earth, non-accountable crust is not required and in fact never existed at all. The rate of increase in surface area with time is then simply a reflection of the increase in Earth radius. What causes this increase will be discussed in a later chapter once all empirical evidence has been introduced and discussed.

The accountability of Earth surface areas for both seafloor and continental crusts can be further understood by comparing the basic premises of the four main global tectonic theories, all of which have been promoted in science at various times during the past 100 years of geological study:

- **Constant Earth radius** (Plate Tectonic Earth): In this theory, excess crust intruded along the mid-ocean spreading centres is continuously being disposed of along subduction zones, displacing and recycling pre-existing crust into the upper mantle. To maintain this view of a constant surface area it must be presumed that non-accountable crust existed prior to subduction. Limited, if any, record of this crust exists and an explanation must also be provided for the accelerating rate of increase in the actual, observed surface areas shown.

- **Partial increase in Earth radius**: In this theory, some excess crust is disposed of along subduction zones during limited Earth expansion.

To maintain a partial increase in surface area, it must be presumed that some non-accountable crust existed prior to partial subduction. The accelerating increase in observed surface area since the early-Jurassic Period must again be explained.

- **Pulsating Earth radius**: Similarly, in this theory crust is generated at spreading centres during a period of increasing Earth radius and is either compressed to form mountains, or disposed of along subduction zones during a reduction in Earth radius. No evidence for a pulsating Earth radius is shown from the available surface area data and new ideas on mountain building negate the requirement for crustal compression to form mountains.

- **Increasing Earth radius** (Expansion Tectonic Earth): For an Expansion Tectonic Earth, new volcanic lava intruded at mid-ocean spreading centres is cumulative with time. The increase in total surface area is a direct result of an increase in Earth radius. On an Expansion Tectonic Earth, non-accountable crust never existed, subduction is not required, and the cumulative surface area is a physical measure of the actual change in Earth radius. The measured change in Earth surface area in turn suggests there has been an accelerating rate of increase in Earth radius over time.

From this brief comparison of the various tectonic theories available it can be seen that both Expansion Tectonic and Plate Tectonic theories represent the extreme end-points of a whole spectrum of global tectonic possibilities. Plate Tectonic theory is currently the accepted theory in science and has been extensively presented in publications elsewhere. Owen in 1976 has adequately presented partial expansion and Milanovsky in 1980 has adequately presented the pulsating Earth theory. It is important to note that each of the Plate Tectonic, partial expansion, and pulsating Earth theories were introduced well before completion of the bedrock geological map of the world and these theories must therefore be able of explain the new phenomena shown in this mapping.

The pattern and distribution of the ancient surface area measurements in Figure 3.3 shows that there is no cause to consider a pulsating Earth during the early-Jurassic to present-day periods, in the style suggested by Milanovsky. It also seriously limits the potential for a partial Earth expansion, as suggested by Owen. The evidence instead shows that any potential Earth expansion process is one of full accelerating expansion, where growth in surface area is a direct result of an accelerating increase in Earth radius through time.

From the measured surface area data it is then a relatively simple means to convert this data to an ancient Earth radius for each time interval measured. This converted Earth radius data is shown plotted as a dashed black line in Figure 3.4, along with a curved red line representing an idealised exponential curve which has been fitted to the measured data. The discrepancy between the measured dashed line and the red curve at the lower left end is interpreted to represent masking of seafloor volcanic crusts by sediments deposited around the margins of the continents. The seafloor data shown in both Figures 3.3 and 3.4 will be combined with measurements from continental surface areas in a later chapter to then derive a formula for calculating Earth radius and a rate of change in Earth radius at any moment back and forward in time.

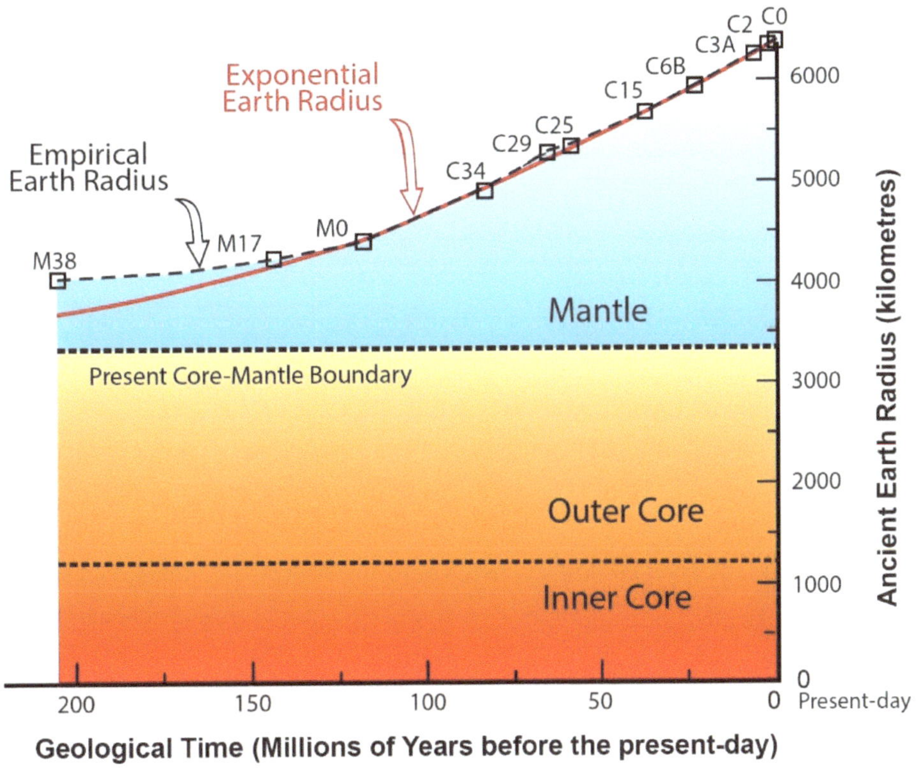

Figure 3.4 Ancient radius of the Earth extending from the early-Jurassic to the present-day calculated from digitised areas of published post-Triassic seafloor mapping data. The red line represents idealised curve fitting of the data.

The implications stemming from these measured surface areas, as well as the derived ancient Earth radii measurements, are that we now have a means to accurately constrain ancient Earth radii for each of the time periods depicted in the coloured seafloor crustal mapping data. From this, by successively removing each coloured seafloor stripe in turn, from each of the modern oceans, and reducing the Earth radius back in time to the early-Jurassic Period, we are also in a unique position to make accurate small Earth models of the continental and seafloor plate assemblages. This accurate mapping was not available before completion of the bedrock geological mapping in 1990.

4 Small Earth Model Construction

"But in the end the rigorous approach has paid off. For it has revealed a discrepancy which had not been apparent. It was not my method that was at fault, but my assumption that the earth of Pangaea was the same size as the earth of today. The assembly of Pangaea is not possible on the earth of the present radius, but on a smaller globe...these difficulties vanish." Carey, 1958

In addition to providing a unique means of measuring an ancient radius of the Earth, the coloured seafloor crustal mapping provides an even more unique means of constraining the precise location of the crustal plates at any moment in time back to the early-Jurassic Period. By returning the seafloor volcanic rocks back to the mantle, from where they originally came from, the surface areas of each of the oceans must, by established protocol, be progressively reduced and each of the continents moved closer together. The uniqueness of adopting an Expansion Tectonic model of the Earth is that there is no need to consider where, or when, pre-existing crusts occurred or similarly where they must go to. All that is required is to let the configuration of the coloured seafloor crustal mapping dictate the precise crustal plate assemblages on a pre-determined smaller radius Earth model.

As previously mentioned, at continental and global scales a sphere cannot be converted into a flat map without first acknowledging the considerable amount of projection distortion involved, especially in the higher polar latitudes. More importantly, we must be acutely aware that information from a flat map cannot be simply cut and pasted or assembled anywhere else on the same map. Doing this would mean that different projection distortions would be transferred from one area of the map into another area that has completely different distortion parameters, even on the same map.

To construct small Earth models of an Expansion Tectonic Earth, spherical polystyrene foam spheres were found to be the best means of assembling the various crustal plates. Cutting and pasting of information from anywhere on a spherical globe is then true-to-scale and distortion free. On these small Earth models, the distribution of crustal plates can then be accurately constrained anywhere on the globe during model construction using standard cartographic techniques.

To test the theory of Expansion Tectonics and, in particular, the ancient Earth radii as determined from empirical seafloor mapping data, a series of spherical small Earth models (Figure 4.1) were initially made for the present-day and the beginning of the Pliocene, Miocene, Oligocene, Eocene, Palaeocene, Late-Cretaceous, Mid-Cretaceous, Early-Cretaceous, Late-Jurassic, and early-Jurassic Epochs and Periods. These small Earth models were constructed using the seafloor geological crustal information displayed on the *Geological Map of the World* sinusoidal base map shown in Figure 3.2, originally digitised with permission from the Commission for the Geological Map of the World and UNESCO (1990).

Figure 4.1 Spherical small Earth models of an Expansion Tectonic Earth extending from the early-Jurassic to the present-day. These models demonstrate that the seafloor crustal plates assembled on an Expansion Tectonic Earth coincide fully with seafloor spreading and geological data and accord with derived ancient Earth radii.

To minimise geological complexity shown by the continental geology, the small Earth modelling presented in this chapter will initially focus on the seafloor crustal mapping only. All continental geology will simply be coloured grey and left blank. The constructed models shown in Figure 4.1 were first used as the basis for my original research into post-Triassic Expansion Tectonics. In later chapters these will be reproduced in more detail in conjunction with modelling of the continental crustal geology. The models shown in Figure 4.1 will now form the basis for discussions on constructing and assembling seafloor crustal plates on Expansion Tectonic small Earth models.

From the outcomes of this early research (Maxlow 1995) I was able to conclude that the seafloor crustal mapping provides a definitive means to test and quantify a potential rate of increase in Earth radius and a means to constrain plate assemblage with a precision not previously available to early researchers. It was also fully acknowledged in this research that if the concept of Expansion Tectonics were wrong then the seafloor crustal mapping will highlight any discrepancies or inconsistencies on each of the small Earth models. This would then require a rethink, revision, or even rejection of the basic assumptions and methodologies adopted here.

4.1 Assumptions

Like most things in life, our actions and achievements must be based on certain assumptions and constructing small Earth models of an Expansion Tectonic Earth is no exception. These assumptions are necessary in order to set guidelines during construction and to also attempt to bridge any gaps in the knowledge base. The main problem with making assumptions, of course, is making sure that these assumptions are realistic, that the limitations are fully understand, and the assumptions do not unduly bias the outcome of the modelling exercise.

In order to accurately quantify any variation in the Earth's ancient radius, and similarly to constrain crustal plate assemblage back in time, it is argued that it is necessary to take into account the area and pattern of the seafloor and continental crusts. What the seafloor crustal mapping shows is the *spatial*—the location of the plates on the Earth's surface—and *temporal*—the location of the plates in time—distribution of seafloor crusts within each of the ocean basins. By moving back in time, successively older coloured seafloor stripes from across the active mid-ocean-ridge spreading zones must then be removed and each of the remaining coloured stripes must be reunited along their common spreading ridges. This is a fundamental premise which is dictated by the mapping data itself.

The main assumptions used to construct and assemble crustal plates on each of the Expansion Tectonic small Earth models presented here are summarised below. These will be further investigated for bias and realism in later chapters when I come to discuss the distribution of additional geological, geophysical, and geographical information on each of the completed models and especially when I come to model the continental crusts.

For the coloured seafloor crusts, on each of the early-Jurassic to present-day Expansion Tectonic small Earth models it is assumed that, for an Earth continuing to undergo an increase in radius and surface area with time:

- The Earth's overall seafloor and continental crustal budget increases with time and all increases in surface area are cumulative with time.

- Once formed, the surface area of seafloor crust, as represented by the bedrock geological mapping in Figure 1.9, is fully fixed in the geological rock-record.

- Seafloor and continental crusts are not removed by subduction processes.

- There are no pre-existing seafloor or continental crusts to account for.

The only stipulation required during cutting and pasting of crustal information between each successive small Earth model is that:

- Continental and seafloor crustal plates must undergo small vertical and surface area adjustments to allow for a progressive change in surface curvature of the Earth with time.

What each of these assumptions, as well as the cutting and pasting stipulation, means is that, for an Earth that is progressively increasing its radius through time, seafloor crust accumulates as a direct result of an increase in Earth's surface area. Once formed, the seafloor crusts then remain firmly fixed in the geological rock-record, the crusts are not removed or returned to the mantle by subduction processes, and similarly there is no need to invoke the existence of non-accountable or pre-existing crusts. This course of action, of course, goes against conventional tectonic wisdom; however we are not dealing with or replicating conventional Plate Tectonic models. We are simply removing the conventional premise that Earth radius has remained constant throughout history and empirically testing the revised premise that Earth radius is increasing with time.

To be consistent with the geological evidence, it is also acknowledged that both the continental and seafloor crusts must undergo some, to variable, crustal distortion, in particular during on-going changes to surface curvature. It is also acknowledged that localised crustal interaction between plates during changing surface curvature, as distinct from plate subduction, may also occur along plate or crustal boundaries but only at a relatively local scale and not to the extent that is required in conventional Plate Tectonic theory.

What this means is that when moving forward in time from the older, smaller radius models, to the younger, larger radius models, the surface curvature of the Earth will progressively flatten during changing Earth radius. As will be shown later, changing surface curvature represents the primary mechanism for most, if not all geological processes observed on Earth's surface today. This includes folding and faulting currently observed in all continental crusts, as well as uplift of the crust along continental margins to form mountain belts and escarpments. These processes also apply to the seafloor crusts where distortion of the new volcanic lava is accommodated for by intrusion and preservation of new lava at ever changing Earth radii and surface curvatures.

These are important points to dwell on. Most previous model makers did not have the seafloor crustal mapping at their disposal to assist in constraining plate assemblages. Instead, they relied on a visual fit-together of the continental crusts to build their models. I can assure the reader that one cannot simply take a fragment of crust from a present-day Earth model and expect to paste it onto a smaller radius globe—it doesn't work because the surface curvatures are vastly different.

In reality, from an Expansion Tectonic Earth perspective, the Earth is increasing in size at an extremely small rate each year. During this process the crusts creep and distort at an imperceptibly small rate in order to keep pace with the ever changing surface curvature. The joints, fractures, faults, and folds in existence on Earth today simply absorb this minute curvature adjustment, generally without us being aware of the processes in action. Catastrophic events, such as earthquakes, volcanic eruptions, and so on are an exception to the rule, however only at the human perception scale. To model back in time, all of these geological processes must then be reversed and the changes in surface curvature must be accommodated for on each of the small Earth models.

I will be discussing change in surface curvature in more detail later, sufficient to say that this progressive flattening of the Earth's surface curvature results in a gradual deformation of the existing crust—what I refer to as a *plastic deformation* process—which occurs over a vastly extended period of geological time. During small Earth model construction, when moving back in time these small crustal changes must be accommodated for and removed. During cutting and pasting of the map fragments onto the small Earth models any wrinkles occurring along the edges of the paper fragments during plate assemblage, for example, are accommodated for by either subtly stretching or compressing the paper fragments during the pinning and pasting stage. This is necessary anyway because each gore still represents a segment of a flat map being pasted onto a curved surface of the globe.

Each of these assumptions, as well as the stipulations used to construct the Expansion Tectonic small Earth models in turn, is consistent with the main assumption. That is, for an Earth undergoing an increase in radius, the actual surface area of the Earth increases as a direct result of an increase in Earth radius with time.

4.2 Model Construction

Model construction was carried out and will be further discussed in two main phases. The first construction phase coincides with the past 200 million year time interval, extending from the present-day back to the beginning of the Jurassic Period. This phase uses the coloured, relatively well age-dated seafloor mapping to constrain the assemblage of seafloor crustal plates and covers the closure of each of the modern oceans. The second construction phase will be dealt with in later chapters and will extend small Earth model coverage further back in time, beyond the early-Jurassic Period, to the beginning of the Archaean Era—around 4,000 million years ago.

This second model construction phase is more difficult to achieve because, while age dating of the continental crusts exists in detail, its use in model construction is complicated by many overprinting phases of sedimentary deposition, erosion, metamorphism, and tectonism. During this extended interval of ancient time there were also no seafloor crusts to guide crustal assemblage. During this second construction phase I will instead be focusing on reducing the surface areas of an identified network of continental sedimentary basins and assembling the remaining, more ancient, continental crusts on pre-determined smaller radius Earth globes.

Because of the complexity of the geological mapping data, it was easier to construct hard copy spherical models rather than attempt to develop digital models. A series of small Earth models were then constructed using high density polystyrene foam spheres. The two main methods used to assemble and model the mapping data include: Method 1) uses spherical scale models pre-cut to the ancient radii as determined from the surface area data; and Method 2) uses foam spheres of the same diameter and then digitally rescaling the sinusoidal base map to represent the actual small Earth model radii. The first method was used to construct the seafloor crustal models shown in Figure 4.1 and the second method will be used later to assemble the continental crustal data.

The construction Method 1 used to assemble the seafloor crustal plates on spherical scale models is very simple and very basic (Figure 4.2). Each coloured seafloor crustal stripe is simply removed from the geological base map in turn and the remaining pieces of map are reassembled by pinning and pasting the crustal plates onto the pre-determined smaller radius Earth model. This, in effect, can be likened to a spherical jigsaw puzzle where the pieces of plates cut from each of the paper gores in the 24-gore sinusoidal base map (Figure 3.2) are reassembled together on a three dimensional scale model sphere.

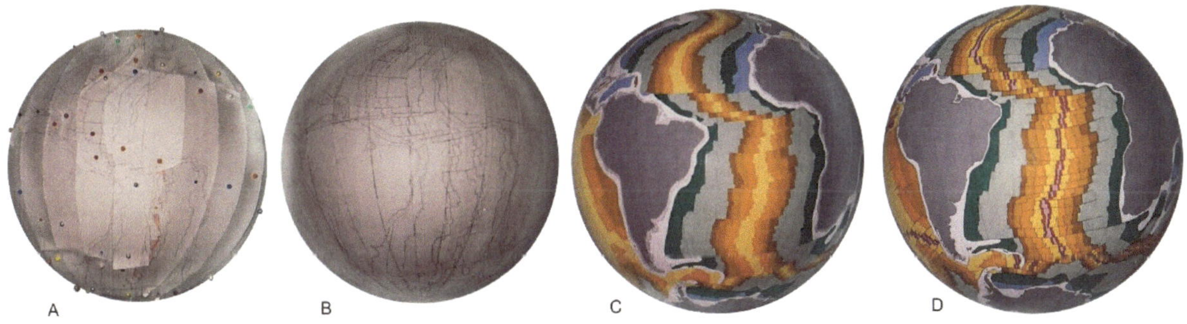

Figure 4.2 Small Earth scale model construction details of various ages and size models: Figure A, preliminary pin-up of plate fragments; Figure B, final paste-up; Figure C, painting of seafloor striping in progress; Figure D, completed present-day small Earth model.

For each model, as the youngest coloured seafloor crustal stripes are progressively removed, the remaining plate boundaries are then reassembled along their common mid-ocean-ridge spreading axes on a reduced radius small Earth model. Of interest is that, on an Expansion Tectonic Earth, the mid-ocean-ridge spreading axes do not move. They instead retain their global spatial integrity throughout time. It is the configuration of the mid-ocean-ridge spreading zones that therefore dictate precisely the assemblage of each adjoining crustal plate boundary. Once the paper fragments are assembled and pinned in place, each piece is then glued into its final assembled position and in this modelling exercise each model was then hand painted (Figure 4.2).

While this outline of small Earth model construction may appear to be brief, the actual construction methodology is also quite basic; each coloured stripe is simply taken away in turn and the remaining crustal plates are re-assembled on a predetermined smaller radius Earth model. The method of constructing small Earth models adopted here then results in a series of models that adhere precisely with the coloured time-scales depicted in the seafloor bedrock mapping. The models also adhere to the ancient Earth radii as determined from measurements of surface areas of this same mapping data. By making more models than required we are also in a far better position to minimise crustal distortion caused by changes in surface curvature, thus achieving the goal of accurately constraining crustal plate assemblages with time.

4.3 Small Earth Models

The early-Jurassic Period to the present-day, as covered by the first phase of small Earth models, represents about 200 million years of Earth history. This interval of time is made up of part of the earlier Mesozoic Era and the later Cenozoic Era. Each of these eras is further subdivided into distinct geological periods and epochs, as presented again in Figure 4.3.

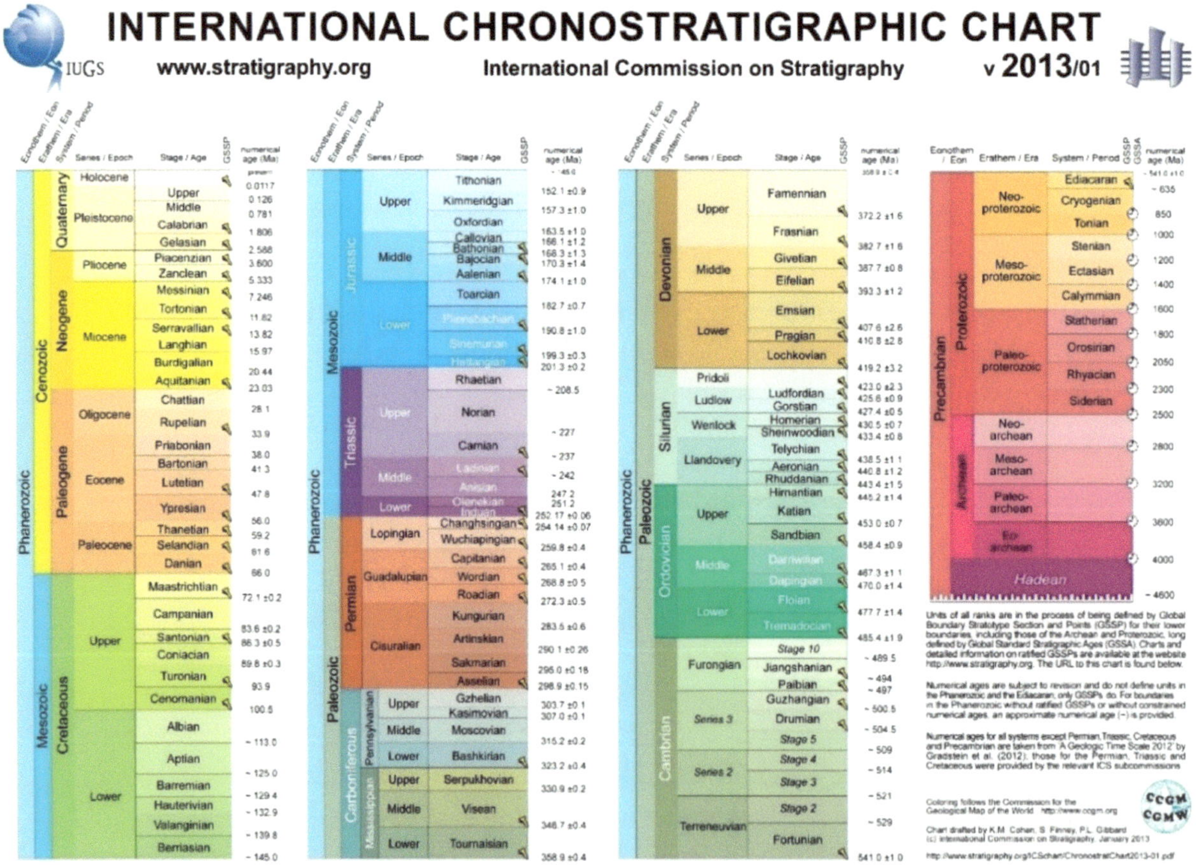

Figure 4.3 The International Stratigraphic Chart, showing the subdivision of geological time into the various eons, eras, periods, epochs and ages, along with their absolute ages (International Commission on Stratigraphy, 2013).

Compared to the total Earth history though, this interval of time represents only about 4 percent of all measured geological time. It is also the interval of time that is most easily modelled from seafloor crustal data and is traditionally considered by model makers, along with all proponents of Expansion Tectonics, to be the period of time most representative of the growth period of Earth expansion.

What we are essentially trying to achieve on each of these small Earth models is a snapshot in geological time which coincides with the beginning of each coloured seafloor stripe. This allows us to proceed back in time in precise, well-defined steps that reflect the empirical mapping data. Modelling is also governed by the need to fully appreciate the influence of distortion of the Earth's surface crust during on-going changes in surface curvature, as well as how distortion influences the assemblage of each model. By moving back in time, rigid fragments of continental or seafloor crusts are not simply reassembled on a smaller radius Earth model. The crustal plates are instead assembled by retaining the integrity of the crusts as they are preserved in the geological rock-record, as well as making allowance for distortion of the crusts during changes to surface curvature, each according to their tectonic histories.

For the completed small Earth models shown in Figure 4.4, as the coloured seafloor stripes are removed in turn it can be seen that the remaining coloured stripes neatly close together on each successive model. Each crustal plate assembles together in a unique, orderly, and predictable manner, during systematic closure of all of the oceans. By removing the coloured seafloor stripes in succession and refitting the plates together on a smaller radius Earth model, each plate is then shown to reunite precisely along their respective mid-ocean-ridge spreading axes, estimated to be at better than 99 percent fit-together for each model. The remaining one percent misfit is simply a combination of map error, cartographic discrepancies, as well as the influence of unaccounted crustal distortion or graphical limitations inherited during model making.

The outcome of this small Earth modelling exercise demonstrates that, for each of the established seafloor crustal plates, there is only one unique plate-fit-option for each of the models shown. From this, it can be seen that there is no need to introduce non-accountable pre-existing crusts or to randomly fragment continental crusts in order to accommodate for the seafloor evidence. What this means is that, apart from local fine-adjustment along the plate margins, there are no alternative assemblage options available. The seafloor mapping in fact dictates precisely where and how the plates are assembled back in time for each time interval shown. It is also emphasised that this modelling represents the first time that the published 1990 bedrock geological mapping has been used to accurately assemble and constrain crustal plates on Expansion Tectonic small Earth models. These models can therefore be considered as unique.

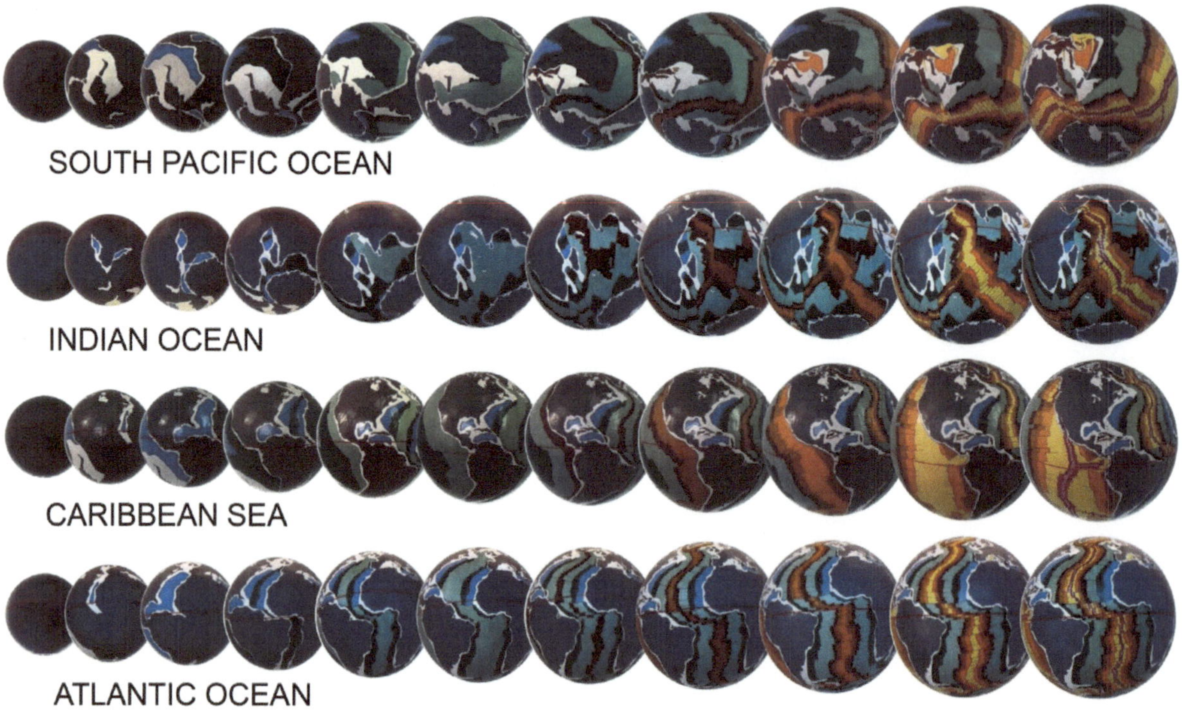

Figure 4.4 Spherical small Earth models of the Jurassic to present-day Expansion Tectonic Earth. Each small Earth model demonstrates that the seafloor crustal plate assemblage coincides fully with seafloor spreading and geological data and accords with the derived ancient Earth radii.

On the early-Jurassic to present-day small Earth models shown, the distinguishing feature of this interval of time is that it covers the break-up and dispersal of the original supercontinent called Pangaea. This break-up is also accompanied by opening of all of the modern oceans. On an Expansion Tectonic Earth the Pangaea supercontinent, which comprised an assemblage of all of the present-day continental crusts, simply wrapped itself around to enclose the entire ancient Earth with continental crust. It is the subsequent rupture and break-up of Pangaea during increase in both Earth radius and surface area that then lead to the formation of the modern continents and opening of the modern oceans. On each of the small Earth models these modern oceans are then shown to progressively increase their surface areas from the early-Jurassic to the present-day in strict accordance with the seafloor mapping data.

As previously discussed, this post-break-up growth history of the modern oceans is now preserved as volcanic rocks within the seafloors and the evidence for this growth history is uniquely shown by the coloured seafloor mapping. As well as the presence of volcanic rocks, it can be seen in Figure 4.4 that sediments, eroded from the exposed lands during initial continental break-up, are also preserved along the margins of each continent and in submerged remnant plateaux within the oceans. These sediments are shown as white areas around the margins of the continents in Figure 4.4. Small, remnant fragments of much older continental crusts have also been identified and mapped by others in a number of areas on the seafloors. These fragments are considered to represent small areas of continental crusts that have either been eroded or fragmented and left behind as the Pangaea supercontinent first ruptured, broke-up, and dispersed.

On Expansion Tectonic small Earth models the continental shelves, marine plateaux, and remnant ancient seafloor crusts are shown to merge during the Jurassic Period to form a global network of marine sedimentary basins. This mergence suggests that during this time the existing network of continental sedimentary basins, plus the marine basins, represents a network of ancient continental seaways. By progressively returning the eroded sediments from these continental and marine sedimentary basins back to the lands where they came from, additional small Earth modelling shows that each continent further re-unites with adjoining continents along their mutual continental shelf margins, at an even smaller Earth radius of approximately 50 percent of the present Earth radius.

It is interesting to note that during this Method 1 construction phase, as the seafloor crusts are progressively removed and each plate and continent is re-united with its adjoining continental neighbour, a network of continent-to-continent *docking-points* is established. The term docking-point is used here to refer to the re-uniting of each continental crustal fragment along a section of coastline or continental landmass, rather than indicating a specific point location. These docking-points represent areas of coastline where two or more previously separated continents precisely touch and lock back together, somewhat like the lugs on a piece of jigsaw puzzle. These docking-points then represent important cartographic and geographic reference points identified during model construction and they are used here to further constrain the positions of the continents on each successive small Earth model. It was observed during model construction that the location and significance of these docking-points continued to remain throughout the Mesozoic Era and their on-going usage will become increasingly important later when even older, smaller radius Earth models are constructed using continental crusts.

From the model studies briefly introduced in this section it is important to again reiterate a very important observation. By globally removing the seafloor crustal geology from each of the coloured stripes in succession and refitting the remaining crustal plates together on small Earth models, all plates are shown to reunite with one, unique assemblage. Each crustal plate is then shown to assemble with a better than 99 percent fit-together, for each model constructed.

If the Earth were not increasing its radius, or was undergoing a partial or pulsating increase in radius, then ***this unique fit-together of all plates and continents would not occur.*** Similarly, if the Earth were not increasing in radius then large gaps or overlaps in the reconstructed plates should occur—as happens in Plate Tectonic assemblages—and the need to fragment continental and seafloor crusts to accommodate for the crustal evidence would become increasingly apparent and necessary. Each of these gaps or overlaps in turn would also outline areas where presumed, pre-existing crusts should have supposedly once been.

Instead, the early-Jurassic to present-day Expansion Tectonic small Earth models demonstrate conclusively that large gaps or overlaps in the crustal plates do not occur, hence the near perfect cartographic match between plates. The fact that large gaps or overlaps do not occur, on any of the models, demonstrates the significance of the coloured seafloor mapping as a valuable tool for reconstructing and constraining the unique fit-together of all past plate assemblages.

5 Origin of the Modern Oceans and Seas

"It is difficult to believe that chance alone can explain this fitting together of the continental margins."
Barnett, 1962

In discussing the origin of the oceans and seas it will be shown that, on an Expansion Tectonic Earth, prior to about 250 million years ago there were no modern oceans, only ancient continental seas. The transition from ancient seas to modern oceans came about when the Pangaea supercontinent first started to break-up to form the modern continents and intervening oceans, thus initiating draining the waters of the ancient continental seas into the newly opening oceans. In this chapter I again make a distinction between the terms *modern oceans* and *ancient continental seas*, and by continental seas I am referring to bodies of seawater that covered low-lying parts of the ancient continental lands—sometimes referred to in geology as *epeiric seas*.

It is also important at this stage to reflect on a number of observations from the bedrock geological map to further understand what is shown in the coloured seafloor spreading data, in particular how it relates to the origin of the modern oceans. In Figure 5.1 it can be seen that the mid-ocean-ridges are centrally located within each of the modern oceans—shown by the centrally located pink Pleistocene stripes. These mid-ocean-ridges include *asymmetric-style* spreading, located along the perimeters of the northeast and southeast Pacific Ocean, and *symmetric-style* spreading as seen elsewhere. A combination of both asymmetric and symmetric-style spreading can also be seen in the Indian Ocean. Here, asymmetric refers to the one-sided, near-continental-margin-spreading zones and symmetric refers to the parallel, two-sided spreading zones that can be seen straddling each of the mid-ocean-ridges proper.

Plate Tectonic studies interpret the asymmetric-style spreading patterns as evidence for subduction, in particular along the west coasts of the Americas. Here seafloor crusts are said to be recycled beneath the continental crusts by convection cells located within the lower crust and mantle. The problem with this argument is that asymmetric-style spreading phenomena are limited to the west coasts of North and South America and east coasts of

Asia and Southeast Asia, both of which are also shown to be opening. In contrast, the symmetric-style spreading was originally interpreted during the 1960s from a limited area in the North Atlantic Ocean, demonstrating that this ocean is opening and increasing its surface area. Completion of the Geological Map of the World has since shown that this symmetric-style spreading occurs in all oceans which suggests, by inference, that all of the oceans are in fact increasing their surface areas.

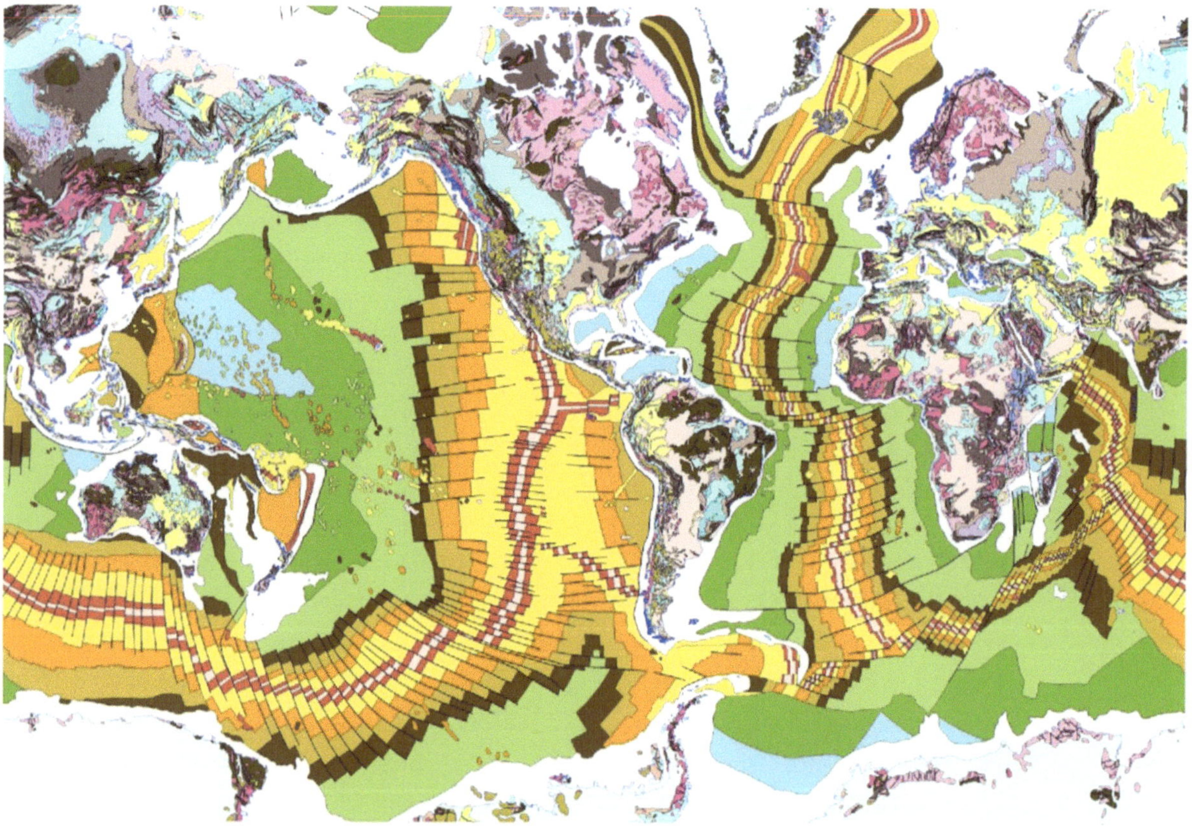

Figure 5.1 Geological Map of the World (digitized with permission from the Commission for The Geological Map of the World and UNESCO, 1990).

The coloured seafloor stripes also make a distinction between the earlier Mesozoic and the later Cenozoic seafloor growth history, where the symmetric pink through to brown coloured stripes coincide with the Cenozoic Era and the greens and pale blue coloured stripes coincide with the Mesozoic Era. In this context, the earlier Mesozoic Era generally shows a more primitive asymmetric-style of mid-ocean-ridge spreading, while during the transition from the Mesozoic to Cenozoic Eras—the transition from green colours to brown in Figure 5.1—symmetric-style mid-ocean spreading ridges become fully established in all of the modern oceans.

In many cases, the symmetric-style Cenozoic spreading patterns can be seen to crosscut the previously established asymmetric-style Mesozoic spreading patterns. From this evidence it can also be seen that the Cenozoic is then characterised by the initiation and propagation of symmetric-style spreading patterns throughout all of the modern ocean basins. Further break-up, rifting, and geographic isolation of the modern continents then accompanied this rapid and unprecedented phase of symmetric-style spreading in all oceans.

In order to commence discussion on the origin of each of the oceans in detail a few observations relating to ancient geography and climate, as will be presented in later chapters, will also be briefly introduced here. When geographical data relating to the distribution of the ancient seas and shorelines are plotted on each of the constructed small Earth models it will be shown that, prior to 250 million years ago, there were no deep oceans, only a network of relatively shallow continental seas generally less than a few kilometres deep.

What this observation implies is that, when moving back in time, as well as returning the seafloor volcanic lava back to the mantle where it originally came from, a proportion of the ocean waters and atmospheric gases must also be returned back to the mantle. This is an empirical observation and must be considered in order to comply with the known distribution of ancient continental seas and shorelines. It will be shown later from this shoreline evidence that there is no justification in inferring that the total volume of ocean waters has remained near-constant over time. If it were constant then this would suggest that at some time in the past the Earth would have been completely covered by water to a depth of many kilometres. The geology preserved in the continental rocks shows that the Earth clearly wasn't completely covered by water during ancient times.

It will also be shown in coming chapters that each of the modern oceans initially started as small marine sedimentary basins forming a series of relatively shallow seaways. Opening of these seaways coincided with where the Pangaea supercontinental crust first started to rupture and break-up during late-Permian times, some 250 million years ago. These marine sedimentary basins and seaways will also be shown to have been located over areas of low-lying ruptured continental crust that had earlier undergone considerable crustal thinning during a prolonged period of continental crustal stretching and extension. These marine basins may have also contained small areas of intruded volcanic seafloor crustal rocks—referred to in geology as *ophiolites.*

During opening of each of these primitive seaways, as well as draining the waters from the ancient continental seas, sediments were eroded from the exposed lands and re-deposited into new low-lying areas. The huge volume of new sediment eroded from the exposed lands initially covered the newly formed seafloor ophiolite volcanic rocks. This erosion of the lands will be shown later to have occurred as a direct result of changing surface curvature of the Earth, where the continental crust will be shown to have initially remained elevated within the central parts of the exposed lands and depressed around the margins. This erosion of the exposed lands, in part, assisted in maintaining a changing surface curvature by lowering the elevated land surfaces and filling the adjoining low-lying sedimentary basins with eroded sediment.

By the mid- to late-Jurassic times—some 165 million years ago—the emerging modern oceans had opened wide enough, beyond the limits of the deposition of sediments, to finally expose the intruded seafloor volcanic lava. These exposed volcanic lava rocks were then subsequently preserved as the coloured seafloor crust shown on the bedrock geological map in Figure 5.1. It is emphasised that it was this exposure and preservation of the seafloor volcanic rocks that gave rise to the coloured seafloor patterns seen in the Geological Map of the World. This process is still continuing today along the present mid-ocean-ridge spreading zones. It should also be noted that this conflicts strongly with the Plate Tectonic requirement for extensive removal of seafloor crust by subduction as well as the presence of pre-existing crusts.

As will be shown on each of the Expansion Tectonic small Earth models, opening of the modern oceans was initially centred in the North Pacific, South Pacific, Arctic, and North Atlantic Ocean regions. Opening of the South Atlantic and Indian Oceans then commenced during mid- to late-Jurassic times, and the Southern Ocean commenced opening during the Paleocene. Opening of each of these modern oceans was also accompanied by marked variations in sea-levels. These sea-level changes were accompanied in turn by climate changes and extinction events occurring during crustal break-up, as well as disruptions to the previously established climate zones, polar ice-caps, species habitats, and migration routes.

The origin of the modern oceans will now be looked at in more detail on each of the constructed Expansion Tectonic small Earth models to see what happens to these oceans when the mid-ocean-ridge spreading zones progressively open as the Earth steadily increases in size. The observations presented in this chapter are taken directly from the Expansion Tectonic small Earth models and can therefore be considered as empirical observations. These models will be elaborated on still further in later chapters as more specific topics, including additional geological, geographical, and geophysical evidence are introduced.

5.1 Modern Oceans

On all Plate Tectonic reconstructions large, ancient, essentially theoretical Tethys, Iapetus, and Panthalassa Oceans are required to maintain a constant radius Earth premise. The hypothetical Panthalassa Ocean was considerably larger than the present Pacific Ocean, having to accommodate for the surface areas of the closed Atlantic, Southern, and Indian Oceans. This Panthalassa Ocean in turn is said to have merged imperceptibly with an equally theoretical Tethys Ocean. With time these, and other lesser oceans, are then depicted as periodically opening and closing. Similarly, ancient supercontinental fragments are also depicted as randomly colliding, amalgamating, re-fragmenting, and dispersing across each of the ancient oceans throughout Earth history.

On an Expansion Tectonic Earth these Panthalassa, Tethys, and Iapetus Oceans cannot be reconciled in their entirety. Instead, these same ancient oceans initially formed a network of much smaller continental seas covering low-lying areas of the ancient Pangaea supercontinental lands. These will be shown here to be represented by a primitive Panthalassa Sea, located between Australia, Asia, and North America; a Tethys Sea, located within the present Eurasian continent; and an Iapetus Sea, located between the West African and North American cratons, possibly extending into South America.

On an Expansion Tectonic Earth the modern oceans initially opened within ruptured areas of continental crust where supercontinental crust had failed to keep pace with changes to the Earth surface area and surface curvature. On small Earth models these rupture zones were initially located within the present northwest Pacific and North Atlantic Ocean regions respectively. The rupture zones then progressively opened and rapidly extended in surface area throughout the Mesozoic and Cenozoic Eras forming what are now the modern Pacific and Atlantic Oceans.

By late-Triassic to early-Jurassic times, rupture of the supercontinental crusts had also occurred in the north-polar Arctic Ocean and south-polar South Pacific and Atlantic Ocean regions. These rupture zones commenced as *passive margin* extensional basins—which means a simple opening process accompanied by extensive deposition of shallow marine sediments. By the late-Jurassic, opening of these rupture zones then shifted from a polar to a more elongate meridional position centred on the North Atlantic and Pacific Ocean regions. On-going changes in surface curvature, accompanied by further rupture and subsequent rifting, then saw the preservation of early-Jurassic seafloor crust within the newly emerging North Pacific Ocean, followed by late-Jurassic seafloor crust in the North Atlantic and Indian Oceans. Similarly, the Southern Ocean commenced opening during the Paleocene Epoch.

Throughout the Cretaceous Period, each of these new oceans continued to increase their surface areas and were accompanied by further break-up and rifting of the modern continents. Within these newly formed oceans, mid-ocean-ridge spreading axes were initiated and continued to develop as either: 1) asymmetric spreading patterns along the perimeters of the North and South Pacific Oceans; 2) as symmetric spreading patterns in the North and South Atlantic and Southern Ocean or; 3) a combination of both asymmetric and symmetric-type spreading patterns in the Indian Ocean. Not until the Mesozoic to Cenozoic transition did modern symmetrical mid-oceanic-ridge spreading patterns become firmly established throughout all of the modern ocean basins. This consistency of timing and sequence of ocean basin development agrees fully with the constraints imposed by the coloured seafloor spreading evidence.

On an Expansion Tectonic Earth it is again interesting note that the Cenozoic Era is characterised by initiation and extension of modern-day symmetrical-style spreading patterns throughout all of the modern ocean basins. This spreading was accompanied by further rifting and isolation of continental areas, such as Australia, Antarctica, and Greenland, as well as breaching of existing land connections between Antarctica and Australia, Antarctica and South America, North America and Europe, and Africa and Europe.

During this 250 million year history, extending from the late-Permian to the present-day, the Earth underwent a considerable and unprecedented increase in radius and surface area which was also accompanied by substantial changes in surface curvature. It should be appreciated that, just like rupturing of the continental crusts, seafloor crusts were also subject to on-going distortion and disruption during changing surface curvature. This is particularly evident in the North Pacific Ocean where rupture first commenced and where extensive areas of older Mesozoic crusts now occur.

These changes to surface curvature of seafloor crust in the Pacific Ocean in particular has given rise to a complex interplay of island-arc volcanism, particularly along what are now referred to as the western Pacific Ocean island-arc and trench margins.

From this brief overview and introduction to Expansion Tectonic small Earth model studies of the modern oceans, the history of each of the modern oceans will now be looked at in more detail. In these descriptions of the modern oceans I emphasise that, when talking about movement or migration of continents on an Expansion Tectonic Earth, this movement refers to an apparent migration of the continental crustal fragments as a direct result of opening and widening of the modern oceans. Any movement of the continents simply accommodates for this opening of the oceans and has nothing to do with conventional continental drift or mantle convection theory.

5.1.1 Arctic Ocean

The Arctic Ocean originated as a very ancient marine sedimentary basin which first formed and commenced opening as the ancient Pangaea supercontinental crust started to rupture around 250 million years ago (Figure 5.2). This early Arctic Ocean basin was originally located in mid- to high-northern latitudes. Successive small Earth models then show that over time, as the ocean continued to open, the ocean basin and surrounding continents progressively migrated into the north polar-regions of today.

In Figure 5.2 it can be seen that the older Arctic Ocean shown in the lower right models is made up of a large expanse of marine sediments, shown as white areas located around the margins of the ancient continents, plus two seafloor basins shown as green colours. These green coloured seafloor basins are called the Amerasia and Eurasia Basins. The basins are Cretaceous in age and are shown to crosscut and displace the extensive white areas of marine sediments which were first deposited along the newly emerging continental shelf margins prior to exposure and preservation of the volcanic seafloor crusts.

On an Expansion Tectonic Earth, opening of the Arctic Ocean occurred as a result of crustal rupture and break-up of the ancient Pangaea supercontinent located between the newly formed North American and European continents. This break-up was then followed by on-going seafloor crustal stretching and opening of the Arctic Ocean within this region. The initially small Arctic Ocean basin progressively increased in surface area and its boundaries continued to extend further to the southeast during the Mesozoic Era.

Figure 5.2 Arctic Ocean small Earth spreading history, extending from the present-day back to the early-Jurassic.

The presence of exposed coloured seafloor crustal rocks within the Amerasia and Eurasia Basins shows that there was an initial period of seafloor spreading in each of these areas during the Cretaceous Period. The lack of any further seafloor crusts suggest that spreading then effectively ceased during late-Cretaceous times. Today, there are no active spreading centres in the older Arctic Ocean regions. Spreading is now located within the adjoining Nansen-Gakkel mid-ocean-ridge, which is a northern extension of the North Atlantic Ocean mid-ocean spreading ridge.

From the late-Cretaceous to the present-day the North Atlantic mid-ocean-ridge continued to extended into the Arctic Ocean region, crosscutting pre-existing marine sedimentary basins in this area. This progression into the Arctic Ocean gave rise to further fragmentation of the ancient continental crust and opening of new oceans located between Greenland, Canada, and Russia. During this opening phase, sediments eroded from the lands were initially redeposited within the early Arctic Ocean basin. Deposition of sediments then progressively changed from deposition within a shallow marine basin to deposition within marine continental shelf settings which then bordered a true deep-ocean-spreading ridge.

Throughout much of this time the present-day Alaskan and Siberian Peninsulas remained joined, forming an important land-bridge between the two continents and effectively isolating the Arctic Ocean from the emerging Pacific Ocean. During this lengthy period of time there was also continental crustal stretching between the two land masses, which was accompanied by faulting along each of the peninsulas during opening of the Arctic Ocean basins. The continental crust that now makes up Greenland and the Canadian Arctic Islands also fragmented during this time and these fragments began to gradually rift apart as the oceans continued to open.

From 65 million years ago to the present-day the Arctic Ocean basin was characterised by an on-going phase of symmetric-style seafloor and continental crustal extension and opening. This phase has resulted in further separation of the Canadian Arctic Islands, opening of Hudson Bay, and further rifting between Greenland and Canada. Separation of the Alaskan and Siberian Peninsulas across the Bering Strait occurred within the past 2 million years and this crustal separation, rifting, and extension is continuing to the present-day.

5.1.2 Atlantic Ocean

The ability to match the east coast of South America with the west coast of Africa has long been recognised by mapmakers for many centuries. This remarkable fit-together now forms the conceptual basis for both early Continental Drift and Plate Tectonic studies. The closing of the Atlantic Ocean and reconstruction of these continents also forms the basis for assemblage of the Pangaea supercontinent. This fitting together of the South American and African continents is further substantiated by an extensive array of geological evidence dating back to the investigations of Wegener.

On conventional reconstructions of the Atlantic Ocean, the corresponding margins of northern Brazil in South America and Guinea in Africa are traditionally fitted together according to their geological matches. But, unbeknown to most people, fitting these coastlines together produces a narrow triangular gap which widens south between the continental margins of South America and Africa south of the Niger Delta region. To minimise this misfit, the margin of South America may also be fitted against southern Africa south of the Niger Delta region. This then produces a narrow triangular gap between the Guinea and north Brazil coastlines, widening northwards. Unfortunately, this also produces a significantly greater area of misfit in the Florida and Central American regions.

It is significant to note that in 1958, when Carey first reassembled these continents on a spherical model representing the Earth's modern dimension, Carey noted these very same misfits. He commented that "*...if all the continents were reassembled into a Pangaean configuration on a model representing the Earths modern dimensions, the fit was reasonably precise at the centre of the reassembly and along the common margins of north-west Africa and the United States east coast embayment, but became progressively imperfect away from these areas.*" Carey concluded from this research that the fit of these ancient continents "*...could be made much more precise in these areas if the diameter of the Earth was smaller at the time of Pangaea.*"

Opening of the North and South Atlantic Oceans on an Expansion Tectonic Earth is shown sequentially in Figure 5.3.

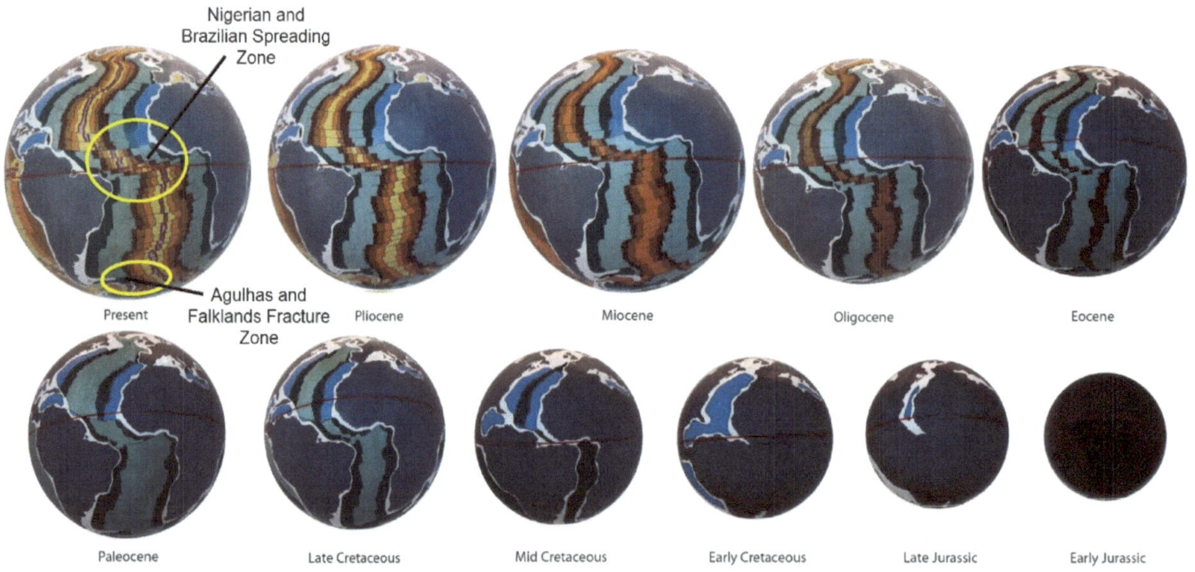

Figure 5.3 Atlantic Ocean small Earth sequential spreading history, extending from the present-day back to the early-Jurassic.

Opening is shown to have commenced in the lower right models during the early- to late-Jurassic in the North Atlantic region, located between North Africa and North America. This opening North Atlantic region later extended south to merge with the opening South Atlantic Ocean, as well as west into the Gulf of Mexico and Caribbean regions.

On Expansion Tectonic small Earth models it is significant to note that, just as Carey concluded, when the North and South Atlantic Oceans are closed misfitting between the continental margins of North and South America, Europe, and Africa is entirely eliminated. Opening of these oceans during the Mesozoic and Cenozoic Eras is then shown to be progressive and symmetrical. Subsequent opening of the Atlantic Ocean is also shown to have occurred in conjunction with opening of both the Arctic and Indian Oceans.

5.1.2.1 North Atlantic Ocean

Opening of the North Atlantic Ocean on an Expansion Tectonic Earth is shown in Figure 5.3 to have commenced as a narrow rift basin, located between the east coast of North America and west coast of Africa. Over time, this rift basin continued to open and progressively extend north into the Arctic Ocean and south into the South Atlantic Ocean. From the early-Jurassic a small counter clockwise rotation of the combined South American and African supercontinent, relative to North America, accelerated opening of the North Atlantic region and also extended the ocean west into what is now the Caribbean Sea. This rotation occurred in sympathy with opening of the Pacific Ocean as each of the adjoining continents continued to adjust for changing Earth radius and surface curvature.

During the early-Cretaceous—about 130 million years ago—the newly formed North Atlantic spreading ridge continued to extend north into the Grand Banks continental shelf, then located between Canada and Iberia. This occurred in response to both continental plate motion and opening of the Caribbean Sea. Rifting between South America and Africa also commenced which then initiated a progressive extension of the spreading ridge into the previously separated North and South Atlantic Oceans.

By the Late Cretaceous—around 80 million years ago—the North Atlantic mid-ocean spreading ridge had continued to extend further north into the Arctic Ocean. This spreading ridge then branched northwest into the Labrador Sea rift zone, located between Canada and Greenland, and northeast into the Mediterranean Sea. The extension and branching of this spreading ridge gave rise to continental break-up and rifting between Canada and Greenland, as well as rifting and rotation of Spain relative to both France and England.

From the late-Cretaceous to the present-day, the North Atlantic mid-ocean-ridge has continued to open in conjunction with both the Nansen-Gakkel spreading ridge within the Arctic Ocean and the South Atlantic spreading ridge. Spreading within each of these areas has resulted in a progressive enlargement of both the North and South Atlantic Oceans and was accompanied by symmetrical seafloor spreading.

5.1.2.2 **South Atlantic Ocean**

The South Atlantic Ocean, as shown in Figure 5.3, commenced opening between the southern South American peninsular and the South African region as a long narrow rift zone and has continued to steadily open to the present-day. The pattern of coloured seafloor stripes shows that this opening occurred in sympathy with opening of the Indian Ocean. Because of this, these two oceans can be considered as extensions of the same continental crustal rupture and opening event.

Separation between the west coast of Africa and east coast of South America commenced during the late-Jurassic—around 155 million years ago. This opening occurred along the now separated Agulhas and Falklands fracture zones, presently located in the southern South Atlantic Ocean. During late-Jurassic times, continental rifting and separation of the two continents progressively extended north. The South Atlantic spreading ridge then eventually merged with the North Atlantic spreading ridge during the early-Cretaceous along the Nigerian and Brazilian rift zone. From that time on the North and South Atlantic Oceans have remained united as a single Atlantic Ocean.

From the late-Cretaceous—around 80 million years ago—to the present-day spreading in both the North and South Atlantic Oceans has continued along a common mid-Atlantic spreading ridge. Subsequent spreading in both regions continued to be symmetrical, with a slow clockwise rotation of South America, relative to Africa, giving a slightly greater spreading rate in the southern South Atlantic Ocean.

5.1.3 Caribbean Sea

The Caribbean Sea is made up of the Mexico, Colombian, and Venezuelan Basins, which are shown in Figure 5.4 to be separated by what is referred to as the Antilles Arc. On an Expansion Tectonic Earth the development and subsequent opening of the Caribbean Sea was intimately related to the continental plate motion histories of both South America and Africa, relative to North America. Bedrock geological mapping of the seafloor shows that opening of each of the Caribbean basins was most active during the Jurassic Period. This was then later reactivated along the

Antilles Arc during the Paleocene—around 65 million years ago—and has continued to open as a relatively restricted basin to the present-day.

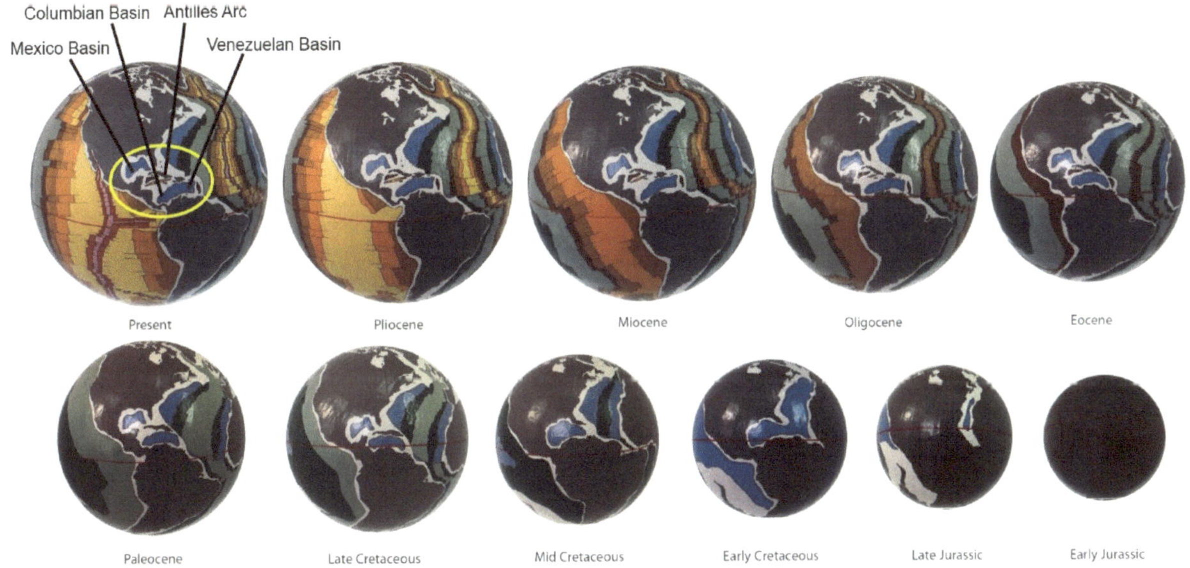

Figure 5.4 Caribbean Sea small Earth spreading history, extending from the present-day back to the early-Jurassic.

Plate Tectonic reconstructions of the Atlantic Ocean traditionally fit the Brazil and Guinean coastlines of South America and Africa together. This fit helps to minimise any misfit in the Caribbean Sea region. The Caribbean region is then seen as a buffer zone between the North American plate, the South American plate, and subducting oceanic plates of the Pacific Ocean. On these reconstructions the Caribbean region is considered to be a preserved piece of the ancient Pacific Ocean, referred to as the Farallon plate, which is inferred to have originated from outside of the Caribbean region.

On an Expansion Tectonic Earth, early development of the Caribbean Sea was intimately associated with opening of the North Atlantic Ocean along with subsequent rifting of Africa away from the Americas. Opening of the Caribbean Sea commenced during the Triassic to early-Jurassic Periods as a result of north-south crustal stretching and extension between the North and South American continents. Subsequent development of the Caribbean Sea was then closely related to a northwest migration of North America, relative to the still joined South American and African supercontinent. The Colombian and Venezuelan basins then opened during the Jurassic to Cretaceous Periods and were later separated by the Antilles island-arc.

This early opening phase lasted until the early-Cretaceous—around 130 million years ago—when rifting and eventual break-up between South America and Africa first began. During this phase, the Nicaraguan and Panama Peninsulas remained joined to South America. After separation of South America from Africa, South America then began to slowly rotate clockwise, relative to North America, in response to opening of the South Atlantic Ocean. This gave rise to further crustal extension and opening within the Caribbean basins, as well as isolation of the Antilles island-arc.

From the Paleocene Period to the present-day, South America has continued to slowly rotate clockwise, relative to North America, and was accompanied by fault movement along the length of the Antilles Arc. This fault movement extended along the western margin of Mexico and has since continued into the Gulf of California where it is known as the San Andreas Fault.

On an Expansion Tectonic Earth, an external origin for the Caribbean crustal region, as proposed in Plate Tectonic studies, is unnecessary. Small Earth model studies show that sourcing a fragment of crust from the Pacific region is untenable and is inconsistent with the established seafloor crustal mapping. Instead, opening of the Caribbean Sea region is shown to be intimately associated with opening of the North Atlantic Ocean as well as on-going crustal plate motion between North and South America and Africa.

5.1.4 Indian Ocean

Reconstructions of the Indian Ocean and assemblage of the surrounding continents is crucial to understanding the assemblage, break-up, and dispersal history of the Plate Tectonic Gondwana and Pangaea supercontinents. These conventional reconstructions place the precursor to the Indian continent within an ancient southern Gondwana supercontinent, located adjacent to Africa, Madagascar, and Antarctica. The Indian Ocean is then inferred to have formed during the subsequent break-up and dispersal of the Gondwana continental fragments during a later Pangaean supercontinental assemblage cycle.

On this conventional reconstruction an inferred ancient Tethys Ocean was located between a southern Pangaea supercontinent, made up of fragments of the Gondwana supercontinent, and a northern Laurasian supercontinent. During break-up of these supercontinents, India is then inferred to have broken away from Gondwana to migrate north as an island continent during subsequent opening of the Indian Ocean. This northward migration of India during the Mesozoic to early-Cenozoic Eras requires subduction of some 5,000 lineal kilometres of inferred pre-existing seafloor crust in order to close the ancient Tethys Ocean. India is then said to have collided with the Asian continent to form the Himalaya Mountains during the Cenozoic Era, leaving behind no trace of this pre-existing ocean crust.

In contrast, on an Expansion Tectonic Earth, small Earth modelling shows that the Indian continent remained firmly attached to the Asian continent throughout the Mesozoic and Cenozoic Earth history (Figure 5.5). The Indian Ocean is then shown to have opened during a post-Triassic phase of crustal rupture of the Pangaea supercontinent, in sympathy with opening of the adjoining Atlantic Ocean basin.

Opening of the Indian Ocean on an Expansion Tectonic Earth is shown in Figure 5.5 to have commenced during the early-Jurassic Period.

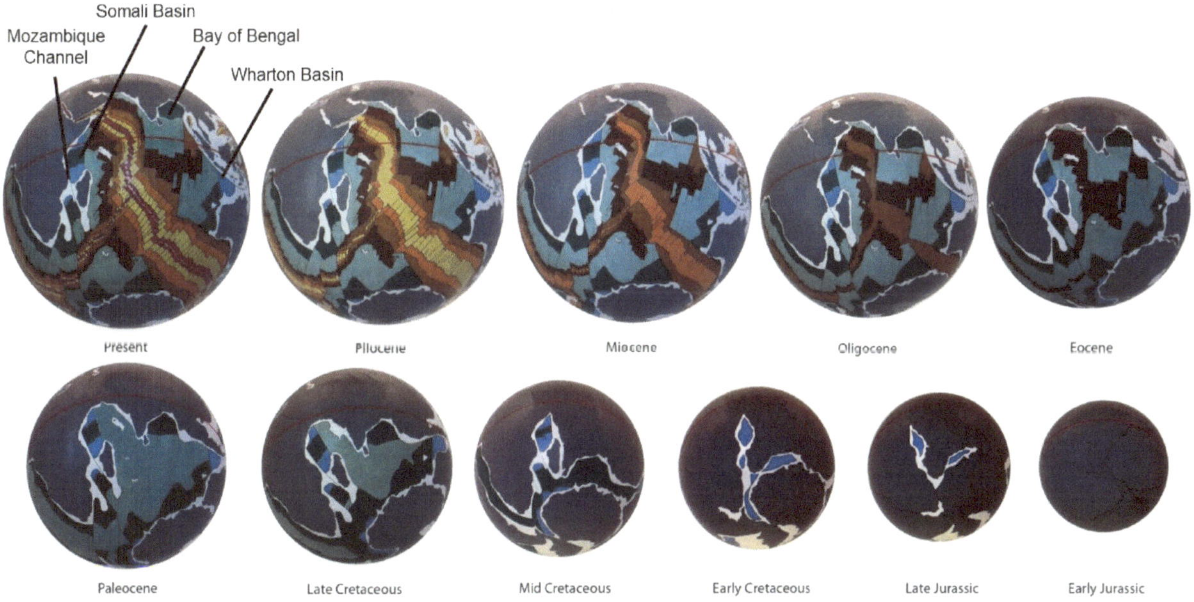

Figure 5.5 Indian Ocean small Earth spreading history, extending from the present-day back to the early-Jurassic

This occurred as a result of initial crustal rupturing within two separate areas, located adjacent to both India and Madagascar, in what are now the Somali Basin and the Bay of Bengal. Opening, and a southward extension of the Somali Basin continued up to the mid-Cretaceous Period which later became the Mozambique Channel located between Madagascar and Africa. During that time the Indian, Madagascan, and Sri Lankan continents and islands continued to remain firmly attached to the ancient Asian continent.

Crustal break-up and opening between Africa, Madagascar, India, and Antarctica then progressively extended south during the early- to late-Jurassic. During that time the Somali Basin and Bay of Bengal areas merged south of Madagascar. The merging of these basins formed what is referred to as a *triple-junction*—the junction of three separate spreading ridges and crustal plates—south of Madagascar and this region now represents the birthplace of the modern Indian Ocean basin.

The Mozambique Channel represents the northern limb of this early Indian Ocean triple junction. The southern limb of the triple junction extended south, around the tip of South Africa, to form an extension of the South Atlantic Ocean, which opened near the present Falkland Islands. The east limb then extended east to form the emerging Wharton Basin, which lay adjacent to West Australia.

By the mid-Cretaceous Period, the Mozambique Channel spreading ridge was abandoned and a new spreading ridge commenced opening between Madagascar and India. Once abandoned, the Somali Basin, Mozambique Channel, and Madagascar continent then remained as part of the African plate as seafloor crustal extension and opening progressively shifted to the east.

From the mid-Cretaceous to Paleocene times seafloor spreading continued between Madagascar, Africa, and India forming the Arabian Sea. During that time the Indian Ocean underwent a rapid north-south and east-west opening between Australia, Antarctica, Africa, and South East Asia, with a rapid northward opening and displacement of the Indian and Asian continents relative to Antarctica.

During the Paleocene—65 million years ago—to the present-day, symmetrical seafloor spreading radiated from a new triple junction located within the central Indian Ocean. This new spreading crosscut and displaced the older, previously established Mesozoic seafloor crust. Spreading then extended into the southern Indian Ocean, east between Australia and Antarctica and northeast into the Gulf of Aden. In the southern Indian Ocean region, the southwest limb of this spreading ridge split across older spreading patterns and now forms an extension of the South Atlantic Ocean mid-ocean-ridge system.

The north-south Ninety East spreading ridge, located within the eastern Indian Ocean, was abandoned during the Eocene. A new spreading ridge then extended southeast from the central Indian Ocean triple junction into a newly opening rift zone located between the Australian and Antarctican continental plates. This rifting initiated separation and northward migration of Australia relative to Antarctica, as well as initiation and relatively rapid opening of the Southern Ocean.

As spreading continued, the Gulf of Aden opened during the Miocene along a northwest extension of the central Indian Ocean and Carlsberg spreading ridges. A second triple junction has since formed at the western end of the Gulf of Aden during the Pliocene. This triple junction is now represented by the actively spreading Red Sea ridge and the East African Rift Valley system.

5.1.5 Pacific Ocean

At present the Pacific Ocean occupies nearly half the surface area of the Earth and can be arbitrarily subdivided into North Pacific, Central Pacific, and South Pacific Ocean regions. In all Plate Tectonic reconstructions, the precursor to the Pacific Ocean was a much larger ancient Panthalassa Ocean. This largely hypothetical early-Mesozoic Panthalassa Ocean is inferred to have possessed an old seafloor crust, which was formed by spreading along ancient mid-ocean-ridge zones during the Palaeozoic Era—the era prior to the Mesozoic Era—extending into the Triassic Period.

The Mesozoic and Cenozoic Plate Tectonic history is then depicted as an east-west and north-south contraction and reduction in surface area of the ancient Panthalassa Ocean in sympathy with opening of the Arctic, Indian, and Atlantic Oceans. This ancient ocean is thought to have eventually contracted in area to the size of the modern Pacific Ocean and is inferred to be still shrinking. During that time, all pre-Mesozoic crust is inferred to have been subducted beneath the continents along subduction zones located around the margins of the Pacific Ocean basins. This inferred subduction must have also included a substantial amount of new Mesozoic and Cenozoic seafloor crust, which is currently being generated within the East Pacific Ocean mid-ocean-ridge zone.

On all Plate Tectonic reconstructions, the North American continent progressively overrides the eastern North Pacific Ocean region. This supposedly occurs during a westward displacement of the North American continent during opening of the North Atlantic Ocean. The North Pacific Ocean spreading ridge is then inferred to have been entirely overridden and subsequently dislocated beneath the Pacific margin of the North American continent.

In contrast, on an Expansion Tectonic Earth an expansive pre-Mesozoic Panthalassa Ocean, and similarly an ancient Tethys Ocean, did not exist. Subduction of between 5,000 to 15,000 lineal kilometres of pre-existing East Pacific seafloor crust beneath the American continent is also not required. In Figure 5.6 the Pacific Ocean is instead shown to originate during early-Jurassic times as two separate marine sedimentary basins. A North Pacific basin was located between northwest Australia, Canada, and China, and a South Pacific basin was located between east Australia, South America, New Zealand and Antarctica.

Both of these marine basins progressively opened to the south and north, along the west coasts of North and South America respectively. These basins then merged to form a single Pacific Ocean basin during the mid- to late-Jurassic Period. Remnants of this early basin history are now preserved as continental margin and marine plateau sediments within the South East Asian and Coral Sea regions—shown as white areas in Figures 5.6 and 5.7.

Figure 5.6 North Pacific Ocean small Earth spreading history, extending from the present-day back to the early-Jurassic.

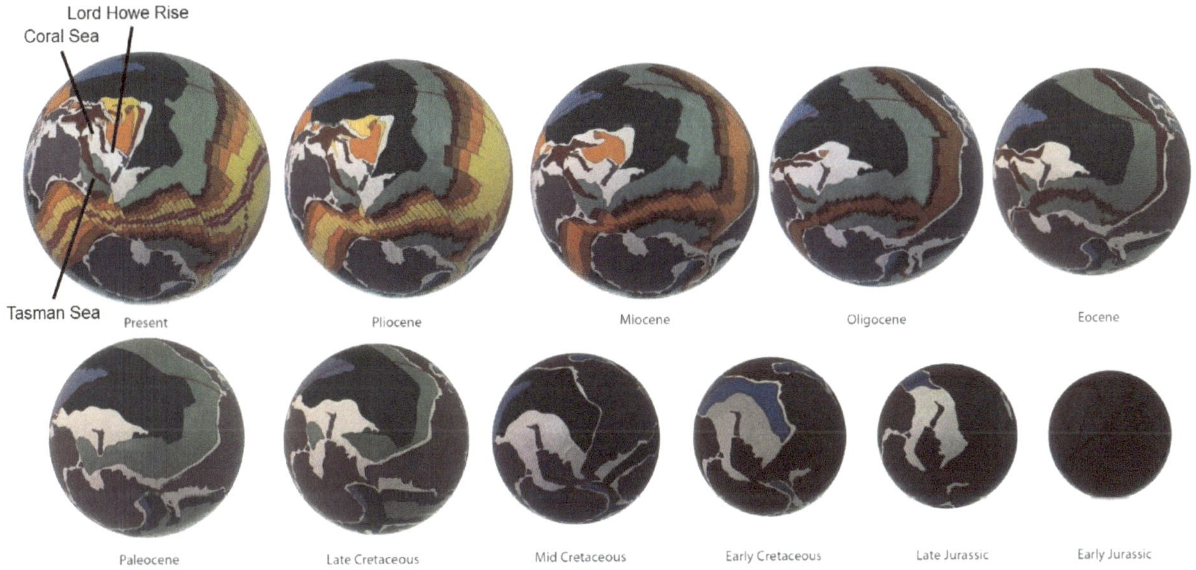

Figure 5.7 South Pacific Ocean small Earth spreading history, extending from the present-day back to the early-Jurassic.

On an Expansion Tectonic Earth, the subsequent evolution of the North and South Pacific Oceans involved a period of rapid northeast to southwest crustal extension and opening between North America, South America, and Australia. By the late-Jurassic, a deep ocean had extended southeast and south along the west coasts of North and South America. This coincided with the initiation, exposure, and preservation of new volcanic seafloor crust in the South Pacific Ocean.

Throughout the Mesozoic Era the North Pacific Ocean underwent a very rapid enlargement, with asymmetric spreading axes extending southeast into the South Pacific region. This spreading and mid-ocean-ridge development curved along the west coasts of North and South America and extended west into the Coral Sea region during the Cretaceous.

During the mid- to late-Jurassic, crustal rifting and opening between New Zealand, New Caledonia, South America, Australia, and Antarctica isolated the Coral Sea plateau, as well as the Lord Howe Rise and New Zealand. This rifting phase then left New Zealand as an island continent, which has remained isolated from surrounding continents to the present-day.

Development of the Pacific Ocean on an Expansion Tectonic Earth during the Cenozoic Era is characterised by the initiation and rapid development of symmetric-style seafloor spreading. This initially commenced within the Tasman Sea region, located southeast of Australia, during the Paleocene and continued to extend east towards South America during the Eocene. From there symmetric spreading continued north, forming the present East Pacific spreading ridge, and then continued to extend along the west coast of North America to its present location adjacent to California.

The opening and formation of the Pacific Ocean on an Expansion Tectonic Earth differs slightly to each of the other oceans. Because of the long period of time involved in opening the Pacific Ocean, the large area of older seafloor crust in the North Pacific region has been subject to considerable crustal stretching and distortion as a result of the changing surface curvature of the Earth. This changing surface curvature was generally absorbed as extension within the thin seafloor crust, but it also gave rise to complex plate interaction and jostling between and along adjoining plates, in particular between the various seafloor and continental plate margins.

Between the North Pacific Ocean plate and the Australian, South East Asian, and Chinese plates, complex crustal interaction has also given rise to the South East Asian island-arc-trench systems, which are now characteristic of the western Pacific region. On an Expansion Tectonic Earth, this region represents a complex interplay of plate motions that were generated during on-going adjustments to surface curvature, especially within the older North Pacific Ocean region.

It should be noted that the development of the Pacific Ocean on an Expansion Tectonic Earth cannot be reconciled with Plate Tectonic reconstructions on a constant sized Earth. On an Expansion Tectonic Earth the restrictions imposed by an early-Mesozoic Panthalassa and Tethys Ocean, as well as circum-Pacific subduction of pre-existing seafloor crust, simply did not exist and were not required during small Earth plate reconstruction. Instead, a process of asymmetric to symmetric evolution of the East Pacific Ocean spreading ridge readily explains the complex crustal patterns shown by the Pacific Ocean seafloor mapping. Likewise, the island-arc-trench systems and extensional back-arc-basins are readily explained by crustal interaction along the plate margin boundaries during relief of surface curvature, especially in the older crustal regions.

5.1.6 Mediterranean Sea

The Mediterranean to Middle East region is a complex and contentious region on Plate Tectonic reconstructions. In order to reconstruct the continents on a constant radius Earth, both Europe and Asia must be fragmented in order to close the Atlantic and Indian Oceans. On all conventional reconstructions the Mediterranean region then forms the western apex of a large triangular-shaped ancient Tethys Ocean. From there, this ocean is inferred to widen eastward towards an even larger ancient Panthalassa Ocean which separated the ancient supercontinents of Gondwana in the south and Laurasia in the north.

The evolution of the Mediterranean to Middle East regions on an Expansion Tectonic Earth is shown in Figure 5.8. Here, the Mediterranean Sea represents the remnants of a more extensive continental Tethys Sea—as distinct from a conventional Tethys Ocean. The Tethys Sea will be shown later to have had an extensive crustal and sedimentary basin history, extending back to the early Precambrian times.

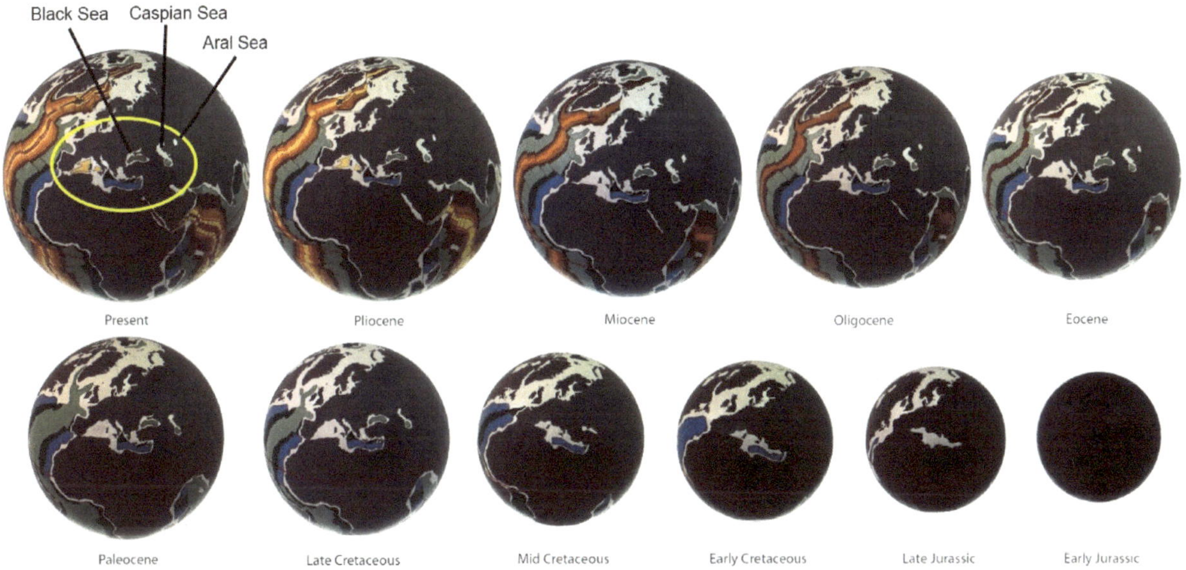

Figure 5.8 Mediterranean Sea small Earth spreading history, extending from the present-day back to the early-Jurassic.

On an Expansion Tectonic Earth the Mediterranean region initially developed during pre-Triassic times as a result of crustal extension between the ancient African and European continents. Crustal rupture within the ancient continental Tethys Sea then commenced during the Jurassic, which initiated opening of the modern Mediterranean Sea and crustal development within the Middle East region. The ancient Tethys Sea, in effect, represents the seabed of what is now the continental region of central Europe, the Middle East, and Asia.

Opening of the Caspian Sea followed during the early-Cretaceous Period. The Black Sea then followed during the mid-to late-Cretaceous and the Aral Sea during the Paleocene. These small seas all represent regions of continental crustal rupture and opening, which was initiated during complex clockwise crustal rotation occurring between the combined African-Arabian plate and central Europe.

Seafloor bedrock mapping shows that crustal extension within the Mediterranean to Middle East region was only active during Jurassic and early-Cretaceous times. From mid-Cretaceous to Miocene times complex crustal motions, including some localised compression between Africa and Europe, formed the Alpine Mountain belts. This event was followed by a continuation of continental crustal extension and renewed basin opening forming the young sedimentary basins presently located within much of central Europe.

The Alpine Mountain building event coincided with a northward extension of the North Atlantic Ocean spreading ridge which was marked by the opening of the Bay of Biscay, located between Spain and France, during the late-Cretaceous. Similarly, opening of the Persian Gulf and Red Sea also commenced during the Eocene and these are continuing to open to the present-day. Seafloor spreading was re-activated in the western Mediterranean region during the Miocene and was accompanied by continental rifting between West Africa and Spain.

On an Expansion Tectonic Earth separation of the Gondwana and Laurasia supercontinents by a large Tethys Ocean, and similarly a north-south closure of the Tethys Ocean during the Mesozoic and Cenozoic Eras, is not required. Instead, opening of the Mediterranean to Middle East region occurred as a result of clockwise rotation and crustal extension between Europe and the Middle East, relative to the adjoining African-Arabian continent. This continental crustal motion, extending from the Iberian Peninsula to the Tibetan Plateaux, also resulted in the rotation of Italy and fragmentation of the Alpine mountain belts during the mid-Cretaceous to present-day times.

These Expansion Tectonic Earth reconstructions also present a straightforward development history for the Black Sea region, the mountain belts of the Balkans, Turkey, and the Caucasus, the development of the southern Russian platform north of the Black Sea and the development of the Aegean Sea during the late-Cenozoic.

5.1.7 South East Asian Seas

The South East Asian region comprises the Philippine, South China, Celebes, Banda, and Java Seas. In this region the present-day South East Asia to New Guinea and Japanese Island chains represent complex island-arc systems. Plate Tectonic reconstructions on a constant sized Earth are particularly vague when it comes to reconstructing South East Asia. On these conventional reconstructions the region is represented by small remnant crustal fragments, which are inferred to have existed at the eastern end of a large ancient Tethys Ocean where it merged with an even larger Panthalassa Ocean. During closure of both the Tethys and Panthalassa Oceans, the South East Asian Seas were then interpreted to represent marginal or back-arc basins. These basins were further separated from the various island-arcs by spreading along zones of inferred crustal subduction and deep trench development.

The development of the South East Asia basins and seas on an Expansion Tectonic Earth is shown in Figure 5.9. This development of the South East Asia region is complex and progressive. Small Earth model studies show that development within this region was intimately associated with the formation and subsequent plate interaction of the North Pacific oceanic plate. On these small Earth models the region is shown to represent the fragmented remains of ancient intercontinental marine and continental sedimentary basins, as well as associated volcanic activity.

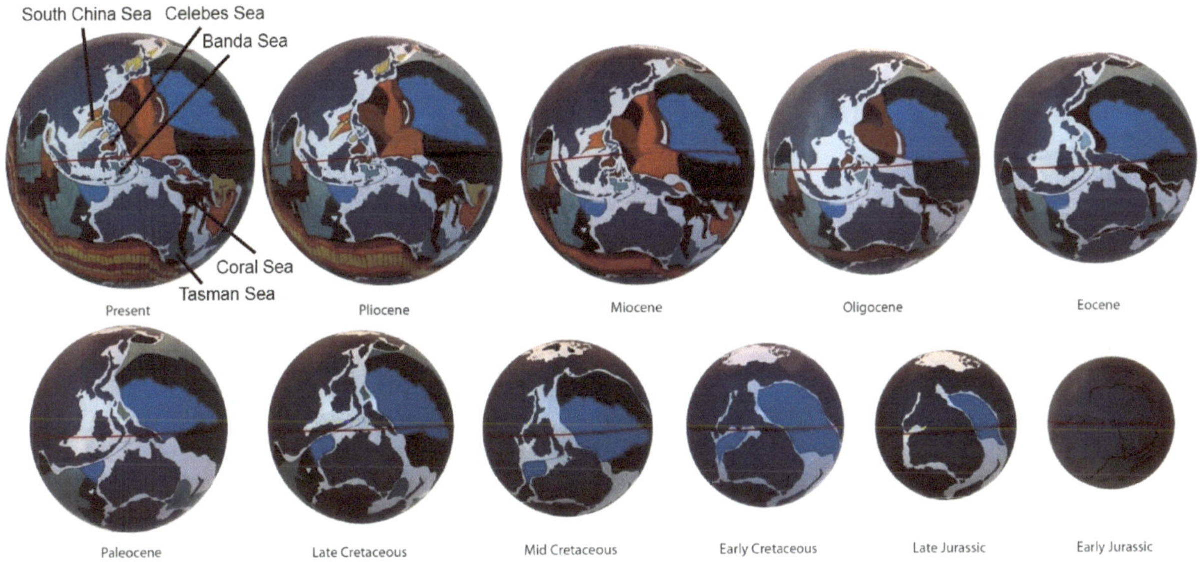

Figure 5.9 Southeast Asian and Southwest Pacific Basin small Earth spreading history, extending from the present-day back to the early-Jurassic.

The South East Asian region initially formed during early-Jurassic times. This region represents an area of marine basin opening, accompanied by deposition of sediments, and was located between the early Australian,

North American, and Chinese continents. These basins first formed in conjunction with the initial rupturing and opening of the North Pacific Ocean. At that time, sediments eroded from the surrounding lands were being redeposited within the newly formed marine basins. These sediments were mixed with volcanic rocks, erupted along the precursors to the modern mid-ocean-ridge spreading zones, to form the early island-arc systems. Remnant sediments from this early opening event are now preserved as both continental shelf and marine plateau sediments, which are exposed as the South East Asian islands and submerged plateaux throughout the South East Asian Sea region.

Development of the South East Asia region has been further complicated by the plate motion history of Australia, relative to Asia, during opening of both the Pacific and Indian Oceans. Progressive crustal extension and opening between each of these continents occurred during late-Cretaceous to Pliocene times which then resulted in opening of the South China, Celebes, and Banda Seas, as well as fragmentation of the early South East Asian island-arc system.

The crustal fragments making up the South East Asian region have since undergone complex clockwise rotation, crustal fragmentation, plate interaction, and on-going island-arc volcanism. These events all occurred during an extended period of southeast to northwest crustal extension between the Asian and Australian continental plates. This interpretation of the South East Asia region contrasts strongly with the Plate Tectonic requirement for continental collision, the closure of pre-existing Tethys and Panthalassa Oceans, and complex subduction of the Australian and North Pacific plates beneath the Asian and Philippine plates.

5.1.8 Southwest Pacific Ocean

The Southwest Pacific Ocean region is also structurally complex and comprises the Coral and Tasman Seas. The region is predominantly made up of remnants of ancient seafloor crust and marine sediments. These sediments were initially deposited and later fragmented during Cretaceous to Paleocene times and again during Miocene to present-day times. The region has been further complicated by crustal plate motion and plate interaction occurring along the margins of the Indo-Australian and Pacific plates.

The southwest Pacific Ocean region is shown on Expansion Tectonic small Earth models in Figure 5.9 above. This region represents the fragmented remains of earlier marine basins and sediments, deposited within an early pre-Triassic South Pacific Ocean basin. Initially, this area opened as a passive marine basin during pre-Triassic times. The newly formed basin then formed part of the early South Pacific Ocean, prior to merging with the North Pacific Ocean during the Triassic Period.

During pre-Triassic times, West Antarctica and New Zealand were assembled adjacent to Australia and South America. Subsequent rifting and opening between South America, New Zealand, and West Antarctica initiated formation of the South Pacific Ocean. During Triassic to early-Jurassic times, New Zealand and New Caledonia were separated from Australia during opening of the Tasman Sea. These remained attached to Ecuador in Central America until final separation from South America during the mid-Jurassic times.

During the mid- to late-Jurassic an early New Zealand and New Caledonian continent comprising the Coral Sea plateau, Lord Howe Rise, and New Zealand island-arc complex were well established. This ancient continent then further fragmented and was partly submerged to form the present Southwest Pacific region. Asymmetric seafloor spreading first developed in the South Pacific Ocean and this spreading extended west during the Cretaceous into the Tasman Sea. This event further isolated the Lord Howe Rise and New Zealand continent from Antarctica and Australia.

On an Expansion Tectonic Earth, the Tongan to South Solomon and New Hebrides trench and island-arc systems represent complex zones of plate interaction between the New Zealand and New Caledonia continental plate and the South Pacific Ocean plate margins. This complex interaction developed during on-going changes in Earth surface area and surface curvature over time. The interaction of these plates resulted in a slow clockwise rotation of the southwest Pacific Ocean region along each of the established trench and arc systems. In addition, the New Zealand and New Caledonian plate was further fragmented and displaced during the Cenozoic Era.

The rapid development of symmetric-style seafloor spreading in the South Pacific Ocean resulted in further opening of the southwest Pacific Ocean region during the Cenozoic. Complex plate interaction and motion along the Kermadec and Tongan trench and arc systems again accompanied this opening during the early-Miocene. This crustal motion was also related to movement along the Alpine Fault system of New Zealand, which continues to the present-day.

On an Expansion Tectonic Earth, the southwest Pacific Ocean basin region cannot be reconciled with conventional Plate Tectonic reconstructions. Instead, this region simply represents a complex interplay of extensional crustal motion and opening between Australia, North America, South America, and Antarctica in conjunction with opening of the North and South Pacific Oceans during on-going changes to Earth radius and surface curvature.

5.1.9 Southern Ocean

The Southern Ocean is located between the present Australian and East Antarctican continents. The ocean arbitrarily merges with the South Pacific and Indian Oceans to the east and west respectively. Rifting and opening of the Southern Ocean first commenced in conjunction with opening of the Atlantic and Indian Oceans during the late-Jurassic. Further opening then initiated final separation of Australia and Antarctica during the Paleocene, some 65 million years ago. Symmetric-style seafloor spreading in the Southern Ocean has since extended west, in conjunction with the eastern arm of the central Indian Ocean spreading ridge, and east to form an extension of the East Pacific spreading ridge.

Opening of the Southern Ocean is a paradox on Plate Tectonic reconstructions and very little mention of this ocean is made in conventional literature. This paradox arises because there are no subduction zones available to absorb the extensive plate motion required to continue opening the ocean.

On Expansion Tectonic Earth reconstructions (Figure 5.10), opening of the circum-polar Southern Ocean forms part of the extensive global network of mid-ocean spreading ridges. The opening and northward migration of the southern continents simply represents a natural consequence of break-up of the Pangaea supercontinent and increase in surface areas of the Southern and adjoining ocean basins during an increase in Earth radius.

Each of the Expansion Tectonic small Earth models shown in Figure 5.10 demonstrate that the Southern Ocean spreading ridge is intimately associated with the global network of ridges exposed within all of the modern oceans. No arbitrary disruption of the Southern Ocean mid-ocean-ridge network is required during subsequent opening. In addition, no pre-existing crust needs to be removed by subduction processes and arbitrary fragmentation of continental crust is also not required.

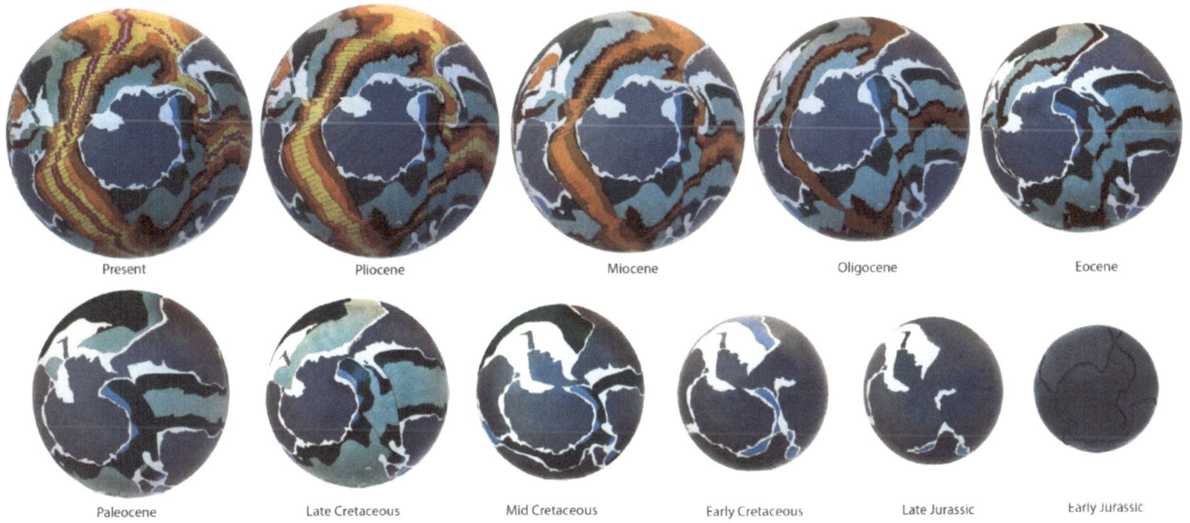

Figure 5.10 Southern Ocean small Earth spreading history, extending from the present-day back to the early-Jurassic.

It is appropriate to also mention that the entire Southern Ocean mid-ocean-ridge has continued to increase in length and has migrated north over time. What this simply means is that the Southern Ocean ridge was once the length of the perimeter of Antarctica, but has since moved away from the margin during opening and is now much longer. This movement and increase in length is a simple reflection of the northward migration of adjoining continents during an increase in the surface area of the Earth over time.

5.2 Summary

The outcome of this discussion on the origin of the modern oceans shows quite clearly that, not only does the published seafloor bedrock mapping provide a unique means of constraining the precise location of the crustal plates at any moment in time back to the early-Jurassic Period, but it also provides a unique framework for discussing the origin of the modern oceans and seas. By moving back in time and returning the seafloor volcanic crust back to the mantle, from where it originated, it is then a simple matter of progressively reducing the surface areas of each of the ocean basins and moving each of the continents closer together.

In contrast to Plate Tectonic reconstructions, the uniqueness of adopting Expansion Tectonic small Earth models means there is no need to consider where, or when pre-existing crusts occurred. The small Earth models also remove the need to invoke hypothetical ancient oceans to maintain a constant surface area Earth, or subduction to dispose of excess crusts. Instead, all that is needed is to let the configuration of the seafloor bedrock mapping dictate the precise crustal plate location and assemblages on a pre-determined smaller radius Earth model.

A further outcome of this discussion is that initial opening of the modern oceans on an Expansion Tectonic Earth consistently occurs during the Permian and Triassic Periods. This opening was intimately associated with break-up of the ancient supercontinents and formation of the modern continents. By continuing to move further back in time and closing off each of the modern oceans in turn, the origin of the modern continents can now be discussed in conjunction with modelling the ancient supercontinents and ancient continental seas.

6 Ancient Continental Crusts

"...it is theoretically possible for the continents, without shelves, to fit together at approximately 40 percent of the present Earth radius by considering that continental shelves were formed after the continental crust had fragmented." Vogel, 1983

As has previously been seen, by continuing to model the seafloor crustal mapping back in time to the early-Jurassic Period, seafloor volcanic lava and sediments deposited around the continental shelves must be progressively removed from each small Earth model in turn. Once all of the seafloor volcanic lava and sediments are removed and returned to where they came from, the remaining continental plates re-unite to form a complete pan-global supercontinental crust on an ancient Earth with a radius of about 55 percent of the present Earth radius. On this small Earth model, as well as the continental crusts assembling together, the remaining continental shelf sediments, located around the margins of the present-day continents, also merge to form a network of shallow seas coinciding with continental sedimentary basins. These seas, in turn, surround and define the geography of the exposed ancient Pangaean land surface.

When Vogel constructed his small Earth models during the 1980s he also suggested that it is *"...theoretically possible for the continents, without shelves, to fit together at approximately 40 percent of the present Earth radius by considering that continental shelves were formed after the continental crust had fragmented."* What he meant by this statement is that the continental shelves are where sediments, originally eroded from the lands, are finally deposited. By moving back in time it is then necessary to return these sediments back to the lands from where they were previously eroded from. By returning the sediments to the lands, it is inferred that any exposed volcanic seafloor crusts beneath these sediments can, just as was done with the younger small Earth models, also be returned to the mantle. It is then conceivable to consider that the ancient Earth could be further reduced in surface area and hence in radius.

What this simply means is, by moving back in time to before the early-Jurassic Period, on an Expansion Tectonic Earth all of the seafloor volcanic lava has been returned to the mantle. Similarly, eroded sediments deposited around the continental shelves must be progressively returned to the continents. In addition to the seafloor volcanic lava and sediments it is also conceivable to further consider that the bulk of the ocean waters and much of the atmospheric gases must similarly be returned to the mantle. Black smokers discharging hot water and gases from along the seafloor mid-ocean-ridges are modern-day examples of this new water and gas discharge process in action and by reversing this, and other discharge processes, back in time the water and gases are simply returned to the mantle where they came from.

Before continuing to model continental crust on an Expansion Tectonic Earth, two important points arising from the small Earth seafloor modelling must be discussed. After having progressively removed all of the seafloor volcanic lava and assembling the remaining continental crustal plates together on a Pangaean small Earth model, the reader must firstly fully understand what the remaining primitive continental crust is made of. Secondly, before attempting to model the continental crust back to the Archaean, an understanding of the mechanics of how this crust would potentially behave during any subsequent change in Earth radius must also be made.

I often think of the assemblage of continental crust on an Expansion Tectonic Earth in context with archaeology. In archaeology, scientists rely on finding fragments of pottery in order to identify where and when the pottery was first made and from this to date the history preserved in the particular archaeological excavation. Being of rigid pieces of fired clay, if enough fragments of pottery were discovered they could be glued together to assemble the broken pieces of pottery back to their original form. From this evidence the precise size, shape, purpose, and type of pottery, as well as the location of manufacture and relative age can be readily determined.

Until relatively recently the present-day continental crusts have been viewed in this same context, that is, the continental crusts were considered to act as rigid fragments that could only be assembled together on a constant radius present-size Earth. In the context of an Expansion Tectonic Earth, early researchers went further to consider that these same rigid fragments may have instead preserved the radius of an ancient Pangaean Earth. From this perspective, the continental crustal fragments were then considered to assemble together to represent the size and shape of the ancient Earth. If these crustal fragments were truly rigid it would then be conceivable to expect them to only assemble on an Earth of either the present-day or ancient Earth radius. The complete assemblage on an ancient small Earth would then adequately quantify an Expansion Tectonic process.

In contrast to rigid crustal fragments, if the fragments were instead made of complexly folded and distorted sediments—unfired clay in the archaeology analogy—there would be no hesitation in unfolding these crustal fragments to fit a model that was dictated by the original shape and size of the fragments, rather than to a preconceived rigid crustal premise. Modern understanding of the continental crusts now tell us that this is precisely how the continental crusts should be considered, as complexly distorted crusts that must adhere to the complex geological history preserved in all rocks.

At this stage, while the fundamental concept of fitting any form of continental crust on a smaller radius Earth may seem incomprehensible, what has been done with the seafloor volcanic lava and sediments so far does not contravene the published seafloor or sedimentary data. Nor does it contravene accepted opinion that: ***...new volcanic magma from deep within the Earth must rise through these weak zones*** [mid-ocean-ridges] ***and eventually erupt along the crest of the ridges to create new oceanic crust.*** It is, of course, the origin of seafloor volcanic lava as well as the waters and atmospheric gases derived from the mantle that is so difficult to comprehend when trying to understand Expansion Tectonics. Where does this enormous amount of material go to when moving back in time and conversely, where does it all come from when moving forward in time? While perfectly valid, these fundamental questions will be suppressed for the moment and considered in more detail later in this book.

6.1 Crustal Stretching

Crustal stretching is a term that refers to the ability of crustal rocks to stretch and thin under an applied extensional force. On an Expansion Tectonic Earth this stretching is eventually followed by crustal rupture and complete failure once the ability of the crust to stretch is exceeded. This applied extensional force is a direct result of an increase in Earth surface area and change in surface curvature during an increase in Earth radius. As a simple example of the processes involved in stretching and rupture, when engineers want to test the strength of a metal bar they clamp the bar between two mechanical chucks and stretch it hydraulically until it breaks. What happens to the test piece under an applied stretching force is that, initially, the metal crystals are able to resist the force. As the applied force continues, the metal crystals begin to very slowly stretch and lengthen in the direction of the force. Ultimately, the metal crystals reach a point where they cannot stretch anymore and they then begin to rupture, tear, and fragment and eventually the metal bar breaks.

Unlike the metal bar, the Earth's outer crust can be visualised as a relatively thin skin or plastic shell where the thickness of the continental crust is known to be about 5 to 10 times thicker than the seafloor crust. This relatively thin continental crust, and similarly the even thinner seafloor crust, in turn covers a much thicker, hotter, and more ductile mantle. I reiterate that the bedrock geological mapping used during the small Earth modelling studies presented here now shows that the seafloor volcanic crusts are actually quenched, cooled, and exposed mantle rocks; not drowned or *oceanised* continental rocks as was previously assumed by scientists.

For the Earth to increase its surface area during a progressive increase in radius, the outer continental crust must be able to distort, stretch, and ultimately rupture, break-up, and fragment, just like the metal bar example. The small Earth modelling completed so far shows precisely this and the continental crustal break-up and fragmentation history is what is now seen preserved in the seafloor crustal rocks. In contrast, the pre-break-up crustal stretching process is what is preserved in the geology of the continental crustal rocks. Only by understanding this stretching, rupture, and break-up process can the reader fully appreciate what the rocks are telling us about how this crustal stretching process works.

Researchers elsewhere have shown that the strength of the Earth's relatively thin outer crusts is controlled mainly by the ability of the crusts to deform, in particular by a process referred to as *plastic deformation*. This plastic deformation process is shown to be dependent on the prevailing temperature gradient existing within the crust and mantle, the composition of the rocks making up the particular crust, and also on how much geological time is available to perform the necessary crustal deformation. These observations were made elsewhere by studying heat-flow data from both active and inactive seismic regions of the Earth. Here, *rifts*—zones of continental crustal extension—were seen to be a common feature of the continents. Within these rift zones, volcanism and seismic activity were found to commonly precede crustal rifting by several million years. This then supported the idea that the crust is heated and weakened prior to rifting and stretching.

What all this means is that, on an Expansion Tectonic Earth, the microscopic amount of crustal stretching occurring each year throughout Palaeozoic and Precambrian times soon accumulated to many hundreds of kilometres of crustal stretching over billions of years of geological time. As will be shown later, throughout Precambrian times it is estimated that the total increase in circumference of the Earth amounted to much less than one millimetre per year, which was distributed around the entire circumference of the Earth. Under these conditions most rocks making up the Earth's crust would have plenty of time to slowly stretch and deform by atomic scale relaxation, plastic flow, or crystal regrowth. If not, local areas of crust may gradually accumulate excess stress and suddenly break during periods of earthquake activity that is more familiar to us.

In this context, the ability of the crust to both stretch during increases in Earth surface area and deform during changes in surface curvature depends on the amount of geological time available, as well as the softening effect that the warmth from the deeper crust and mantle provides. The various rock types that make up the continental crusts then determines how the crust will stretch and how much time is available determines how successfully they will deform. It will be shown later that a global network of continental sedimentary basins located on each of the small Earth models represents the primary focus for crustal stretching, which in turn may also be accompanied by jostling and deformation of adjoining continental crusts during on-going changes to surface curvature.

In the context of an Expansion Tectonic Earth, volumetric expansion of the Earth occurs deep within the mantle and results in what I refer to as *mantle swell*. Mantle swell is then transferred to the outer crust as crustal stretching, which is localised along zones of crustal weakness within the established global network of sedimentary basins. Within these zones of weakness crustal stretching is also accompanied by high heat flow, as currently demonstrated by seismic heat-flow data, and on the small Earth models this also coincides with the accumulation of sediments within a network of established sedimentary basins.

In contrast, within the seafloor crusts this stretching process is now occurring mainly along the mid-ocean-rift zones. This activity is accompanied by mantle-derived high heat flow and intrusion of volcanic lava along the full length of the mid-ocean-ridges. This then gives rise to the observed seafloor spreading and accumulation of new volcanic seafloor crusts, as discussed in previous chapters. Within the continents, crustal stretching may also be accompanied by intrusion or extrusion of igneous and volcanic rocks as well as on-going seismic activity.

The main processes and outcomes that affect crustal stretching during the progressive increase in Earth surface area are schematically summarised in Figure 6.1.

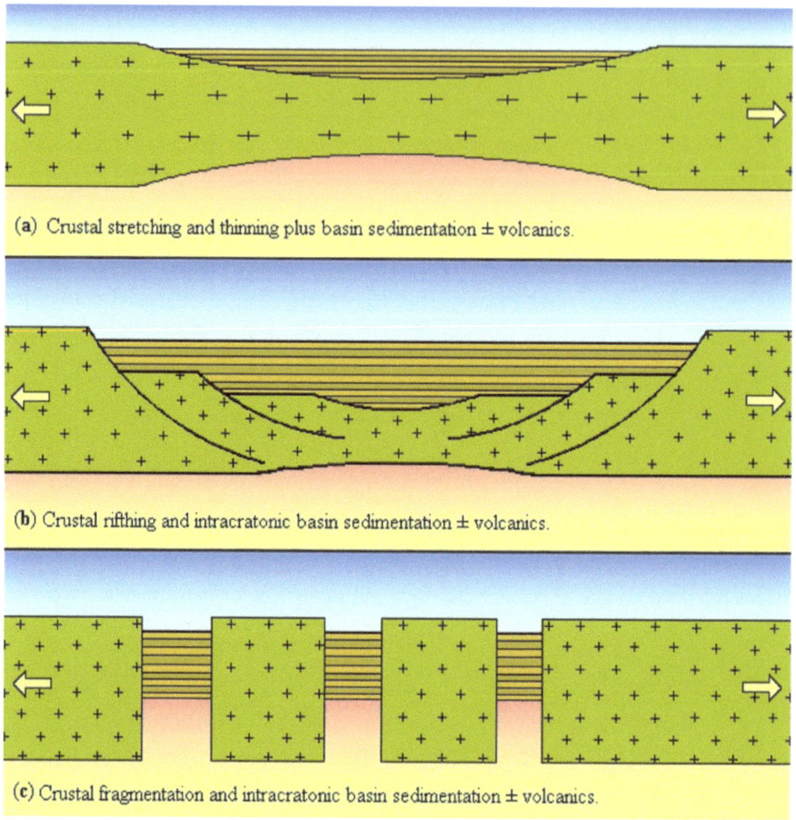

Figure 6.1 Primary mechanisms for crustal stretching during Earth expansion. Figure a) Simple crustal stretching; b) crustal rifting; c) crustal fragmentation. In each figure continental crust is shown as green, the mantle as pink shading and basin sediments are shown as horizontal stripes.

These processes and outcomes include:

1. **Simple crustal stretching:** This process occurs as stretching, thinning and down warping of the continental crust to form shallow sedimentary basins (Figure 6.1a). These basins are subsequently filled with sediment eroded from the surrounding elevated land surfaces. Continuation of this process eventually leads to crustal failure and may also be accompanied by localised crustal fragmentation at the point of rupture.

2. **Crustal rifting:** This process occurs mainly within the continental crustal regions as the crust starts to fragment and separate during tectonic activity (Figure 6.1b). The process forms what are referred to as *horst and graben* rift structures, with curved, *listric* faulting

occurring along the margins of the rift zones. On the continents this process may then form a rift basin, which is in turn filled with sediment eroded from the surrounding elevated land surfaces plus intrusion of new volcanic lava. The East African rift zone and the Red Sea are good modern examples.

3. **Simple brittle fragmentation:** This process is a simple version of crustal rifting and may be more applicable to the ancient Precambrian granite crusts (Figure 6.1c). Here, the thick, brittle crust simply fractures, fragments, and separates and would again be accompanied by deposition of basin sediments as well as accumulation of intrusive and extrusive volcanic lava within any newly formed basins.

While schematic and simplistic, the examples shown in Figure 6.1 can be likened to a piece of toffee as simulating a fragment of continental crust. If the toffee is held with two fingers between both hands and pulled apart nothing much will happen. If we are strong enough, or the toffee brittle enough, we may then be able to break the toffee, as shown in Figure 6.1c. Alternatively, if we were able to apply and maintain a steady pull over an extended period of time we would eventually notice an atomic scale relaxation of the toffee molecules and after a while we would detect an atomic scale stretching of the toffee fabric. Over a prolonged period of time this stretching process would both accumulate and steadily increase and would eventually stretch and thin as in Figure 6.1a, in particular if the toffee is heated and softened from below. This atomic scale stretching is well known from testing metal products in civil engineering. Further stretching of this example will then eventually result in failure, initially as shown in Figure 6.1b and finally as complete rupture or fragmentation. Rupture and fragmentation is here considered the primary mechanism for break-up and rifting of the continental crusts to form the modern continents.

By moving back in time, regardless of the toffee's current stretched or fragmented form, it is easy to comprehend and understand that the pieces of toffee in this hypothetical example must be returned to their original form, size, thickness, and shape, regardless of what the pieces look like after stretching. The analogy with Earth's continental crusts is identical. By moving back in time, in particular during the vastly more extended intervals of geological time available, the geological processes involved in stretching, deformation, faulting, fragmentation, as well as erosion of the continental crusts must be fully reversed and returned to their original pre-deformation states. These crusts must then be precisely reassembled into their pre-stretched, pre-deformed, pre-rifted, or pre-fragmented configuration.

Objections to crustal stretching mechanisms generally revolve around a long-standing misconception in geology that if we were to dig or drill deep enough into the Earth's crust we would eventually intersect much older bedrock. My reaction to this is, of course older bedrock will be intersected. In Figure 6.1a the age of the basement rocks will be as old as the surrounding crustal rocks that are being stretched. By digging or drilling through the overlying younger sediments the basin will be shown to be founded on a much more extensive area of ancient rocks than what is shown by the original surrounding rocks outcropping at surface. Similarly, for Figure 6.1b and Figure 6.1c the basement rocks will be as old as either the remnant ancient rocks or as old as the mantle rocks exposed at the time of crustal break-up. By moving back in time, the crusts are simply moved back to where they originated from by undoing the geological stretching processes that had acted on the crusts during this extended interval of time. This, of course, does not in any way contravene what is preserved in the geology.

On an Expansion Tectonic Earth, each of the examples of crustal thinning and rifting shown in Figure 6.1 may be accompanied by elevated heat flow originating from the mantle. This heat flow may also be accompanied by intrusion of mantle-derived igneous rocks in the form of granites or intrusive dykes or by volcanic eruptions at surface to form lava flows. Vast quantities of hot fluids and gases may also be involved, which would then accumulate in low-lying areas to form the ancient seas or rise to form the ancient atmosphere over time.

This heat flow is derived from deep within the mantle and on an Expansion Tectonic Earth its transfer to the surface occurred throughout an extensive global network of crustal weaknesses, which are the precursors to the modern mid-ocean-ridge spreading zones. This elevated heat flow also represents an important mechanism for the observed geological processes of *metasomatism*: the alteration of rocks by hot fluids, *metamorphism*: the alteration of rocks by heat and pressure, and *granitization*: the alteration of rocks by extremes in heat or pressure to form granites.

6.2 Continental Crust

In contrast to the relatively simple seafloor crusts, continental crust is made up of a diverse range of present-day to ancient rocks dating back to the earliest Archaean times. These rocks include ancient granite and *greenstone*—altered volcanic rocks—deformed and physically altered sediments eroded from the more ancient lands, intrusive and extrusive magmatic rocks, as well as multiple layers of overlying younger sediments deposited in past low-lying regions. These younger rocks in particular generally cover vast areas of older rocks (Figure 6.2, repeated). Studies also show that the average composition of the continental crust is that of granite. In other words, rocks rich in silica and aluminium in the form of quartz and feldspar minerals. This contrasts with the seafloor crust which has an average composition of the volcanic lava called basalt—a lava rich in iron and magnesium. Average composition simply means that if all of the continental rocks were re-heated and melted their composition would approximate that of granite whereas the seafloor crustal rocks would be basalt, hence the big difference between continental crust and seafloor crust.

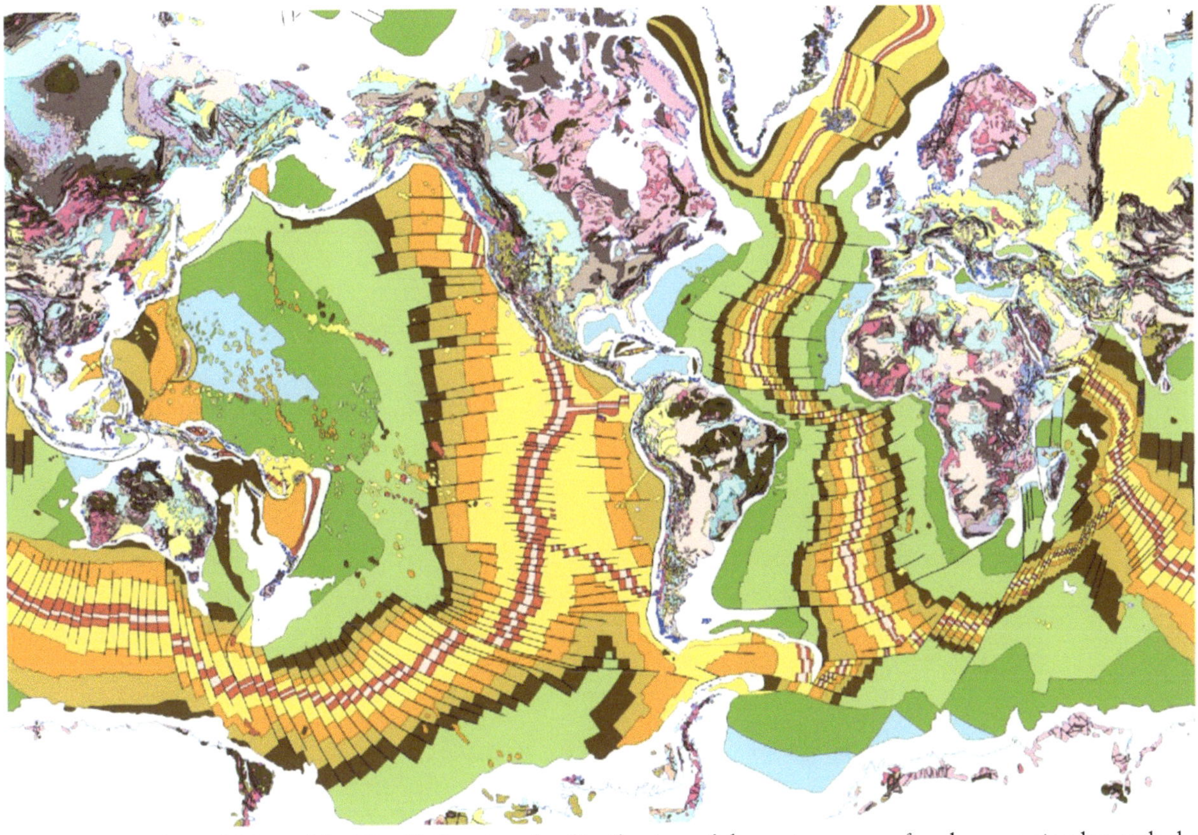

Figure 6.2 Geological Map of the World showing the distribution of the various ages of rocks occurring beneath the layer of surface soils.

As has previously been discussed, the continental and seafloor crusts displayed in Figure 6.2 represent time-based bedrock geology. What this means is that this geology is not specific to any one rock type, but displays all rocks according to their ages. These rock ages can then be broadly correlated with the geological history, in particular for each of the areas of continental crust. The continental geology shown in this figure is further complicated by subsequent deposition of many layers of young sedimentary and volcanic rocks, which generally cover and overprint older crustal rocks lying below them. These in turn are complicated still further by many periods of metamorphism, folding, faulting, weathering, and erosion that may have occurred intermittently throughout Earth history.

In reality then, the Earth's continental crustal geology is far more complex than that depicted in Figure 6.2 and well beyond the needs of this explanation. Like the coloured seafloor crustal patterns, this map simply shows the ages of continental crustal growth and therefore the expansive growth history of the Earth's outer continental crustal skin. It should be further appreciated that very little of this geological information was available to early researchers during the 1950s and 1960s when Continental Drift, Plate Tectonic, and Earth Expansion theories were being developed, in particular, the global distribution of rocks shown in this map.

On this bedrock geological map the most ancient continental crustal rocks are shown as areas of pink and red colours, mainly occurring in Canada, Africa, the Baltic region and Australia. The distribution of these most ancient continental crustal rocks shown within the modern continents can be likened to islands of ancient rocks surrounded by younger sedimentary rocks—analogous to the remaining pieces of toffee in the stretched toffee analogy above. Like the modern seafloor lava, by progressively moving back in time, the eroded sediments shown on this map must be returned to these ancient rocks from where they originated and the sedimentary basins progressively closed. From an Expansion Tectonic perspective, these ancient crustal islands may then potentially represent remnant fragments of a more ancient and primordial Archaean supercontinent.

In Figure 6.2 these islands of ancient crustal rocks are in turn shown to be surrounded by deformed, less ancient and often metamorphosed sedimentary rocks shown as khaki to brown, blue, and yellow colours. These sedimentary rocks can be likened to the distribution of continental shelf sediments currently surrounding the modern continents. Just like the modern continental shelf sediments, when moving back in time the geological processes operating must be reversed and the sediments returned to the ancient lands where they originated from.

Put a different way, at the global scale shown in Figure 6.2, the continental rocks can be visualised as a mosaic of continental crustal fragments which are broadly made up of three dominant crustal types. In geology, these crustal types are referred to as *cratons*, *orogens*, and *basins*. These terms vary considerably in their usage and definition throughout the world and also within the published geological literature. To avoid confusion I will first define these terms so that all readers will understand what is being discussed.

- A **craton** is defined as a part of the Earth's crust that has attained relative crustal stability and the rocks have been little deformed for a prolonged period of time. By definition, cratons must have reached crustal stability by about 2,400 million years ago (the end of the Archaean Eon) and since then have undergone little deformation compared to adjacent parts of the crust.

- An **orogen** refers to a belt of rocks characterised by regional folding, metamorphism, and intrusion of magmatic rocks. The rocks of an orogen can include deformed, eroded, and reworked parts of older, early-formed cratons, as well as young volcanic and sedimentary rocks. A distinct tectonic phase of Earth movement, over a relatively short period of time, first establishes an orogen. It is also possible for an orogen to become re-activated during subsequent tectonic events and the belt normally remains as a permanent zone of relative weakness within the Earth's crust.

- A **basin** refers to an area that is underlain by a substantial thickness of sedimentary rocks. These rocks possess unifying characteristics of both sediment type and deformation history. Within a basin, sediments are deposited during a regionally restricted period of time, often extending for tens to hundreds of millions of years, during crustal depression or a related sequence of such events. The term basin is usually synonymous with the term *sedimentary basin* and it represents a regional topographical down-warp of the Earth's surface, generally filled with water.

From these definitions it should be appreciated that not all continental crusts are the same and not all rocks were formed by the same process or during the same interval of time. Cratons, for instance, are by definition the most ancient crustal rocks and are older than 2,400 million years. It is conceivable to imagine that an orogen may have started out as an ancient sedimentary basin, but later may have been deformed and metamorphosed and perhaps intruded by molten igneous rocks at any time during the subsequent Earth history. If older than 2,400 million years, these rocks may now be stable enough to be considered a craton.

During modelling of the continental crusts the physical characteristics of each of these various types of crust, as well as their distribution and tectonic histories, must be recognised and strictly adhered too. During model construction, the general shape and configuration of each of these crustal regions must also be retained throughout Earth history in strict accordance with their known distributions and ages.

It will be shown that construction of Expansion Tectonic small Earth models, extending from the early-Jurassic Period back to the early-Archaean, involves the progressive removal of all younger basin sediments and magmatic rocks and simply returning these to the ancient lands or back to the mantle where they came from. Each basin is then restored to a pre-stretching or pre-rifting configuration on a smaller radius Earth. By moving back in time, adopting this process will enable us to progressively move the adjacent margins of each sedimentary basin closer together as part of one seamless ancient supercontinental crust, while still preserving the spatial integrity of the adjacent more ancient crusts.

This fundamental understanding of continental crusts is necessary to provide a mechanism for crustal modelling back in time in order to justify determination of an ancient primordial Earth radius during small Earth modelling studies.

6.3 Primordial Earth Radius

By measuring the surface area of seafloor crustal rocks from each of the oceans it was shown that this information can be used to derive ancient Earth radii for the various geological time intervals back to the early-Jurassic—about 200 million years ago. Similarly, by progressively removing young seafloor crust from each of the modern oceans and reducing the Earth radius accordingly, each of the remaining crustal plates can be reassembled precisely on small Earth models. These models demonstrate that the remaining seafloor crustal plates can be accurately reassembled together along their respective mid-ocean-ridge spreading axes, without the need to invoke pre-existing crusts, to remove crusts by subduction, or to arbitrarily fragment the continents.

The ultimate aim of this exercise though is to establish a radius for the most ancient primordial Earth and, similarly, to establish a formula for calculating Earth radius. This formula will then be used to calculate the radius of the ancient or future Earth at any moment back or forward in time.

The presence of ancient pink and red continental rocks shown in Figure 6.2 suggests that a primordial small Earth may have existed at the beginning of geological time when the oldest known rocks on Earth were first formed—around 4,000 million years ago. Since there were no, or very limited, volcanic seafloor crusts in existence during pre-Jurassic times evidence for this primordial Earth radius must then be established from the surface areas of ancient continental crusts found on each of the present-day continents.

Unlike volcanic rocks preserved on the seafloors, the continental crust is much more complex and generally much more ancient. Again, continental crust is a mixture of sedimentary, igneous, metamorphic, and volcanic rocks, which have been folded, faulted, eroded, and buried many times over throughout Earth history. In addition, young rocks generally cover the more ancient rocks and the presence of ancient rocks can often only be inferred from deep drilling or geophysical seismic surveys. Apart from surface or near-surface rock exposures, it is then extremely difficult to accurately measure the total surface area of each rock type existing on the present-day continents. This applies in particular to the buried bedrock crustal rocks that are used throughout this study and especially where the process of crustal stretching, as introduced in the previous section, has been active.

A number of possible scenarios for establishing a radius of the most ancient primordial Earth from measured areas of continental crusts are graphically shown in Figure 6.3. These scenarios range across a whole spectrum of possibilities, ranging from adoption of the present total continental surface area (Curve A), through to an estimate of the total surface area of preserved Archaean crusts (Curve D). Additional constraints also include the approximate relationship between sedimentary rock age and amount of outcrop area on the present land surface (Curve C), plus an intermediary curve determined by removing areas of young sedimentary rocks from around the continental margins (Curve B).

Unfortunately, while each of these scenarios has some merit in constraining a primordial Earth radius, none of these can substantiate, with any degree confidence or accuracy, what a primordial Earth radius may have in fact been. The estimates shown in this figure vary from between 1,500 kilometres to 3,500 kilometres radius for the early-Precambrian and the wide variation of estimates shown make them very inconclusive.

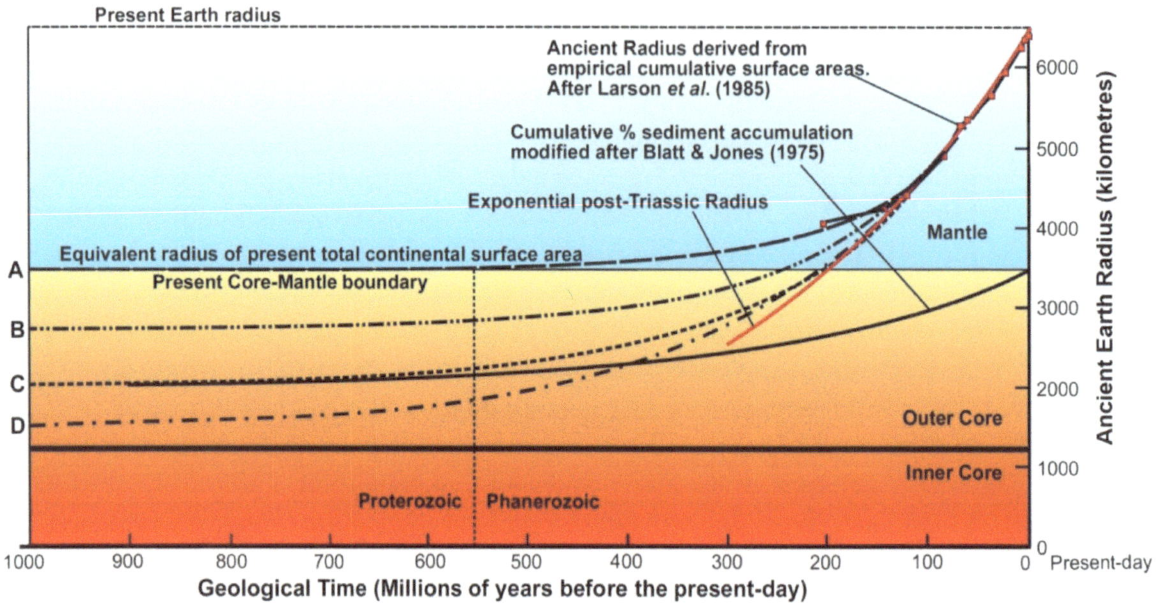

Figure 6.3 Constraints to determining a primordial Precambrian Earth radius. Curves depict Earth radius constrained by, Curve A: the present total continental surface area; Curve B: present total continental surface area less marginal basins; Curve C: published estimates of cumulative sediment areas and; Curve D: total estimated area of Archaean crust.

Because of this uncertainty, for the Expansion Tectonic small Earth modelling presented here, a primordial Earth radius was established during model construction by simple, empirical, trial and error. This may appear somewhat unscientific but, given the surface area data limitations outlined above, it was found to be the only viable means available with sufficient flexibility to constrain or to reasonably eliminate each of the four possible scenarios shown. I must also emphasise that modelling continental crustal data back to the Archaean has never been done before, for any tectonic theory, and, as such, the modelling presented here represents the first time that this has ever been attempted.

The method adopted here to determine a potential primordial Earth radius is basically very simple. During model construction, by moving back in time the radius of each preceding model is progressively reduced in small, incremental stages. An equivalent area of the youngest sedimentary and magmatic rocks remaining is then removed from the primary bedrock geological base map. The global network of sedimentary basins is then progressively reduced in surface area until only the most ancient Precambrian continental crustal rocks remain.

This method is justifiable because the younger crustal rocks represent sedimentary and magmatic rocks that were deposited, intruded, or extruded after the older crustal rocks were first formed. During progressive removal of the younger rocks, each of the sedimentary basins and rift zones are then restored to a pre-stretching or pre-rift configuration and all sediments and magmatic rocks are simply returned to the exposed land surfaces or mantle where they originated from.

As will be elaborated on later, by using this rather subjective construction method an Archaean small Earth with a primordial Earth radius of about 1,700 kilometres can be readily constructed. This primordial small Earth is made up of an assemblage of the most ancient pink and red coloured Archaean crustal fragments, plus remnant khaki coloured early-Proterozoic crustal rocks (as previously seen in Figure 6.2). In this construction method all young sedimentary and igneous rocks are simply returned to their original crustal or mantle sources during a process of simple, step-wise reduction in Earth radius.

While this primordial Earth radius, which is very similar in radius to the present Moon—about 27 percent of the present-day Earth radius—is very difficult to comprehend, it is reiterated that the crustal reconstruction is readily achieved by using a very basic construction method. By using this reconstruction method, no fundamental physical law has been violated. The primordial Archaean Earth simply comprises an assemblage of the most ancient crustal components, where all younger rocks have been returned to their places of origin.

This 1,700 kilometre Earth radius, while subjective, represents an approximate limiting radius for the most ancient primordial Earth. This primordial radius can then be used to establish a mathematical formula to calculate Earth radius at any moment back or forward in geological time. From this primordial small Earth radius, mathematical modelling studies of Archaean to present-day Earth radius, as outlined in detail in Appendix A, shows that the change in Earth radius increases in accordance with an exponential rate of increase in Earth surface area.

Derivation of this Earth radius formula is given in Appendix A and for the mathematically minded is expressed as:

$$R_a = (R_0 - R_p)e^{kt} + R_p$$

Where: R_a = the ancient Earth radius at time t, R_0 = the present-day mean Earth radius, R_p = the primordial Earth radius = 1,700 kilometres, e = base of natural logarithm, k = a constant = 4.5366 x 10^{-9}/year.

The exponential increase in Earth's radius throughout geological time, as determined from empirical model studies, is graphically displayed in Figure 6.4.

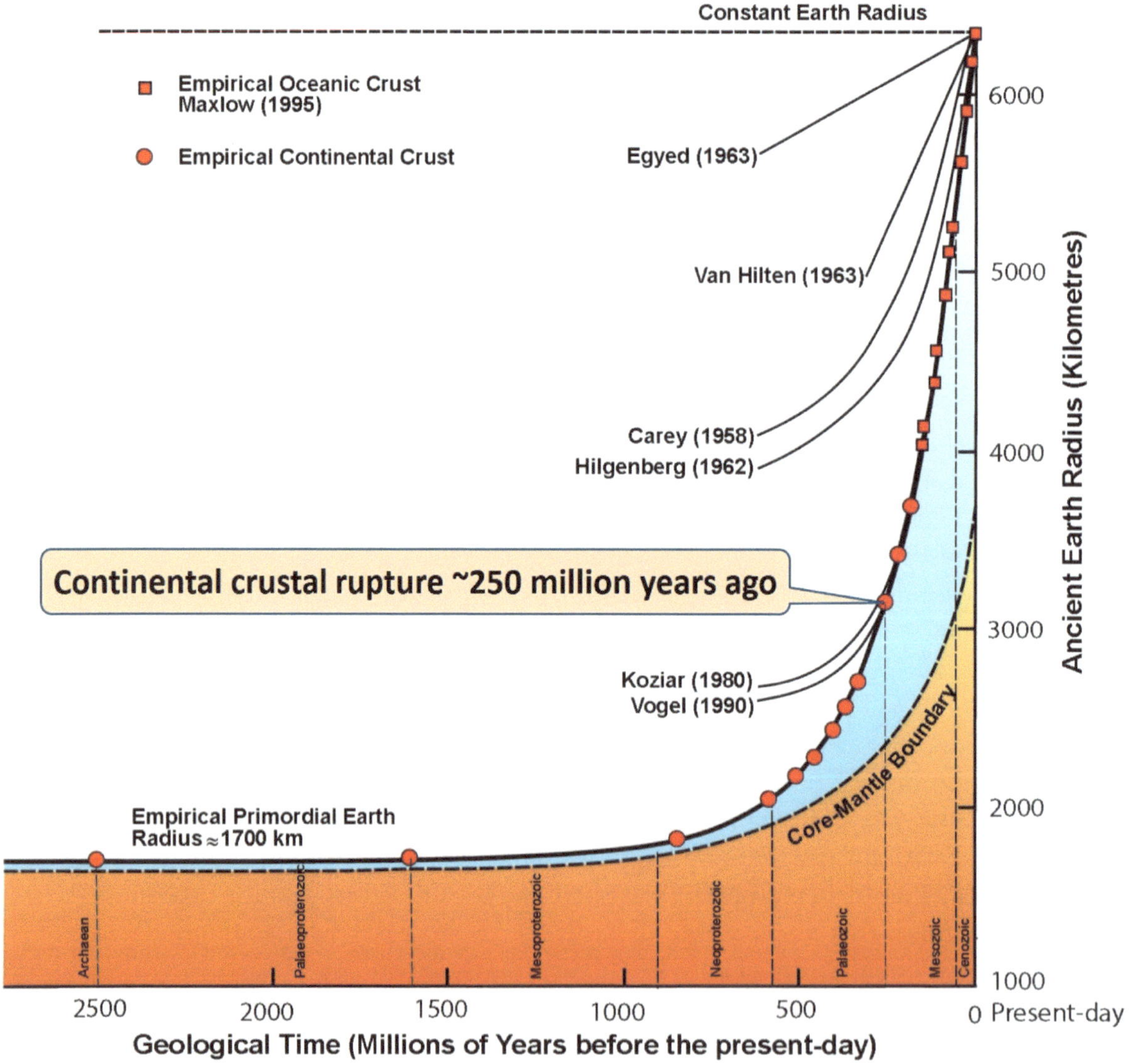

Figure 6.4 Exponential increases in Earth radius extending from the Archaean to present-day. Graph shows post-Triassic increase in radius derived from seafloor mapping and pre-Triassic change in radius derived from an Archaean primordial Earth radius of 1700 km. Small Earth models constructed are shown as red coloured circles and squares. A suggested core-mantle boundary remains speculative.

The location of the Archaean to present-day Expansion Tectonic small Earth models used to constrain this graph are shown as red dots and squares, along with historical estimates of ancient Earth radius by others based on both geophysical methods and visual reconstructions of the continents. The radius curves determined by Koziar and Vogel are also shown for comparison. The curves of Koziar and Vogel coincide precisely with the estimates of ancient Earth radius established here to about 200 million years ago. Their curves were, however, limited to early versions of seafloor mapping data available at the time. They also used estimates of pre-Jurassic continental crustal areas only, without considering the potential for further continental crustal stretching.

As can be seen in this figure, the available data has improved considerably since the early geophysical estimates of ancient Earth radius. It also shows that there is considerable agreement with previous estimates made using the seafloor surface area data back in time to at least 200 to 250 million years ago. This graph, as it stands, suggests that during the early-Archaean to mid-Proterozoic times—extending from about 4,000 to 1,600 million years ago—the Earth's ancient radius remained relatively static, increasing in radius by approximately 60 kilometres during the entire 2,400 million years of ancient Earth history. From the mid-Proterozoic—about 1,600 million years ago—the Earth then underwent a steady to accelerating increase in radius to the present-day. This accelerating increase is now reflected in the break-up of the Pangaea supercontinental crust, the subsequent opening of the modern oceans, and the relatively recent development of modern seafloor crusts.

This is an interesting outcome. A civil or structural engineer testing the tensile strength of a particular metal or the compressive strength of concrete or rock would get a similar curve to that shown in Figure 6.4. The curve would then be discussed in detail with respect to the molecular-scale stretching or compression of the particular test product prior to molecular failure, final rupture, and breaking of the test piece. The only difference between the engineering scenario and the small Earth models is the scale difference between the test piece and the size of the Earth; as well as human comprehension.

The graph above suggests that, during Precambrian times, an increase in both radius and surface area of the Earth was accommodated for by molecular-scale redistribution and stretching of the crusts. This is evidenced in the rock-record by linear rock fabrics so often seen in the most ancient rocks exposed at surface today—such as schist and gneissic rocks. The ability for the continental crust to absorb this molecular-scale crustal stretching then began to rapidly decrease, resulting in an overall weakening and subsequent thinning of the crusts, until the ability to stretch was finally exceeded during late-Permian to Triassic times. During these times the continental crust then began to fail, rupture, and break apart to ultimately form the modern continents and open to form the modern oceans. By removing the inherent strength of the overlying continental crust, the Earth then underwent a rapid to accelerating increase in Earth radius to the present-day and by inference this process will continue well into the future.

Until more research effort is undertaken, it is acknowledged that this primordial Earth radius determination and Earth radius formula must remain speculative for the Palaeozoic and Precambrian times. The ultimate test of the methodology behind determining an ancient primordial Earth radius though is in the ability for the continental crusts to assemble back in time. As was discussed in the previous chapter on seafloor crustal modelling, if the mechanisms, and similarly the assumptions, behind this construction method are incorrect then ancient crustal fragments will fail to assemble. If they fail to assemble then consideration will have to be given to alternative methods, or even to abandoning the quest.

6.4 Changing Earth Parameters

Now that a formula for changing Earth radius through time has been established, it can be used to further investigate what the annual changes in the physical dimensions of the Earth would be if the Earth were expanding. As well as changes to Earth radius, these variables include circumference, surface area, and volume. Consideration can also be given to the annual changes in mass, density, and surface gravity, but, at this stage, these changes remain speculative and cannot be determined directly from ancient radius.

The calculated changes in the annual rate of change in Earth radius, surface area, circumference, and volume throughout the ancient and future geological times are graphically shown in Figure 6.5. The values used to generate this graph show that, prior to about 2,200 million years ago, the annual rate of increase in Earth radius was less than one micron—one millionth of a metre—per year. By about 1,200 million years ago this annual rate had then increased to approximately 100 microns per year, which is about the thickness of a coarse human hair.

What these changes mean is that for the first 2,800 million years of known ancient Earth history the required increase in radius of the Earth per year amounted to much less than the average thickness of a human hair. By the beginning of the Palaeozoic Era—around 560 million years ago—radius had then increased to just 1.5 millimetres per year and by the late-Permian Period—250 million years ago—the annual rate of increase was about 8.5 millimetres per year. The current measured rate of increase in Earth radius is now 22 millimetres per year.

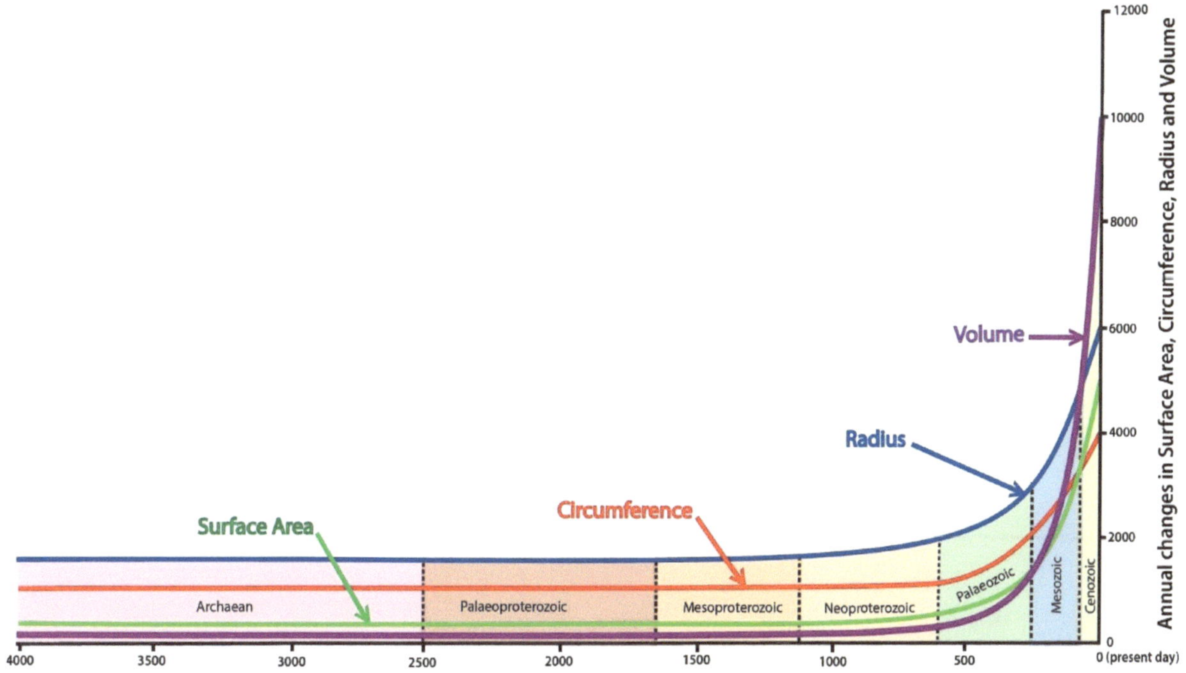

Figure 6.5 The annual changes in Earth radius, circumference, surface area and volume over time. The graphs range from the Archaean to the present-day and are derived using the established Earth radius formula. Annual change in Radius is in kilometres, Circumference is times 10 kilometres, Surface Area is times 100 million square kilometres, and Volume is times 100 million cubic kilometres.

These figures can be compared to historical estimates of annual rates of change in Earth radius obtained by geological methods and published by the various authors shown in Table 6.1. It is interesting to note that these results are all very similar and the mean of each of these published values is 21.9 millimetres per year; essentially the same as the 22 millimetres per year as measured here from seafloor mapping.

As can be seen in Figure 6.5 a steady to rapidly accelerating increase in Earth radius began sometime during the mid-Proterozoic Era—called the Mesoproterozoic—and this increase has continued through to the present-day. This accelerating increase in radius can be visualised as reflecting a progressive weakening of the original supercontinental crust which then ultimately lead to crustal failure, rupture, continental break-up, and opening of the modern oceans. This trend and growth rate can also be readily projected to the near future, with some degree of confidence.

Author	Year	Rate [cm/yr]	Method
Koziar	1980	2.59	Increase in the Earth's surface area (Phanerozoic)
Blinov[1]	1983	1.99	Present annual increase in the surface area of oceanic lithosphere
Blinov[2]	1983	>1.91	Present annual increase in the Earth's circumference
Osipishin & Blinov	1987	1.96	Increase in the Earth's surface area (Meso-Cainozoic)
Koziar[3]	1996	2.7	Present annual increase in the Earth's circumference
Maxlow	2002	2.2	Increase in the Earth's surface area (from the Archean)
Koziar	2011	>2.0	ratio of the lengths of Atlantic Ridge and its African parent margin

[1] correct interpretation of the result obtained by Steiner (1977)
[2] correct interpretation of the result obtained by Kulon (1973)
[3] correct interpretation of the result obtained by Le Pichon (1968)

Table 6.1 Published historical estimates of the present-day rate of increase in Earth radius. Table from Koziar, pers. com.

As well as radius, the present annual increases in circumference, surface area, and volume can also be determined using the established Earth radius formula. The present-day annual rates of increase in these variables are calculated as:

Radius annual change = 22 millimetres/year
Circumference annual change = 140 millimetres/year
Surface Area annual change = 3·5 square kilometres/year
Volume annual change = 11,000 cubic kilometres/year

It should be realised that these present-day annual rates of increase represent the actual rates that are occurring now and these are continuing to increase exponentially with time. The values determined here compare favourably with estimates derived from published data, which were based on spreading values from early versions of the seafloor mapping program. It should also be noted that, while an increase in seafloor spreading and surface area was acknowledged by many of these previous authors, it was presumed that this increase was then balanced by removal of excess crust by subduction in order to maintain a constant Earth radius.

For example, in 1977 Steiner estimated a rate of global increase in seafloor spreading during the past 5 million years of 3·19 square kilometres per year. This equates to a calculated rate of increase in radius of 20 millimetres per year. Also, in 1975 Garfunkel calculated an increase in seafloor spreading of 3·15 square kilometres per year and in 1982 Parsons calculated 3·45 square kilometres per year, which equate to rates of increase in Earth radius of 20 and 23 millimetres per year respectively. These figures are further substantiated by early space-based geodetic measurements of intercontinental distances between various observation sites. A rate of increase in Earth radius was calculated by Parkinson (in Carey, 1988) to be 24±8 millimetres per year and a published global mean value of 18 millimetres per year was subsequently calculated from space-based data in 1993 by Robaudo and Harrison.

A rate of increase in circumference of the Earth was calculated by Blinov in 1983 to be 120 millimetres per year by considering the relative plate motions in the Pacific, Atlantic, and Southern Oceans. This equates to an annual increase in Earth radius of 19·1 millimetres per year. The mean value of each of these published surface area and circumference estimates is then 21.2 millimetres per year, which again compares favourably with the value of 22 millimetres per year measured here from the surface areas of modern published seafloor mapping.

6.5 Mass, Density and Surface Gravity

At this stage, investigating rates of change in Earth mass, density, and surface gravity is speculative, however, it is very important to have some understanding of how these variables may behave when discussing a potential cause and mechanism for Earth expansion. This causal mechanism will be discussed in more detail much later in the book. Both mass and density are mathematically bound and are calculated using the standard formula:

Mass = Density divided by Volume.

Similarly, surface gravity is calculated using the standard formula:

Surface gravity = the product G.M divided by Earth radius squared

Where G is the value for universal gravitation and is assumed to be a constant, and M is the mass of the Earth.

In the mass and density equation, volume can be simply calculated using the ancient Earth radius formula introduced above, however, the variables mass and density are both unknown, making calculation of these variables very difficult. Because of this, three possible scenarios exist for an expression representing potential changes to Earth mass, density, and surface gravity with time. These include:

- Mass remains constant requiring density to decrease as volume increases exponentially throughout geological time.

- Density remains constant requiring mass to increase as volume increases exponentially throughout geological time.

- Both mass and density are variable as volume increases throughout geological time.

At this stage, each of these unknown variables are difficult to determine and only by constraining either mass or density to constant values can they be investigated. Potential changes in these mass and density variables range from a constant Earth mass with a decreasing density scenario, to a constant density with an increasing Earth mass scenario. Between these two end points a whole spectrum of possibilities exist, where both mass and density may be variable at the same time.

Within the present Solar System the planets can be divided into the inner rocky planets, with densities ranging from 3.3 to 5.5 grams per cubic centimetre, and the giant gaseous planets, with densities ranging from 0.7 to 1.7 grams per cubic centimetre. While all of the planets are different, they do show a marked decline in density values between the rocky planets and the giant planets. These limits to the range of densities may then potentially place constraints on ancient Earth values.

To determine changes to surface gravity it must again be assumed, for the purpose of this exercise, that the value for universal gravitation "G" has been constant or near constant throughout Earth's history. This is not necessarily so, but, any rate of change in this value is potentially very small and may also be indeterminate at this stage.

6.5.1 Constant Earth Mass Scenario

For a hypothetical Earth increasing its radius under conditions of a constant Earth mass, density is shown in Figure 6.6 to decrease exponentially. This decreases from a peak value of approximately 52.5 times the present value during Archaean and Early Proterozoic times, to a present value of 5·52 grams per cubic centimetre. Density then approaches zero by about 300 million years into the future.

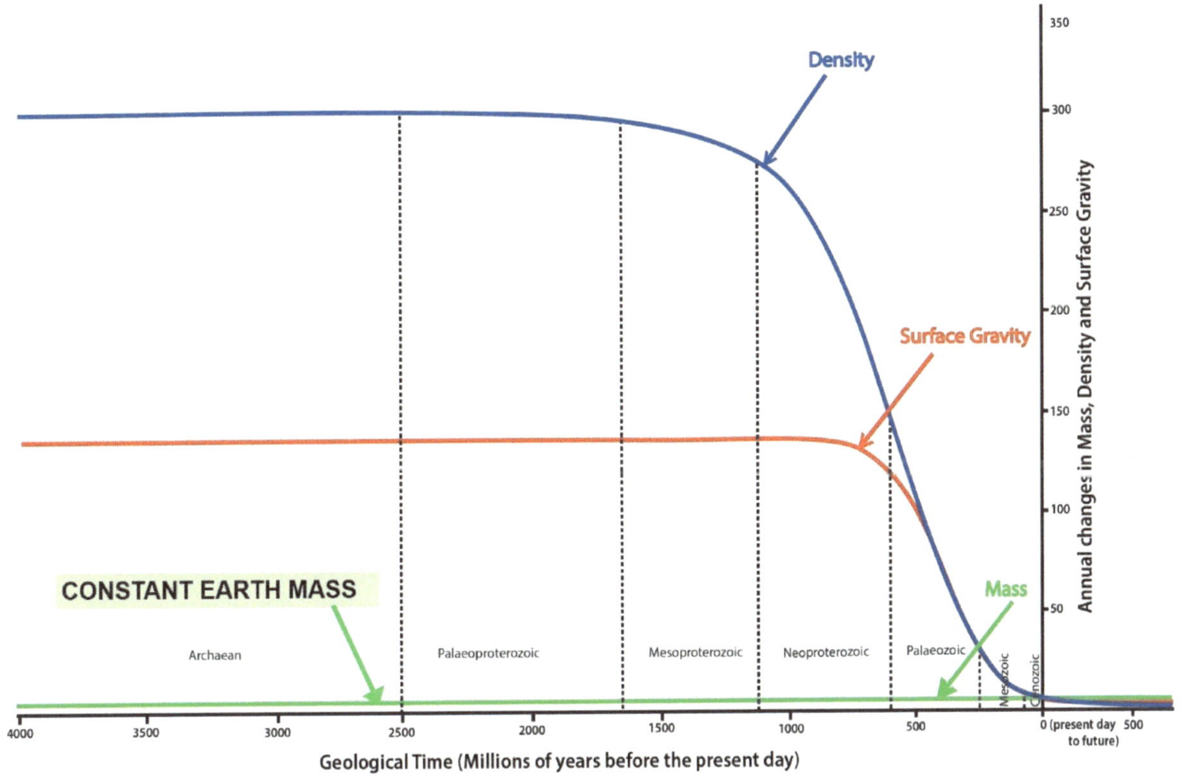

Figure 6.6 Changes in Earth Mass, Density, and Surface Gravity assuming a constant Earth Mass scenario. The graphs show that, for a constant Earth mass scenario both Density and Surface Gravity would be excessively large during the Precambrian Eras. Annual change in Density is in grams per cubic centimetre, Surface Gravity is in metres per second per second, and Mass is kilograms times 10 raised to the power of 24.

For this same constant Earth mass scenario, surface gravity is also shown in this figure to decrease exponentially. This decreases from a peak value of approximately 14 times the present value during Archaean and Early

Proterozoic times, to 9.5 times the present value during the Cambrian Period. Surface gravity is then shown to approach a value of zero by about 300 million years into the future.

What all this means is that, by assuming that the Earth's mass remains constant, or near constant, since Earths' formation, it must be accepted that for the ancient Earth everything was over 50 times heavier during the Precambrian. Surface gravity must have also been over 14 times stronger than it is now.

Put simply, for Archaean times this means that fifty cubic metres of present-day material must be squashed into one cubic metre and no gases would escape into the atmosphere. To approach zero density and zero surface gravity by about 300 million years into the future means that at that time the Earth would evaporate and cease to exist, which, as the reader can appreciate, is a daunting scenario.

6.5.2 Increasing Earth Mass Scenario

In contrast, for a hypothetical Earth increasing its radius while retaining a constant density, mass is shown in Figure 6.7 to increase exponentially. This increases from a primitive Archaean and Mid-Proterozoic value—shown as Mesoproterozoic in Figure 6.7—somewhat similar to the mass of the present Moon, to a present mass of about 54 times the ancient value. This increase then continues to about 32 times the present Earth mass by about 300 million years into the future.

Similarly, surface gravity is shown in this graph to increase exponentially. This increases from approximately one third of the present value during Archaean and mid-Proterozoic times and reaches approximately 3 times the present value by about 300 million years in the future.

Again, what all this means is that, by assuming that the Earth's mass has been steadily increasing since formation of the Earth and density has remained constant or nearly constant, then surface gravity on an Expansion Tectonic Earth will progressively increase with time. It can then be seen that, in this increasing Earth mass scenario, surface gravity during Precambrian times would have been about one third of the present value and about one half of the present value during the early-Mesozoic. The Mesozoic Era, of course, was the age of dinosaurs, those very large, very long bodied creatures that could very well have benefited from a much reduced surface gravity.

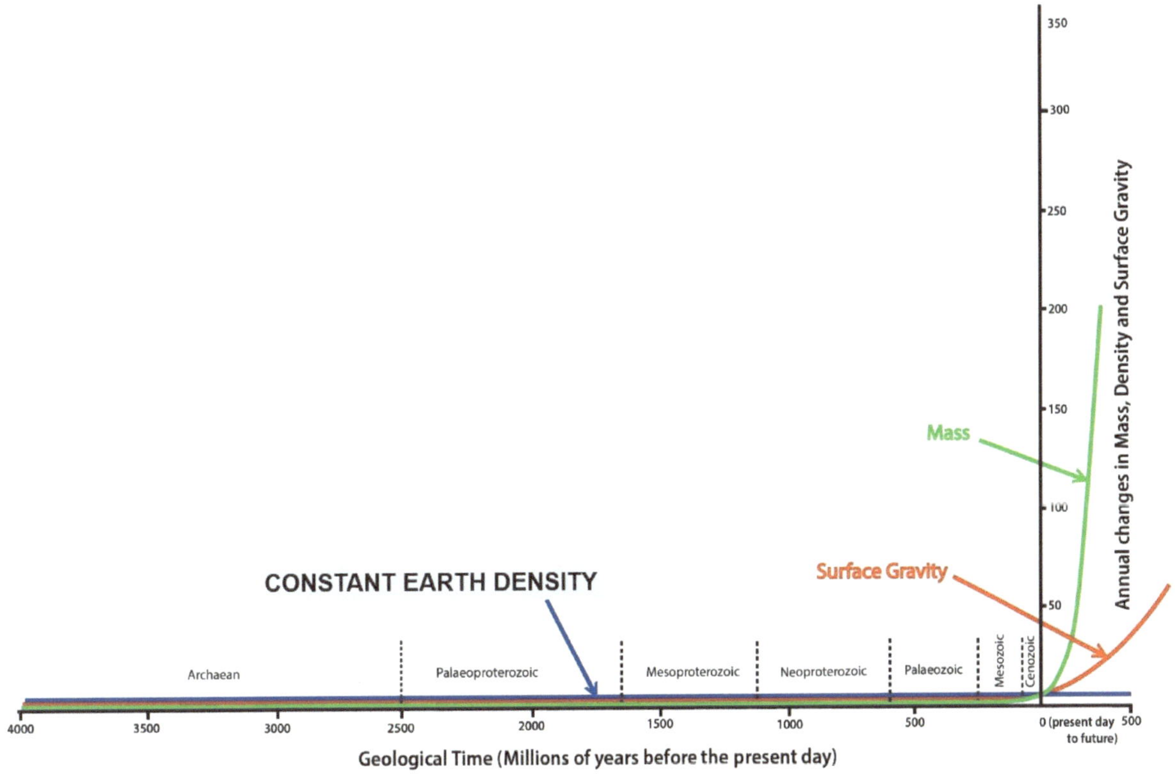

Figure 6.7 Annual changes in Earth Mass, Density and Surface Gravity assuming a constant Earth Density scenario. The graphs show that an increasing Earth Mass scenario is more in keeping with values from the other planets within the present Solar System. Annual change in Density is in grams per cubic centimetre, Surface Gravity is in metres per second per second, and Mass is kilograms times 10 raised to the power of 24.

Projecting this increasing Earth mass scenario to the future suggests that the Earth would approach the same size as Jupiter or Saturn within about 500 million years. During this interval of time the Earths' oceans, as well as the gaseous elements presently retained in the crust and mantle, may simply evaporate to form a thick atmosphere or ring structure similar to those of the present giant planets. This scenario would be subdued slightly if a density gradient were superimposed on the data, declining in time from higher rocky planet densities to lower giant gaseous planet densities.

This increasing mass scenario may then suggest that the transition from an inner rocky planet to a giant gaseous planet may be a natural evolutionary planetary process in our Solar System. If the Earth were increasing in radius it may mean that our Earth is currently in a transitional phase and may end up as either another giant planet or perhaps a failed planet, such as the asteroid belt might resemble.

While the changes in mass, density, and surface gravity shown in these figures represent end-points of a whole spectrum of possible causes for Earth expansion, they are speculative. For a hypothetical Earth with a constant mass however, the densities and surface gravities prevailing during Precambrian times would have been excessively high, particularly when compared to the density and surface gravity of the other planets within our present Solar System.

Carey in 1995, concluded from his own research that, because of the limitations shown from estimates of our ancient surface gravity, there may be no alternative but to consider an exponential increase in Earth mass with time as both the primary cause and effect of Earth expansion.

By adopting this constant density and increasing mass scenario for the cause of Earth expansion, as first proposed by Carey, the present approximate annual rates of increase in mass and surface gravity are then calculated to be:

Mass annual change = 60×10^{12} tonnes/year
Surface gravity annual change = $3{\cdot}4 \times 10^{-8}$ metre sec/sec/year

Without additional cosmological input and constraints, however, at this stage the exact changes, and hence primary cause of Earth expansion, cannot be determined with any degree of certainty from the changing Earth radius formula. There is though, no reason to suspect that whatever mechanism is involved it has not been on going through the entire history of the Earth, nor did it necessarily commence at a specific period of time due to some specific event. While an increase in Earth mass appears to be the most acceptable scenario by most researchers, it also begs the very pertinent and natural question as to where does all this new mass come from?

6.6 Important Assumptions

As was discussed during the seafloor crustal modelling studies, prior to carrying out model construction it is important to firstly acknowledge and define any adopted assumptions. Acknowledging these assumptions will then help understand the limitations of our knowledge. These assumptions will also help define the direction our current understanding of the process will potentially take the modelling. By understanding the limitations of these assumptions, subsequent quantification studies can also be better understood in the context of the adopted limitations.

In addition, to avoid confusion, the term subduction, when used, will be restricted here to its classical Plate Tectonic usage. In order to maintain a static radius Earth model, Plate Tectonic usage implies that thousands of square kilometres of excess seafloor crust must be disposed of beneath the continents. Similarly, subsequent closure of oceans during continental drift gives rise to continental collision to form mountains. In contrast, subduction does not occur on an Expansion Tectonic Earth and observed subduction-related phenomena are basically related to crustal interaction processes during changing surface curvature. This will be discussed in detail in later chapters.

For the reconstruction of continental crust on each of the Archaean to Triassic Expansion Tectonic small Earth models presented here it is assumed that, for an Earth undergoing an exponential increase in surface area and radius through time:

- Sediments deposited in continental sedimentary basins, as well as magmatic intrusions and volcanic eruptions, represent new crusts that accumulate primarily within down-warped areas of continental crust.

- These continental basin sediments, magmatic intrusions, and volcanic eruptions are not removed from the Earth's surface by subduction processes and, in conjunction with additional on-going geologic processes, are essentially preserved in the rock-record.

- Moving back in time, all younger sediments and intruded igneous rocks must be returned to their respective places of origin.

- Moving back in time, the surface area of sedimentary basins must be progressively reduced to their pre-rift, pre-stretching, or pre-break-up configurations.

What these Expansion Tectonic Earth assumptions mean is that, by moving back in time, all sediments must simply be returned back to the ancient lands where they were eroded from and similarly, all magmatic and volcanic rocks must be returned back to the mantle or lower crustal regions where they originally came from. After doing this, the surface area of each of the sedimentary basins or rift zones must then be reduced to simulate the return of the crust to a pre-stretching or pre-rifted state when moving back in time.

Once again, during model construction the only stipulation required during cutting and pasting of crustal information between each successive small Earth model is that:

- Continental crustal plates must undergo small vertical and surface area adjustments to allow for a progressive change in surface curvature of the Earth with time.

6.7 Model Construction

By now it should be well and truly appreciated that there are significant fundamental differences between the composition of the geologically relatively ancient continental and relatively modern seafloor crusts. Only by recognising this fact can the process of Expansion Tectonic small Earth modelling be extended back to the primitive Archaean Eon. What will now be done is to progressively remove continental crust from within the identified network of continental sedimentary basins and restore the remaining crusts to their pre-basin configuration on smaller radii Earth models.

The series of 24 small Earth models shown in Figure 6.8 represent the earlier seafloor crustal models—models younger than about 200 million years—recreated to include continental crustal geology, plus additional pre-Jurassic continental crustal models extending back in time to the early-Archaean. One model has also been extended to 5 million years into the future. Shown on these models is both seafloor and continental crustal time-based bedrock geology. The models in this particular figure show the closure of the Indian Ocean after removing all seafloor crusts. These models also show the relative sizes of the ancient Earth through time after progressively removing seafloor volcanic rocks, igneous rocks, and eroded continental sediments and returning these rocks and sediments back to either the mantle or ancient lands where they originally came from.

The step-wise construction of the additional Jurassic to Archaean models presented here is continued back in time until removal of all seafloor volcanic and continental sedimentary basin sediments and magmatic rocks is complete. At that time the most ancient Archaean continental crustal rocks, along with any remnant Proterozoic basement rocks, are then assembled together as a complete, primordial, pan-global Archaean small Earth model.

Modelling studies suggest that this primordial Archaean small Earth model had a radius of approximately 1,700 kilometres, which is about 27 percent of the present Earth's radius and of a similar size to the present Earth's Moon.

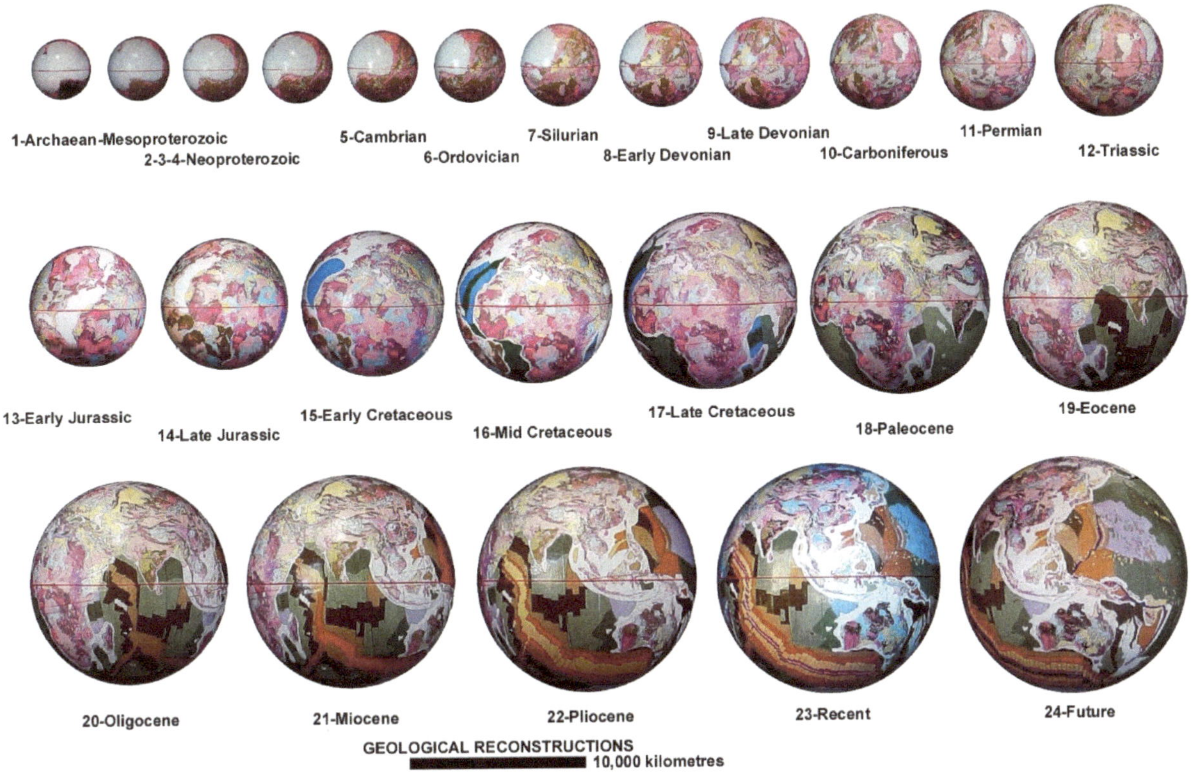

Figure 6.8 Spherical Archaean to future Expansion Tectonic small Earth geological models. Models show relative increase in Earth radii over time showing both continental and seafloor geology. Models range in age from the Archaean to present-day, plus one model projected 5 million years into the future.

As was shown in the earlier seafloor crustal modelling studies, as the oceanic plates were progressively re-assembled a network of continent-to-continent docking-points was established. These initial docking-points formed important positional constraining points for assembling each of the continental plates, and were further complimented by constraints from seafloor mapping. These docking-points were maintained throughout the post-Triassic seafloor assemblage times and their usage will be extended here to include the pre-Triassic continental crustal constructions.

Examples of continental docking-points used on each of the pre- and post-Triassic models (Figure 6.9) include: south-west Australia to Wilkes Land of East Antarctica; Tasmania to both south-east Australia and Victoria Land of East Antarctica; South Africa to New Schwabenland of East

Antarctica adjacent to the Weddell Sea; west Africa to South America; India to Queen Maud Land of East Antarctica; Siberia to Alaska; the Canadian Northwest Territories to Greenland; Greenland to Norway; Spain to Morocco; and Arabia to Africa.

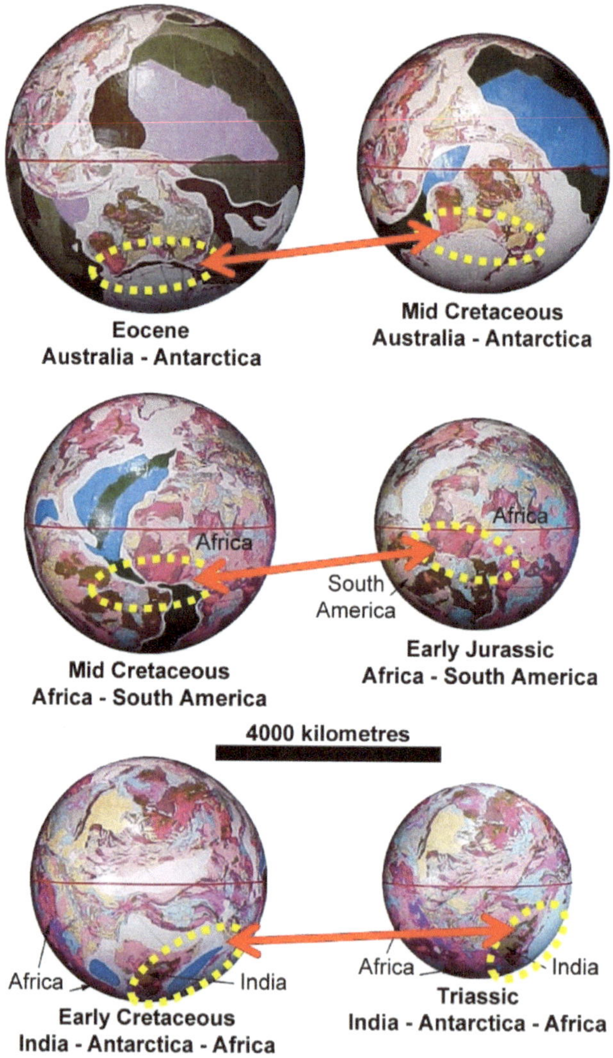

Figure 6.9 Examples of continent-to-continent docking-points used to constrain continental reconstructions. These include: Australia and Tasmania to Antarctica; Africa to South America; and India to Antarctica, Madagascar, and Africa.

Recognition and use of these docking-points allows each of the remaining continental crustal plates, in particular the ancient cratons and orogens, to be accurately assembled relative to each other. For each successive small Earth model, this recognition then enables the relationship between each of the docking-points to be strictly maintained across each crustal fragment until assemblage of the most ancient crusts is finally complete during the earliest Precambrian times.

While not immediately evident at the scale shown, each of the Triassic to Archaean Expansion Tectonic small Earth models in Figure 6.8 show that crustal stretching, thinning, and rifting of continental crust occurred within a global network of crustal weakness. This global network coincides with the previously established network of sedimentary basins and is shown to surround each of the most ancient continental orogenic crustal rocks. This network of crustal weakness also represents the precursor to, and is exactly coincident with, the known early-Jurassic to present-day mid-ocean-rift zones. The presence of magmatic, intrusive, and extrusive volcanic rocks within these areas of crustal weakness, as well as the close proximity of Precambrian metamorphic rocks, also suggests that these zones were associated with high heat flows emanating from the mantle, as previously predicted. Again, this distribution is a reflection of and was a precursor to the modern mid-ocean-rift zones.

By moving forward in time from the primitive Archaean small Earth model, successive models in Figure 6.8 show that the distribution of ancient continental sedimentary basins, continental magmatic rocks, and zones of crustal weakness progressively evolve into the better established network of geologically more recent sedimentary basins. This demonstrates that the zones of crustal weakness, as represented by these basins and mobile zones, have been important sites for progressive crustal stretching, basin formation, and seawater accumulation throughout Earth's history. These zones of crustal weakness eventually form the loci for continental crustal rupture, as well as complete continental break-up and opening of the modern ocean basins during late-Permian and younger times.

7 Origin of the Ancient Supercontinents

"The fit of the continents on a smaller Earth appears to be too good to be due to coincidence and requires explaining." Creer, 1965

The most important outcome of the bedrock seafloor mapping of each of the modern oceans is the recognition that seafloor crust is intruded volcanic lava derived from the mantle, not drowned continental crust as had previously been assumed by scientists. While this is obvious now, this shows that there is a complete contrast between the seafloor and continental crusts. It is even more important to appreciate that essentially all of the modern continental crusts are now exposed as dry land, where something like 66 percent of the exposed lands are actually sedimentary rocks, originally deposited in shallow ancient seas quite unlike modern oceans. The point I am making here is that if *the present is the key to the past,* as modern Earth science insists, then why aren't large areas of the modern oceans underlain by sedimentary basins as was common in the ancient past?

From an Expansion Tectonic Earth perspective, this conundrum is quite natural. For approximately 94 percent of Earth's history all of the continents were once joined together as a complete supercontinental crustal shell on a smaller radius Earth. During that time the distribution of ancient seas coincided with the continental sedimentary basins. It was only during the past 250 million years that the ancient Pangaea supercontinental crust then ruptured and fragmented to form the modern continents. In addition, during that same time the intervening areas between each of the modern continents opened to form the modern oceans, which were in turn accompanied by intrusion of seafloor volcanic lava along the mid-ocean-ridges. Hence, it is easy to understand why the sedimentary basins making up the modern continental crusts were, in fact, originally part of the ancient supercontinental crusts, not the modern oceans.

It is also important to fully appreciate that, prior to crustal break-up, the ancient supercontinental crust had undergone an extremely long period of crustal stretching and complex distortion during on-going changes in surface area and surface curvature. This change in surface curvature, in particular, has given rise to all of the geological features, such as folding and faulting, that are now preserved in the present-day crustal rocks. Change in surface curvature is also the prime mechanism for more grandiose geological features, such as mountain building and tectonism. It will be shown later that this complex crustal stretching and distortion mechanism also explains why and how the sediments were first deposited, where they now occur, and why they now occur as exposed lands and not as shallow continental seas.

Modelling studies have shown that, for each of the Archaean to late-Permian Expansion Tectonic small Earth models, all of the present-day continents remained assembled together as a complete pan-global supercontinental crustal shell encompassing the entire Earth at less than 50 percent of the present Earth radius. It wasn't until the late-Permian Period—around 250 million years ago—that rupture and eventual break-up of this supercontinental crust first began in earnest, resulting in formation of the modern continents and opening of the modern oceans. Prior to that time none of the modern continents existed in their present geographic outlines. Instead, these modern continents were preceded by ancient supercontinents surrounded by a network of ancient continental seas.

Within each of these ancient supercontinents, all of the separate crustal fragments that now make up the modern continents did exist. Throughout that time each of these fragments remained assembled together and continued to maintain an intimate spatial relationship with each other. What I mean by this is that prior to the late-Permian at no stage were there separate isolated crustal fragments or a need for random crustal break-up, collision, or dispersal events, as is mandatory in Plate Tectonic assemblages. Instead, the ancient Expansion Tectonic Earth history shows that each of the supercontinents remained firmly attached to each other as a complete pan-global continental shell, which then explains why there was no room or need for random break-up, collision, or dispersal.

As previously discussed, the crusts making up each of the modern continents are made up of three primary crustal elements, namely: cratons, orogens, and basins. On each of the small Earth models the distribution of ancient sedimentary basins coincides with a global network of shallow seas, which in turn coincide with low-lying regions located around the margins of the more elevated ancient lands. These sedimentary basins were previously seen to occur within areas undergoing crustal stretching and down-warping during changes in Earth surface curvature. Throughout this early part of Earth history, these basins were then periodically or progressively filled with sediments eroded from the surrounding elevated land surfaces.

Fragments of the most ancient supercontinental crusts, as well as evidence for the assemblage and physical characteristics of the supercontinental lands and seas, are now preserved within the ancient rocks exposed throughout the present-day continents. Not all of the original evidence is preserved of course as there has been an extremely long history of weathering, erosion, and metamorphic change since these rocks were first formed. There is, however, enough evidence left preserved in these rocks to give a good understanding of what the ancient Earth may have been like during these early times.

On an Expansion Tectonic Earth, throughout the ancient Precambrian times the network of shallow sedimentary basins and seas represented an interconnected global network of crustal weakness. This network of crustal weakness is where the ancient continental crusts first began to stretch, thin, and slowly break apart during on-going changes in surface area. The network of crustal weakness also represented the focus for on-going crustal instability—that is, where fault movement and seismic activity were prevalent—as well as where high heat flow, volcanic lava, and magma found their way to the surface via fractures and faults. Because of the focussed crustal instability in these areas, the sedimentary basins within this network were also periodically subjected to more localised crustal movement. This crustal movement then formed the precursors to the ancient long linear fold mountain belts, remnants of which remain on many of the modern continents.

Remnants of the ancient sedimentary basins that existed during Precambrian times occur on most of the modern continents, in particular large remnant areas within Europe and Asia, plus related areas within North and South America and Australia. The European and Asian areas are also referred to as the *Tethys region* in both Plate Tectonic and Expansion Tectonic Earth assemblages. In this Tethys region sediments are shown to have been deposited within an extensive ancient continental *Tethys Sea*, which has since been drained during falling sea levels and the sediments are now mostly exposed as dry uplifted lands.

Throughout the following Palaeozoic Era—extending in time from about 560 to 250 million years ago—each of the previously established Precambrian sedimentary basins continued to increase in surface area during on-going but rapidly increasing Earth radius. These established sedimentary basins extended in surface area to also include major basins now located in North America, Africa, Asia, and Russia. By the end of the Permian Period the ability for these continental sedimentary basins to continue to stretch was then exceeded, which resulted in crustal failure, rupture, and initiation of continental crustal break-up. As a result of this break-up the modern continents and modern oceans were first formed, where the growth history of the oceans is now faithfully preserved as the coloured seafloor bedrock mapping as previously discussed and modelled.

7.1 Four Phases of Crustal Development

In the following discussions it must be appreciated that crustal development on an Expansion Tectonic Earth is an evolving process which is intimately related to changing Earth radius, surface area, and surface curvature. This process commenced with an extremely long phase of slow to steady continental crustal stretching, followed by a phase of steady to rapidly increasing crustal stretching leading to crustal rupture and finally to break-up of the ancient supercontinental crust and formation of the modern continents and oceans. This entire process was unimaginably slow and protracted, matched only by the unimaginably long span of geological time available since formation and stabilisation of the Earth some 4,000 million years ago.

This evolving crustal development process is graphically summarised in Figure 7.1, shown in relation to the exponential radius curve as introduced in the previous chapter. This crustal development process can be considered as four distinct phases of Earth history.

The four phases of continental and seafloor crustal development include:

- **Phase 1**: An Archaean to Mesoproterozoic–times older than 1,600 million years ago–phase of minimal increase in Earth radius. The increase in surface area during this phase was readily absorbed during the first beginnings of continental crustal stretching and was marked by early sedimentary basin development and intrusion of granite and volcanic rocks. Deposition of sediments eroded from the exposed lands was confined to a global network of continental sedimentary basins which coincided with relatively shallow continental seas. Break-up and subtle jostling of each of the ancient plates during changes in surface curvature was first initiated during this phase, in particular within the established network of sedimentary basins, giving rise to long linear zones of crustal weakness. Because of the prolonged period of time involved in this phase, the continental crust had a subdued featureless topography for much of the time and continental seas were relatively shallow. Single to multi-celled organisms first became established and evolved during this phase.

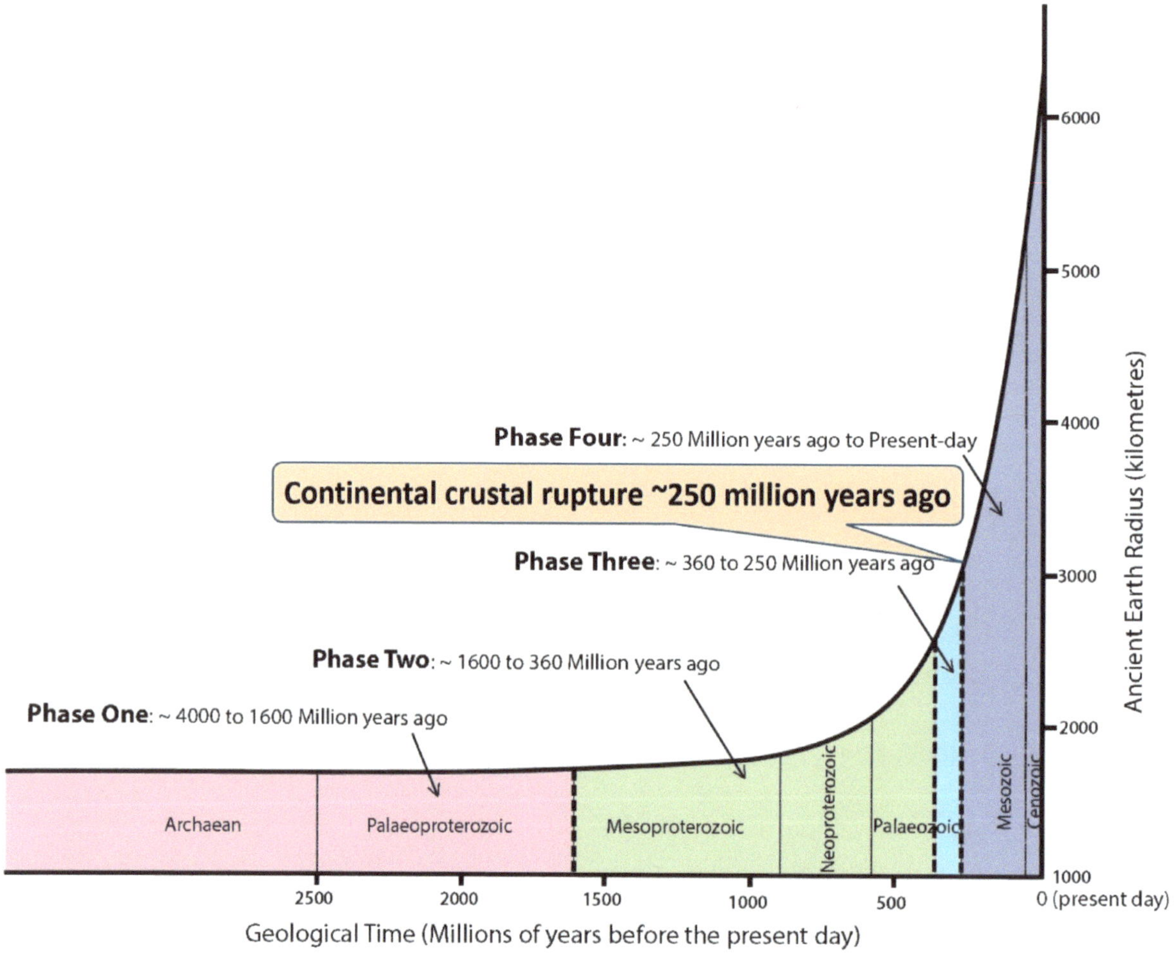

Figure 7.1 Four phases of Expansion Tectonic Earth development history. Phase 1): an early phase of minimal increase in Earth radius extending from 4000 to 1600 Ma; Phase 2): followed by a phase of steady to rapidly increasing Earth radius extending from 1600 to 360 Ma; Phase 3): a short, late-Palaeozoic transitional phase extending from 360 to 250 Ma where continental crustal stretching is exceeded and crustal rupture is initiated; Phase 4): a post-Permian phase extending from 250 Ma to the present-day where expansion is predominantly manifested within the modern oceans. [Ma = millions of years before the present].

- **Phase 2**: A Mesoproterozoic to Carboniferous—from 1,600 to 360 million years ago—phase of steady to rapidly increasing Earth surface area, crustal stretching, and changing surface curvature. The increasing surface area during this time continued to be absorbed within the continental crusts as on-going crustal stretching, accompanied by magmatic and volcanic intrusions. Crustal stretching was again confined mainly within the coincident network of crustal weakness, sedimentary basins, and shallow seas. Over time, the network of sedimentary basins and seas continued to increase their surface areas throughout the early- to mid-Palaeozoic Era—560 to 360 million years ago—as crustal stretching also increased. Movement of the crustal plates during adjustments to changing surface curvature resulted in multiple and often reoccurring phases of crustal jostling and tectonic activity, followed by renewed basin extension as stretching continued. The topography contrast between the low-lying sedimentary basins and exposed continents gradually increased during this phase, in particular during periods of crustal interaction and early mountain building. Hard shell and invertebrate life forms and fishes were well established during this phase.

- **Phase 3**: A Carboniferous through to late-Permian—360 to 250 million years ago—transitional phase, where continental crustal stretching was exceeded and stretching changed to crustal rupture, rifting, and initiation of continental break-up and opening of the modern oceans. Continental seas commenced draining and deposition of sediments then progressively shifted away from the established continental sedimentary basins into newly opening marine basins. Evidence for this early marine basin phase is now preserved seawards of the continental shelf margins and as remnant ocean plateaux within many of the modern oceans, such as the Lord Howe Rise in the Tasman Sea. During this phase reptiles, dinosaurs, and mammals evolved and dominated the lands.

- **Phase 4**: A post-Permian—250 million years ago to the present-day—phase of relatively rapid increase in Earth surface area and rapid opening of the modern oceans. Here, crustal stretching predominantly occurred within the modern oceans along well-defined mid-ocean-ridge spreading zones. During this phase, the increase in surface area and volcanic activity within the mid-ocean-ridge zones gradually exceeded the input of sediments from the continents. By the early-Jurassic Period the seafloor volcanic rocks were then exposed and preserved as seafloor crust. All seafloor crusts formed since then have remained preserved within each of the oceans. Two sub-phases are also recognised during this time, with each phase corresponding to the distribution of Mesozoic and Cenozoic seafloor crusts. During this phase dinosaurs became extinct and were superseded by the mammals.

On an Expansion Tectonic Earth each of the four phases of crustal development are fundamentally governed by the rate of increase in Earth radius, which is a direct result of an increase in Earth surface area and is, in turn, accompanied by increasing changes to surface curvature. The interval of time represented by the single, Phase 1, Archaean to Mesoproterozoic small Earth model shown in Figure 7.2 was so long and so imperceptibly slow to develop that this one model actually represents about 60 percent of all Earth history—50 percent more time than the rest of the models combined.

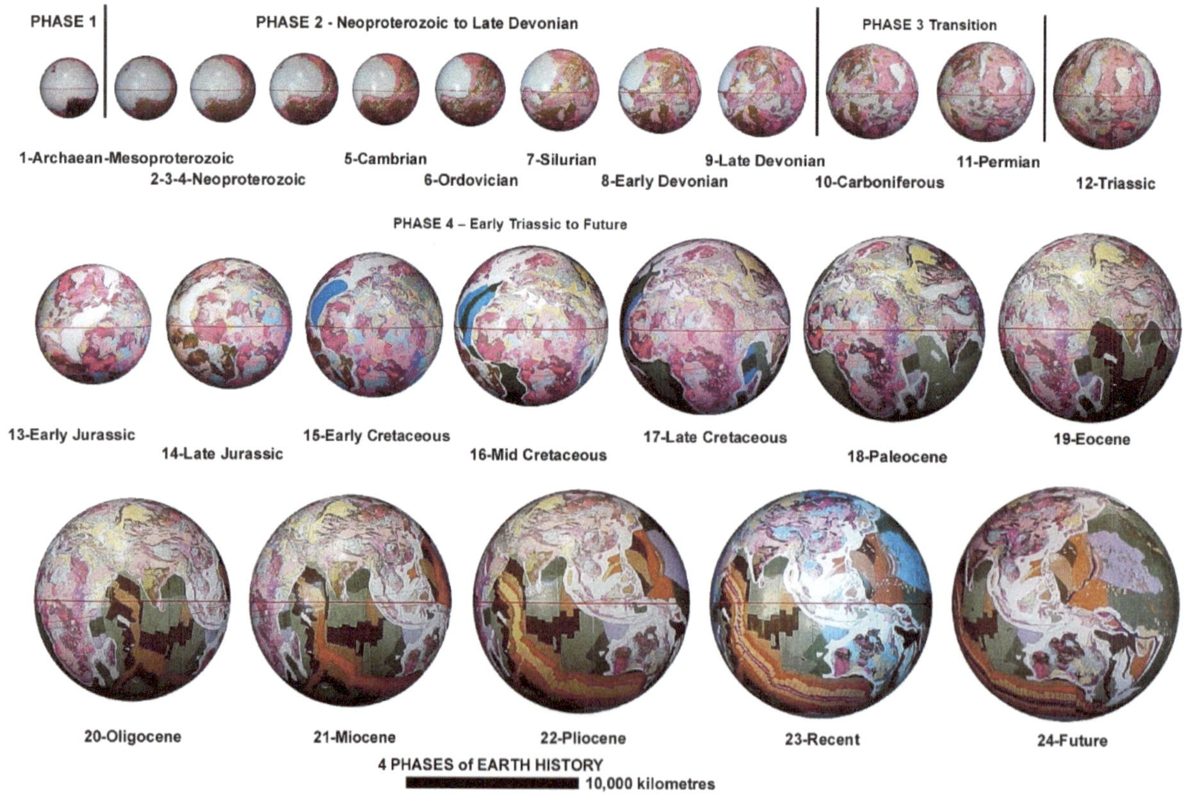

Figure 7.2 Four phases of Expansion Tectonic Earth development shown in relation to small Earth models extending from the early-Archaean to 5 million years into the Future.

The two Archaean to Carboniferous phases—times older than 360 million years ago—then represent prolonged periods of steady to steadily accelerating increases in Earth surface area focussed primarily within a network of established sedimentary basins. Once crustal rupture commenced during the late-Permian Period, this rupture initiated a rapidly accelerating phase of increase in surface area accompanied by continental break-up, draining of the continental seas, and modern ocean development to the present-day, extending seamlessly into the future.

This unique supercontinental crustal development and break-up process is shown by the small Earth models to be a steadily evolving and, more importantly, a predictable process that will now be elaborated on still further by investigating the development of each of the ancient supercontinents on an Expansion Tectonic Earth. This development extends in time from the primordial Archaean Earth model through to the late-Permian Period—about 250 million years ago—when the supercontinents first began to break-up and disperse as the modern continents. Modelling back in time to the Archaean has never been done before, for any tectonic theory, so it will be a very interesting exercise.

The observations that will be discussed here are, out of necessity, global scale and no attempt is made to incorporate complex local geology or tectonic detail. The outcomes of these observations will be further discussed in later chapters when dealing specifically with additional evidence from geophysics, ancient geography, ancient climate, and ancient biogeography.

7.2 Ancient Supercontinents

As was briefly introduced in Chapter 2, in Plate Tectonic studies the ancient supercontinents represent the assemblage of fragmented remains of more ancient supercontinents. The present continents in turn represent the assemblage of more recent fragmented crustal remains. On these conventional tectonic reconstructions there are a multitude of identified ancient supercontinents, most notably called Pangaea, Gondwana, Laurentia, Baltica, Laurussia, and Rodinia, each with its own separate crustal assemblage, break-up, dispersal, and subsequent collisional histories. The fundamental tectonic assumption is that the radius of the Earth has remained constant throughout Earth history. These assemblage processes are then depicted as being random and non-predictable and crustal assemblages are often portrayed as having a multitude of plate-fit options and ill-defined plate histories. While the primary continental crustal components of each of these conventional Plate Tectonic supercontinents are essentially the same as presented here, the configuration and assemblage of the ancient supercontinents on an Expansion Tectonic Earth, as well as the reasons why, differ markedly from conventional assemblages.

On all Expansion Tectonic small Earth models there is no requirement for random, non-predictable, or multiple plate-fit options or ill-defined plate histories. Nor is there a requirement for extensive, largely hypothetical ancient oceans to comply with a constant surface area premise or for the fragmentation of any of the modern continents to comply with palaeomagnetic *apparent-polar-wander* studies—the study of the apparent migration of the ancient poles from studies of remnant magnetism preserved in rocks.

Instead, at the reduced Earth radii shown by the small Earth models in Figure 7.2, these ancient supercontinental crusts are shown to tightly wrap around each of the ancient small Earth models. These crusts are then shown to assemble against continents that were otherwise not previously considered to be related on conventional assemblages. In this setting, on an Expansion Tectonic Earth, it is the distribution of the ancient continental seas, as well as the changes in coastal outlines and sea levels, that define the development and configuration of these ancient supercontinents, not because of random break-up, collision, or assemblage processes. All crustal fragments making up the ancient supercontinents are instead shown to retain an intimate relationship throughout Earth history, lasting until crustal break-up was initiated and the modern oceans began to open during the late-Permian Period.

In the following sections, discussion will start with the earliest primordial Archaean supercontinent and then move forward in time through to Rodinia, Gondwana, and finally to Pangaea. It should be noted that there are many published variations of these conventional supercontinental assemblages in the literature, as well as a multitude of lesser continental fragments. These lesser fragments are generally poorly defined, with a strong emphasis on variations in their locations rather than their definitive crustal shapes or assemblages. I must also mention that the term supercontinent, as used here, differs somewhat from Plate Tectonic studies where it is used to imply an assemblage of ancient crustal fragments. The term supercontinent is instead used here purely and simply to distinguish between the ancient continental crusts and the modern continents. As such, I reiterate that on an Expansion Tectonic Earth it is the distribution of continental seas that defines the ancient supercontinents, not an arbitrary assemblage of fragmented crusts.

The earliest Archaean supercontinent does not have a formally recognised name in conventional studies, so it will be introduced here as *Archaea,* representing the most primordial small Earth model assemblage.

7.2.1 Archaea

As mentioned earlier, the ages of the oldest known rocks found on Earth today represent the moment in time when these rocks first formed and, more importantly, the moment in time when these rocks were first stable enough to be preserved to the present-day. Evidence for the early-Archaean crusts will be discussed in later chapters, sufficient to say that crustal temperature and pressure indicators—minerals preserved in the ancient metamorphic and igneous rocks used to define the prevailing crustal temperature and pressure conditions—suggest that global crustal temperatures during the early-Archaean where very high. This suggests that the Earth was potentially molten during pre-Archaean times.

The significance of this statement is that, if global temperatures were excessively high during pre-Archaean times, the 4,000 million year age of the Earth possibly represents that moment in time when surface temperatures had cooled sufficiently enough for the ancient igneous rocks to crystallise. These rocks were then preserved to this day, without being re-melted or recycled back into a molten crust or mantle. In modern stratigraphy, the pre-Archaean time is referred to by the informal term Hadean Era.

The primordial Archaea crustal assemblage, as determined from the small Earth modelling studies, is shown in Figure 7.3 along with the location of the Archaean equator and poles. Establishing the location of the equator and poles was achieved via palaeomagnetic studies and will be discussed in Chapter 13. The named remnants of the present-day modern continents are also highlighted in this figure as black outlines. It should be noted that this figure represents an assemblage of the most ancient continental crusts that are currently preserved on the present-day Earth. On this model, continents, such as Antarctica, that have large amounts of ancient crusts will appear to be much larger than what they are now. This is only apparent and over time new sedimentary and volcanic crusts will be added to these original ancient crusts prior to break-up and forming the modern continents during late-Permian times.

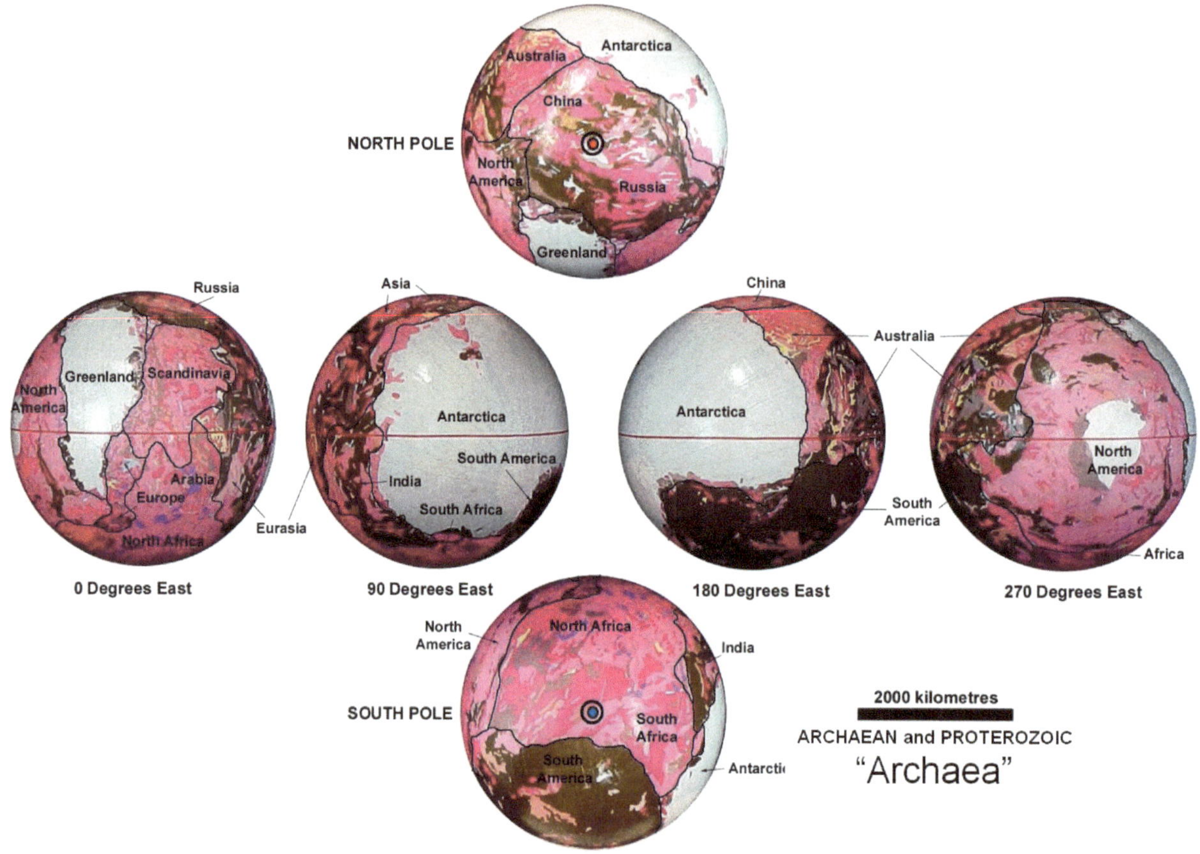

Figure 7.3 The primordial *Archaea* Expansion Tectonic small Earth model. The model shows the assemblage of ancient early-Precambrian continental crusts. The cratons are shown as pink and red, Proterozoic basement rocks are shown as khaki and the location of the present-day Antarctic and Greenland ice-sheets (covering Precambrian crustal rocks) are shown as pale blue. The ancient equator is shown as a horizontal red line and the poles are shown as dots. Ancient remnants of each of the present-day continents are outlined in black.

Geological evidence today shows that between 10 to 40 kilometres of rock material has been eroded from each of the ancient crusts. In the introduction to this chapter I mentioned that approximately 66 percent of the exposed continental crusts are actually sedimentary rocks, which once formed the seafloor of the ancient continental seas. All that has been done to create the Archaea model shown in Figure 7.3 is to then return these eroded sediments back to the ancient land surfaces, as well as return any intruded or extruded volcanic and magmatic rocks back to the mantle or lower crustal regions where they originally came from. It is then a relatively simple process to progressively close off the remaining sedimentary basins until each of the most ancient crusts dock together as a complete primitive supercontinent. In essence, these ancient continental crustal fragments are analogous to assemblage of the modern continents into the Pangaea supercontinental assemblage after returning the seafloor crusts back to the mantle.

After having progressively removed all younger sediments and reducing the surface areas of each of the surrounding sedimentary basins back in time, Figure 7.3 shows that these ancient crusts assemble together with a relatively high degree of precision. It should be noted that this assemblage is made without the need to arbitrarily fragment continental crusts, or the need to invoke pre-existing crusts to account for any shortfalls in surface area or crustal configuration. This is, in fact, the first time that the primordial Archaea Earth assemblage has been contemplated and achieved.

The size of this Archaea supercontinent was about 27 percent of the present Earth radius, about the size of the present Moon. Because of the extremely low expansion rates existing during the Precambrian times, the Archaea small Earth model shown in Figure 7.3, which is also the first model (top left) shown in Figure 7.2, represents the period of time extending from the early-Archaean through to the Mesoproterozoic—middle-Proterozoic—Eras. This one model then represents about 2,400 million years, or two thirds of all known geological history.

Modern geological mapping of the ancient Archaean crusts also show that the vast majority of these ancient crusts comprise granite and volcanic rocks, with lesser sedimentary rocks. These crusts are often referred to as *granite-greenstone* terranes—where the term greenstone refers to the distinctive dark green colour of the volcanic rocks. The sediments themselves are also distinctively different from the later and more modern sediments in that they are what are loosely referred to as *dirty sediments*. The term dirty sediment simply refers to the large amount of rock fragments and volcanic material—such as volcanic ash—that occurs in these sediments. Under modern weathering and erosion conditions this ash and rock fragment material would quickly decompose to form relatively clean clays, silts, and sands. This sedimentary evidence then implies that the atmospheric conditions existing during the early-Archaean times were vastly different to those of today, confirming that it may have been quite acidic and devoid of oxygen.

During the very earliest Archaean times it is envisaged that once the primitive granite and volcanic crust had cooled and solidified, onset of Earth expansion was first initiated as global-scale cracking and fracturing of the crust to form a network of crustal weakness surrounding relatively stable lands. This network of crustal weakness, in effect, would have been made up of a network of cracks which were subsequently intruded by renewed granite activity or primitive volcanic lava—the greenstones—similar to the lava being intruded along the modern mid-ocean-ridges. As time very slowly progressed the influence of changing surface curvature, although minimal during that time, assisted in maintaining a distinction between the elevated land surfaces and the low-lying areas located around the margins. Eroded sediments were then deposited in these low-lying areas. It is considered that this elevation contrast was instrumental in establishing a distinction between the first Archaean supercontinental lands and the first ancient seas and sedimentary basins.

The emergent Archaea land surface would have been vastly different to what is familiar to us now. It would have been a barren, possibly windswept rocky landscape devoid of all forms of life and exposed to erosion of rock and dust particles by winds and maybe some acidic rain. There would have also been strong chemical weathering and erosion of the rocks themselves, in particular erosion by hot and potentially acidic waters from volcanic eruptions localised along the newly established network of crustal weakness. This weathering and erosion then gave rise to deposition of the very first sediments, deposited in low-lying regions, along with any extruded volcanic lava.

In Figure 7.3, remnants of this network of low-lying regions can be seen by the distribution of khaki coloured, mainly sedimentary rocks. These early sedimentary basins would have been small and isolated, becoming progressively more extensive after a very long period of time as the low-lying basins progressively filled with sediment. This early granite-greenstone and sedimentary crust now represents the oldest preserved crustal remnants on Earth today. The prevailing acidic atmosphere and waters also enabled metals rich in sulphur as well as elemental carbon to accumulate within the sedimentary basins, which in turn may have formed the basis for bacterial life forms to emerge and evolve during the later Precambrian times.

The geological evidence suggests that during the latter part of this Archaean supercontinent phase, very large stable sedimentary platforms—very large flat basins filled with sediments—covered with relatively shallow seas had formed. These had a very low elevation contrast between the dry lands and the seas. In other words, at that time there was a relatively flat landscape covering the entire ancient Earth. These large platforms simply represent sedimentary basins that had been filled to capacity over the extremely long period of time operative during the ancient times. Any further erosion of the lands and deposition of sediments was limited to mainly chemically-precipitated siliceous chert, carbonate, and banded iron formation rocks within the relatively shallow seas. Remnants of these chemically-precipitated rocks, along with the aerially extensive granite, volcanic, and associated sedimentary rocks making up the ancient lands are now preserved on many of the modern continents.

During this extended period of time, the ancient North and South Poles are shown in Figure 7.3 to have been located within what are now Northern China and West Africa respectively. Similarly, the ancient equator is shown to have passed through what is now North America, East Antarctica, Australia, Greenland, and Scandinavia; a strong contrast to where it is located now. Over this extended length of time the primitive Archaea supercontinent also underwent many changes to its network of seaways and sedimentary basins. These geological changes eventually evolved imperceptibly into the better known supercontinent called *Rodinia,* some 1,000 million years ago. Again, it is emphasised that on an Expansion Tectonic Earth it is the changes to sea-level and the distributions of the network of sedimentary basins that define the progression from one supercontinent to the next; not as a result of fragmentation or re-assemblage of prior supercontinental crustal fragments.

7.2.2 Rodinia

The conventional *Rodinia* supercontinental assemblage relies heavily on palaeomagnetic evidence to constrain a multitude of plate-fit options that are currently published in the literature. These assemblages are further supported by ancient climate, fossil, and geographic evidence as required. These assemblages are, however, seriously hindered by the longitudinal limitations imposed by palaeomagnetic constraints—as will be discussed in Chapter 13. Beyond these limitations, the presence of largely theoretical ancient oceans also requires that the continental crustal fragments making up Rodinia must be sourced from remote and equally theoretical continents, which have in turn also been arbitrarily fragmented and dispersed. This fragmentation and dispersal process results in a confused multitude of assemblage options for each supercontinental fragment, usually where fragmentation is not supported by local geological or geographical evidence.

Rodinia on an Expansion Tectonic Earth is represented by the late-Proterozoic—about 800 million years ago—small Earth model shown in Figure 7.4.

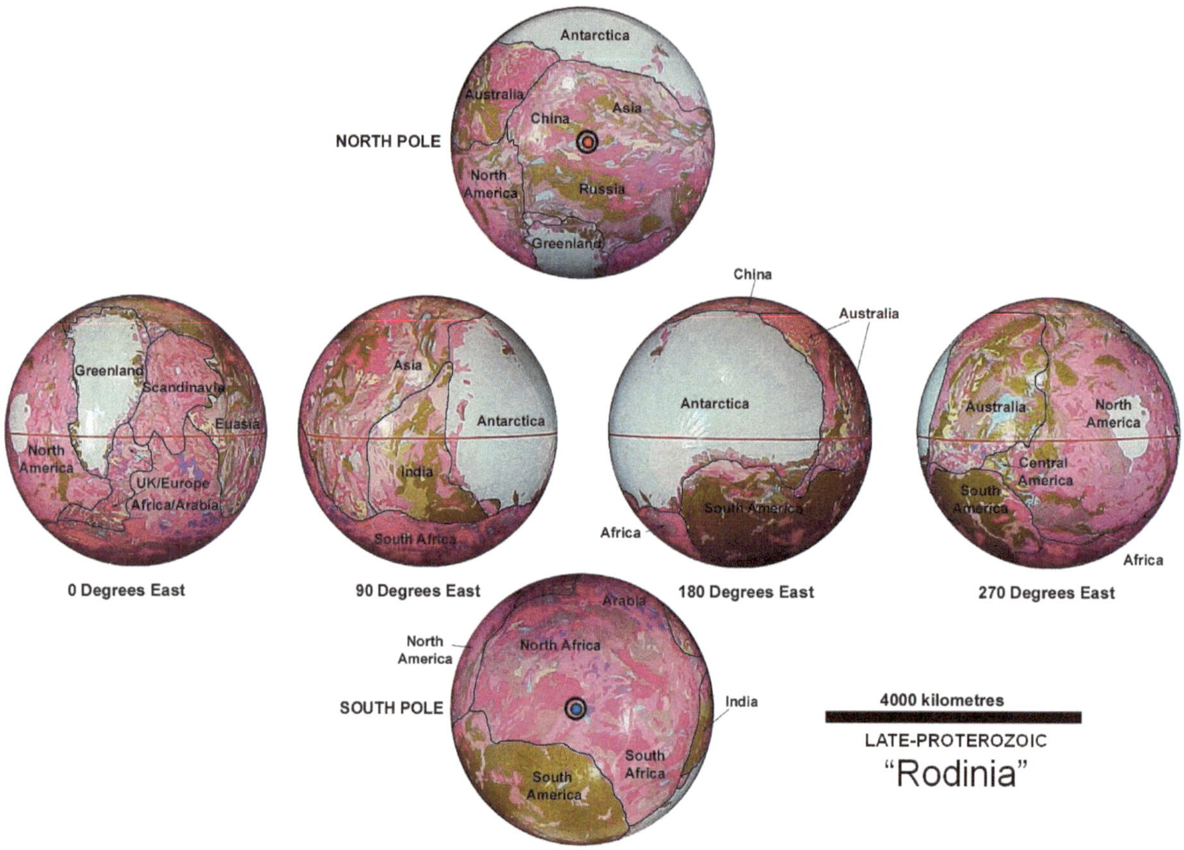

Figure 7.4 Late-Proterozoic Rodinia Expansion Tectonic supercontinent. The model shows the distribution of Precambrian cratons (pink and red) inferred to represent exposed land, and a network of basins (khaki and brown) inferred to represent continental seas. The black lines represent outlines of remnant present-day continents.

The assemblage and distribution of the remnant modern continental outlines are also shown as black outlines, along with the Proterozoic magnetic poles and equator. Unlike younger small Earth models to come, there is no published ancient coastal information available for the Precambrian times to properly define the distribution of the ancient continental seas. The reason for this is not because there were no seas or coastlines present during the Precambrian times but because conventional tectonic assemblages are inconclusive for these times. The outline of the Rodinia supercontinent and seaways on an Expansion Tectonic Earth is therefore inferred from the distribution of Precambrian continental sedimentary basins.

It is also inferred that the assemblage of each of the ancient Precambrian cratons shown on this model–red and pink coloured rocks–represent the exposed lands and similarly, the network of sedimentary basin rocks–khaki and brown coloured rocks–represent the ancient continental seas. These sedimentary basins are quite justifiable considering the very close agreement between coastal outlines and continental sedimentary basins which will be shown later for the younger small Earth reconstructions. The exposed lands shown in Figure 7.4 agree in part with the conventional Rodinia continental assemblage except that extensive ancient oceans, as distinct from continental seas, are simply not required at the reduced Earth radius shown by this small Earth model.

Comparison of the Rodinia small Earth model in Figure 7.4 with the Archaea model in Figure 7.3 shows that there is not a lot of difference between these two models. The location of the poles and equator are essentially the same and the distribution of the ancient cratons is also much the same. The final phase of the Archaea supercontinental development history was the formation of very large, stable sedimentary platform basins and shallow seas with a low elevation contrast between the lands and seas. This, in essence, is the only difference between Archaea and Rodinia where these basins simply represent mature sedimentary basins that have been filled to capacity. Any erosion and input of sediment was then limited to mainly chemically-precipitated rock—chemical precipitation of various elements is a common feature during the Early-Precambrian Eras—which requires very stable crustal conditions.

The distinguishing feature of the Rodinia supercontinental history is that it coincides with the second phase of steadily increasing surface area and Earth radius (as shown graphically in Figure 7.1). What this means is that, as a result of the increasing Earth radius, increasing changes to surface curvature were starting to take effect which began to elevate the exposed lands and, in turn, also increased the elevation contrast between the lands and seas. This elevation contrast initiated an increased rate of erosion of the exposed lands, which also influenced changes to the distribution of seas and corresponding coastal outlines of the supercontinents. Once again, in Expansion Tectonics, it is these changes to the coastal outlines of the shallow seas that defines the shape and configuration of both the ancient seas and supercontinents.

Another distinguishing feature about the Rodinian times was the changing atmospheric conditions. These conditions changed from reducing conditions throughout much of the earlier Archaean times, to the accumulation of atmospheric oxygen during the Proterozoic. This transition was marked by an increasing accumulation of banded iron formation rocks—important sources of iron ores—and chemically precipitated calcium and magnesium carbonate rocks during this time. The transition also coincided with development and preservation of the earliest life forms, soon to explode in diversity during the following Gondwanan supercontinental times.

During late-Proterozoic Rodinian times, the ancient North and South Poles continued to be located within what is now Northern China and West Africa respectively. The ancient equator continued to pass through what is now North America, East Antarctica, Australia, Greenland, and Scandinavia, as well as Europe, Asia, and India; essentially the same as during the Archaea supercontinent times.

On subsequent, younger Expansion Tectonic small Earth models the early Rodinia assemblage represents a precursor to the better known *Gondwana* supercontinent and ultimately the *Pangaea* supercontinental configurations. As shown on the Archaea and Rodinia small Earth models, the change from one supercontinent to another is progressive and evolutionary and this theme will follow through to the Gondwana and Pangaea supercontinents. Small Earth modelling again shows that this progression is intimately related to the variation in coastal outlines during change in sea levels; not by random continental break-up and reassemblage processes. On an Expansion Tectonic Earth it is only Pangaea that eventually breaks up and disperses to form the modern continents.

7.2.3 Gondwana

The Plate Tectonic *Gondwana* supercontinental assemblage is made up of an assemblage of the present-day South America, Africa, Arabia, Madagascar, India, Australia, and Antarctican continents, as well as crustal fragments from Florida, southern and central Europe, Turkey, Iran, Afghanistan, Tibet, and New Zealand. This conventional assemblage shows that these present continents were assembled into large, separate northern and southern Gondwana supercontinents plus smaller sub-continents separated by a number of largely inferred ancient oceans. This assemblage is said to have formed during a *Pan-African tectonic event*. There is no need to talk about this event, but it represents a geological event that occurred at the end of the Precambrian Era and evidence is still preserved in rocks from the various modern-day continents. From its initial assemblage, the conventional Gondwana assemblage is said to have remained intact throughout the Palaeozoic to Early Mesozoic times. As a number of researchers have noted though, *...past movements of the Gondwana supercontinent, based on ancient magnetic, biogeographic and climatic evidence, are both equivocal and contentious*, and this contention still remains despite the addition of new data from Africa and Australia.

Gondwana on an Expansion Tectonic Earth is shown reconstructed on the Ordovician small Earth model—about 460 million years ago—in Figure 7.5. In this figure the distribution of the ancient coastal outlines is also shown for the first time as blue lines and the continental seas as shaded blue areas. These coastal outlines, which simply represent the global distribution of the ancient beaches existing at that time, were plotted from published geographical data and are in turn based on various geological mapping studies. This distribution of ancient coastlines, in effect, defines the outlines and locations of the ancient supercontinents and lesser sub-continents as well as the distribution of continental seas existing at that time.

In this figure, the distribution of the ancient coastlines now enables the various ancient continental seas to be located and named. As has already been mentioned, and as will be elaborated on in Chapter 14 when ancient geography is discussed, on the Expansion Tectonic small Earth models prior to the early-Triassic there were no ancient oceans.

The need for ancient oceans is an artefact of conventional tectonic studies, whereby largely hypothetical ancient oceans are required in order to maintain a constant surface area Earth. While seeming to be a contradiction of terminology, on an Expansion Tectonic Earth a clear distinction is made between the terms modern oceans and ancient seas. To comply with conventional naming, rather than refer to the oceans as ancient oceans, I again simply call them continental seas. As shown in Figure 7.5, this usage will be limited to the more familiar ancient Tethys Sea, Panthalassa Sea, and Iapetus Sea.

During this Gondwanan time the Earth surface underwent a steady to rapid increase in area and was also fast approaching crustal rupture—the moment in time when the continental crusts first begin to fail and break-up. This Gondwana crustal assemblage also retained the exact same configuration of cratons, orogens, and basins as seen in both Rodinia and Archaea. The only difference being the greater surface area of surrounding sedimentary basins and hence aerial distribution of continental seas. This crustal assemblage was retained still further until initiation of crustal rupture and break-up of the pan-global continental crust to form Pangaea occurred during the late-Permian.

The coastal information shown on the Ordovician small Earth model shows that, at that time there were distinct elevated ancient land surfaces. These land surfaces were in turn surrounded by a network of equally distinct, relatively shallow continental seas. Gondwana on this model was subdivided into a North and South Gondwana, separated in part by an early *Panthalassa*

Sea—the precursor to the modern Pacific Ocean. North Gondwana was made up of Australia, East and West Antarctica, and India and also possibly included Tibet and Afghanistan. South Gondwana, joined at Madagascar, comprised Africa, Arabia, and South America. The ancient sub-continents, called *Laurentia* and *Baltica*, were made up of North America, Greenland, and Scandinavia, as well as smaller Precambrian fragments now represented by Scotland and Ireland. Similarly, the ancient *Laurussia* sub-continent was centred on the Precambrian regions of Mongolia and northern Russia.

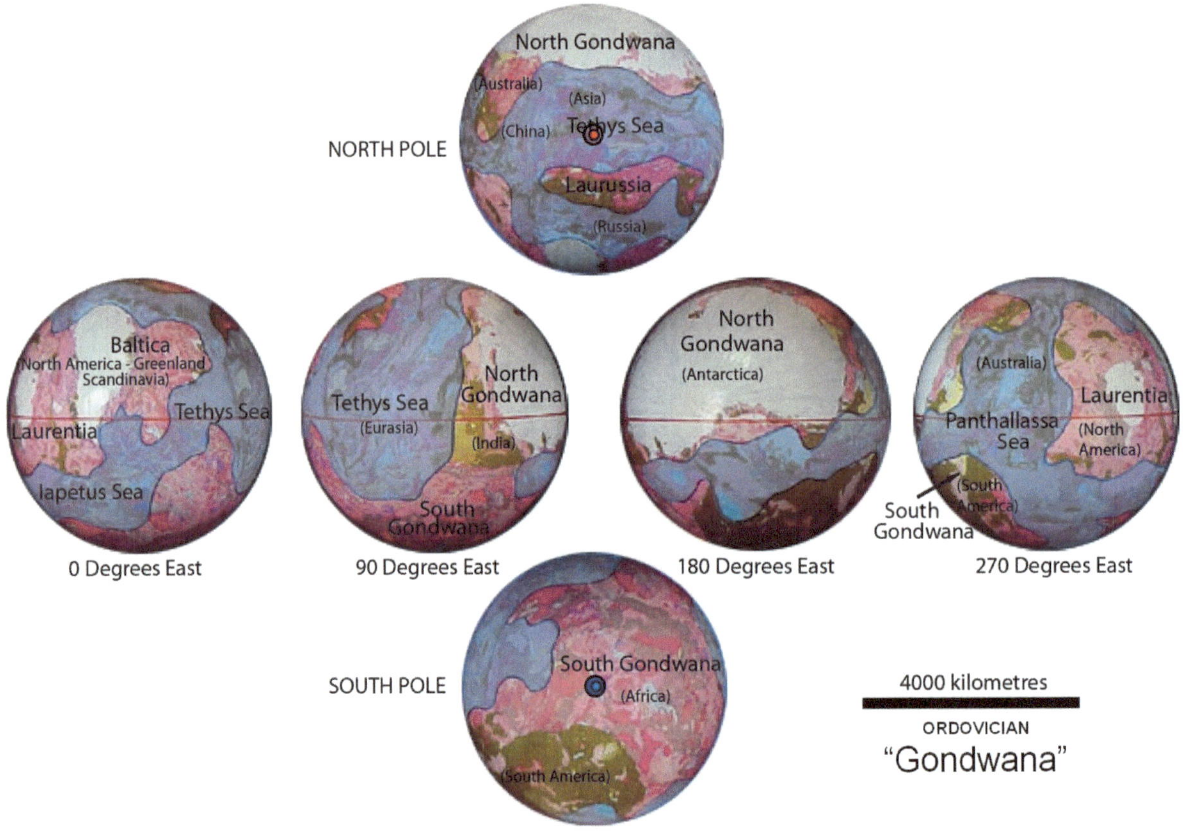

Figure 7.5 The Ordovician Gondwana Expansion Tectonic small Earth assemblage. This model shows the ancient coastline distribution (blue lines) defining North and South Gondwana in relation to Laurentia, Baltica, and Laurussia. The ancient Tethys, Iapetus, and Panthalassa Seas form part of a global network of continental seas (shaded areas) surrounding each of the exposed supercontinents.

Each of these exposed Gondwana land surfaces were in turn surrounded and interconnected by the ancient continental Tethys, Iapetus, and Panthalassa Seas. Remnants of these are now preserved and represented by many of the ancient sedimentary basins that are located in Eastern Australia, North and South America, Europe, Asia, and Africa.

This Expansion Tectonic small Earth Gondwana assemblage differs markedly from Plate Tectonic reconstructions. On an Expansion Tectonic Earth Eastern Australia, for instance, is assembled adjacent to both North America and South America and is separated by a narrow Panthalassa Sea, as distinct from the need for a very much wider conventional Panthalassa Ocean separating each of these continents.

The existing physical and geological similarities between the Australian and South American continents have long been acknowledged in a number of publications. These similarities include the distribution of marine and terrestrial plants and animals, which link Australia with New Zealand across to Central and Southern America, and also rock and fossil types that link Australia directly with South America. It should be appreciated that these physical links between each of these modern continents can only be maintained on an Earth that had a much-reduced ancient radius. In contrast, because of the Plate Tectonic requirement for a very large ancient Panthalassa Ocean, conventional reconstructions generate enormous separations which cannot be justified or reconciled with the factual geological or geographical evidence.

It is significant that the interval of time that Gondwana existed on an Expansion Tectonic Earth coincides with the later part of the second crustal development phase shown in Figure 7.1. This is characterised by a rapidly accelerating phase of increasing surface area and accompanying changes in surface curvature. These accelerating changes gave rise to a marked increase in erosion and deposition of sediments and marked changes to the distribution of continental seas and coastal outlines. The increasing changes in surface curvature also initiated localised compression of the sedimentary basins to form long linear fold mountain belts, further disrupting established seaways. It is significant to note that these times also coincided with the rapid development and evolution of all life forms on Earth. Evolution of species is often seen as driven by the need to keep pace with environmental change and similarly extinction is seen as a by-product of not being able to keep pace with environmental change. The degree of crustal change during this interval of time certainly had the capacity to markedly influence evolutionary change in all life forms.

During Gondwanan times, the Expansion Tectonic small Earth South Pole was located within central West Africa in what was then South Gondwana. The North Pole was located within Northern China in what was part of the Tethys Sea. The ancient equator passed through East Antarctica, central Australia, North America, central Eurasia, and India, through what was then North Gondwana. This geographic configuration approximates Plate Tectonic reconstructions in part, but differs substantially in the South Pacific region because Plate Tectonics require the presence of a wide expanse of Panthalassa Ocean.

The Expansion Tectonic coastal outlines also show that continental land connections existed between north Australia and North America and between north Africa-Arabia and Scandinavia. The existence of these land connections formed important migration routes or barriers for terrestrial and marine life forms. A change in configuration of the various continental seas may have also resulted in periods of relatively rapid sea-level changes in these areas, with disastrous consequences for plant and animal species existing at the time.

The late-Palaeozoic to early-Mesozoic times eventually coincided with the initial break-up of the Gondwana supercontinent, which was accompanied by draining of the continental Tethys, Panthalassa, and Iapetus Seas as the modern oceans began to open. As a result of this draining of the seas each of the Gondwana and related supercontinents geographically merged with the smaller Laurentia, Baltica, and Laurussia sub-continents to form the more familiar *Pangaea* supercontinent. On an Expansion Tectonic Earth this evolution of continents and seaways was again reflected in the progressive changes to coastal outlines and sea levels, as displayed in each of the small Earth models.

7.2.4 Laurentia, Baltica, Laurussia

Laurentia, Baltica, Laurussia and numerous smaller fragments, such as Kazakhstania and fragments making up Southeast Asia are described in Plate Tectonic literature as sub-continents, or simply continental fragments. These are routinely described in the literature as fragments of pre-existing supercontinents that periodically dispersed and amalgamated between the larger supercontinents during continental break-up and subsequent reassemblage of the various continental fragments.

On the Ordovician Expansion Tectonic small Earth model shown in Figure 7.5 above, each of these sub-continents were geographically connected to the supercontinents. Their presence and coastal outlines depended primarily on the variation in the sea levels existing at the time and the sub-continents were often separated geographically from the main supercontinents by the global network of continental seas.

As the continental crusts began to rupture during the late-Permian, the established continental seas then began to drain into newly formed marine sedimentary basins as the modern oceans began to form and open. The previously submerged seafloor surrounding each of the sub-continents was then exposed as dry lands and the surface areas of the sub-continents progressively increased to eventually form part of the more familiar Pangaea supercontinent.

7.2.5 Pangaea

The assemblage of a large *Pangaea* supercontinent was initially promoted by Alfred Wegener in 1915. In his proposal, he presented geologic evidence to suggest that the large continental areas of the modern world were originally united late in the Palaeozoic Era prior to subsequent break-up and dispersal of the continental fragments. This has since been adopted by Plate Tectonic studies, whereby Pangaea is said to have assembled from crustal fragments originating from the break-up of Gondwana. During the late-Palaeozoic Era the Pangaea supercontinent then comprised two contrasting geographical provinces: a southern Gondwana province and a northern Laurasian province. The Pangaea supercontinent was then inferred in Plate Tectonic studies to have fragmented within the Tethyan region, where continental break-up began in earnest during mid-Jurassic times.

Early Pangaea on an Expansion Tectonic Earth is shown on the Permian small Earth model in Figure 7.6. This figure again shows the coastal outlines as blue lines and continental seas as shaded blue areas, and also names the various supercontinents and intervening seas existing at the time. This assemblage coincides with the Carboniferous to Permian transitional crustal development phase 3, shown in Figure 7.1. In this figure, increasing Earth surface area had progressed from advanced continental extension to true crustal rupture prior to continental crustal break-up and dispersal as the modern continents.

The presence of Pangaea on an Expansion Tectonic Earth represents a simple, progressive, although relatively rapidly changing evolution of the Gondwanan coastal outlines prior to continental break-up and formation of the modern continents. During this Pangaean time rupture and break-up of the continental crusts also initiated draining of the continental seas, which was in turn accompanied by a shift in where the eroded sediments were being deposited. This shift changed from sediments being deposited within the existing network of continental sedimentary basins, to being deposited within newly opening marine basins and along the continental shelf margins of the newly formed continents. This influx of sediment, along with intrusion of new volcanic and magmatic rocks, is now commonly preserved within submerged marine plateaux and continental shelves surrounding most of the modern continents.

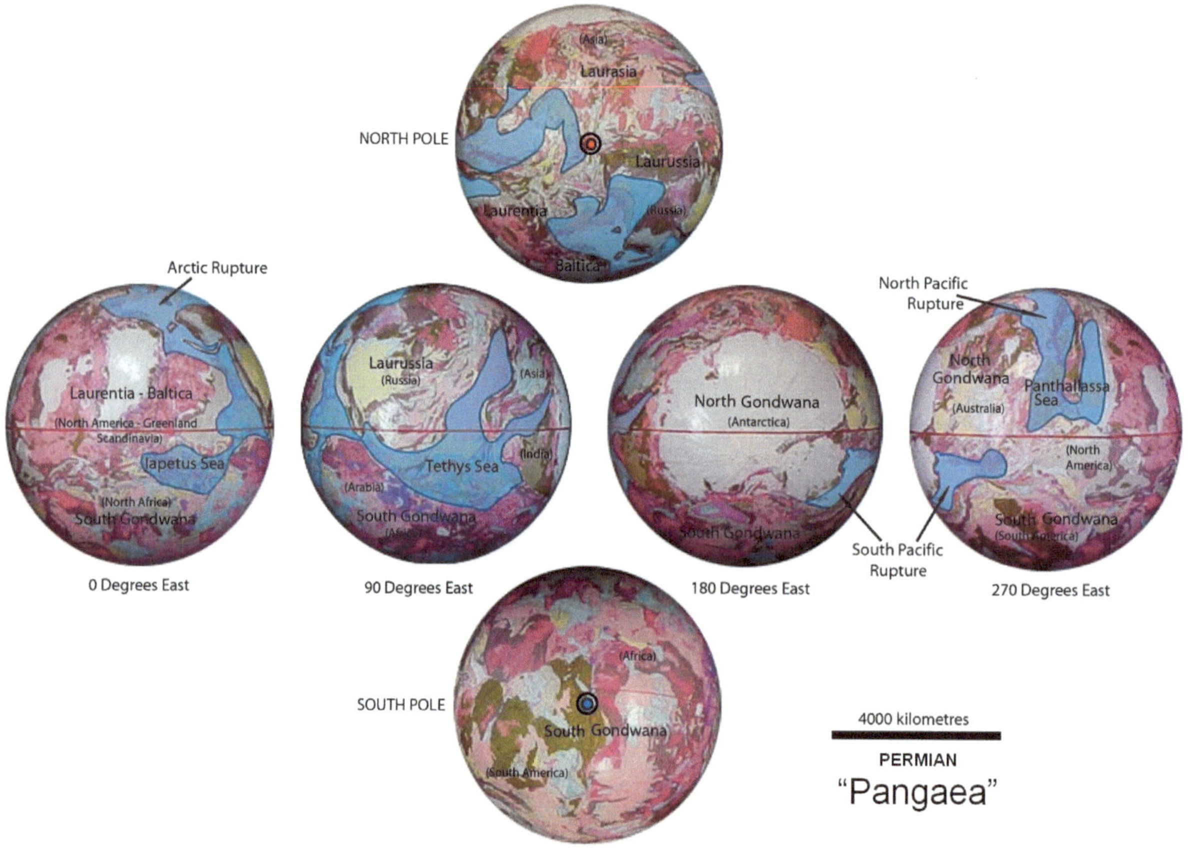

Figure 7.6 Permian Pangaea Expansion Tectonic small Earth crustal assemblages. The model shows the ancient coastline distribution (blue lines) as well as the ancient Tethys, Iapetus, and Panthalassa Seas (blue shaded areas) forming part of a global network of continental seas. The figure also shows the locations of continental rupture commencing in the north and south Pacific and Arctic Ocean regions to form the modern oceans.

Following rupture of the supercontinental crust, Pangaea on an Expansion Tectonic Earth eventually broke-up during the late-Permian and dispersed as the modern continents during Triassic and Jurassic times. The subsequent seafloor crustal history is now preserved as intruded volcanic seafloor lava throughout all of the modern oceans.

This post-Pangaea interval of time also saw large apparent shifts in the location of the North and South Poles and equator, occurring as a direct result of opening of the modern oceans and an apparent shift in the location of each of the modern continents. These shifts, as well as geographic isolation during break-up, in turn led to plant and animal evolutionary changes which increased markedly during post-Jurassic times. Changes were also accompanied by increased extinction of many of the established life forms as well as increased isolation of many species as a number of the modern continents separated to form island continents. The unique fauna of Australia, as well as the lack of unique fauna in India despite long isolation in Plate Tectonic constructions, are classic examples of this important event.

8 Origin of the Modern Continents

"...terrestrial expansion brought about the splitting and gradual dispersal of continents as they moved radially outwards during geological time." Vogel, 1983

As has been shown on each of the empirical small Earth models, on an Expansion Tectonic Earth the ancient supercontinents existed as a complete continental crustal shell encompassing the entire Earth for 94 percent of geologic history. This supercontinental phase lasted for some 3,750 million years and culminated with break-up of Pangaea about 250 million years ago. During this ancient time, continental crust covered the entire Earth with no intervening oceans. The origin of the supercontinents then simply involved a progressive, evolutionary crustal process during a prolonged period of crustal stretching and changes to both Earth surface area and surface curvature through time. The outlines and configurations of the supercontinents were then controlled and dictated by changes to the ancient sea-levels and coastal shorelines.

In strong contrast, the modern continents only existed in their current form after break-up of the ancient Pangaea supercontinental crust started some 250 million years ago. These modern continents simply represent the fragmented remains of the ancient Pangaea supercontinental crustal shell. This fragmentation and break-up occurred because the ability of the supercontinental crusts to continue to stretch during on-going increases in Earth surface area was finally exceeded during late-Permian times. Once crustal stretching was exceeded the Pangaea supercontinental crust then ruptured, broke apart and fragmented to form the modern continents and opened to form the intervening modern oceans.

Discussion on the origin of the modern continents overlaps and follows on from the development of the more ancient supercontinents. During Precambrian and Palaeozoic times, ancient continental crusts making up the ancient supercontinents had already undergone a considerable amount of stretching and distortion during changes to both surface area and surface curvature. This stretching process originated from early-Archaean times where, once the ancient crust had cooled and stabilised, it started to crack and fracture during onset of Earth expansion. These fractures were focussed within a global network of crustal weakness and were also the loci for on-going intrusion of volcanic lava and formation of a network of low-lying sedimentary basins. On-going stretching and extension of the crust was then primarily focussed within this network of crustal weakness.

As will become apparent in the following figures, the modern continents often have a nucleus of most ancient crusts which are surrounded in turn by younger sedimentary basins or orogens. The orogens generally comprise older sedimentary basins that have been heated and folded. This global network of ancient sedimentary basins is analogous to the modern oceans whereby the network of crustal weakness contained within these basins represents the precursor to the modern mid-ocean-ridge spreading zones. Once rupture and separation of the supercontinents began, part of these sedimentary basins then remained attached to one continent and other parts remained attached to adjoining continents.

When describing the development and subsequent history of these modern continents in this and subsequent chapters, geographical orientations relative to the ancient equator will be given in lower case, for example, the long axis of Australia was orientated **north-south** throughout the Precambrian—as distinct from its current **east-west** orientation. In contrast, current geographical descriptors and names of continents relative to the present-day equator will be given in upper case, for example **North** America.

8.1 Australia

The Australian continent, shown in Figure 8.1, has large areas of ancient crusts preserved throughout what is now Central, Northern, and Western Australia. The rest of present-day Australia is made up of relatively young sedimentary rocks which were originally linked to similar sedimentary basins now located in China and North and South America. The older parts of the Australian continent had their beginnings as part of the ancient Archaea supercontinental crustal assemblage. This ancient assemblage later extended in surface area as new sedimentary basins opened and extended around its margins.

Throughout the various Precambrian and Palaeozoic Eras most of ancient Australia was located in the northern hemisphere and the long axis of the primitive Australian continent was orientated north-south relative to the ancient equator. Once the ancient Pangaea supercontinent started to break-up and the modern Pacific Ocean commenced opening during the Mesozoic Era, the newly formed Australian continent then rotated counter clockwise to its present east-west orientation and migrated south into its present location in the southern hemisphere.

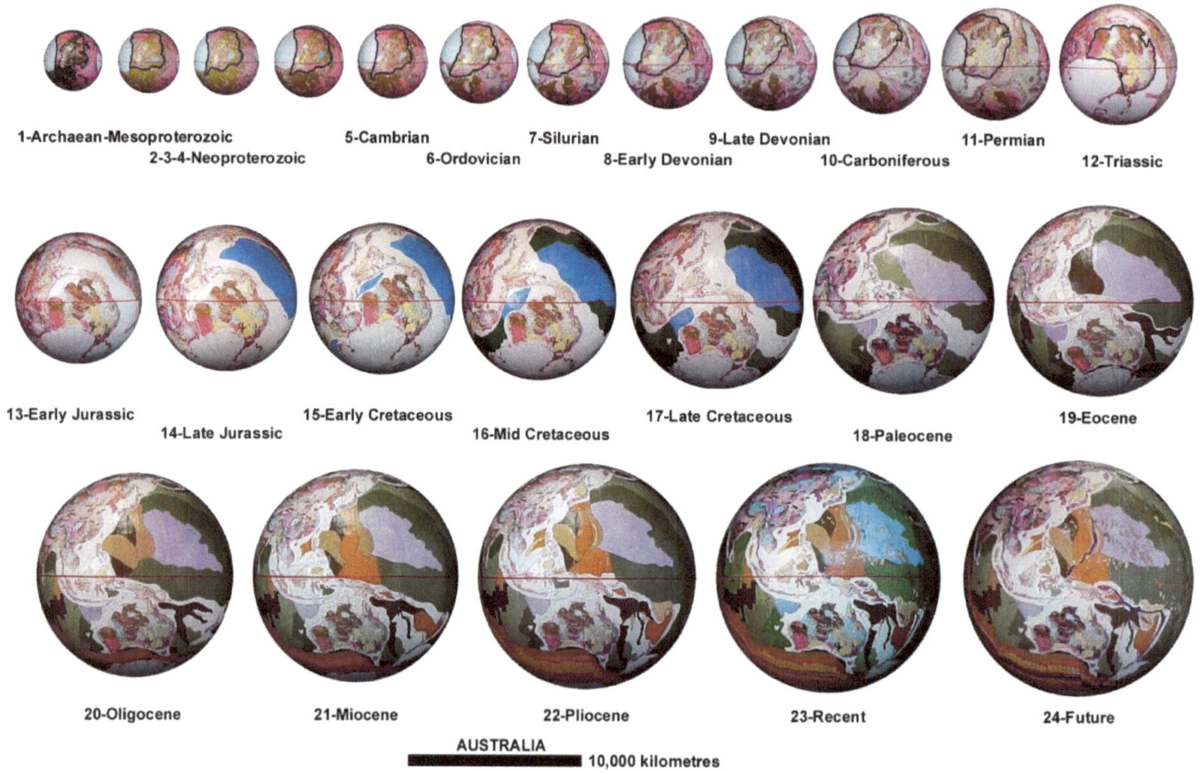

Figure 8.1 The continental crustal development of Australia. The outline of the pre-Jurassic Australian crust is shown as a black line. During the Precambrian to late-Palaeozoic Eras Australia was orientated north-south prior to rotating counter clockwise to the present east-west orientation. This occurred during opening of the modern Pacific Ocean. The horizontal red line represents the location of the ancient equator.

During these ancient times the Australian crusts abutted directly against crusts from primitive China to the north, Canada and North America to the east, South America to the south, and East Antarctica to the west. At that time ancient Australia was located within mid- to high-northern latitudes relative to the ancient equator and the ancient equator passed through what is now Central and Northern Australia.

During this pre-break-up supercontinent time the Proterozoic sedimentary basins of Northern and Central Australia formed part of an extensive global network of sedimentary basins. These basins extended north into the Proterozoic basins of Alaska, Canada, Northern Russia, and Asia, and to the east and south these basins were also linked to the Proterozoic basins of North America, Central America, and South America. Deposition of sediments within these ancient Australian sedimentary basins was most active to the south—within what is now Eastern Australia—and this deposition extended into adjoining regions in primitive New Zealand, South America, North America, and Antarctica. Break-up of this extensive sedimentary basin occurred during Permian times during initial opening of the South Pacific Ocean. Remnants of this basin are now preserved in East Australia, New Zealand, Central and South America, North America, and Antarctica

Crustal movement during this time was accompanied by ancient mountain building events, mainly within the network of sedimentary basins in what is now Central and Eastern Australia. Remnants of these same mountain building events are also preserved as the Andean mountain events of South America, the Appalachian and Grenville Mountain events of Eastern North America, and the Cordilleran mountain event of Western North America and Canada. Events in Central and Northern Australia were also associated with periods of crustal movement and jostling of the various ancient crusts located between Australia and North America. This jostling occurred as each of the ancient crusts adjusted for changing surface curvature. The mechanism for this jostling process will be discussed in later chapters.

Continental crustal rupture to form the modern Australian continent first commenced during Permian times in areas located to the northeast of Australia, relative to the ancient equator, adjacent to what is now the Pilbara and Kimberley regions of Western Australia. Opening also occurred in the south, adjacent to Eastern Australia, separating Australia from New Zealand. The outline of the modern Australian continent then began to take shape in these areas once the North and South Pacific and similarly the Indian Ocean began to open.

After slowly opening during the Permian and Triassic Periods, the early North and South Pacific Oceans then started to rapidly open, separating Australia from North and South America, while initially retaining a brief land link between Queensland in East Australia and California in North America. These previously separate North and South Pacific Oceans then merged during the Jurassic Period, finally separating Australia and New Zealand from North and South America.

Similarly, the Indian Ocean commenced opening during the Jurassic Period and was located adjacent to what is now Northwest Australia. As a result, Australia separated from China and South East Asia as the Indian Ocean continued to open and extend southwards along the west coast of Western Australia. At that time lands connecting Australia with adjoining continents remained attached to South East Asia to the north and East

Antarctica to the south allowing plant and animal species migration between each of these continents.

Rifting between Australia and East Antarctica commenced during the Paleocene—about 65 million years ago—during opening of the Southern Ocean. Australia and Antarctica are continuing to drift apart to the present-day leaving both continents as separate island continents. During this rifting process, Australia migrated south from mid-northern latitudes crossing the ancient equator into its present-day mid-southern latitude location. Bear in mind that migration is apparent and it is the opening of the oceans that causes this apparent migration.

Rocks exposed throughout Australia now reflect this geographic migration history, with extensive coral reef deposits located along the full length of Eastern Australia reflecting its original equatorial location throughout the Palaeozoic Era. This was followed by a prolonged period of tropical weathering as Australia rotated and crossed the equator, which is marked by extensive deposits of *laterite* rocks—rocks that have undergone deep chemical weathering from tropical rains—throughout present-day West and Northeast Australia. This was then followed by a progressive drying and *desertification* of the landscape to the present-day as Australia continued to migrate south away from the equator. Once the Southern Ocean began to open, Australia then remained geographically isolated from its neighbouring continents and has continued to progressively migrate further south into low- and mid-southern latitudes.

This southern migration of Australia is at odds with Plate Tectonic studies whereby Australia is instead said to be migrating north, out of more temperate to polar climate zones, and is said to be colliding with South East Asia. This interpretation is based primarily on palaeomagnetic studies—more on this in later chapters—and this interpretation is in fact contrary to what the rock-record shows. Irrespective of what Plate Tectonics insists, coral reef deposits and tropical weathering simply cannot occur in high polar latitudes; they are diagnostic of equatorial latitudes.

8.2 Africa and Arabia

Africa and Arabia contain extensive areas of ancient Precambrian crusts. Both originated from the Archaea supercontinental crustal assemblage located adjacent to the Archaean and Proterozoic regions of primitive South America, North America, Central Europe, Scandinavia, India, and East Antarctica. This early continental crustal configuration, shown in Figure 8.2, has since remained spatially intact throughout Earth history relative to each of the surrounding continents.

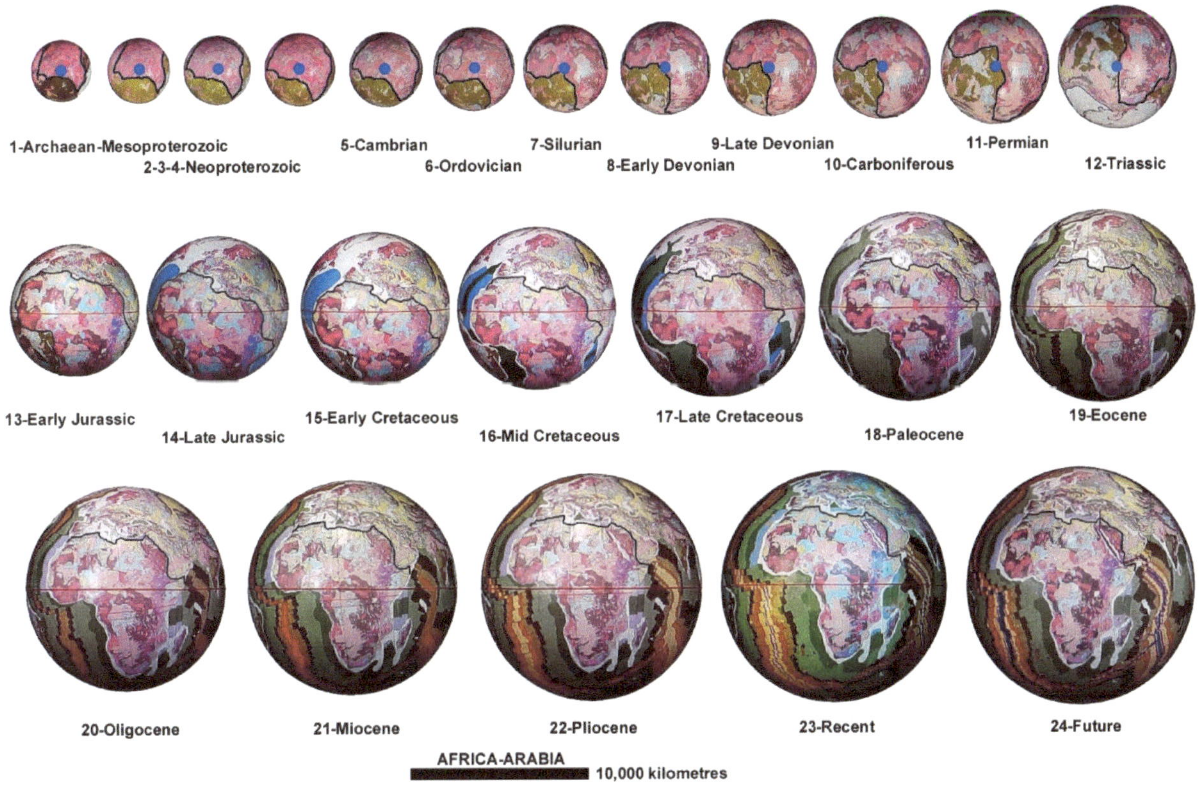

Figure 8.2 The continental crustal development of Africa and Arabia. In each of the Precambrian and Palaeozoic models North Africa is shown located south of the ancient equator and the South Pole is located in west Central Africa. The Archaean to Triassic models are centred on the South Pole (blue dots). The horizontal red line represents the location of the ancient equator and the black lines represent the African continental crustal outlines.

Throughout Precambrian and Palaeozoic supercontinent times the ancient South Pole was located within what is now Central West Africa and during that time a south-polar ice-sheet extended from there into Arabia and South America. Once crustal break-up commenced, this ancient South Pole is shown to have made an apparent migration south along the present West African coastline during the Mesozoic Era as the African continent migrated north.

The crustal development of primitive Africa and Arabia during Precambrian and Palaeozoic supercontinent times involved extensive continental crustal stretching and formation of sedimentary basins. Each of these areas was, in turn, accompanied by crustal jostling between each of the South, East, and West African cratons during changing Earth surface curvature. Throughout North Africa, crustal stretching and deposition of sediments continued into the Palaeozoic and later eras in conjunction with similar crustal stretching and jostling within what are now the Mediterranean, Central European, and Asian Tethys regions.

Precambrian and Palaeozoic crustal movements and mountain building occurred along what is now the present East African coast, associated with similar events in primitive Madagascar and India, as well as along the West African coast associated with the Grenville, Appalachian, and Hercynide Mountain events in Eastern North America and Europe. Remnants of events associated with the younger Alpine and Himalaya Mountain building episodes also occur within what is now North Africa.

Break-up of the ancient Pangaea supercontinent to form the modern African and Arabian continents was first initiated between West Africa and North America during the Permian and Triassic Periods during opening of the North Atlantic Ocean. Rifting and separation between South America and Africa, as well as between Africa, Antarctica, Madagascar, and India, commenced during the Jurassic adjacent to what is now South Africa. This rifting continued north, within the opening South Atlantic Ocean, to eventually join with the North Atlantic Ocean. Similarly, rifting continued north along the East coast of Africa during opening of the Indian Ocean which eventually gave rise to the modern African continental outline.

Rifting and separation of Africa from Arabia, and similarly Africa from the Mediterranean and Middle East regions, commenced during the Cretaceous Period during opening of the Mediterranean Sea and this rifting is continuing to the present-day within the opening Red Sea.

During the Mesozoic and Cenozoic Eras, Africa and Arabia continued to slowly migrate north, relative to the South Pole, in conjunction with opening of the Atlantic and Indian Oceans. From its central West Africa location throughout Precambrian to Palaeozoic times, the ancient South Pole is shown to have made an apparent migration south—due to the northward movement of Africa away from the southern polar region toward to equatorial region—along the west coast of South Africa. It then crossed the newly opening South Atlantic Ocean during the Mesozoic Era and continued onto the Antarctic continent during the later Cenozoic Era. This apparent polar migration severely disrupted existing plant and animal species development within Africa as the polar climate zone also experienced an apparent migration south across the land surface.

Once the ancient South Pole crossed the Atlantic Ocean, much of Africa and Arabia maintained a centrally-located equatorial position through to the present-day with the ancient equator experiencing a slow apparent migration from a North African position to its current Central African position.

It is interesting to note that on an Expansion Tectonic Earth a direct geographical connection across to Europe and Asia is maintained throughout Earth history and only relatively recently are the continents shown to be separating and rifting along what is now the Mediterranean and Red Seas. This, of course, is again at odds with Plate Tectonic assemblages where Africa and Arabia are said to be migrating north and colliding with Europe and Asia to form the Alpine Mountain belt. This conventional migration and collision, however, cannot explain the opening of the Mediterranean Sea during the late-Cretaceous times without invoking subduction of pre-existing Tethyan crusts, nor can it explain the present opening of the Red Sea and rifting between Arabia and Africa.

8.3 Antarctica

The continental development of both East and West Antarctica is speculative because of the masking effect of the present-day ice-sheet coverage. This ice cover, as well as the ice-sheet covering Greenland, is part of the Bedrock Map of the World base map used to construct the small Earth models and could not be removed during model construction. The Antarctic icecap is known to be about 33 million years old, representing a considerable part of the Cenozoic Era. Reconstructions and limited published field evidence also indicates that the East Antarctica continent comprises mainly Precambrian crusts and this crust has remained relatively intact throughout Earth history.

The continental reconstruction of Antarctica is shown in Figure 8.3. This figure shows that, unlike present-day East Antarctica, during the Precambrian Eras West Antarctica was made up of small remnants of Proterozoic crusts. The reconstructions also show that during Precambrian and early-Palaeozoic times the primitive crusts making up both East and West Antarctica straddled the ancient equator. These continental crusts were located adjacent to primitive Australia, South America, South Africa, India, and Central and Southeast Asian crusts.

Ancient Proterozoic and Palaeozoic sedimentary basins located in what are now India, Central and Southeast Asia, Australia, and South America surrounded the ancient Antarctican continent and extended into West and East Antarctica beneath the present ice-sheet. At that time mountainous regions in West and East Antarctica also formed extensions of similar mountainous regions located in Eastern Australia, as well as the Andean regions of South America.

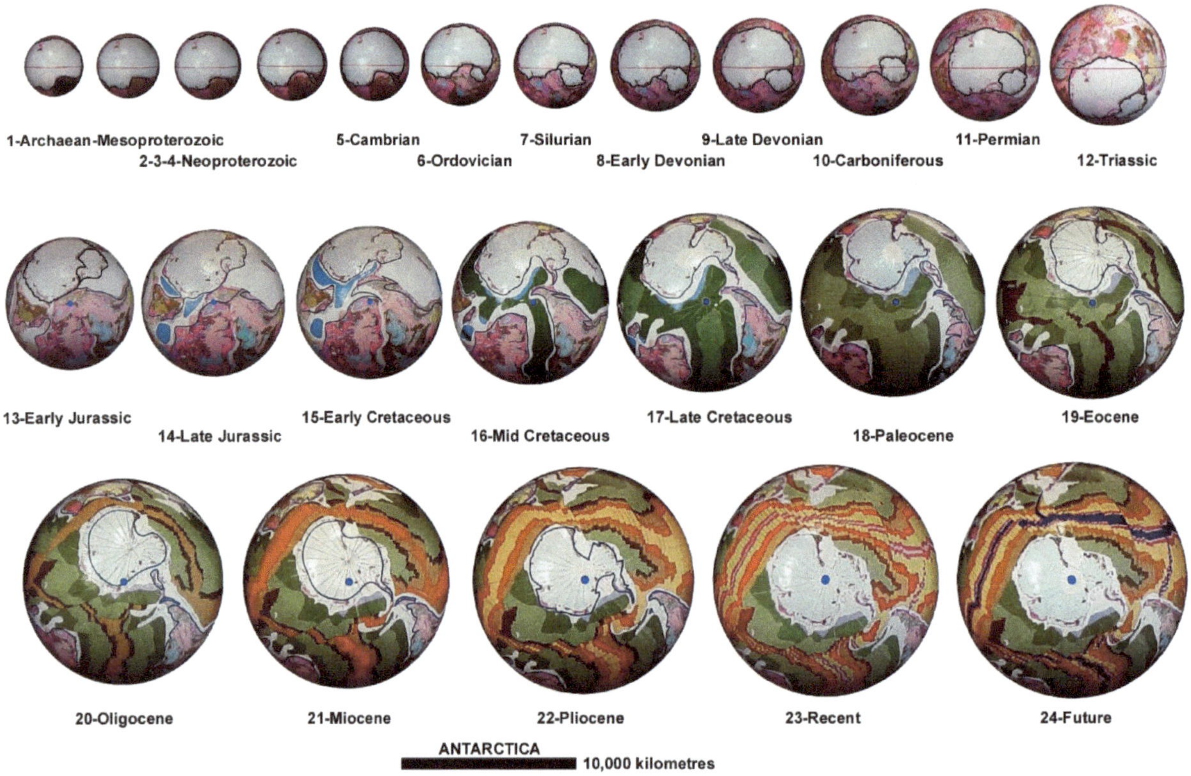

Figure 8.3 The continental crustal development of East and West Antarctica (shown as pale blue—the present-day south-polar ice shelf). During the Archaean to late-Palaeozoic Eras Antarctica straddled the ancient equator. The continent then rapidly migrated south, as the modern oceans opened, to its present location at the South Pole. The horizontal red line represents the location of the ancient equator and blue lines represent the ancient coastlines. Note: models 13 to 24 are centred on the South Pole (shown as blue dots).

Break-up of the former Pangaea supercontinent to form the modern Antarctican continent commenced during early-Permian times during a period of crustal rupture and opening of the South Pacific Ocean—the ancient South Panthalassa Sea. By the Late Jurassic, opening of the Indian and South Atlantic Oceans had also commenced leaving a land connection between Australia and East Antarctica and similarly between South America and West Antarctica. Final break-up, rifting, and separation of modern Antarctica from Australia and East Antarctica commenced during the Paleocene, some 65 million years ago.

On an Expansion Tectonic small Earth, the development of West Antarctica relative to East Antarctica was closely related to opening of the

South Pacific Ocean. By the Late Cretaceous West Antarctica had separated from East Antarctica and had begun to rotate clockwise, relative to East Antarctica, as the South Pacific Ocean continued to widen and open. During that time the West Antarctican Peninsular remained joined to the southern South American Peninsular. This peninsular continued to remain joined until separation and rifting between the two continents began during the Miocene leaving modern East and West Antarctica as separate and isolated continents through to the present-day.

Throughout Precambrian and Palaeozoic times the Antarctican continent straddled the ancient equator. Once break-up was initiated the combined East and West Antarctican continent is then shown to have commenced an apparent migration south, relative to the ancient South Pole. It then continued migrating further south to its present south-polar location throughout Mesozoic and Cenozoic times, establishing and preserving the present south-polar ice-sheet.

This reconstruction and migration history shows that Antarctica evolved throughout most of its history within a warm tropical environment prior to migrating south into high southern and then into south polar-regions. A close look at the various small Earth models also shows that Antarctica is currently not stationary over the South Pole, as convention suggests, but is continuing to migrate slowly across the south-polar region. It is also important to note that seafloor mapping shows that the surrounding Southern Ocean contains a single, continuous mid-ocean-rift zone. All of the surrounding continents are, in fact, migrating north away from Antarctica as new seafloor crust continues to be intruded along this mid-ocean-rift zone. This is a fatal flaw in Plate Tectonic reconstructions, whereby a suitable explanation cannot be given to explain why there are no subduction zones available to absorb this huge amount of Southern Ocean seafloor crustal spreading surrounding Antarctica.

8.4 Europe, Russia and Asia

The European and Asian continent, inclusive of Russia, is a vast area and constitutes the largest land mass on Earth today. The continental crustal development of Europe and Asia is shown to be a complex and prolonged history of Precambrian to recent crustal stretching and extension, sedimentary basin formation, mountain building, and intrusion of magmatic rocks. Much of the present European and Asian continent now represents the uplifted and exposed seafloor of the former Tethys Sea region, with relatively minor areas of more ancient crusts.

In Plate Tectonic studies a large Tethys Ocean is depicted as being located between the fragmented remains of the present European and Asian continent and is inferred to have covered many small crustal plates, as well as Cretaceous *island-arcs* and small continents. Parts of Central and Eastern Europe are also said to have been covered by a northern branch of the Tethys Ocean. This branch was separated from the Tethys by the formation of the Alps, Carpathians, Dinarides, Taurus, and Elburz Mountains, before it gradually disappeared during the late Miocene Period, becoming an isolated inland sea.

In contrast, on an Expansion Tectonic Earth, during Archaean and early-Proterozoic times the European and Asian continent was largely made up of a network of Precambrian sedimentary basins—the early Tethys Sea basin in particular. It also included small fragments of Archaean crusts that are now dispersed throughout Europe, as well as the larger ancient Mongolian and Chinese crusts. These crusts were once united as part of the ancient Archaea supercontinent. After fragmenting during the Precambrian, these crusts now form small remnants embedded within the much larger Tethys sedimentary basins.

Throughout the Precambrian and into the following Palaeozoic times, the Northern Asian region of the ancient European and Asian Tethys Sea was centred over the ancient North Pole. Similarly the European region, in what is now the Mediterranean region, was centred on the ancient equator. East Antarctica was located along the eastern margin of this large continent, Australia was located to the north, Canada to the northwest, Greenland to the west, Scandinavia and Arabia to the southwest and India was located to the south (Figure 8.4).

During Precambrian and Palaeozoic times, sediments were being eroded from each of the exposed surrounding continental land surfaces and deposited within the extensive Tethys Sea. This north-south orientated Tethys Sea basin in turn formed part of a much more extensive global network of continental seas. In this context, the ancient Tethys Sea also included Precambrian sedimentary basins that are now exposed in Central Australia and India and further afield also included ancient basins located within South America.

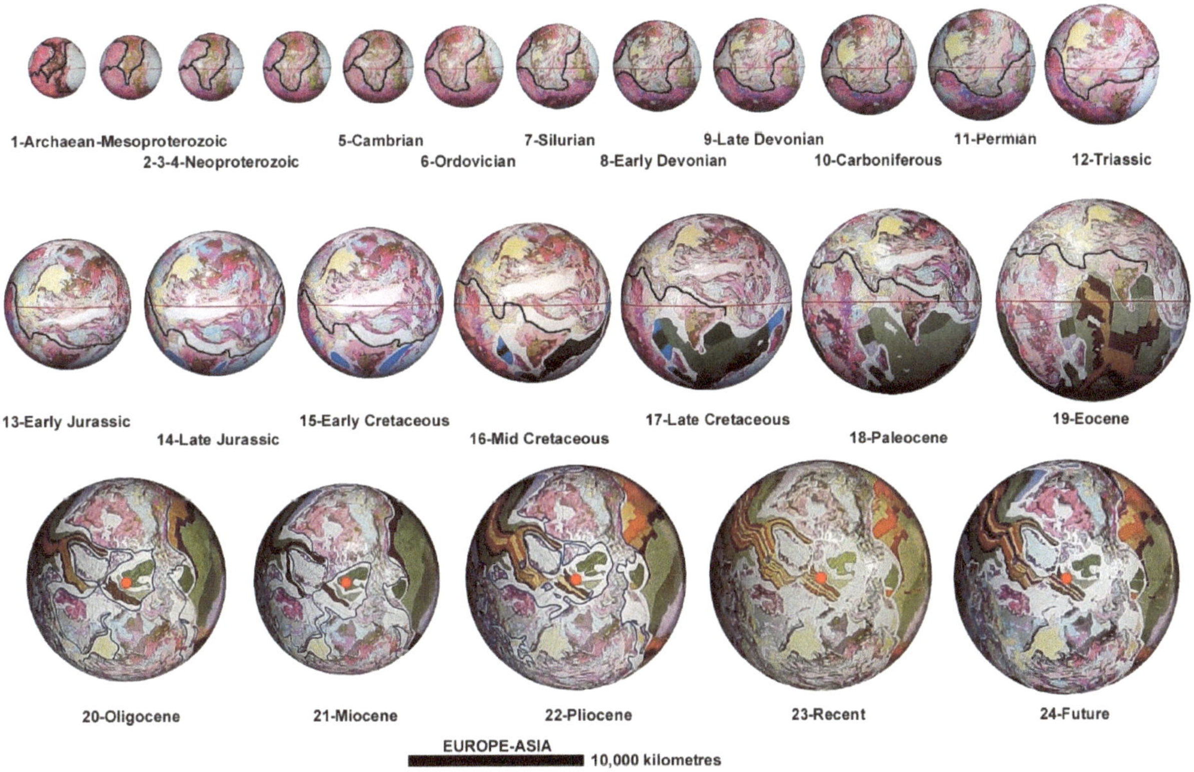

Figure 8.4 The continental crustal development of Europe and Asia. Models 20 to 24 are centred on the North Pole (shown as red dots). The horizontal red line represents the location of the ancient equator, blue lines represent the ancient coastlines, and the black line represents the outlines of the ancient European and Asian continent.

Break-up of the Pangaea supercontinental crust first began in the ancient Arctic, North Atlantic, and North Pacific Oceans during the Permian which then initiated break-up and separation of the combined European and Asian continent. During that time, as each of the surrounding modern continents began to drift apart and the modern oceans began to open, the existing Tethys continental sea was then disrupted and began to progressively drain. During the Mesozoic, the ancient Tethys Sea gradually became exposed as dry lands as the waters slowly drained into the opening modern oceans. Deposition of sediments within the ancient Tethys Sea region was then disrupted and deposition of eroded sediments shifted into the newly formed marine basins, now located around the margins of the modern continents.

During that time the on-going development of Europe, Asia, and Russia was strongly influenced by ever changing sea levels and changing coastlines. The Tethys Sea was completely drained during the Cenozoic, exposing Europe, Asia, and Russia as the elevated continent it is today. Because of its large size, Europe and Asia straddled many climate zones, ranging from north-polar to equatorial, with parts extending into low southern latitudes. Today, the entire European and Asian continent is located in the northern hemisphere but still extends from the equator through to the North Pole.

Throughout Earth history, crustal stretching and mountain building associated with changes in Earth surface curvature played an important role in shaping the European and Asian continent. Precambrian and Palaeozoic events in Western Europe were associated with the ancient Grenville, Appalachian, and Hercynide Mountain building events now preserved within Eastern North America and Scandinavia. Similarly, during the Mesozoic and Cenozoic Eras, the Alpine and Himalaya Mountain belts were formed during opening of the Mediterranean Sea and were accompanied by renewed stretching and crustal extension between Europe and Asia relative to Africa.

This Expansion Tectonic continental crustal history differs markedly from Plate Tectonic reconstructions. On an Expansion Tectonic Earth fragmentation of former supercontinents and inclusion of an extensive ancient Tethys Ocean is not necessary in order to close off the North Atlantic Ocean to conform to seafloor bedrock mapping data.

8.5 India

The Indian continent is traditionally shown on Plate Tectonic reconstructions to be an island continent migrating north during the Mesozoic Era, moving across a vast pre-existing Indian Ocean until it collided with Asia during the Cenozoic. Collision with Asia is then thought to have resulted in formation of the Himalaya Mountain belt.

In contrast, the crustal development of India on an Expansion Tectonic Earth is shown in Figure 8.5. During Precambrian times, the ancient Indian continent initially formed a southern extension of the European and Asian Tethys Sea basin. During that time, and extending into later Palaeozoic times, India was located adjacent to East Antarctica to the northeast, Madagascar and South Africa to the southwest, and Arabia to the west. The ancient crust making up the present Indian continent was originally located within mid-southern to equatorial latitudes throughout the Precambrian and Palaeozoic times.

During the Mesozoic, India briefly migrated into mid-southern latitudes before returning to equatorial and low-northern latitudes during the later Cenozoic Era. This apparent migration of India was related to its proximity to the rapidly extending European and Asian Tethys Sea basin. It was also influenced by the migration of adjoining continents away from the ancient South Pole, which was then located in central West Africa, and to the newly opening Indian Ocean.

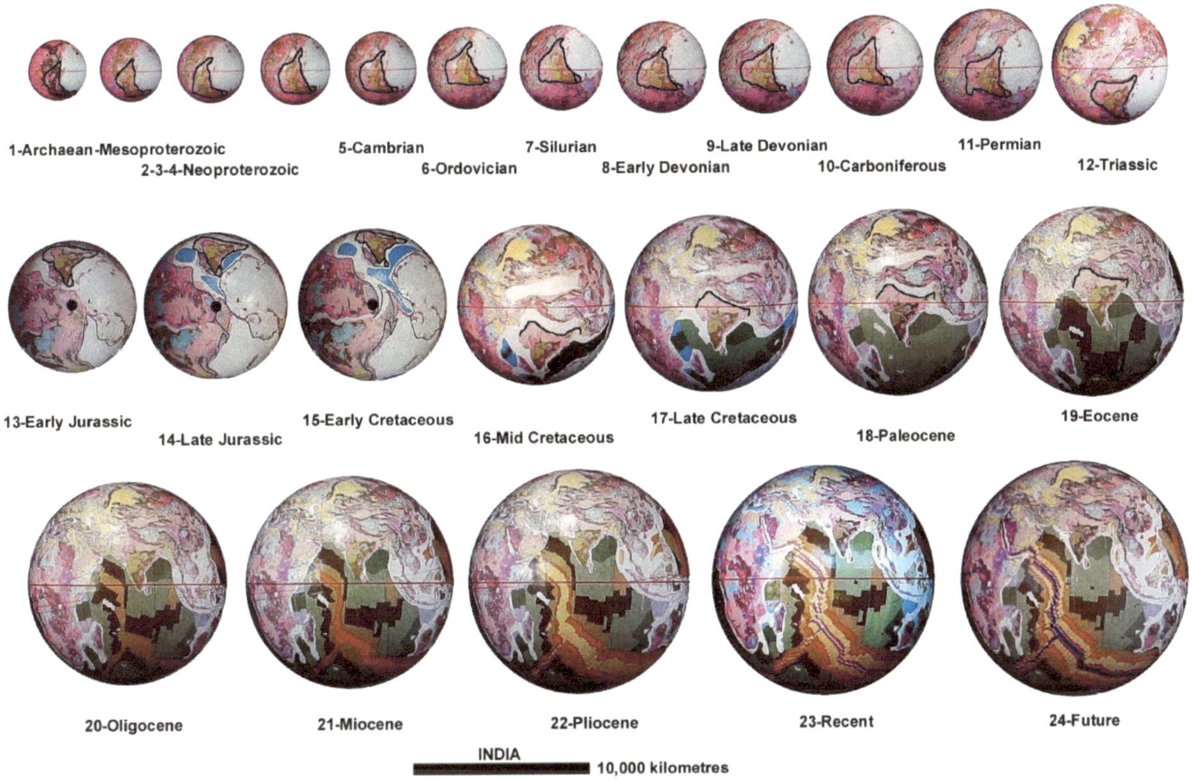

Figure 8.5 The continental crustal development of India. Each of the early-Jurassic to early-Cretaceous models are centred on the South Pole (shown as black dots). The horizontal red line represents the location of the ancient equator, blue lines represent the ancient coastlines, and the black line represents the Indian continental crustal outline.

Crustal extension between the ancient north and south Indian crustal regions occurred during the Proterozoic Era in conjunction with related crustal development in the Tethys region. This period of crustal extension continued into the Mesozoic Era. Crustal movement and mountain building in India was associated with crustal motion relative to Antarctica, Madagascar, and Africa, as well as a number of Palaeozoic to Cenozoic mountain building events along the northern Himalaya contact with Europe and Asia. In this context the Himalaya Mountain chain was intimately associated with changes to surface curvature focussed along this northern Himalaya contact.

Continental separation and rifting of India, Madagascar, and Sri Lanka from Antarctica and Africa commenced in the Jurassic during initial opening of the Indian Ocean. Madagascar and Sri Lanka then began drifting away from India during the early- to mid-Cretaceous, with Sri Lanka continuing to remain in close proximity with India.

In contrast to Plate Tectonic reconstructions, on an Expansion Tectonic Earth the Indian continent has remained geologically attached to the Asian continent throughout all Earth history. Because of the proximity of India to the European and Asian Tethys Sea region, India was geographically, but not geologically, isolated from Asia for much of that time by the presence of shallow continental seas. As the European and Asian Tethys Sea progressively drained during the Cenozoic Era, India and Asia were then fully exposed as one continuous continental plate with no requirement for a separate Indian sub-continent or collision event.

8.6 North America

The development of the North American continent is shown in Figure 8.6 to be intimately related to the ancient Archaea supercontinent crustal assemblage and its ultimate break-up to form the ancient Canadian, Greenland, and Scandinavian crusts. This cluster of ancient continental crusts is discussed together and is here referred to as the North American cluster. On Plate Tectonic reconstructions this cluster is referred to as the Laurentia and Baltica supercontinents and these have been extensively studied and referenced throughout North American literature.

On the Expansion Tectonic small Earth models shown in Figure 8.6 the ancient North American cluster of crustal fragments remained intact throughout Palaeozoic and Precambrian times. This cluster is shown to straddle the ancient equator, extending from mid-southern to high-northern latitudes. The ancient cluster was, in turn, assembled against the European and Asian Tethys region to the north, the Australian Proterozoic basins to the west, South America to the southwest, and West Africa to the south and southeast, relative to the ancient equator. Small Precambrian crustal fragments now located within Europe were also clustered adjacent to and southeast of Greenland and Scandinavia.

In effect, this ancient North American cluster formed a nucleus for surrounding crustal development. The cluster represented exposed elevated lands throughout these ancient times and supplied eroded sediments to surrounding sedimentary basins. The Precambrian and Palaeozoic development of the North American cluster then involved crustal extension and basin sedimentation within a surrounding network of continental sedimentary basins. These basins included links to the Tethys Sea basin to the southeast, as well as basins linked to what is now Russia, China, Australia, and South America to the north, west, and southwest.

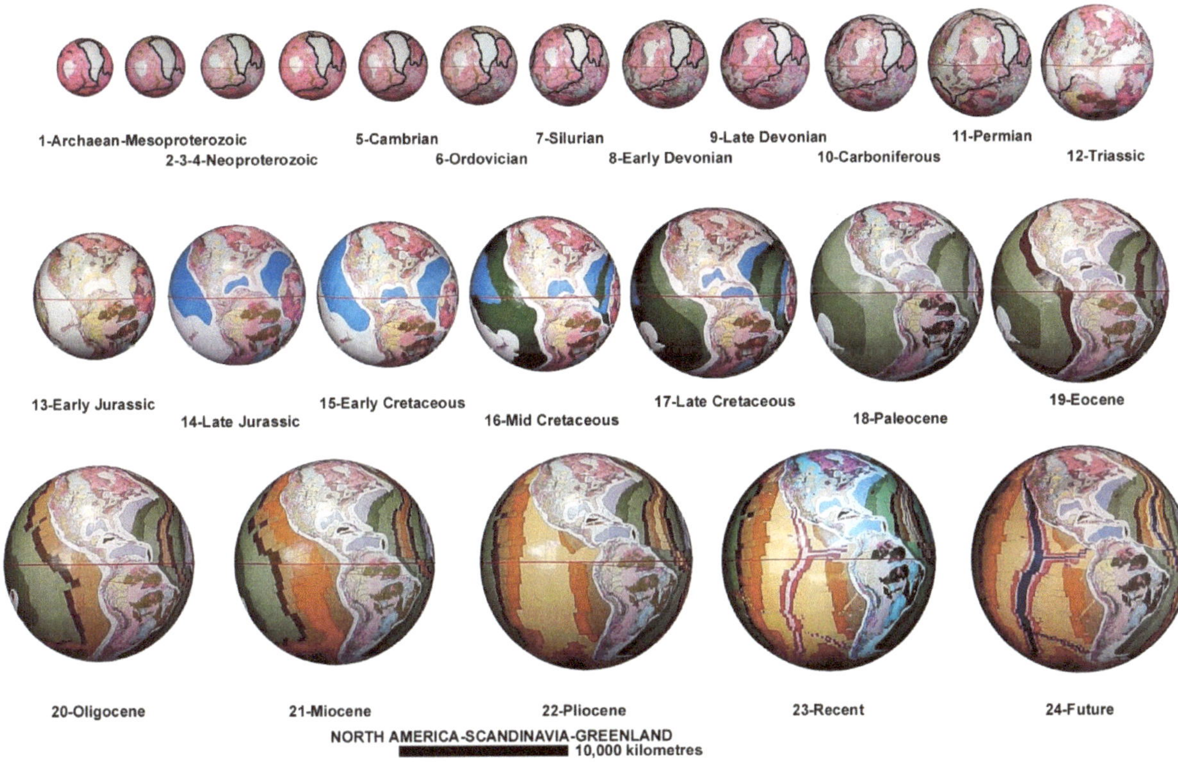

Figure 8.6 The continental crustal development of North America-Greenland-Scandinavia. The horizontal red line represents the location of the ancient equator and black lines represent the continental outlines.

This North American cluster remained intact throughout the Precambrian and Palaeozoic times until the Pangaea supercontinental break-up and opening of the North Atlantic Ocean commenced during the late-Permian Period. During opening of the North Atlantic Ocean, Scandinavia and the Baltic region were separated from North America and have since remained attached to Europe during further opening of the North Atlantic and Arctic Oceans.

Crustal jostling and mountain building events occurred during Precambrian and Palaeozoic times as a result of on-going changes to surface curvature. These events formed long linear mountain belts located around the margins of the North American cluster, forming the precursors to the Cordilleran, Grenville, and Hercynide Mountain belts.

Break-up and fragmentation of the ancient Pangaea supercontinent began during the late-Permian and by the Triassic the early Arctic, North Atlantic, and North Pacific Oceans had also commenced opening. This break-up and opening of the modern oceans then effectively defined the modern North American continental outline. By the early-Jurassic, break-up had continued to extend into the Arctic and North Atlantic Oceans and also into the Caribbean and Labrador Seas. Greenland was then separated from Canada and has remained in close proximity. Similarly, South America began an apparent drift away from North America. Fragmentation of the Northern Canadian Islands also occurred during the Jurassic Period, which was intimately related to rifting occurring between Canada and Greenland. Further rifting within the Northern Canadian region has continued to the present-day.

During break-up of Pangaea and opening of the Pacific, Atlantic, and Arctic Oceans, each of the established Precambrian, Palaeozoic, and Mesozoic mountain belts were then fragmented. Remnants of these mountain belts are now separated as far away as Australia, Africa, South America, Russia, and Europe. Much of the Grenville and Appalachian fold-mountain belts remained attached to Eastern North America and the Cordilleran Mountain belt remained attached to Western North America. The Hercynides remained attached to Europe, the Caledonides to Scandinavia, and the Mauritanides to West Africa. The northern extension of the Cordilleran Mountain belt continued via Alaska into Asia and continued as the Andean Mountain belt into South America. Fragments of this Andean belt also include the New England fold belt of Eastern Australia and remnants can also be seen in New Zealand.

Throughout late-Jurassic to Miocene times, north-south stretching of the Cordilleran and Andean Mountain belt maintained a continuous continental link extending from South America through to North America to Siberia. The Siberian connection was then severed during the Pliocene Epoch during opening of the Bering Strait and the Central American link still remains attached today.

From its original equatorial location, the North American continent and continental cluster then slowly rotated clockwise as a result of crustal break-up and opening of the Atlantic, Pacific, and Arctic Oceans. Each of the North American, Scandinavian, and Baltic continents has since migrated north into mid- and high-northern latitudes, relative to the ancient North Pole.

8.7 South America

The development of the South American continent is shown in Figure 8.7. This continent is closely associated with the development of Africa and the South America to Africa assemblage has long been recognised in Plate Tectonic reconstructions. The closing of the Atlantic Ocean and assemblage of the American and African plates forms the basis of the Gondwana supercontinental assemblage, as well as the basis for both Continental Drift and Plate Tectonic theory.

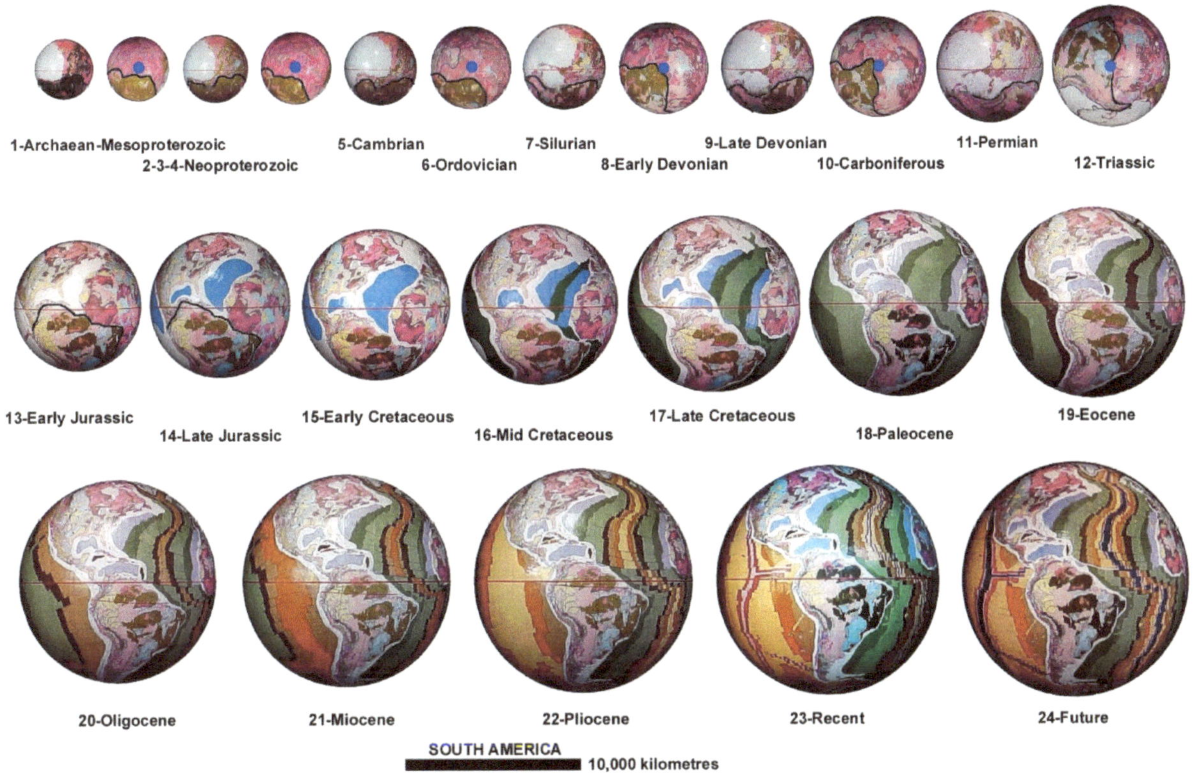

Figure 8.7 The continental crustal development of South America. The horizontal red line represents the location of the ancient equator, blue dots represent the South Pole, and black lines represent the South American continental outline.

In Expansion Tectonic studies the South American crusts originally formed part of the ancient Archaea supercontinent, which in turn formed part of an extensive network of ancient Precambrian sedimentary basins. East Antarctica and Precambrian remnants of West Antarctica and New Zealand were located to the northwest, Australia was located to the north, North

America to the northeast, and Africa to the east, south, and west. Subsequent development of South America involved an extended period of crustal extension and fragmentation of the ancient crusts during the Proterozoic Eon and Palaeozoic Era. This development occurred in conjunction with similar Pan-African events in South and West Africa and was also associated with events in ancient Australia and New Zealand.

During Precambrian and Palaeozoic times, the ancient South American continent extended from low equatorial to high south polar latitudes. During the late-Palaeozoic to mid-Cretaceous, as the Atlantic and Indian Oceans progressively opened, the South American continent then slowly migrated north, in conjunction with Africa, relative to the ancient South Pole. From its Precambrian and Palaeozoic southern hemisphere location, South America then rotated clockwise, in sympathy with opening of the Atlantic Ocean, and migrated north to straddle the present-day equator. During the Palaeozoic, crustal extension and basin formation also gave rise to extensive deposition of basin sediments located between what are now South America, Antarctica, Australia, New Zealand, and North America.

Continental break-up between South America, New Zealand, Australia, and Antarctica commenced during late-Permian to Triassic times. This occurred in conjunction with opening of the South Pacific Ocean, with New Zealand retaining a brief link to Mexico. During the Jurassic, opening of the South Atlantic Ocean commenced in the south and this opening then migrated north to merge with the opening North Atlantic Ocean. During this time, South America separated from North America during opening of the North Atlantic Ocean and Caribbean Sea. A land connection with North America still remains along the South American and Central American peninsulas. A land connection between the West Antarctican Peninsular and southern South America also remained until final separation during the Miocene during opening of the Southern Ocean.

Mountains developed during the Mesozoic and Cenozoic as long linear belts along the West Coast of South America. This occurred in conjunction with further opening of the South Pacific Ocean and formed a Southern extension of the Cordilleran event in North America. Fragments of these mountain events also occur in New Zealand and Eastern Australia.

This crustal history of South America is in strong contrast with Plate Tectonic assemblages, whereby opening of the South Atlantic Ocean is compensated for by subduction of the South Pacific plate beneath the west coast of South America to form the Andean Mountain belt. As has been outlined previously, closing of the Pacific plates by subduction against the

Americas is not required on an Expansion Tectonic Earth. Instead, reconstruction of each of the oceans is shown to adhere with the seafloor crustal evidence preserved on the seafloors.

9 Into the Future

"All inferences from experience suppose...that the future will resemble the past." David Hume, 1777
"The result, therefore, of our present enquiry is that we find no vestige of a beginning, and...no prospect of an end." Hutton, 1788

In cosmological thinking it is a commonly held belief that the eventual demise of planet Earth will involve a fiery end, related to a steady expansion and decay of the Sun in the far distant future. As the Sun expands and decays scientists tell us that it will eventually envelop and consume each of the planets in turn to form a red giant star. This is certainly nothing for any of us, or future mankind to worry about as all life forms will have long since become extinct or evolved into something more adaptive by then.

From the continental and seafloor descriptions of Earth's past and present geological history discussed in the previous chapters, the concept of an Expansion Tectonic Earth can now be projected forward in time. From this, an account of what may eventually happen to not only the modern continents and oceans but also the Earth itself will be seen. On Expansion Tectonic small Earth models this projection into the future is readily achieved by simply calculating a future Earth radius at any moment in time and extrapolating opening of the mid-ocean-rifting throughout all of the oceans.

When the rate of increase in Earth radius established from the small Earth models is projected forward in time, both Earth surface area and radius is shown to increase to the size of the planets Jupiter and Saturn by about 500 million years into the future. For an Expansion Tectonic Earth, it is envisaged that one of two things may happen to the Earth during that time. Firstly, the Earth may fragment and disintegrate to form a second asteroid belt or, secondly, it may simply continue growing to become another giant gaseous planet. How this relates to an expansion of the Sun is uncertain, but again this does not concern the wellbeing of present mankind.

Because the present Earth is essentially a wet planet—because it has an atmosphere and hydrosphere—when compared to the dry inner rocky planets of the Solar System, to me the giant planet scenario seems the most likely outcome. As the Earth's mantle continues to swell, it is envisaged that more entrapped fluid and gas will continue to be expelled from the mantle to form a dense gaseous atmosphere in the far distant future. This may then extend in time to form planetary ring structures as lighter gases are progressively lost into space.

A reconstruction of an Expansion Tectonic Earth at five million years into the future is shown in Figure 9.1—which coincidently still looks very similar to the present Earth. Reconstructing crustal plates on this future small Earth model is readily done by simply calculating the predicted Earth radius from the established radius formula and then adding new seafloor crust along each of the mid-ocean-ridge axes—precisely as is happening today. Apart from the increased distances between the various continents and subtle changes to the coastlines, on this model the distribution of continents and oceans on the future Earth is shown to be essentially the same as it is today.

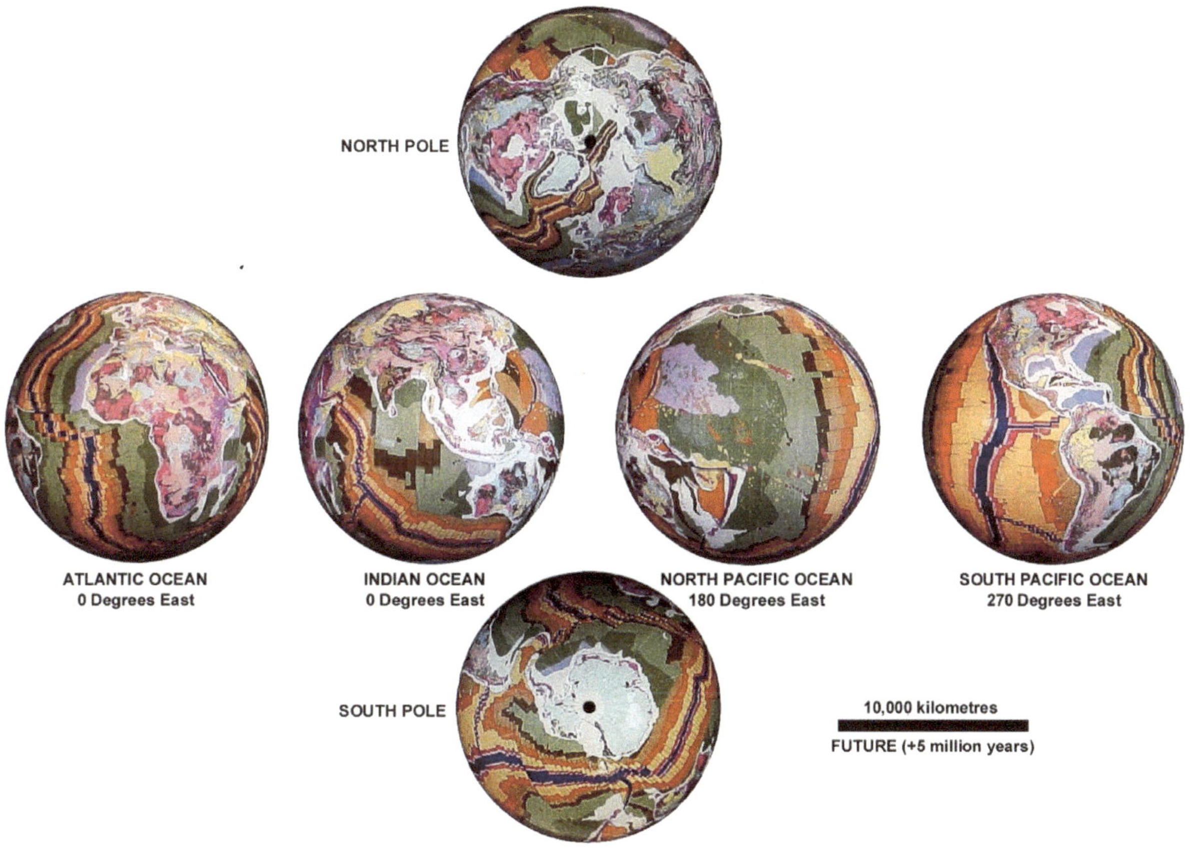

Figure 9.1 The Expansion Tectonic Earth projected to 5 million years into the future. The dark blue spreading ridge represents spreading along each of the mid-ocean-ridges for the next 5 million years. The model shows an extension of all mid-ocean-ridge axes into seismically active areas such as Turkey, Japan, California, and New Zealand.

During this geologically short interval of time, it is calculated that Earth radius will increase by 107 kilometres to 102 percent of the present radius. The series of images in Figure 9.1 show that the increase in Earth radius to 5 million years into the future is consistent with a continued increase in surface area of each of the oceans and lengthening of each of the present-day mid-ocean-ridge spreading axes. This process of lengthening of the mid-ocean-ridge spreading axes is a direct result of the increase in circumference of the Earth during an increase in Earth radius. The mid-ocean-ridge lengthening process then represents an important mechanism for future crustal development during on-going Earth expansion.

On an Expansion Tectonic Earth, as the mid-ocean-ridge spreading axes lengthen and the ridges open they can be visualised as being huge propagating cracks in the seafloor—which is precisely what they are. As the cracks propagate and lengthen they also continue to break-up the continental and seafloor plates into ever smaller fragments. This is currently occurring within each of the major earthquake prone areas of the world today. In these earthquake prone areas the seismic and earthquake activity occurs as a result tensional cracking and breaking apart of the crusts, not due to collision as portrayed in Plate Tectonics. This break-up is also accompanied by intrusion of volcanic lavas, elevated heat flow, and expulsion of new water and gases from the mantle.

Lengthening of the East Pacific mid-ocean-ridge spreading axis on an Expansion Tectonic Earth is currently occurring as a northward extension of the spreading ridge passing through the Gulf of California. On an Expansion Tectonic Earth this gulf will eventually rift and separate the Californian Peninsula from North America to form an island. A northward extension of the Red Sea Rift zone through the Gulf of Aqaba and Dead Sea region into Turkey will eventually result in the rifting and separation of the Sinai Peninsula from Arabia. A northern extension of the Marianas spreading ridge is shown to be continuing towards Japan and a southern extension of the Tongan spreading ridge is also continuing through New Zealand.

Elongation of the mid-ocean-ridge spreading zones within these areas contrasts strongly with the Plate Tectonic requirement for plate convergence, continental collision, and subduction of vast areas of oceanic crust. On an Expansion Tectonic Earth, an increase in surface area accompanied by elongation of the mid-ocean-ridges is considered to better represent the break-up and separation of the continents and opening of the existing oceans when moving into the future. These mid-ocean-ridge opening processes, as projected well into the future, also continues to comply with the established seafloor bedrock geological mapping.

10 Summary of Part One

"If 50 million believe in a fallacy, it is still a fallacy." Carey, pers. comm. 1995

Until now, heavy reliance has been made on the bedrock geological mapping of the oceans and continents to both measure the ancient radii of the Earth and to establish a formula to determine an ancient Earth radius at any moment in time. This bedrock mapping and measured ancient radius data was then used to construct small Earth models extending from the Archaean Era, some 4,000 million years ago, to the present-day plus one model extended to 5 million years into the future.

Throughout the Triassic to present-day times it was found that the seafloor crustal plates reassemble together on each small Earth model with a better than 99 percent global plate-fit without the need to fragment continents, subduct large amounts of seafloor crusts beneath the continents, or invoke pre-existing crusts in order to maintain a constant surface area Earth. Similarly, when progressively older sediments and intruded magmatic rocks from continental sedimentary basins are returned to the ancient lands or mantle respectively, it was found that remnants of the most ancient Precambrian continental crusts also reassemble together with a high degree of global fit-together on ancient small Earth models.

Throughout this extensive small Earth modelling study, the history of each of the ancient supercontinents and seas was also discussed, as well as the modern continents and oceans, in the context of an Earth progressively increasing its surface area through time. In order for the Earth to increase its surface area, it was stressed that the relatively thin outer continental and seafloor crusts must stretch and distort as the surface curvature of the Earth progressively flattens throughout history. It is considered that this distortion of the crusts during change in surface curvature has given rise to all of the known geological and geographical features that are now familiar to us on Earth today.

A side-line to this small Earth modelling exercise was to test the validity of past models of an Expansion Tectonic Earth, as previously constructed and published by Ott Hilgenberg in 1933, Klaus Vogel in 1983, and numerous others. Seafloor mapping was not available to Hilgenberg; hence his reliance on a visual fit-together of continental crustal fragments was far from convincing. This contrasts with the models of Vogel who had access to, and was able to use, early versions of the seafloor mapping data. I personally visited Vogel's home in East Germany where together we compared and contrasted each other's models. In both cases the models of Vogel shown in Figure 1.4, and the author's models shown in Figure 4.1, both of which were constructed independently of each other, are essentially the same and differ only in fine graphical detail. Completion of the Geological Map of the World in 1990 is therefore seen as a very important milestone towards quantifying Expansion Tectonics. This bedrock geological mapping is now considered to be a valuable tool for anyone to accurately constrain plate assemblages on small Earth reconstructions, as well as a means to accurately measure the ancient Earth radius against time.

The conclusions of the various historical researchers regarding observations and impressions from their own modelling studies are worth briefly mentioning. Barnett in 1962 said, "*...it is difficult to believe that chance alone can explain this fitting together of the continental margins.*" Creer in 1965 said, "*...the fit of the continents on a smaller earth appeared to be too good to be due to coincidence and requires explaining,*" and Vogel was "*...continually amazed that the simplicity with which earth expansion answers so much of the earth's evolution has been so delayed in universal adoption.*" From their various works, these researchers concluded, "*...if all of the earth's continents were fitted together they would neatly envelope the earth with continental crust at approximately 55 to 60 percent of the present earth size.*" This led Hilgenberg to conclude "*Earth expansion resulted in the break-up and gradual dispersal of continents as they moved radially outwards during geological time.*" Similarly, Vogel suggested it was also "*...theoretically possible for the continents, without shelves, to fit together at approximately 40 percent of the present earth radius by considering that continental shelves were formed after the continental crust had fragmented.*" Vogel concluded from his modelling studies that, "*...the earth has expanded with time, from an early Pangaean configuration to the Recent, with continental separation caused by a radial expansion of the earth.*"

It is also worth mentioning again that during the 1950s, professor Carey, in researching and modelling the concept of continental drift, demonstrated "*...if all the continents were reassembled into a Pangaean configuration on a model representing the Earth's modern dimensions, the fit was reasonably precise at the centre of the reassembly and along the common margins of north-west Africa and the United States east coast embayment, but became progressively imperfect away from these areas.*" Carey concluded from this research that, "*...the fit of the Pangaean reassembly could be made much more precise in these areas if the diameter of the earth was smaller at the time of Pangaea.*"

It is unfortunate that since acceptance of Plate Tectonics as the ruling tectonic theory during the 1960s, none of this alternative research is considered, acknowledged, or probably even known about today. It is also unfortunate that the basic observations and conclusions of Vogel, Carey, and others continue to be totally ignored or simply discarded by mainstream science. Unbeknown to most Earth scientists today, these observations are in fact extremely important and valid in our modern understanding and perception of global tectonics. It is the little meme that is firmly embedded in the back of our minds—the constant Earth radius one—that is unfortunately still hindering geoscientific research today and the one that continues to get in the way of rational scientific thinking.

It is emphasised that, although Plate Tectonic reconstructions of individual regions on a constant sized Earth can achieve a high degree of fit-together, in most cases these reconstructions obscure the fact that crustal assemblage, development, and displacement in one region of the Earth globally affects all other areas. What this means is that problems of misfit in one area of the Earth on conventional reconstructions cannot, and must not, be conveniently transferred to an adjacent region and then simply ignored. Global tectonics is a holistic phenomenon, not a local phenomenon. Similarly, modern surface geological mapping also shows that the continents are not rigid plates, as is often portrayed in conventional tectonic studies, but are instead a complex array of crustal features each with their own equally complex crustal histories that must be taken into consideration when undertaking any modelling studies of the Earth.

In contrast, the modelling studies outlined in each of the previous chapters utilise the modern bedrock crustal mapping of the Commission for the Geological Map of the World and UNESCO (1990) to constrain plate assemblage on small Earth models of an Expansion Tectonic Earth. This mapping was not available to early researchers into Continental Drift, Plate Tectonics, or Earth Expansion. It is important to note that this map was originally commissioned and completed with the primary intent of quantifying the plate motion histories and prior plate assemblages on a Plate Tectonic Earth model. However, it is rarely used in Plate Tectonic studies today simply because it failed to achieve its intended aim—in other words, the factual data did not fit the preconceived static radius Plate Tectonic model.

By successively removing young seafloor crustal geology from along the mid-ocean-ridge spreading axes shown on this map and reuniting the remaining continental and seafloor crustal plates on small Earth models, the modelling studies of each of the Earth Expansion researchers mentioned above are indeed validated. In addition, the bedrock mapping demonstrates that all crustal plates do indeed assemble precisely, with one unique global fit-together on small Earth models. Again, as Creer so astutely noted, "*...the fit of the continents on a smaller Earth appeared to be too good to be due to coincidence and requires explaining.*"

The Expansion Tectonic modelling studies presented here demonstrate conclusively that the crustal plates, when reconstructed on small Earth models, coincide fully with the seafloor spreading and geological data and accord precisely with the derived ancient Earth radii for each model constructed. This coincidence applies not only to the more traditional oceans, such as the Atlantic Ocean where conventional reconstructions agree in principle, but also to the Pacific Ocean where the necessity for subduction of all or part of the seafloor crusts generated at spreading ridges is refuted. The small Earth modelling studies instead demonstrate that the mechanism of seafloor spreading, as highlighted by the seafloor mapping, provides a definitive means to accurately quantify an Expansion Tectonic Earth process.

It is emphasised that this bedrock geological mapping is factual empirical evidence and the Expansion Tectonic research undertaken can therefore be justly considered as empirical research. Empirical research, by definition, is a way of gaining knowledge by means of either direct or indirect observation. In Expansion Tectonics, this empirical evidence is the published seafloor and continental crustal mapping data preserved in each of the oceans and continents. This evidence was analysed here both quantitatively and qualitatively during model studies, without imposing prior premises or assumptions about Earth radius. By quantifying the seafloor mapping evidence and making sense of it in qualitative form, a number of empirical questions were then able to be asked in order to further knowledge and understanding within the bounds of the evidence available.

The most fundamental question considered during this modelling exercise was: ***Does the seafloor mapping really reflect an increase in Earth radius with time?*** This was dually answered and tested by measuring each of the seafloor surface areas in turn and from this, calculating ancient Earth radii for each time interval involved. The results were then further quantified by using the radii data to create small Earth models. These models were in turn used to test whether or not the remaining seafloor and continental crustal data will reassemble back in time to ultimately assemble as Pangaea and primordial Archaea small Earth models respectively. The construction of each of these small Earth models relied on the fundamental premise that seafloor volcanic lava intruded along the mid-ocean-ridge zones, as well as sediment deposited within marine and continental sedimentary basins, are cumulative with time. It was argued that, by moving back in time, both the lava and sediment must be returned to the mantle or ancient lands respectively, from where they originated.

By progressively removing seafloor volcanic lava from each of the small Earth models in turn it was shown that the plate fit-together along each mid-ocean-ridge plate margin achieved a better than 99% global fit for each model constructed. This unique fit-together was considered to empirically demonstrate that a post-Triassic Expansion Tectonic Earth is indeed a viable process and therefore justified extending modelling studies back further to the Archaean. This experiment empirically demonstrated that all remaining continental crusts do indeed assemble as a complete Pangaea Earth at approximately 50 percent of the present Earth radius during the late-Permian. Similarly, by extending this philosophy back in time to the Precambrian times, remnant ancient Proterozoic and Archaean continental crusts were shown to assemble together as a primordial Earth at approximately 27 percent of the present Earth radius.

From this small Earth modelling exercise it was shown that, prior to the Triassic Period, the ancient supercontinents had existed as a complete continental crustal shell for 94 percent of early Earth history, lasting for some 3,750 million years. This supercontinental crust covered the entire Earth with no large intervening oceans. During that time large bodies of water were instead restricted to a network of relatively shallow continental seas. The evolution of the supercontinents during pre-Triassic times simply involved a progressive and evolutionary crustal process during a prolonged period of crustal stretching, accompanied by changes in both Earth surface area and surface curvature through time. The outlines and configurations of the supercontinents were then dictated by changes to the ancient sea-levels and coastal shorelines, primarily as a result of changes to the surface areas of each of the ancient seas.

In strong contrast, the modern continents simply represent the fragmented remains of the ancient Pangaea supercontinental crustal shell. This fragmentation and subsequent break-up occurred because the ability for the continental crusts to continue to stretch during on-going increases in Earth surface area was finally exceeded during the late-Permian Period. As a result, during the late-Permian the continental crust then ruptured and broke apart to form the modern continents. During that time the ancient continental seas were also disrupted and progressively drained from the continents into the newly opening modern oceans.

Quantification of an Expansion Tectonic Earth process back to the early-Archaean required an extension of the fundamental cumulative seafloor volcanic crust premise to include continental crusts. Continental crust was reconstructed on pre-Triassic small Earth models by considering the primary crustal elements cratons, orogens, and basins. In order to achieve this, consideration was given to an increase in Earth surface area occurring as a result of crustal stretching and extension within an established network of continental sedimentary basins.

Moving back in time, this crustal extension was then progressively restored to a pre-extension, pre-stretching, or pre-rift crustal configuration by simply removing young sedimentary and intruded magmatic rocks and reducing the surface areas of each of the sedimentary basins in turn, consistent with the empirical data shown on the Geological Map of the World. During this process, the spacial integrity of all existing ancient cratons and orogens was retained until restoration to a pre-orogenic configuration was required. By removing all basin sediments and magmatic rocks, as well as reducing the surface area of each sedimentary basin in turn, an ancient primordial small Earth with a radius of approximately 27 percent of the present Earth radius was then achieved during the early-Archaean. This primordial Earth comprised an assemblage of the most ancient Archaean cratons and Proterozoic basement rocks; all other rocks were simply returned to their places of origin.

Mathematical modelling of both the seafloor and continental crustal surface area data demonstrate that the Earth is undergoing an exponential increase in surface area and radius, commencing from a primordial Earth of approximately 1,700 kilometres radius during the early-Archaean. From this, a current rate of increase in Earth radius is calculated to be 22 millimetres per year. Increase in Earth radius throughout the Archaean to mid-Proterozoic times was shown to increase by approximately 60 kilometres during 3 billion years of Earth history, prior to a steady to rapidly accelerating increase in Earth radius commencing during the late-Proterozoic and continuing to the present-day. Extrapolating this increase to the future demonstrates that expansion to 5 million years in the future is consistent with a continued spreading along all present-day mid-ocean-ridge axes.

Crustal assemblages on each of the small Earth models showed that large Panthalassa, Tethys, and Iapetus Oceans were not required during reconstruction. These oceans were instead replaced by continental Panthalassa, Iapetus, and Tethys Seas, which represent precursors to the modern Pacific and Atlantic Oceans as well as ancient sedimentary basins located on many of the present-day continents. Similarly, emergent land surfaces during the Precambrian and Palaeozoic Eras were shown to equate to the conventional Rodinia, Gondwana, and Pangaea supercontinents and smaller sub-continents. On each small Earth model, supercontinental development was shown to be progressive and evolutionary, with no requirement for random continental dispersion-amalgamation cycles. Supercontinent configuration was then shown to be related to progressive crustal extension within an established network of continental sedimentary basins, changes to Earth surface curvature, and changes to sea levels. Also, as will be shown in the following chapters, it is important to appreciate that on an Expansion Tectonic Earth the poles and equator remain stationary while the land masses make an apparent traverse across them. It will also be shown that this does not support conventional palaeomagnetic apparent-polar-wandering of the ancient poles.

While the extensive small Earth spherical model studies presented in this first part of the book empirically demonstrate that the concept of an Expansion Tectonic Earth is viable, the problem of where the excess Earth volume comes from and similarly where does it go to moving back in time has not been addressed and remains a very real enigma. I can assure the reader that this enigma will be discussed at length in Part Two. Despite this enigmatic origin for the excess volume required to quantify an Expansion Tectonic process, it is emphasised that the empirical evidence presented so far shows that the ability to globally reassemble Archaean to present-day crusts with such a high precision cannot be mere coincidence and therefore needs further explanation.

From this statement I maintain that scientists cannot continue to justify imposed ad-hoc premises or assumptions about Earth radius simply to maintain a perceived Plate Tectonic consensus of opinion. It is considered that this consensus of opinion, based on an untested static radius Earth meme, must therefore be rejected in order to allow true science to correctly consider and assess the empirical evidence presented here.

At this stage, the outcomes of this empirical crustal research are more than adequate to quantify the validity of an Expansion Tectonic Earth crustal process. However, in the eyes of many, it is still seen as largely subjective even though small Earth models were able to be readily constructed by simply assuming that the Earth's surface area is increasing with time as a direct result of an increase in Earth radius, with nil crustal consumption. Part Two of this book will now progressively introduce new geological, geophysical, and geographical evidence to further test the validity of these small Earth modelling studies.

PART TWO

The Earth's Rock-Record

"Ultimately world reconstructions must be congruent not only with the data from geology and geophysics, but also with palaeobiogeography, palaeoclimatology, and palaeogeography." Shields, 1997

11 Geological Implications

"Why should a body which is expanding develop huge diapiric extensional basins with a very deep source, as testified by the earthquake foci pattern, while elsewhere, very shallow extensional ridges with an associated shallow seismicity implies a great complexity of the global expansion process?"
Scalera, 1990

In each of the chapters in Part One of this book the adopted approach has been to use published bedrock geological mapping of the oceans and continents to construct small Earth models of the past crustal assemblage on an Expansion Tectonic Earth. This completed mapping was not available to early researchers, or small Earth model makers, prior to the 1980s. Because of this, early model makers were limited to visual refitting of the continents across vast areas of ocean. Even though this was feasible, otherwise they would not have considered doing it, the continental assemblage's on their models were not always convincing. Most other researchers of the time were only just accepting the fact that continents were capable of moving–as proposed in Continental Drift–so to go the next step and also consider that the Earth is expanding was a very difficult concept to accept. History shows that the less than convincing small Earth models of these early researchers all contributed to rejection of the concept of Earth expansion leading to subsequent adoption of the seemingly more convincing theory of Plate Tectonics.

By far the single most important contribution to modern scientific understanding of the concept of Expansion Tectonics has been completion of the bedrock geological mapping and age dating of all the continental and seafloor crusts during the early 1990s. This bedrock mapping has now enabled assemblage of crustal plates to be accurately constrained on models of an Expansion Tectonic Earth, and, for the first time, has enabled small Earth modelling studies to be extended back to the earliest Archaean times. This mapping has also provided a means to mathematically define an increase in Earth radius over time, which, in turn, has provided a means to investigate additional related physical phenomena.

Each of the Expansion Tectonic small Earth models presented in previous chapters demonstrate that, by using this modern bedrock geological mapping, an Archaean to present-day Expansion Tectonic Earth process is indeed viable. From this small Earth modelling, it was then intimated on a number of occasions that an increase in Earth radius is simply transferred from the mantle to the outer crusts as continental and seafloor crustal stretching and extension. This crustal stretching process was shown to have been operative throughout Precambrian and Palaeozoic times prior to supercontinental crustal failure and rupture during the late-Permian. Crustal rupture then gave rise to break-up of the Pangaea supercontinent to form the modern continents and rifting and opening to form the modern oceans.

As Shields so astutely wrote in 1997, *"Ultimately world reconstructions must be congruent not only with the data from geology and geophysics, but also with palaeobiogeography, palaeoclimatology, and palaeogeography"*—where the word *palaeo* simply means ancient. What Shields was referring to here is that it doesn't matter how well or how convincing a tectonic theory or model is, if the theory doesn't agree with the physical evidence preserved in the rock-record, it is the theory that is incorrect, not the physical evidence.

In the following chapters I will now move on to investigate the geological implications of what the seafloor and continental crustal data are showing, in particular, by way of additional geological, geophysical, geographical, and biogeographical evidence. That is, evidence that is still preserved in the rock-record from all continents and seafloors throughout the world for all to see and study. This evidence is also collectively referred to in conventional plate studies as *global tectonic data*.

When correctly interpreted, this physical evidence can provide an extensive amount of information about the past geologic history, past geography, biogeography, and ancient climate, along with additional physical parameters such as atmospheric conditions, sea levels, location of the ancient equator and poles, Earth radius, and so on. For Expansion Tectonics to be seen as a truly viable concept, the evidence preserved in this rock-record must therefore fully support and substantiate the empirical small Earth crustal evidence that has so far been investigated in earlier chapters.

11.1 Historical Overview

Acceptance of the theory of Earth expansion—Expansion Tectonics—was, and still is, regarded by many researchers as being thwarted by major obstacles. These obstacles were originally considered by the researchers to *outnumber the evidence in favour*. It must be stressed, however, that these concerns stem from times dating back to the 1950s, well before the advent of modern tectonic studies and well before modern understanding of these perceived obstacles. Included in these concerns was the need to explain the existence of two very different extensional structures observed on the seafloors. These structures included the mid-ocean-ridges and the *trench-arc* and *back-arc zones*—Plate Tectonic terms relating to the various island-arcs, mainly located within the Pacific basin—characterised by very different seismic activity and volcanism. As well as this, concerns were raised regarding an explanation for the perceived problem of accumulating the atmosphere and ocean waters on an Expansion Tectonic Earth, as well as the adaptation of *palaeomagnetics*—the study of remnant magnetism preserved in rocks—to a constantly changing Earth radius.

Brunnschweiler in 1983, in a paper dealing with the evolution of modern tectonic concepts, considered that "*Earth expansion was essentially a radial movement and therefore its tangential plate displacements are only apparent, not real.*" In 1990 Scalera asked, "*...why should a body which is expanding develop huge diapiric extensional basins with a very deep source, as testified by the earthquake foci pattern, while elsewhere, very shallow extensional ridges with an associated shallow seismicity implies a great complexity of the global expansion process?*" Similarly, the possibility of orogenesis—the compression and folding of rocks to form mountain belts—developing under conditions of radial expansion was also discounted by Rickard as early as 1969 because he considered, "*...the necessary vertical movements did not appear to explain the observed compressional features.*"

What should be realised is these early discussions about crustal mechanisms on an Expansion Tectonic Earth relate to times when it was considered that continental crusts acted as rigid bodies and hence any subsequent break-up of these crusts only involved simple crustal fragmentation. This consideration continues, to some extent, through to present thinking where conventional Plate Tectonic modelling studies continue to treat supercontinental assemblages as amalgamations of previously fragmented and accreted crusts. Since these early times though, modern global geology now shows that the continental crusts have, in fact, had a long and complex geologic history, a history that must be strictly adhered to during all crustal modelling and theoretical studies.

Bailey and Stewart in 1983 further considered that, "*...for an Earth undergoing expansion with time, the bulk of the oceans would have to be outgassed since the Palaeozoic, requiring fundamental changes in atmosphere, climate, biology, sedimentology and volcanology.*" In 1986 Weijermars also considered that, "*...for a pre-Jurassic small Earth with a continuous continental crust, a large expansion process implies that the entire Earth would have been covered by an ocean with an average depth of 6.3 kilometres.*" This implication of Weijermars is contrary to the evidence preserved in rocks, hence his concern. In fact, Weijermars introduces yet another meme whereby he assumes that the volume of ocean water has been constant, or near constant with time.

Palaeomagnetic studies have long been considered the cornerstone of Plate Tectonic studies. These studies have supplied an immense amount of data about past locations of continents and oceanic plates. These data have also provided evidence about motion histories of the various continental crusts and continental growth, as well as measurements of the Earth's ancient radius. Analyses of ancient Earth radii were made by van Andel and Hospers in 1968 and similarly McElhinny and Brock in 1975. McElhinny and Brock considered "*...within the limits of confidence, theses of exponential Earth expansion, or even moderate expansion of the Earth, are contradicted by the palaeomagnetic evidence.*" This then led McElhinny and Brock to conclude, "*...there has been no significant change in the ancient radius of the Earth with time.*"

As mentioned, many of these perceived concerns stem from times dating back to the 1950s, well before the advent of modern tectonic studies and well before a full and modern understanding of these phenomena. Most of the problems envisaged by these researchers have now been resolved, rejected, or accepted as common knowledge within Plate Tectonic studies. It is only through additional research that these same problems will be resolved and accepted from an Expansion Tectonic Earth perspective.

Subsequent literature, such as Smiley in 1992, indicates there is also an increasing awareness of a number of "*perceived problems*" confronting Plate Tectonics, as modelled on a fixed radius Earth. Smiley considered that these perceived problems continue to be simply ignored, or are overcome by invention of new ad-hoc "*fixes.*" This is similar to what Kuhn had to say in 1970 where he talks about "*normal*" versus "*extraordinary*" science. In so-called normal science, a problem tends to be discounted as probably explainable in terms of what is known but not at the current state-of-the-art. The outcomes of empirical small Earth modelling studies detailed in previous chapters then lend substance to these concerns that the present concepts of Plate Tectonics, including Continental Drift and palaeomagnetic based polar wandering studies, may indeed need to be "*...re-evaluated, revised, or rejected*" as Smiley suggests.

The intention of the following sections of this chapter is to reconsider the evidence and proposals presented by a number of these early authors in order to see whether their ideas and concerns still apply to the modern global evidence. Much of this evidence and proposals have also been extensively researched and published in Polish by Jan Koziar and Stephan Cwojdzinski. This evidence will then be further discussed in the following chapters by incorporating some of the evidence and ideas in understanding relief of surface curvature on an Expansion Tectonic Earth.

11.1.1 Changing Earth Surface Curvature

Understanding how both the continental and seafloor crusts behave and adjust for changing surface curvature during an increase in Earth surface area throughout time is crucial towards understanding the mechanisms behind Expansion Tectonics. During the 1960s, there was a lot of active debate in the literature regarding a perceived incompatibility of the continental crusts during changes in Earth curvature through time. This incompatibility reflected the prevailing view that the Earth's continental and seafloor crusts acted as rigid bodies. It was further considered that the continental crusts could not change shape without undue fragmentation or distortion. These considerations were, in reality, artefacts of the presumption that Earth radius has remained constant throughout time and this view has since been used in the literature to justify rejecting any theories that consider otherwise.

Other researchers though were a little more imaginative. Some considered that crustal distortion on an increasing radius Earth should be possible because the complex folding, faulting, and shearing that are now seen in the present-day continental rocks would lead one to expect continental fragmentation and distortion in any event. Others suggested that the outline of continental margins may have changed in time as a result of *accretion* of the continents, by loss of continental crust, or both. What these authors were talking about is a Plate Tectonic concept whereby new crusts are added to or removed from the margins of old crusts during periods of continental collision or break-up.

On an Expansion Tectonic Earth, in particular during pre-Triassic supercontinent times, accretion by continental collision could not occur because during pre-Triassic times the supercontinental crusts covered the entire Earth as a single continental crust. Instead, a variation of this accretion theme occurred within intervening sedimentary basins during an increase in surface area. In these areas new sediment was deposited in low-lying areas as each of the basins continued to enlarge or rift apart. In contrast, loss of continental crust on an Expansion Tectonic Earth can only occur during post-Triassic crustal break-up times. However, this is only an apparent phenomenon as crust lost from one continent was generally retained on its adjoining continental margin—a phenomenon referred to in Plate Tectonics as *displaced terranes*.

These perceived problems revolve around not only a lack of appreciation of what crusts are made of, but also a lack of appreciation of the extensive amount of time involved during an expansion of the Earth. In addition, there was also a lack of appreciation of the extremely small annual increases in radius that have occurred each year with respect to the dimensions of the Earth itself. It must be further understood that during the pre-1970s there was a general lack of consensus on how much the Earth radius could potentially increase each year. This is understandable and in the eyes of many, regardless of the potential for Earth expansion, a rigid crust was a justifiable conclusion. Rocks are solid objects, so how could they be anything other than part of solid rigid crusts?

As shown in Chapter 6, it is now known that the Earth's continental crust is made up of stabilised ancient *cratons*, complexly folded and metamorphosed *orogens*, and a network of sedimentary *basins*. The ancient cratons comprise mainly granite and volcanic rocks with lesser sediments and can, by definition, be considered as *rigid* crusts. The orogens comprise mainly metamorphosed sedimentary rocks—rocks that have been folded and recrystallized as a result of strong Earth pressures and temperatures—and when originally formed these rocks were essentially *flexible* crusts. However, these orogenic rocks have since increased their rigidity during metamorphism. The orogenic rocks can then be considered as having an intermediary strength between the ancient cratonic rocks and the younger basin sediments. The rocks forming the sedimentary basins are, by comparison, truly flexible rocks. By flexible, this means that over the extended period of geological time available they will readily flex, fold, fault, shear, stretch, and distort, as highlighted in many rock exposures throughout the world today.

In addition to the three main types of continental crustal rocks, there is also the completely different seafloor crust to consider. This crust has been shown from the seafloor geophysical mapping to consist predominantly of intruded volcanic lava. This lava was intruded directly from the mantle and can be considered as quenched and solidified mantle rocks. In other words, the seafloor crust is quenched and exposed mantle rock with nothing to do with continental crusts. The style of intrusion along the mid-ocean-ridge spreading centres—essentially intrusion as vertical dykes within open fractures and faults—imposes an inherent weakness to the seafloor rocks and, as such, these crusts can also be considered as flexible.

The point being made here is that there is a very strong contrast between the various types of continental crustal rocks and the seafloor crusts. In particular, there is a major difference between the rigid ancient cratonic continental crusts and the younger flexible continental sedimentary basin and seafloor crusts. Each of these crusts will behave completely differently during changing Earth radius, surface area, and surface curvature. The rigid crusts will tend to *brittle fracture* and fragment, whereas the flexible continental and seafloor crusts will tend to stretch and deform.

In order to visualise the effects of changing surface curvature on an ancient rigid continental crust, imagine fitting broken fragments of a 150 millimetre diameter ceramic sphere onto a 151 millimetre diameter sphere. This would be perfectly feasible, in particular if the broken fragments were small enough. It would also be feasible to consider that all of the 150 millimetre diameter fragments would refit neatly on the 151 millimetre diameter sphere, albeit with loose-fitting joints. It is envisaged that this scenario is an approximation of what may have occurred to the ancient primordial Earth crust during early-Archaean supercontinental times. During these early-Archaean times a rigid ancient cratonic crust covered the entire primordial Earth. Any increase in Earth radius would simply fracture and separate the fragments of crust as a myriad of joints, faults, fractures, and so on. In turn, any major openings between each of the crustal fragments may then heal and seal with either new volcanic lava intruded from below or filled by sediment eroded from above. This initial fragmentation of the ancient supercontinental crust would then set up a precursor for on-going increases in Earth radius. Any subsequent increases in radius and surface area would simply be focussed within these already existing fractures, faults, and joints.

During the pre-1970s, what the early researchers, and also many of the small Earth model makers, were considering was that the entire modern continental crust was originally rigid. The implications of this consideration can be further visualised by trying to assemble large rigid crusts together, similar to say assembling a fragment of a broken ceramic sphere of 150 millimetres diameter onto a sphere of say 300 millimetres diameter—similar to the proportions of a Permian small Earth model when compared to a present-day Earth model. This cannot be done and no one in his or her right mind would insist otherwise. The broken fragments would have to be broken into ever-smaller pieces and even these smaller fragments would not assemble together accurately on the larger diameter model, nor would they be able to accommodate for the huge changes in surface curvature.

Next, consider the effects of changing surface curvature on the flexible crusts, such as the sedimentary basins and seafloor volcanic rocks. This effect can be visualised by placing your hand on the surface of the 150 millimetre diameter sphere. In this instance your fingers will be tightly closed together to simulate the assemblage of fragments of more rigid crusts. Next place your same hand on the 300 millimetre diameter sphere. In this instance the fingers become separated and, if humankind were more aquatic, webbing between the fingers—simulating the sedimentary basins—would be stretched and exposed. By using a soft clay sphere rather than a rigid ceramic sphere, fitting more flexible 150 millimetre diameter fragments of clay onto a 300 millimetre sphere would then be a perfectly feasible proposal. These fragments would accommodate for the changing surface curvature by simply stretching or distorting and there would be no need to further fragment the crust into smaller pieces.

Applying these various scenarios to the Earth, rather than considering the continental and seafloor crusts as acting as rigid bodies, now imagine them behaving as a combination of both rigid and flexible crusts acting over an extended period of geological time. In particular, visualise the ancient rigid primordial crust as firstly being fragmented, then the intervening sedimentary basins being stretched and distorted by microns to millimetres per year. This minute amount of crustal stretching and distortion is then distributed over distances of many hundreds to thousands of kilometres of continental crust, with stresses acting over billions of years of Earth history. The resulting crustal distortion then becomes perfectly feasible, especially, for example, when viewed in relation to the amount of vertical, tide-related crustal distortion that the Moon imposes on the modern crusts each day. These tidal effects alone would go a long way towards helping relieve any changes to surface curvature and allow the crusts to remain at least partially equilibrated to the prevailing new Earth radius.

Once the ability of the ancient supercontinental crusts to continue fragmenting, stretching, and distorting during increasing Earth radius was exceeded, the supercontinental crusts would then rupture and fragment to form the modern continents, precisely as happened during the late-Permian times. Once ruptured, any gross changes in surface area and curvature of the Earth would be simply transferred to the seafloor crusts as extension along the mid-ocean-rift zones as well as on-going fragmentation and distortion within the continental sedimentary basins.

The relief of surface curvature on an Earth undergoing a progressive increase in radius with time was first considered at length by Rickard in 1969 and was modelled extensively by Koziar during the 1980s and 1990s. The associated implied stress relief was also discussed by Weijermars in 1986. The latter author discounted Earth expansion on the grounds of a perceived *"spatial incompatibility"* resulting from a changing surface curvature. This spatial incompatibility was previously emphasized by Jeffreys in 1962 when commenting on Barnett's 1962 small Earth model. Jeffreys pointed out that crust on Barnett's 115 millimetre diameter model could not be simply placed on his 76 millimetre diameter model without undue distortion—precisely as suggested in the analogy above. The reconstruction of continents on a small Earth model was then considered by Jeffreys to be dependent on the distribution of continental distortion. Again this is, in effect, a roundabout way of saying what I have just discussed above that the continental and seafloor crusts are not all rigid crusts and must instead be allowed to stretch and distort over time.

Within the extended period of geological time available on an Expansion Tectonic Earth, the flexible crusts instead behave *plastically*. This plastic deformation process involves minute atomic-scale changes to the rocks which are in turn distributed over vast areas of the Earth surface. These atomic-scale changes gradually accumulate over time as each of the crustal fragments and sedimentary basins re-equilibrate to the ever changing surface curvature. Carey compared the features seen in metamorphic rocks [and by inference sedimentary rocks] with the plastic flow of glaciers. Here, over time the slow atomic-scale creep of the ice enables the ice to *flow* under the influence of gravity.

The mechanism for plastic-flow and distortion processes, as well as the characterization of a material's behaviour during changes to an applied stress, is known as *rheology*. Rheology is the study of the flow of matter, primarily in the liquid state, but also as solids under conditions in which they respond during plastic flow rather than deforming elastically in response to an applied force. A material that behaves as a solid under low applied stresses may start to flow above a certain level of stress, which is called the yield stress of the material. The term plastic solid is often used when this plasticity threshold is rather high, while yield stress is used when the threshold stress is rather low. Geophysicists also include the flow of molten lava and debris flows such as mudslides and volcanic ash in the study of rheology.

In 1962 Dennis, in reply to Jeffreys' comment about crustal spatial incompatibility, suggested that the implied crustal distortion could be met because of what he called "*pseudo-viscosity of the crust.*" He further commented that distortion by normal geological processes, such as folding, faulting, and shearing, would lead one to expect continental distortion in any event.

Van Hilten in 1963 considered the problem of crustal distortion and break-up on an increasing radius Earth during his early investigations into palaeomagnetism. Van Hilten suggested that "...*during a potential Earth expansion process, a complete continental crust would distort and eventually be displaced by forming a number of radial tears in the crust.*" He likened this process to the radial peeling of an orange and introduced the term "*orange peel effect*" to describe such a tearing process (Figure 11.1).

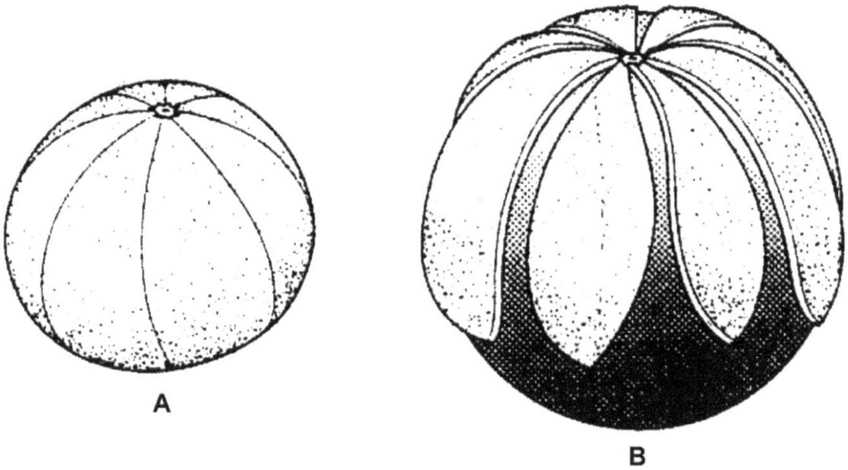

Figure 11.1 Van Hilten's "*orange peel effect*" model for Earth expansion. Van Hilten suggested that, during expansion of the Earth from A to B, the continental crusts would distort and displace by forming radial tears (from van Hilten, 1963).

This *orange peel model*, which suggests that the continental outlines were formed by merely *pumping up* the size of the Earth to split the continental crust like an orange peel, was never seriously accepted by researchers at the time. Yet, as shown on the Expansion Tectonic small Earth models, once the continental crust first began to rupture during the late-Permian Period, this orange peel model gives a very simple but effective visual impression of how the complete Pangaea supercontinental crust eventually broke-up and how each of the modern oceans opened and spread apart. Note in particular the large opening and separation around the bottom of the sphere in Figure 11.1B.

This orange peel effect is shown in the Geological Map of the World (Figure 11.2), displayed in Mollweide equal area projection, where the Americas and Africa, along with India, Australia, and Southeast Asia can be likened to four of van Hilten's radial tears. This is also supported by observations of Carey in 1963, where he commented that "...*the greater separation of the southern continents, and general northward migration of all continents, has resulted because of a greater expansion of the southern hemisphere than in the northern hemisphere,"* identical to what van Hilten portrays in Figure 11.1B.

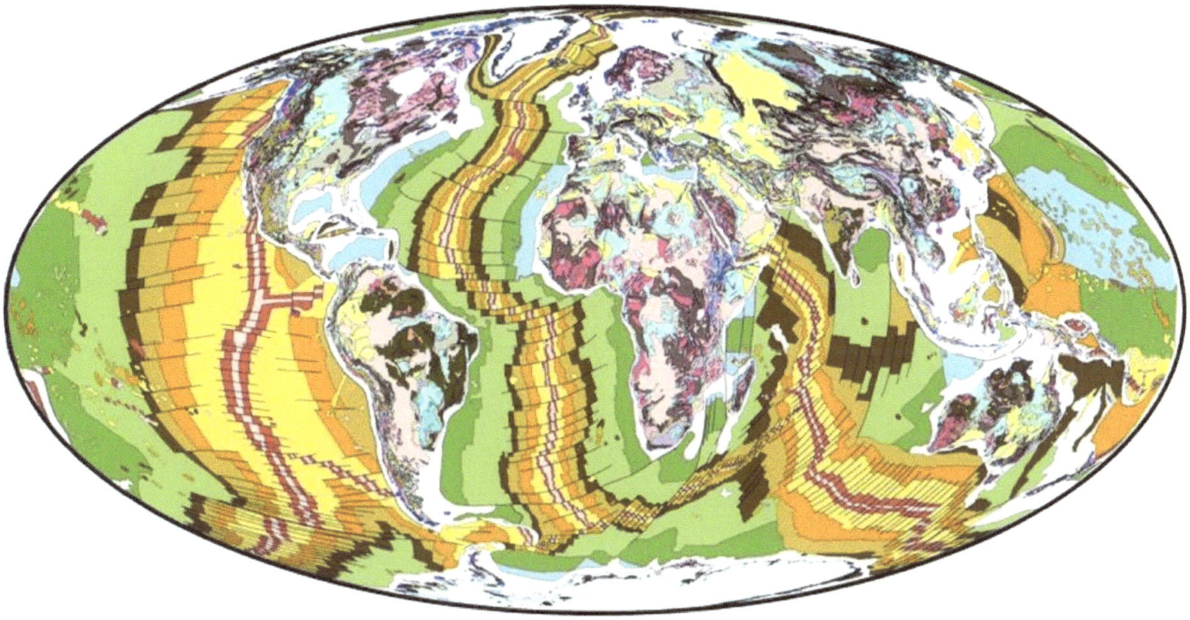

Figure 11.2 The Geological Map of the World displayed in Mollweide equal area projection highlighting both van Hilten's *"orange peel effect"* model for Earth expansion and Carey's comment regarding the greater separation of the southern continents and general northward migration of all continents.

In 1969, Rickard further suggested that relief of surface curvature during an increase in Earth radius was a possible mechanism for both *"geosynclinal"* formation and *"orogenesis."* The term geosyncline has now been replaced by additional Plate Tectonic terminology, but it basically refers to a subsiding low-lying linear trough or down warp in the Earth's crust where sediments accumulate. Geosynclines can be viewed in a similar context to continental crustal extension and development of the network of sedimentary basins on each of the small Earth models, as previously discussed in Chapter 5. Geosynclines on an Expansion Tectonic Earth will be discussed in a later section.

In contrast to comments by others, where it was suggested that continental crust adjusted to the new surface curvature by plastic flow and cracking during increasing Earth radius, Rickard suggested that the Earth's crust may be sufficiently strong to ensure a considerable time lag before complete adjustment for continental curvature was achieved. Rickard argued that, during an increase in Earth radius, geosynclines–sedimentary basins–

would initiate as furrows along the margins of continental cratons (Figure 11.3). He considered that this occurred because of a perceived *"differential radial expansion"* operating during opening of the oceans. The effective increase in the slope of continental crusts by a few degrees was then considered to increase the rate of erosion of the exposed lands, giving rise to a rapid accumulation of new sediment within the low-lying basins.

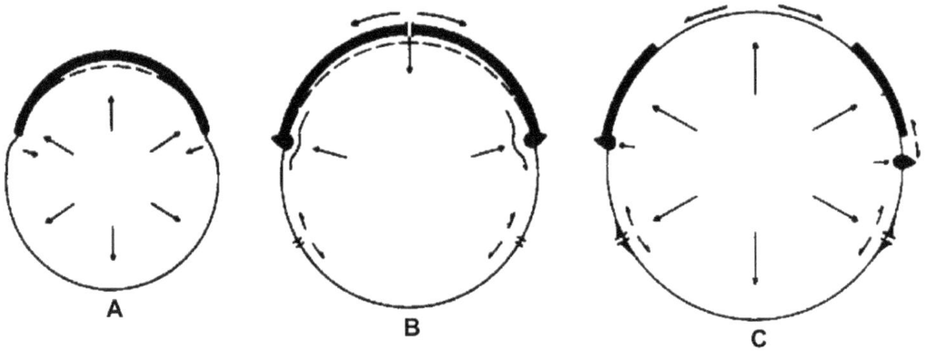

Figure 11.3 Rickard's 1969 model for relief of surface curvature during an increase in Earth radius. Figure A represents an *"initial stage"* of *"marginal geosynclinal furrow"* development; Figure B represents a *"critical stage"* in the development of orogenesis, rifting, and initiation of sea-floor spreading; and Figure C represents the *"relieved stage"* of curvature readjustment, isostatic uplift, block faulting, sea-floor spreading, continental separation, and development of island-arcs (from Rickard, 1969).

The delicate state of balance between the opposing forces involved in relief of surface curvature was considered by Rickard as adequate to account for the complex variety of vertical movements required during early geosynclinal activity. Rickard further considered that, eventually a critical stage would be reached where magmatic activity and rising heat may accompany orogenesis—essentially the crumpling of the rocks by compression and folding movements.

Carey was critical of Rickard's model because of his assumption of a significant *"...enduring strength in the continental crust and an implied super-elevation of the continents."* Carey considered instead that, because of the *"...rapid adjustment of isostatic inequalities in the lower crust the required super-elevation could never come about."* Carey appreciated that the central sector of the continental crust must rise *"...because of the megatumour beneath it,"* but maintained that it would never depart far from isostatic equilibrium, nor would there be any *"...lateral gravitational force beyond that arising from hydrostatic equilibrium."*

What all this simply means is that, rather than the continental crust doming upwards by super-elevating the central part of the crust, as shown in Figure 11.3A, the crust would naturally adjust for changing surface curvature by distortion and fragmentation but would retain some slight elevation within the central part of the crust due to a limited retention of an enduring strength of the crust. It will be shown later that this is precisely what must happen; otherwise the entire Earth would have a flat featureless surface devoid of mountains and rivers.

Carey further considered that *"...the Earth consists of a crystalline mantle some 3,000 kilometres thick over a fluid core."* The first result of an increase in Earth radius would be rupture of the whole mantle—plus overlying crust—into polygonal blocks of a few 1,000 kilometres across (Figure 11.4). He considered this would then be surrounded by accreted seafloor crust added during the last 160 million years. As the Earth radius continued to increase, Carey envisaged that the first adjustment to a decreasing surface curvature would occur at the primary mid-ocean-ridge spreading zones. These would extend to second rank basins and swells as the crust continued to adjust. Further adjustment would continue down through a hierarchy of fractures and ultimately to ordinary joints in consequence of a final adjustment to the changing surface curvature. These primary polygonal blocks would then be patterned by second-order polygonal basins and swells, extending through both the continental and developing seafloor crusts.

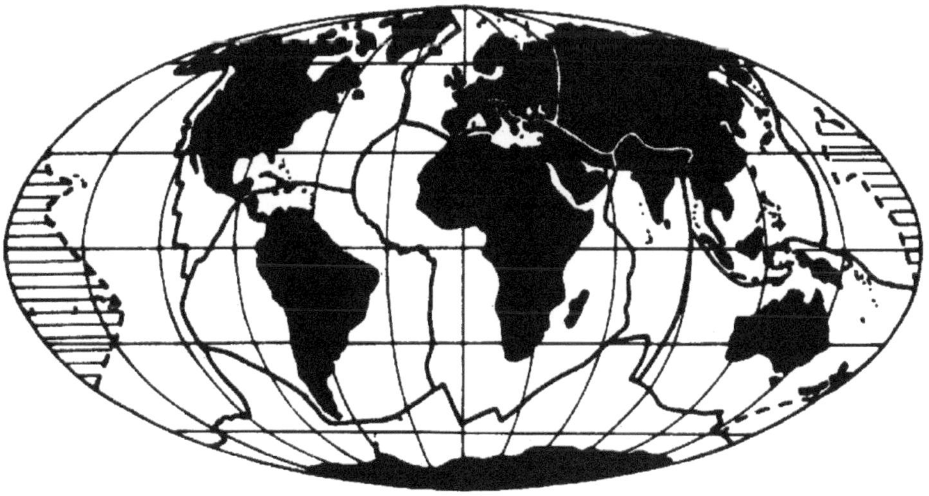

Figure 11.4 Carey's first-order primary crustal plate polygons surrounded by circum-continental spreading *"diapirs"*. Each numbered primary polygon consists of a continent surrounded by its accreted oceanic crust, added during the past 160 million years—now called plates in all tectonic studies (modified after Carey, 1996).

While the proposals above are feasible, Carey failed to fully appreciate that Earth expansion has been operative for the entire 4,000 million years of known Earth history, not just for the past 160 million years as he implied. This is understandable because modelling of the continental crusts on small Earth models prior to the Triassic period has never been done before, nor was it possible before publication of the modern bedrock geological mapping in 1990. The small Earth modelling studies outlined in the first part of this book are, in effect, the first time this pre-Triassic modelling has been achieved and formerly presented.

In this context, Carey's consideration of primary polygonal crustal blocks must be shifted well back in time to when the initial fracturing of the ancient crust first occurred during early-Archaean times. As intimated in the ceramic sphere analogy above, during an increase in early Earth radius, this mainly crystalline ancient magmatic and volcanic crust may have initially simply fractured and fragmented. Periodic intrusion of new granite magma and volcanic lava then re-stabilised the primordial crust prior to further fracturing over time. During Proterozoic times, these fractured ancient crusts were in turn *accreted to*—added to—by an on-going crustal extension process operating within the network of established sedimentary basins and crustal weakness. This accretion occurred mainly during crustal stretching and extension along the margins of each of the fragmented crusts. Only during the late-Permian Period did true rupture and break-up of the ancient supercontinental crusts occur to form the modern continents, as well as opening of the modern oceans along each of the newly formed mid-ocean-ridge spreading centres.

Within the context of Expansion Tectonics, when considering the mechanism for changing Earth's surface curvature, it is important to again appreciate that continental crusts are made up of a broad distribution of cratons, orogens, and sedimentary basins. Each of these continental crustal elements has their own definitional characteristics, which differ fundamentally from the modern seafloor crusts. Previous authors did not address this crustal hierarchy, considering instead that continental and even seafloor crusts were basically similar and simply acted as rigid plates.

By only considering these crusts as rigid plates, it is understandable why the various authors discounted Earth expansion on the grounds that continental plates could not adjust for changing radius while still retaining the curvature of an earlier smaller radius Earth. Without acknowledging Earth expansion during pre-Triassic times it is also understandable why it was further considered by these authors that a large majority of the continental crust must have already been in existence during these pre-Triassic times and why growth of continental crusts throughout these times was considered to have been negligible?

11.1.2 Geosynclines

While the term geosyncline is now rarely used in conventional tectonic studies, I am going to persist with its usage here to remain in context with what Rickard and others discussed during the 1960s and 1970s. Additional Plate Tectonic terms also include *subduction, island-arcs,* and *back arc basins,* which, strictly speaking, cannot be reconciled in their entirety here. These terms and usage on an Expansion Tectonic Earth will be discussed later. On an Expansion Tectonic Earth the term geosyncline is, in fact, better suited to the extensional sedimentary basin settings that the small Earth model studies show.

Rickard's 1969 model for geosynclinal development, shown modified slightly in Figure 11.5, implies that an increase in Earth radius is essentially a radial process. The way the figure is drawn also implies that basin formation, orogenesis, and mountain building are post-continental break-up phenomena. This is again a reflection of the lack of appreciation that Expansion Tectonics has occurred throughout earlier pre-break-up Earth history, rather than being simply confined to the past 160 to 200 million years or so.

In this model, Rickard envisaged that a geosynclinal trough formed along the outer margin of a continental crust—left end of the continental crust shown in Figure 11.5A—whereby relief of surface curvature is balanced by the strength of the crust and the downward acting weight of the sediments deposited in the trough. During a radial expansion of the Earth from A to B, sediments within the geosynclinal trough undergo compression and orogenesis during ongoing change in surface curvature. This may have also been accompanied by intruded granite magma. It is interesting to note how Rickard depicts the continental crust as being rigid and hence fragments during change in surface curvature, and also how the geosynclinal sediments are considered to be flexible crusts.

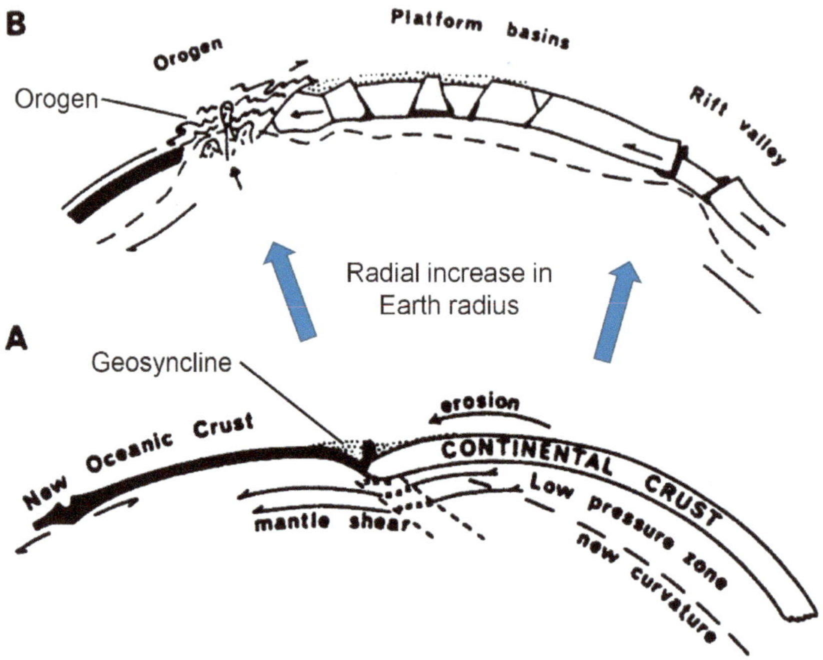

Figure 11.5 Rickard's cross-section of the Earth model for development of a *"geosynclinal trough"* prior to orogenesis. Figure A represents a *"critical stage"* of development where relief of surface curvature is balanced by the strength of the crust and downward acting weight of the sediments. Figure B represents an *"orogenic stage,"* where crustal compression is induced by *"isostatic uplift"* and accompanied by intruded granite magma (modified after Rickard, 1969).

Rickard further considered that compressional forces leading to orogenesis of the crust during an increase in Earth radius resulted from "*basin inversion.*" This basin inversion term refers to the uplifting and exposure of basin sediments during change in surface curvature as distinct from slowly sinking during deposition of the sediments. He also considered the orogenic process involved an interaction of the continental and seafloor plates as the Earth proceeded to adjust its surface curvature. This was radical thinking for the time and he must be credited for being at least conscious of the effects of changing surface curvature during crustal development. Rickard went on to suggest that a lateral outwards movement of the continental plate caused "*...compressional buckling and over-thrusting of the sedimentary fill within a narrow geosynclinal trough.*" Similarly, he also considered that continental-style geosynclines were developed where continental plates were fractured internally, giving rise to lateral compression as the two adjacent plates moved together during vertical relief of surface curvature.

The delicate state of balance between the opposing forces involved in relief of surface curvature was then considered by Rickard to account for the complex variety of vertical movements during early geosynclinal activity. This eventually led to what he called the critical stage, whereby magmatic activity and rising *"geo-isotherms"*—heat flow—was accompanied by orogenesis.

In contrast to Richard's model, the Expansion Tectonic small Earth models show that prior to 200 million years ago there were no volcanic seafloor crusts. During that time supercontinental crust completely enclosed the entire ancient Earth as a single Pangaea supercontinent with sedimentary basins coincident with relatively shallow continental seas. What this means is that Rickard's model shown in Figure 11.5, even though it is schematic, should not be seen as continents and large oceans in isolation with geosynclinal and orogenic activity only occurring after supercontinental break-up. This model should instead be seen as geosynclinal and orogenic development acting on a complete supercontinental crustal shell. The network of continental seas then coincides with the location of geosynclines which ultimately formed the loci for orogenesis. This will be discussed further in Chapter 12.

The model for geosynclinal development and orogenesis, as put forward by Carey in the 1970s and 1980s, for an Earth undergoing an exponential increase in Earth radius is shown in Figure 11.6. In Carey's model, the development of a simple *"diapiric orogen"*—an upward rising crust—resulted from initiation of primary stretching in the continental crust due to *"radial expansion."* Carey further considered this upward rising process would then lead to *"necking"* or thinning of the crust.

The top and bottom surfaces of the continental crust were then said to *"...converge towards zero with time at some 5 kilometres below sea-level"* during which time *"...the mantle rises some 30 kilometres."* Carey considered that, although the surface of the thinning continental crust subsided steadily, the bottom of the crust, as well as the presence of a *"mantle diaper"*—comprising magmatic rocks such as granite—below it must also rise. This process continued to do so throughout the period of orogenesis. Carey further considered that *"...crustal thinning, caused by the expanding interior, resulting in gravity drive towards isostatic equilibrium, causes all the motions necessary for orogenesis."*

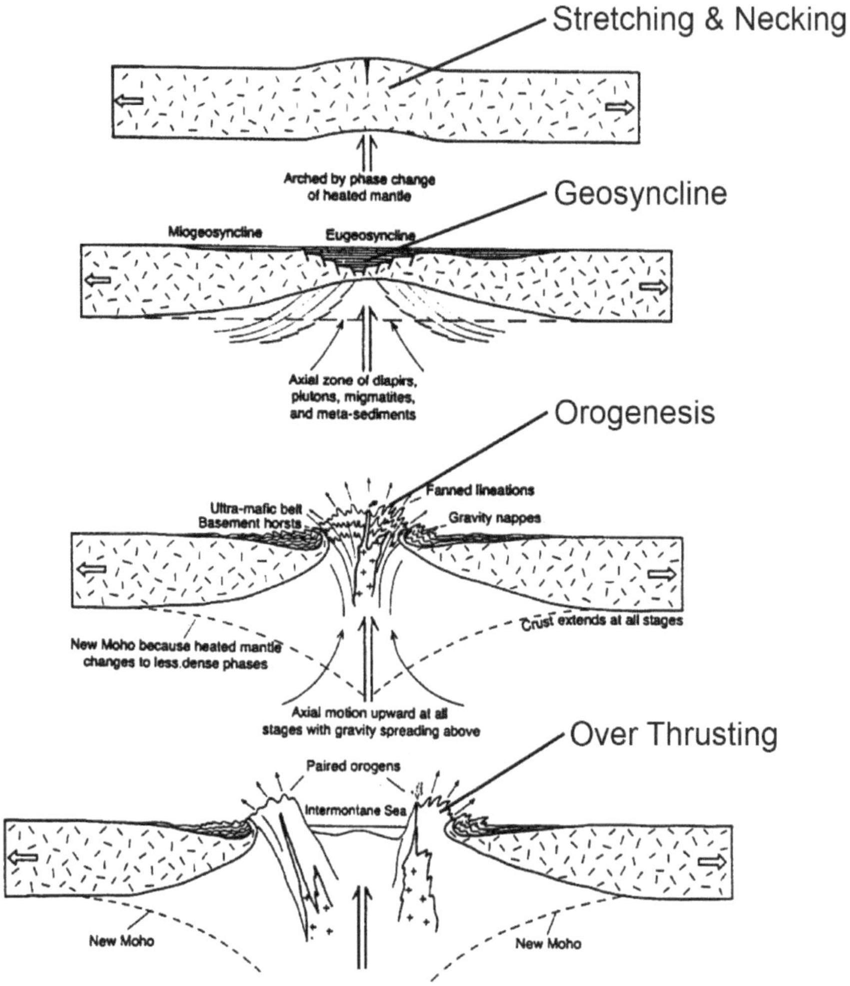

Figure 11.6 Carey's (1996) model for a symmetrical single phase orogen rising about 100 kilometres. The top figure shows initiation of primary stretching in the continental crust leading to "*necking*" or thinning. The bottom of the crust and mantle diapir below it rises and continues to do so during orogenesis. The middle figures show two contrasting sites of geosynclinal sediment deposition. The lower figures show a continuing and accelerating rise of the diapir signaling the onset of orogenesis and over-thrusting of the geosynclinal sediments to form folded mountains (from Carey, 1996).

11.1.3 Mountain Formation

Whenever we talk about mountains, for most people visions of grandiose snow-capped alpine scenery generally comes to mind. What must be appreciated is that all of this scenery, along with the foothills and surrounding plains, is actually the result of prolonged erosion which, in itself, creates the topography but has very little to do with the actual formation of the mountains. In this respect a mountain is geographically defined in the Oxford English Dictionary as *"a natural elevation of the Earth surface rising more or less abruptly from the surrounding level and attaining an altitude which, relative to the adjacent elevation, is impressive or notable."*

What must be further appreciated is that this impressive mountainous topography can be, and generally is, relatively young whereas the original formation of the mountains is generally much older and, over time, the mountains may have been eroded and uplifted many times. It is now known that mountains have been formed at various times throughout much of geological history. Over an extended period of time, once formed they then begin a natural cycle of erosion, decline, and reactivation. What we currently see as mountains today will eventually be eroded down to simple foot hills in the distant future—geologically speaking.

In conventional Plate Tectonics the formation of mountains is a random process primarily associated with convergence and collision of the various supercontinental crusts, where rocks are said to be pushed upwards during collision to form mountains. In Plate Tectonic studies, mountains form along what are referred to as convergent boundaries where the tectonic plates move towards each other and collide. In this process, the collision is said to trigger deformation and thickening of the crust. This, in turn, leads to uplift of the crust to form elevated mountains locally raising the surface of the Earth by over 3,000 metres. Over time, these mountains slowly erode through the action of rivers, weather conditions, and glaciers. The material eroded away is then washed into surrounding sedimentary basins or along the continental shelves and the reduction in weight of the mountains may result in still further uplift of the mountainous terrain.

This continental collision process is considered in Plate Tectonic studies to represent a horizontal compression mechanism that leads to deformation, crumpling, and faulting of the various rocks into folds or thrusts along the various convergent plate boundaries. To balance the weight of the mountains formed at surface, much of the compressed rock is also considered to be forced downward producing deep crustal roots. It is considered in Plate Tectonic studies that compression due to convergence of the plates can be the result of either continental to oceanic crustal convergence, oceanic to oceanic crustal convergence, or continental to continental crustal convergence processes.

These processes include:

- When a plate of continental crust converges with a plate of oceanic crust, the heavier oceanic crust is considered to move under the continental crust by a process called subduction. In conventional Plate Tectonics this is the main process through which mountains and volcanoes are formed where the subducted oceanic crust is then melted and recycled to the surface at a later time.

- When a plate of oceanic crust converges with another plate of oceanic crust, the older crust is said to be subducted under the newer crust leading to formation of volcanic island arcs, for example the Southeast Asian islands.

- When two plates of continental crusts come into contact with each other, neither will subduct beneath the other due to their excessive thickness. This type of collision is then said to form long fold mountain belts extending for thousands of kilometres, often containing remnants of oceanic sediments in the highest peaks. For example the Alps in Europe, and the Himalaya.

It is important to also appreciate that formation of mountains by collision on a Plate Tectonic Earth is a random process which is intimately linked to the equally random process of formation and break-up of the various supercontinents. Because of this random formation process, it is conceivable to suggest that very few of the known mountains on Earth today would be inter-related. What I mean by this is that, on a Plate Tectonic Earth, mountains in Australia may not be related to mountains in the Americas simply because they may not have been involved in mutual collision. In Plate Tectonics this random formation and break-up process has been operative throughout much of Earth history and generally had a long history of initial uplift and formation, followed by ongoing erosion to a subdued topography.

The most common types of mountains recognised in geology today include fault-block mountains, dome mountains, volcanic mountains, and plateaux mountains.

Fault-block Mountains, or simply block mountains, are formed when faults or cracks in the Earth's crust force large blocks of rocks—hundreds of square kilometres scale blocks—up or down. The uplifted blocks are called horsts and the intervening down-faulted blocks are called grabens. The largest of these mountain processes can form rift valley systems such as the East Africa rift valley. Movement of these blocks of rock may also stack on one another usually with a steep sided front and a sloping back. The Sierra Nevada Mountains in North America and the Harz Mountains in Germany are good examples.

Dome Mountains are formed when large amounts of molten rock or magma push up the Earth's surface rocks from deep within the crust. The magma, though, rarely reaches the surface of the Earth. This intruded molten rock then cools and forms a dome shaped mountain terrain. The Black hills of South Dakota and the Adirondack Mountains of New York in the USA are examples of Dome Mountains.

Volcanic mountains are created when molten magma pushes its way from beneath the Earth's crust and when it reaches the surface it erupts as lava, ash, and volcanic gases. These erupted materials accumulate and build up around the volcano as a large shield-like volcanic terrain. Volcanic mountains are then shaped by further eruptions, lava flows, and crustal collapses. Mount Fuji in Japan, Mount Rainer in the USA, as well as Mauna Loa and Mauna Kea on the Big Island of Hawaii are examples of volcanic mountains.

Plateaux mountains are formed when the land is uplifted and exposed to ongoing erosion. These plateaux initially form large areas of elevated flat land, often over 600 metres above sea-level. Over millions of years, the reactivated rivers and streams cut deep into the plateaux forming elevated mountainous topography. The remnants of the original plateaux can often be seen on the horizon as a common, relatively flat elevation of the various mountain peaks. The mountains in New Zealand and the Catskills of New York are examples of plateaux mountains.

During uplift and exposure of each of these mountainous terranes, the newly formed mountains are then subjected to the agents of erosion, such as by water, wind, ice, and gravity, which proceeds to gradually wear the uplifted mountains down. Each of these erosion processes then form characteristic mountain systems ranging from high glaciated peaks to desert landforms and low undulating foot hills.

This brief summary of conventional thinking on the formation of mountains will be elaborated on further in Chapter 12 when dealing with mountain building and orogenesis on an Expansion Tectonic Earth.

11.1.4 Orogeny

The word *orogeny* comes from the Greek word *oros* for mountain plus *genesis* for creation or origin. Gilbert originally defined the term orogeny in 1890 as representing a period of mountain building. Burg and Ford in 1997 went on to define orogeny as a concept referring specifically to the "*...folding of rocks in fold belts.*" This folding was further considered to be "*...characterised by intense internal rock deformation along long linear mountain belts, and, at places, by associated regional synkinematic metamorphism.*" Here, *regional synkinematic metamorphism* simply means the alteration of rocks by heat and pressure occurring at the same time as the folding or deformation event.

Burg and Ford also considered that "*...orogenesis encompasses all processes that result in crustal thickening causing uplift and hence high topography*" which may occur over variable intervals of time and may consist of a number of distinct tectonic events. With changing emphasis to the terminology, in particular with respect to Plate Tectonic usage, orogeny now refers to forces and events leading to a large structural deformation of the Earth's crust and upper mantle due to the interaction of tectonic plates. Deformation of the crust then results in the formation of long tracts of highly deformed rock called orogens or simply, orogenic belts.

Plate Tectonic theory incorporates two profoundly different explanations for orogeny, referred to as *non-collisional* and *collisional* models. The non-collisional model requires an on-going and comparatively continuous event to produce a discontinuous, relatively short duration event called orogeny. Cebull & Shurbet in 1992 were concerned, however, that the model itself contains no mechanism by which these continuous-to-discontinuous transformations could be accomplished. The compressive strength of rock was also considered too low to be able to transmit the stresses required. Because of this, and also because of a failure to express a means of transmitting the stresses required for orogenic deformation, the non-collisional model is now little used in the interpretation of ancient orogenic belts.

In this context, the Plate Tectonic collisional model is now used to explain most orogenic structures in terms of crustal shortening occurring as a result of continental collision and compression. In addition, it is also used to explain how subduction can be considered continuous and orogeny as a discontinuous event during periods of continental collision.

Cebull & Shurbet concluded, however, that neither model in this two-fold explanation for orogeny depicted in Plate Tectonic theory is sufficiently comprehensive to explain the wide range of complexities inherent in orogenesis. They also concluded that both models contain inadequate mechanisms to drive subduction.

In 2000, Ollier and Pain noted in their book on mountains that a distinction must be made between orogeny, which deals with the formation of fold belts, and uplift to form mountains. This distinction is not always clear in conventional tectonic usage. In their book, Ollier and Pain showed that in contrast to folding of rocks to form mountains, modern mountain building instead results from vertical uplift to form "*plateaux*"—large areas of relatively flat uplifted land surfaces—rather than from horizontal compression. Ollier and Pain provided many examples from major mountain ranges throughout the world to show that most mountains are, in fact, relatively young eroded plateaux, with very little to do with the earlier folding event, or events, as preserved in the rocks below the plateaux. These plateaux, shown schematically in cross section in Figure 11.7, were seen as once being relatively flat, low-lying eroded surfaces, such as now seen along many continental margins. As a result of vertical uplift, the ancient rivers and streams of the time were then reactivated and new erosion surfaces proceeded to cut across older folded bedrock structures to form new plateaux and eroded escarpments.

This observation of Ollier and Pain contrasts strongly with the Plate Tectonic view that vertical crustal movement to form mountains can only occur from horizontal movement during continental collision or crustal subduction. In considering this, Ollier and Pain pointed out that any conventional tectonic explanation must account for the extended amount of time needed, as well as the stable conditions required to allow enough time for erosion of the land surfaces to form the observed plateaux, in particular where multiple and overlapping plateaux events occur. Also, Ollier and Pain considered that it must provide an explanation for the rapid vertical uplift that exposed the plateaued land surfaces to renewed erosion, as distinct from the completely separate periods of folding and uplift proposed by conventional tectonics.

In their book Ollier and Pain went further to present a model for what they referred to as "*...passive margin continental development and erosion*," again based on their extensive field-based observations. They concluded from this model that, rather than continental collision, early break-up and rifting to form continents started with a *"rift valley stage"*–a period of crustal extension and faulting.

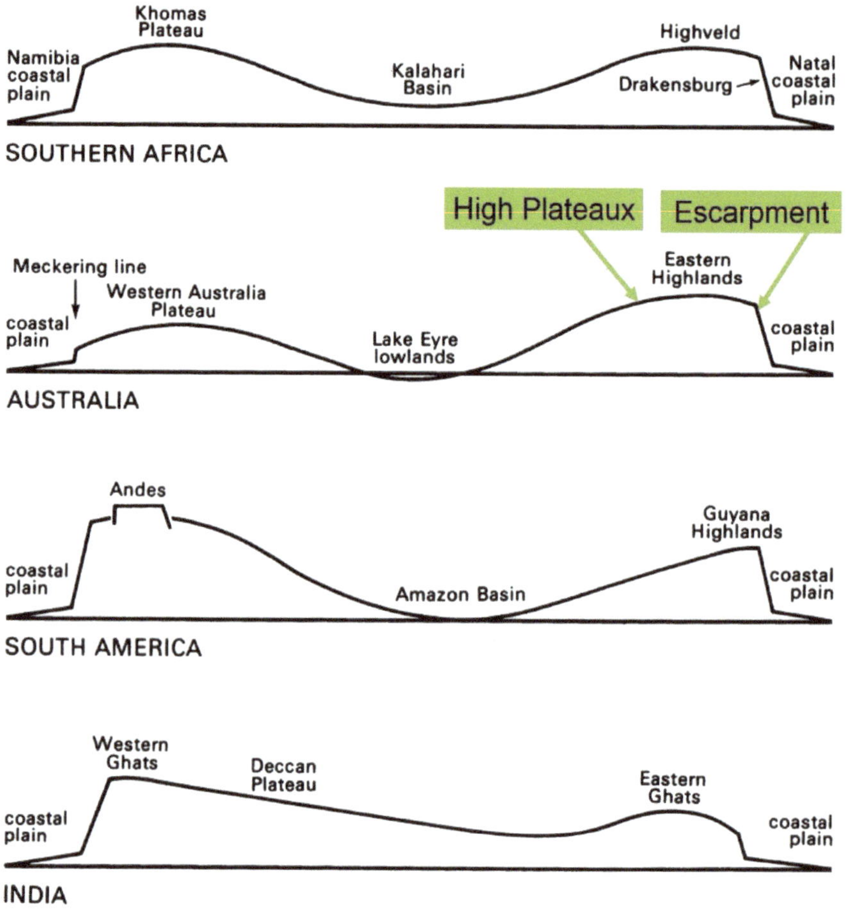

Figure 11.7 Schematic vertical cross-sections of mountains from various continents. In each example great escarpments parallel the coasts and separate a high plateau from the coastal plain. Each continent also has a central depression, bounded by gentle rises to high plateaux, beyond which are abrupt outward-facing escarpments and coastal plains (after Ollier, 1985).

During this stage an ancient coastal plain, located along the edge of the continent, was simply down-warped. On-land changes to the existing drainage pattern then occurred and, as seafloor spreading began, part of the coastal plain sank beneath the sea. Erosion was then concentrated on the steeper slopes of the exposed land surface, located between the axis of the warping and the coast, which may, in time, have been uplifted to form a great escarpment. Sediments eroded from the exposed land were then deposited offshore, covering the sunken plain in the process.

On an Expansion Tectonic Earth this coastal down warping of Ollier and Pain occurs naturally during changing Earth surface curvature. Relief of surface curvature then eventually reverses this process, resulting in uplift of the continental margins exposing the plateaux and forming an escarpment, followed by further erosion and planation.

While seemingly complex and sometimes outdated, what each of the authors in the previous sections were attempting to do was accommodate for crustal extension and thinning of the crust to form geosynclines, as well as compression of the crust to form orogenesis and mountains. The mechanism behind these phenomena is not immediately apparent in any of the schematic models shown above and will now be further discussed in context with an Expansion Tectonic Earth model.

12 Relief of Surface Curvature

"The appeal of the [Plate Tectonic] *collision model as a unique explanation of orogeny is diminished partly by its failure to rationalise the nature of contemporary orogenesis along such chains as the Andes and around much of the rest of the Pacific Rim."* Cebull and Shurbet, 1992

The expanding Earth proposals for changing surface curvature, orogenesis, and geosynclines, as well as conventional mountain formation introduced in the previous chapter were each considered adequate for their times and intended usage. From an Expansion Tectonic perspective, when each of these proposals were first suggested, the ancient radius of the Earth was only thought to have increased from a Pangaean supercontinent of around 55 percent of the present Earth radius. Expansion was then thought to have only occurred during the past 160 to 200 million years or so. There was very little consideration in the literature for changes to the continental crust or for any increase in Earth radius prior to that time. Because of this, the primitive Earth was considered by these researchers to have been fully developed, or at least mostly developed, by 200 million years ago, implying that it had also been fully developed since early-Archaean times.

Since these proposals were put forward there has been a wealth of new global tectonic data collected, as well as completion and publication of the Geological Map of the World in 1990. Based on this new data and geologic mapping, Expansion Tectonic small Earth modelling studies presented here now show that Earth radius has been increasing exponentially throughout the entire history of the Earth, extending from early-Archaean times through to the present-day. Where the term exponential refers to growth rates starting quite slowly and only building to the rapid exponential growth rates after considerable time has passed.

From these small Earth modelling studies, crustal extension and relief of surface curvature within the ancient supercontinental crusts was then shown to have occurred at a very much slower rate than the seafloor crusts. Extending over a far greater period of time than that currently shown by the modern seafloor crustal spreading.

Within the context of Expansion Tectonics, the primordial Archaean granite and volcanic supercontinental crust was initially formed and stabilised as a primitive Earth with an ancient radius of about 27 percent of the present Earth radius. The ancient Earth curvature at that time differed markedly from, for example, that shown by rocks deposited within a sedimentary basin during the much younger Permian Period. Both of these Earth curvatures were, in turn, markedly different to that now preserved in the seafloor volcanic rocks intruded during opening of the modern oceans. The bedrock geology previously shown in Figure 11.2 therefore represents a collage of rock types and rock ages, each of which were deposited, emplaced, or intruded at very different geological times and under completely different Earth curvature conditions. This appreciation, of course, is unique to an Expansion Tectonic Earth.

What the crustal evidence from the Expansion Tectonic small Earth models now show is that there have been four distinctly different, but inter-related, crustal forming episodes during Earth history. These episodes are distinguished primarily by the differing styles of relief of surface curvature and include:

1. A primordial Archaean Earth episode representing formation and stabilisation of the most ancient granite, metamorphic, and volcanic cratonic crusts.

2. A second more extensive Proterozoic episode marked by onset and fragmentation of the ancient Archaean crust, followed in turn by crustal extension and deposition of sediments within an established network of sedimentary basins. Relief of surface curvature during that time was maintained by on-going erosion of the land surface, reducing it to a flat featureless landscape.

3. The third Palaeozoic episode represents a time when supercontinental crustal extension and relief of surface curvature had exceeded the ability to erode the landscape to a flat profile. This episode was characterised by an increasingly elevated to mountainous landscape prior to eventual supercontinental crustal rupture and break-up during the late-Permian.

4. Finally the fourth Mesozoic and Cenozoic episode is represented by rupture and break-up of the supercontinental crust to form the modern continents and opening to form the modern oceans. This, still current, episode is characterised by intrusion and preservation of mantle-derived seafloor volcanic lava along each of the mid-ocean-ridge spreading centres.

Formation of the Earth prior to 4,000 million years ago is a highly speculative topic, which has only been studied from a static radius Earth perspective and may never be known definitively. However, geological evidence in the form of high temperature minerals and rock-types, in particular their distribution on a primordial Expansion Tectonic Earth, strongly suggests that the pre-Archaean Earth—times older than 4,000 million years ago—may have been hot and molten. The sheer volume of granite and volcanic rocks making up the primordial small Earth at that time suggests that high temperatures must have existed anyway. Only by 4,000 million years ago were surface temperatures sufficiently cool to enable the various minerals making up the ancient continental crustal rocks to crystallise and form a stable Archaean supercontinental crust.

This observation is contrary to some current conventional opinions where the Earth is said to have been formed by cold accretion of interstellar material which then subsequently heated, melted, and fractionated to form the crust, mantle, and core as is currently known, at the current Earth radius. While this may or may not have happened prior to 4,000 million years ago, the evidence presented here shows that the Earth was at least cool enough by 4,000 million years ago to have formed a stable crust.

On an Expansion Tectonic Earth the most primitive early-Archaean crust can be visualised as being analogous to slag floating on the surface of molten metal within a furnace. On the ancient Earth, this slag represents the ancient crust and comprises mainly lighter silica- and alumina-rich elements which, when cooled, crystallised to form the silica- and feldspar-rich minerals common to the continental crusts. In contrast, the mantle, located below the continental crust, comprises the mainly heavier iron- and magnesia-rich elements and minerals that are common to the intrusive volcanic rocks including those that now make up the modern seafloor crusts.

It is envisaged from a geological perspective that once this most ancient primordial Archaean Earth crust first cooled and stabilised, the mainly crystalline granite and intruded volcanic rocks remained as a rigid supercontinental crust and covered the entire primitive Earth. It is further envisaged that onset of very subtle Earth expansion was accompanied by simple global-scale fracturing of this brittle supercontinental crust. Fracturing then gave rise to further intrusion of granite magma and volcanic lava along faults and fractures within a newly established network of crustal weakness. This network of crustal weakness also formed the precursor to the primitive network of low-lying sedimentary basins. The extremely small rate of change in both Earth radius and surface curvature occurring during these ancient times was enough to maintain a very limited super-elevation of the central parts of the fragmented crust and to also establish a subtle elevation contrast between the locally elevated lands and low-lying sedimentary basins.

The terms *super-elevation* and similarly *enduring strength* and *relief of surface curvature* will be elaborated on later. Super-elevation can be visualised as being vaguely similar to the cross-section of a road pavement, where the central part of the road is raised higher than the outer margins and the ditches either side of the pavement represent the low-lying sedimentary basins. In relation to an Expansion Tectonic Earth, super-elevation simply refers to the raised central part of a fragment of rigid crust. Similarly, enduring strength simply refers to the ability of the continental crust to remain rigid and so retain its original surface curvature and hence its ability to super-elevate during Earth expansion. Relief of surface curvature of the crust then first begins when super-elevation fails and the crusts further fragment, deform, and gravitationally collapse to re-equilibrate to the changing surface curvature.

The second Proterozoic crustal episode was marked by retention of these fragmented primordial crusts along with on-going crustal extension localised between each of the ancient crustal fragments. This crustal extension process formalised the global network of crustal weakness and accompanying sedimentary basins, which are now common to the ancient orogenic belts on Earth today. It is further envisaged that relief of surface curvature during these times was accompanied by localised super-elevation and fragmentation of the rigid primitive crusts as well as depression around the margins of the land surface within the relatively flexible sedimentary basins. The extremely long period of time involved during this episode enabled the elevated lands to be continually eroded to a flat featureless landscape. This allowed the rigid crusts to still maintain a limited, but apparent, super-elevation while erosion continued to reduce the elevation contrast between the lands and seas.

While seemingly straight forward, this second episode differs from what most geoscientists currently believe. The exposed rocks within the known ancient Precambrian crusts are generally complexly folded and strongly metamorphosed—rocks that are recrystallized due to high pressures and temperatures. In conventional tectonic studies this complex folding would imply that the rocks have been subjected to continental collision to form high fold mountain belts. But, contrary to this belief and as will be discussed later, folding and metamorphism of rocks does not necessarily imply crustal collision, it only requires sustained localised compression.

Because of the extended period of time available during this second crustal episode, it is envisaged that downward relief of surface curvature of the fragmented crusts would have initiated the required prolonged and sustained compression and folding—estimated to be of the order of microns per year—extending over many hundreds of millions of years. The extensive amount of time available to fold these crustal rocks would also mean that erosion of the folded rocks exposed at the land surface would have continued to maintain an on-going flat landscape. The ancient rocks exposed today do indeed show preservation of the complex folding of these rocks but not necessarily any evidence of the mature eroded land surface.

During the third Palaeozoic episode Earth radius and surface area began to steadily to rapidly increase. Relief of surface curvature during that time was not able to keep pace with the accelerating changes to the land surfaces. These accelerating changes also lead to initiation of orogenesis and formation of high elevated mountain ranges. As a result of these changes, a feature of this episode was the presence of increasingly active drainage systems where erosion of the landscape and deposition of eroded sediments became prominent. It was also during this time that the Earth began to acquire the varied landscapes that are now more familiar to us today—albeit devoid of vegetation during much of that time. Older sediments deposited during the more ancient, smaller Earth radius times were then subject to periods of renewed folding, faulting, erosion, and orogenesis. All of this was accompanied by an increasingly elevated and disrupted landscape prior to eventual crustal failure and rupture during the late-Permian to form the modern continents.

This Palaeozoic episode was succeeded by the fourth post-Permian episode of crustal break-up and opening of the modern oceans, extending over the past 250 million years. This fourth episode is characterised particularly by on-going intrusion and preservation of the modern mantle-derived seafloor volcanic crustal rocks. Of importance is that during this episode each of the seafloor volcanic crusts shown as coloured stripes in Figure 11.2 were intruded and preserved at ever-changing Earth curvatures. Relief of surface curvature during this post-Permian episode then shifted from extension-style mechanisms involving sedimentary basin formation and erosion, to completely new continental and seafloor crustal styles. These new styles mainly involved crustal uplift and erosion along the margins of the fragmented modern continents to form mountain plateaux, plus on-going mid-ocean-ridge extension and transform faulting within the oceans.

During each of these four fundamental crustal forming episodes, both the sedimentary rocks and seafloor volcanic lava represent *time-dependent* deposition or intrusion. What I mean by this term is that both the sediments and seafloor volcanic rocks were not all deposited at the same time or at the same Earth curvature. Instead, their deposition or intrusion was spread over different periods of time and were formed *at the prevailing Earth curvature existing at the time of deposition or intrusion*. During the small Earth modelling studies outlined in previous chapters, when moving back in time, it was this time-dependent process that represents the primary mechanism adopted when returning these rocks to where they originated from and to then assemble each of the remaining crusts on smaller radius Earth models.

What the distribution of exposed ancient lands, as well as the network of ancient continental seas, show on the small Earth models is that the ancient supercontinents, in particular those making up the most ancient cratons and orogens, did indeed show an "*enduring strength in the continental crust*" but not necessarily a prominent "*super-elevation*" of the continents as suggested by Rickard. Super-elevation on the ancient Expansion Tectonic Earth only occurred within the fragmented ancient cratons but, because of the imperceptibly slow rate of change in Earth radius during these times, erosion was able to maintain a flat featureless landscape. What this means is that during these ancient times, because of the rigidity of the ancient crusts, the crusts did indeed partially super-elevate but were in turn continually eroded. Because of this, they were then never able to truly super-elevate above a flat and featureless landscape.

It should be appreciated that some form of super-elevation is considered necessary during these ancient times in order to maintain some form of elevation of the exposed lands, which, in turn, was necessary for erosion to occur. The subtle enduring strength and elevation contrast between the cratons, orogens, and basins then persisted throughout the early Earth history increasing to include periods of orogenesis and formation of fold mountain belts during Palaeozoic times as the Earth surface curvature rapidly changed. Super-elevation during these ancient times was, in turn, accompanied by a downward adjustment for changing surface curvature within each of the continental sedimentary basins, a process that gave rise to formation of geosynclines and ultimately to orogenesis and mountain formation.

12.1 Primary Crustal Mechanisms

In order to fully appreciate the processes involved in relief of surface curvature on an Expansion Tectonic Earth it is necessary to introduce a number of primary mechanisms to visualise the way the various crusts may behave during Earth expansion. The basis of these mechanisms is common knowledge in structural and civil engineering studies and is used extensively for designing and testing structural components for engineering projects. These mechanisms are firstly introduced here at an engineering scale in the form of a beam and will in turn be adapted and modified to the scale of the Earth's crust.

Consider what happens to a straight horizontal beam supported at either end, such as a steel beam in a high rise building. If a heavy load is applied at the mid-point of the beam it will tend to flex down due to the combined weight of both the load and the beam, shown exaggerated in Figure 12.1. Two things will then happen to the beam. The physical properties of all beams tells us that, due to downward flexing caused by the applied load, the top half of the beam will be in compression and the bottom half will be in tension regardless of what material the beam is made of.

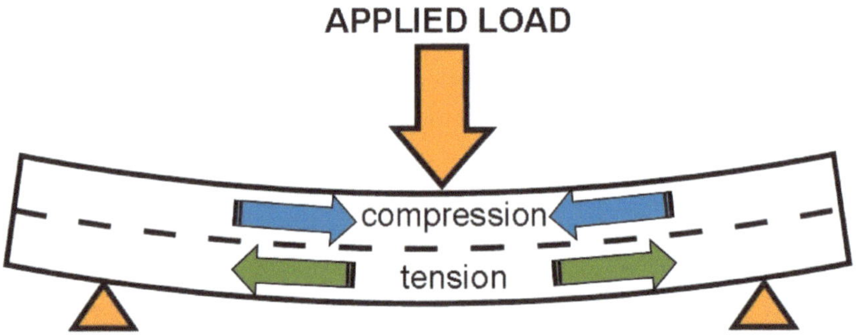

Figure 12.1 A simple horizontal beam supported at both ends, with a centrally located applied load.

Taking this a step further, depending on the physical strength and applied load acting on the beam, a rigid beam, for example one made of concrete, will be able to resist compression in the top half but will easily break by cracking across the bottom of the beam. In contrast, a flexible beam, for example one made of metal, will be strong in tension but may simply wrinkle or buckle as it sags and compresses along the top surface. A combination of these two properties is the basis of reinforced concrete, where bars of steel reinforcing stop concrete from cracking under tension and the compressive strength of the concrete stops the steel bars from buckling. Similarly, for a beam made of a soft material, such as clay, where it is weak in compression as well as in tension, the bottom half of the beam will simply stretch and the top half will wrinkle and compress.

Applying this engineering principle to the scale of the ancient Earth, Figure 12.2 is a schematic cross-section of a small portion of a brittle primordial Earth crust. In this example the crust, which is analogous to the beam in Figure 12.1, is upwardly curved because of the curvature of the Earth. The crust is also part of a stabilised ancient craton and comprises brittle crystallised granite magma. During the early-Archaean, this primordial crust covered the entire ancient Earth surface and can be considered analogous in physical characteristics to the concrete beam in the previous analogy. During an increase in Earth radius from A up to B, the increase in radius generates localised tangential compression at the surface interaction of each crustal fragment and a tangential extension at the base of the crust.

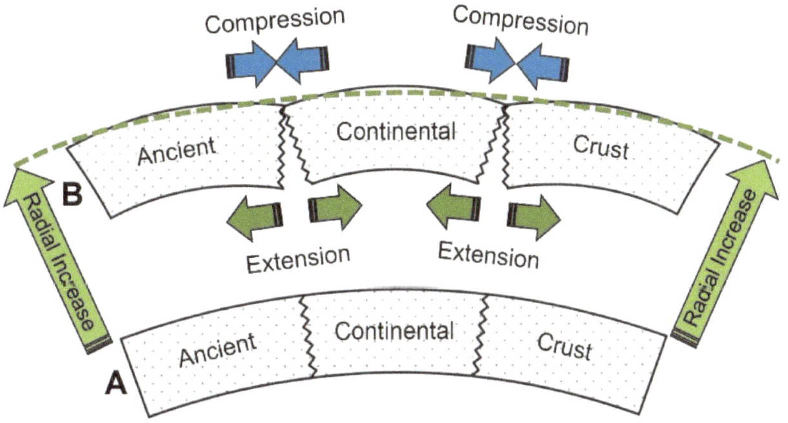

Figure 12.2 A schematic cross-section of a fragment of an ancient supercontinental crust undergoing break-up and fracturing during a change in Earth radius and curvature from Figure 12.2A to Figure 12.2B. Note: crustal thicknesses are not to scale.

The Earth's crust shown in the lower Figure 12.2A can then be visualised as a simple, albeit convexly curved beam. In this case, the applied load is the downward force of gravity plus the weight of the crust itself. Because the crust is brittle, and hence weak in tension, during an increase in Earth radius the crust will initially start to super-elevate and rise in the central part of the crustal fragment, but over time it will eventually fracture and fragment. The smaller, fragmented crusts will then gradually collapse under the influence of gravity and adopt the new surface curvature existing at the time—shown as a green dashed line in Figure 12.2B. In geology this process represents a primary mechanism for compression of rocks at the surface plus extension focussed within the lower crust, allowing melting of the lower crust to form new granite as well as intrusion of volcanic lava and magma as dykes. The *tonalite-trondhjemite-granodiorite* group of granite rocks common to all Archaean terranes throughout the world may be related to this fundamental compression-extension crustal mechanism and crustal fragmentation process.

Geological mapping evidence from continents around the world shows that the ancient primordial crust continued to fragment and stabilise until such time as it reached a size where it was able to retain its integrity and resist further large-scale fragmentation. In other words, fragmented ancient crusts of a number of hundred kilometres in diameter are now commonly preserved in the rock-record. It is envisaged that the ancient crust would be further stabilised during intrusion of new granite magma and volcanic lava, as well as new material added at the base of the crust sourced and crystallised from the mantle below. As will be further discussed later, the extremely long time-frame involved in this early-Archaean expansion phase—of the order of billions of years—would also see the super-elevated crusts being continually eroded and the eroded sediments being washed away and deposited in surrounding low-lying sedimentary basins. This erosion would have the added effect of thinning and weakening the central parts of the elevated crusts, leading to further cracking and crustal fragmentation extending down to the ordinary faulted and jointed rocks as proposed by Carey.

As well as the cross-sectional view of this hypothetical crust it is also important to understand what happens to the fragmented crusts when looking down at the surface of the primitive Earth. This consideration is only relevant to an Expansion Tectonic Earth where fragmented crusts also undergo what I refer to as *crustal jostling*. This crustal jostling process occurs as the fragmented crusts adjust their positions relative to each other during changes in surface area and curvature. As the Earth radius increases, surface area must then increase and surface curvature flattens with time. Crustal jostling is then seen as a global compensating process of adjustment of each of the crustal fragments. A schematic example of this crustal jostling process is shown in Figure 12.3.

Figure 12.3 is a hypothetical plan view of the surface of the Archaean Earth showing a number of fragments of ancient supercontinental crust, each with dimensions of a number of hundred kilometres across. The dashed areas between each of these crusts are a mixture of intruded volcanic lava, as well as sediments deposited in a network of low-lying sedimentary basins.

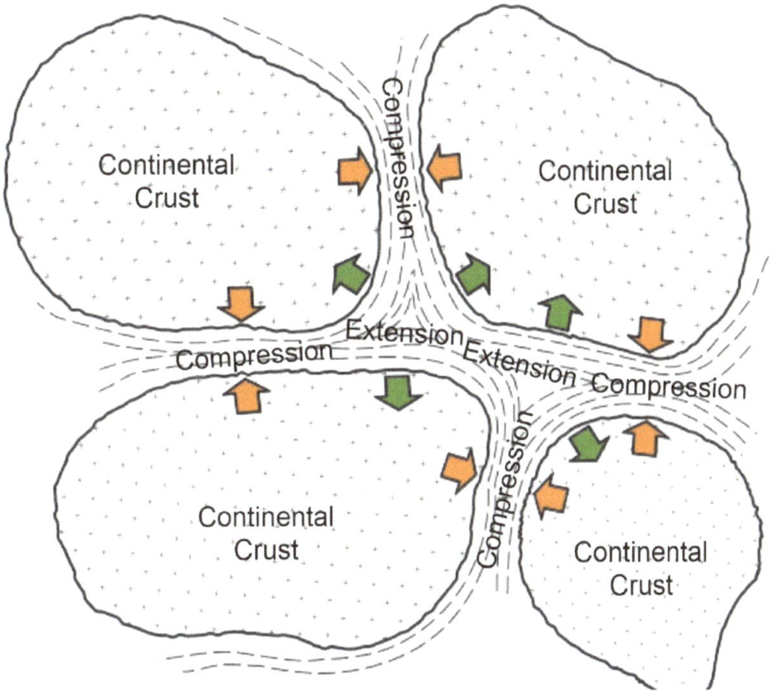

Figure 12.3 Schematic representation of the Earth surface of a brittle fractured primordial supercontinental crust, accompanied by crustal jostling, intrusion of magmatic rocks, and formation of a network of sedimentary basins. Orange arrows show localised compression and green arrows show crustal extension.

What is shown in this figure is the effects of crustal collapse, previously shown in cross-section in Figure 12.2B, as the change in surface curvature is distributed amongst the fragmented crusts. As each fragment of crust collapses due to the increasing Earth radius and changing surface curvature, local points of contact between the crustal fragments will be under compression—shown as orange arrows in Figure 12.3. These points of contact act somewhat like hinge points. In contrast, the intervening areas undergo crustal extension—the green arrows in Figure 12.3—and give rise to formation of low-lying sedimentary basins. This localised compression-extension phenomenon represents the focal points for both on-going folding of sediments and intrusion of volcanic lava plus further opening and deposition of new sediments within the low-lying sedimentary basins.

Both Figures 12.2 and 12.3 are an attempt to show the process of fragmentation and changing surface curvature of part of the Earth's crust in three dimensions. These figures need to be visualised together, one in vertical cross-section overlapping the other as a surface plan view. In addition to the localised crustal jostling shown in Figure 12.3, on-going changes in surface curvature will also involve regional-scale rotation of the fragmented crusts—which represents more advanced crustal jostling. The outcomes of this rotation and crustal interaction process are now preserved in the rock-record as fault and shear movement, as well as folding and metamorphism of the sediments and volcanic rocks as they are heated, compressed, and jostled during orogenesis.

12.2 Enduring Strength and Super-Elevation

The problem of an enduring strength and super-elevation of the supercontinental crusts, as originally proposed by Rickard, will now be further investigated on an Expansion Tectonic Earth. As previously mentioned, super-elevation simply refers to the raised central part of a large fragment of a rigid, or semi-rigid, continental crust. Similarly, enduring strength simply refers to the ability of the continental crust to remain rigid and retain its original surface curvature and hence its ability to super-elevate during increase in Earth radius. Once the enduring strength of the crust is exceeded then gravity-induced relief of surface curvature takes over and, over an extended period of time, the crust collapses and re-equilibrates to the new surface curvature.

On the Expansion Tectonic small Earth models it has been shown that the rate of increase in Earth radius prior to about 2,400 million years ago was of the order of microns, increasing to tens of microns per year. During these ancient times the increase in Earth circumference also amounted to tens, increasing to hundreds of microns per year—which is equivalent to the thickness of a human hair—where one micron is equal to one millionth of a metre, or one one-thousandth of a millimetre. These imperceptibly small Earth circumference measurements were in turn distributed around the entire 10,600 kilometre circumference of the ancient primordial Earth, which represents atomic-scale adjustments to the crust per kilometre per year.

With this in mind, prior to about 2,400 million years ago the ancient supercontinental crust can be considered a mixture of rigid ancient crusts surrounded by a network of relatively flexible sedimentary and volcanic crusts. The increase in Earth radius during that time is considered to have initially been absorbed as simple crustal fragmentation and on-going stabilisation of the ancient crusts. This was, in turn, accompanied by crustal stretching and extension between each of the crustal fragments. This process may have also been accompanied by localised faulting and fracturing, which extended down the structural hierarchy to movement within normal surface fractures and joints—the cracks and flat surfaces seen on everyday rocks.

Small Earth modelling studies suggest that some form of enduring strength and super-elevation of the supercontinental crusts must have been present during these ancient times. This was necessary in order to maintain an elevation contrast between the lands and shallow seas to maintain erosion and deposition of sediments, as now preserved in the rock-record. This enduring strength and super-elevation was, in turn, accompanied by a steady, downward gravitational adjustment to surface curvature as the crusts further fragmented and distorted.

This whole process would have also been accompanied by extensive erosion of the super-elevated land surfaces and deposition of eroded sediments within surrounding low-lying sedimentary basins. This super-elevation and erosion process was maintained because of the imperceptibly small changes to Earth radius occurring during these Precambrian times. The style and extensive distribution of chemical sediments present during Proterozoic times, which require very stable conditions to form, also suggests the landscape remained relatively flat and featureless during these times. The presence of an enduring strength and super-elevation of the ancient crust can then be visualised as a delicately balanced process comprising crustal fragmentation, change in surface curvature, erosion, and deposition of sediments within extensive sedimentary basins.

Since these ancient times, on an Expansion Tectonic Earth the rate of increase in Earth radius and similarly the rate of increase in Earth circumference and surface curvature has increased markedly. These changes have increased exponentially from microns per year during the ancient times to 22 millimetres per year increase in radius and 140 millimetres per year in circumference for the present-day. While these current figures may seem quite large, the 140 millimetres per year increase in circumference is in fact distributed over the entire 40,075 kilometres circumference of the present-day Earth, which again represents atomic-scale adjustments to the crust per kilometre per year. This present-day increase in circumference is now mainly confined to crustal extension within each of the seafloor crusts, where comparable extension rates of between 10 to 120 millimetres per year are currently recorded from each of the mid-ocean-ridge spreading zones.

The effects of enduring strength and changing super-elevation of a rigid continental crust during changing surface curvature of the Earth can be visualised by way of an example. Consider a fragment of an ancient craton of say 500 kilometres diameter that has remained relatively rigid and stable during an increase in Earth radius from an ancient radius to the present-day

(Figure 12.4). The calculated difference in mid-point elevation of this craton between the surface curvatures of the ancient Earth of 1,700 kilometres radius—the radius of the Archaean small Earth model—and the present-day Earth is approximately 27 kilometres. What this means is that the central part of the crust must, over time, either fragment and gravitationally collapse by 27 kilometres during changing surface curvature, or it must be eroded by 27 kilometres to maintain the prevailing surface curvature through time. Or, more realistically, a combination of both collapse and erosion must occur. This order of magnitude in super-elevation is well within the realms of presently observed erosion of these ancient cratons as still preserved in the rock-record. Erosion of between 10 to 40 kilometres of vertical crustal thickness, as measured from geological observations, is common within each of the ancient cratons present throughout the world today leaving behind the presently observed continental crustal thicknesses varying between 35 to 50 kilometres thick.

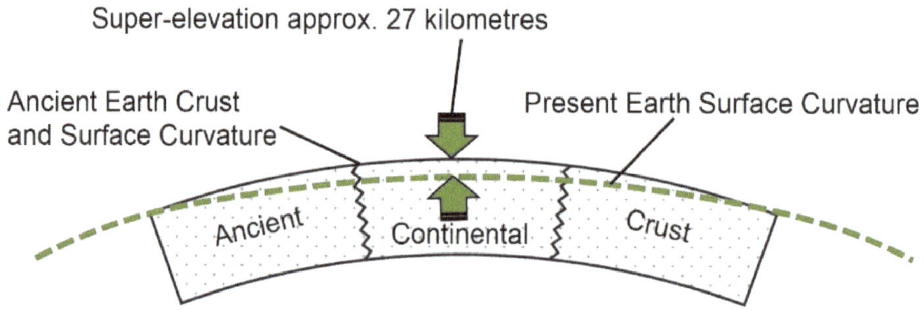

Figure 12.4 Cross section of the Earth showing the magnitude of change in surface curvature during Earth expansion from the primordial Archaean Earth to the present. The mid-point super-elevation of the 500 kilometre diameter craton shown amounts to approximately 27 kilometres. Crustal thickness not to scale.

Of interest also is that the measured surface area of the ancient crustal rocks occurring on Earth today represents about 33 percent of the total surface area of the modern continents. By eroding between 10 to 40 kilometres of ancient continental crustal rocks and depositing the sediments in surrounding sedimentary basins, the surface area of the supercontinents then theoretically increases during the Precambrian times to an estimated 70 to 80 percent of the present continental surface area—a process analogous to what is known as crustal accretion in Plate Tectonic studies. The remaining 20 to 30 percent surface area of continental crust is easily accommodated for by on-going fragmentation of the ancient cratons, increases in surface area during crustal extension within the sedimentary basins, as well as intrusion of new granite magma and volcanic lava.

What the example shown in Figure 12.4 suggests is that fragmented parts of the ancient supercontinental crust may have remained rigid enough to undergo relatively simple super-elevation during a very subtle, but prolonged increase in Earth radius. Fragments of crust up to say 500 kilometres diameter may have had enough enduring strength to remain in an on-going super-elevated state, with enough time available for steady on-going erosion to maintain the prevailing Earth surface curvature. Elsewhere, localised fragmentation, folding, crustal extension, volcanic intrusion, and deposition of sediments are also abundantly evident in the rock-record. These processes are particularly evident in the intervening areas between and surrounding the more rigid crustal fragments, which are now represented by the folded orogenic crustal rocks. This suggests that a combination of localised crustal rigidity, fragmentation, and gravity-induced collapse, plus extensive erosion of the crusts was aptly maintained during these ancient times.

In contrast to the most ancient supercontinental crusts, throughout Proterozoic and into the following Palaeozoic times the supercontinental crusts were dominated by crustal extension localised within an established network of sedimentary basins (Figure 12.5).

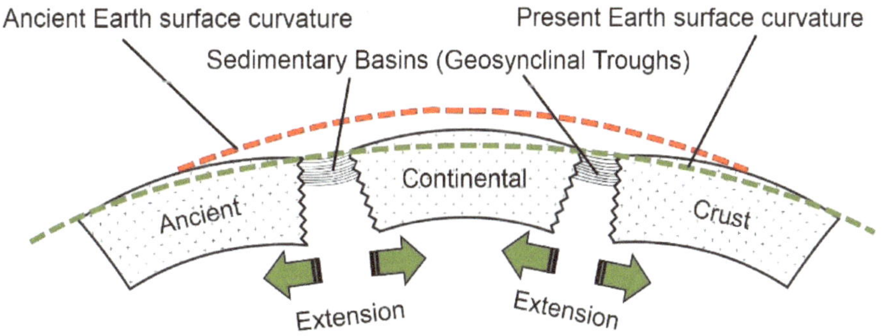

Figure 12.5 A schematic cross-section of an idealised fragment of an ancient supercontinental crust showing geosynclinal sedimentary basin formation during ongoing increase in surface area and Earth radius. Crustal thickness not to scale.

Figure 12.5 is shown for the purpose of emphasis and illustration only and represents simple fragmentation of an ancient crust. In reality, this fragmentation would also include a considerable amount of plastic deformation of the lower crust as well as faulting and rifting of the brittle crusts. In this context, the sediments deposited within the basins shown represent flexible *time-dependent* crusts. In other words, the sediments were deposited during periods of crustal extension at the prevailing surface curvature existing at the time of deposition, so there was no requirement for super-elevation of the sedimentary basins. The relative flexibility of the sediments also allowed the basins to continue to re-equilibrate to the changing surface curvature through time by simple, gravity-induced down-warping, extension, or folding of the sediments within geosynclinal troughs.

Following break-up of the supercontinents and opening of the modern oceans during the late-Permian Period, crustal extension then shifted from within the continental sedimentary basins to within each of the newly opening mid-ocean-ridge spreading centres. Preservation of this extension and seafloor growth process is now seen as the zebra-stripe-like magnetic growth patterns on the seafloor, paralleling each of the mid-ocean-ridges within each of the modern ocean basins. It is worth noting again that these seafloor crusts represent exposed and quenched mantle-derived lava, not continental rocks. Like the sediments deposited within sedimentary basins, deposition and preservation of this lava was at the ancient Earth surface curvature prevailing at the time of intrusion.

It is now known that intrusion of this seafloor volcanic lava occurs via near-vertical faults and fractures localised along the full length of the mid-ocean-ridge spreading axes. In effect, as new faults and fractures open, lava is intruded and then quenched by seawater when it reaches the seafloor. Over time, the lava may accumulate and extend in surface area as a series of stacked vertical dykes analogous to cards in a pack of playing cards placed on edge. Because of the physical properties of these stacked vertical dykes, the strength of the seafloor lava is then severely compromised. Relief of surface curvature during changing Earth radius is easily accommodated for by simple vertical movement along the margins of each intruded dyke.

The lack of extensive exposure of seafloor crusts on Earth today suggests that the inherent flexibility of these volcanic rocks is quite high resulting in a much subdued enduring strength and super-elevation of these time-dependant seafloor crusts. Larger-scale adjustments to surface curvature can also be seen as fracturing and faulting of the seafloor crusts in all of the modern oceans and these are shown in the Geological Map of the World as an extensive array of cross-cutting transform faults.

Over the past 250 million years the flexibility of this seafloor crust has then maintained a steady downward change in super-elevation and relief in surface curvature in keeping with changes in Earth radius. The mechanism for this seafloor crustal distortion process is also aided by the presence of elevated heat flow from the mantle below, which assists in relief in surface curvature by softening the basal regions of the seafloor crusts, as well as the weight of the overlying ocean water. This, in turn, encouraged additional plastic deformation, crustal stretching, and further intrusion of volcanic lava.

12.3 Relief of Surface Curvature

In addition to fragmentation, folding, and extension of the crusts, another equally important phenomenon occurs to the crusts during relief of surface curvature. In Figure 12.6 an ancient crust has fragmented and has undergone relief of surface curvature from the ancient radius to the present-day Earth radius, as shown by the red and green dashed curved lines respectively. In addition to the relief of surface curvature, as the crust flattens the chord length–the straight line distance between the outer margins of the crust–as shown by the long horizontal ancient red and present-day green arrows, lengthens. For example, for a crust with a curved surface length of 1,000 kilometres this increase in chord length is calculated to have been approximately 15 kilometres since early-Archaean times.

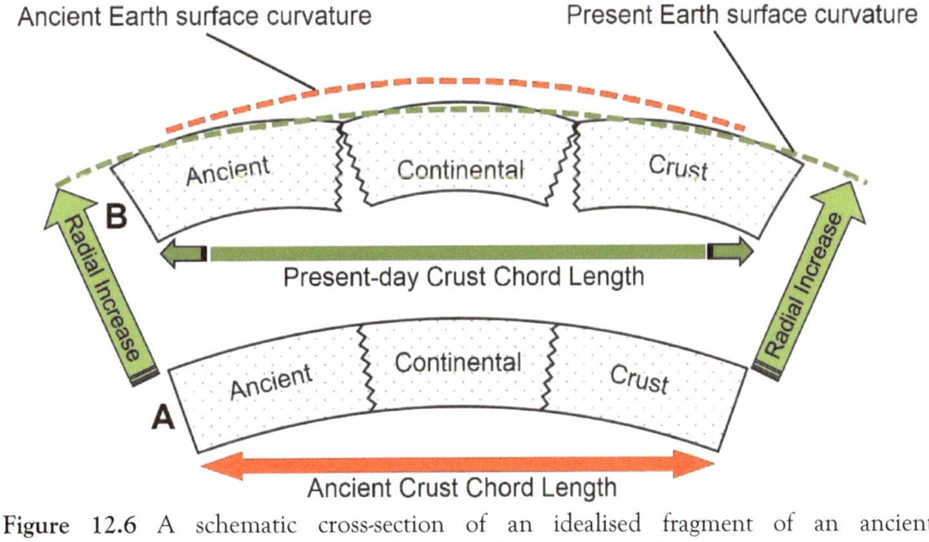

Figure 12.6 A schematic cross-section of an idealised fragment of an ancient supercontinental crust undergoing relief of surface curvature during an increase in Earth radius from Figure 12.5A to Figure 12.5B. During this relief of surface curvature process the chord length of the fragmented crust is shown to increase over time. Crustal thicknesses are not to scale.

This increase in chord length, in particular where multiple adjoining crustal fragments are involved, provides a prime mechanism for both ongoing crustal extension within surrounding sedimentary basins plus periods of localised crustal compression and shortening leading to orogenesis and fold mountain building. It must be emphasised that mountain building though is more complex than just compression leading to an upward movement of the crust. The scenario in Figure 12.6 is more applicable to formation of the ancient, pre-supercontinental break-up, fold-mountains. Modern, post-supercontinental break-up, mountains generally involve the uplift of a plateaux along the margins of the modern continents, followed by extensive erosion to form the varied elevated topography that is more familiar to us.

The fundamental mechanisms involved in crustal relief of surface curvature on an Expansion Tectonic Earth are highlighted schematically as coloured vector arrows at the left and right ends of Figure 12.7.

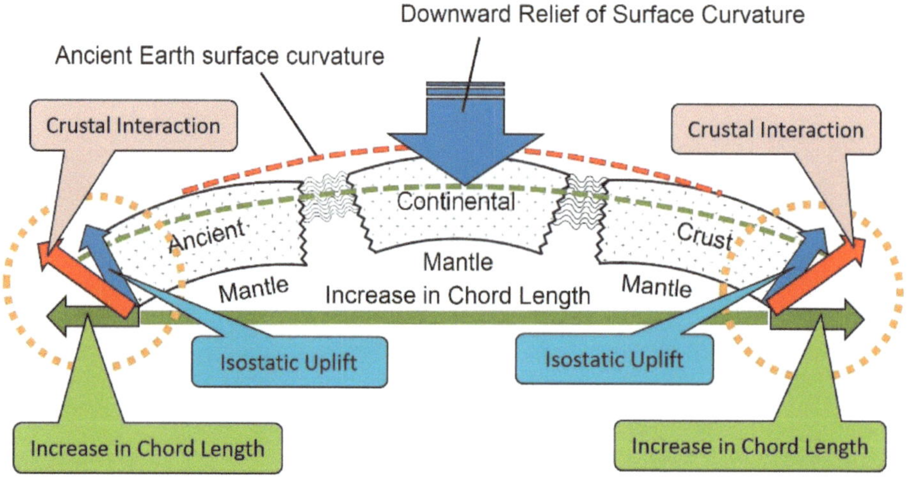

Figure 12.7 A proposed mechanism for relief of surface curvature on an Expansion Tectonic Earth. The large blue arrow represents relief of crustal curvature during changing surface curvature. The small red, blue, and green arrows represent the crustal movement vectors along the outer margins of a continent which simulates crustal interaction and orogenesis. Crustal thickness not to scale.

In this figure a hypothetical cross-section of a fragmented ancient supercontinental crust is shown located at the present-day Earth surface. This supercontinental crust has undergone fragmentation of the ancient cratonic crusts plus extension within a number of intervening geosynclinal sedimentary basins during relief of surface curvature. Folding of the sediments during compressional orogenesis is also evident within the basins. In this scenario the supercontinental crust now forms part of a modern continent and is separated from other continents by intervening modern oceans and seafloor crusts.

In order for the ancient continental crust shown in this figure to equilibrate to the present-day surface curvature, a downward relief of surface curvature, shown by the large blue arrow, is compensated for by an increase in the chord length, as shown by the long green arrow between the outer margins of the continent. Isostatic uplift of the continental crust has also occurred around the margins of the continent to form an escarpment and associated mountain plateaux. The relative crustal movements associated with this relief of surface curvature are highlighted by the red, blue, and green vector arrows at the left and right sides of the figure. These vector arrows show a green horizontal movement related to an increase in chord length, a blue vertical movement related to isostatic uplift of the outer margin of the crust, and a red net upwards and outwards crustal interaction movement at the interface between the continental crust and the seafloor crust.

This red net crustal interaction vector is seen as a prime mechanism for Wadati-Benioff-related seismic phenomena along the outer margins of the continental crusts. In addition, chord-related movement along the lower crust-mantle interface is seen as consistent with known crustal movement along the Mohorovičić discontinuity.

The Wadati–Benioff zone was named after two seismologists, Kiyoo Wadati of the Japan Meteorological Agency and Hugo Benioff of the California Institute of Technology, who independently discovered the zone. In conventional Plate Tectonic usage the Wadati–Benioff zone is a deep active seismic zone located within a subduction zone. Motion along this zone produces deep earthquakes, the foci of which may be as deep as 700 kilometres. This exact same zone and seismic phenomena on an Expansion Tectonic Earth is considered to be related to the red crustal interaction vector and crustal movement shown in Figure 12.7.

The Mohorovičić discontinuity, usually referred to as the Moho, is the boundary between the Earth's outer crust and the mantle and is named after the pioneering Croatian seismologist Andrija Mohorovičić. This Moho discontinuity is located some 5 to 10 kilometres below the seafloor and between 20 to 90 kilometres beneath continents, with an average depth of about 35 kilometres. In this context, the Moho basically separates both the seafloor crust and continental crust from the underlying mantle—bearing in mind that seafloor crust is simply cooled and stabilised upper mantle. On an Expansion Tectonic Earth the Moho represents the discontinuity between the crust and mantle during relief of surface curvature and increase in chord length acting over time.

While schematic, Figure 12.7 shows that there is adequate mechanism for crustal interaction on an Expansion Tectonic Earth during downward relief of surface curvature giving rise to currently observed seismic phenomena. The increase in chord length of continental crusts is again seen as a prime mechanism for generating both Wadati–Benioff-related phenomena along the outer margins of the continental crusts as well as Moho-related discontinuity movement along the base of the crusts. These phenomena, in turn, also represent a mechanism for the long linear mountain escarpments and plateaux of Ollier and Pain, as well as island-arc and back-arc basin phenomena commonly seen around the western Pacific Ocean rim.

12.4 Orogenesis and Mountain Building

The definition of orogeny, as proposed by Burg and Ford in 1997, is defined as the *"folding of rocks in fold belts,"* which are, in turn, *"characterised by intense internal rock deformation along long linear mountain belts, and, at places, by associated regional synkinematic metamorphism."* Again, synkinematic simply means at the same time. This Plate Tectonic definition for orogeny and mountain building is now used to explain these structures in terms of continent to continent collision.

I take this opportunity to reiterate that this definition of continental collision is both overly simplistic and confusing. The reader may recall that the original Continental Drift model of Wegener suggested that continental crusts simply ploughed their way through the basaltic seafloor crusts, hence providing a potential mechanism for both continent to continent break-up and collision. Modern seafloor mapping now shows that the seafloor crusts, in fact, form part of a much larger global array of crustal plates. These plates are the basis of both Expansion Tectonic and Plate Tectonic theories and comprise complementary continental and seafloor crusts. Rather than simply ploughing their way through the seafloor crusts, it is now known that both continental and seafloor crusts must act in unison in order preserve the observed geology of these identified plates over time.

In Plate Tectonic usage, the terms collision and compression are generally considered synonymous implying that compression of rocks to form fold belts and uplift to form mountains can only occur as a result of plate to plate, plate to continent, or continent to continent collision. These terms are not necessarily synonymous. As an example, clapping your hands is analogous to collision whereas simply holding hands is compression. Two crusts do not need to be separated to then come together in collision to compress, fold, or uplift rocks. Compression can just as easily take place between two mutually adjacent or interacting crusts.

On an Expansion Tectonic Earth, the definition of orogenesis is instead seen as *a sustained compression of rocks between mutually adjacent crusts giving rise to folding of the rocks to form long linear fold mountain belts, and, at places, by associated regional synkinematic metamorphism.*

The fundamental mechanism for orogenesis on an Expansion Tectonic Earth involves an extension of the primary relief of surface curvature mechanisms outlined in Sections 12.1 and 12.2. During a downward gravity-induced re-equilibration of the ancient crust to the changing surface curvature, two fundamental processes are seen as contributing to sustained compression which may then contribute to orogenesis (Figure 12.8).

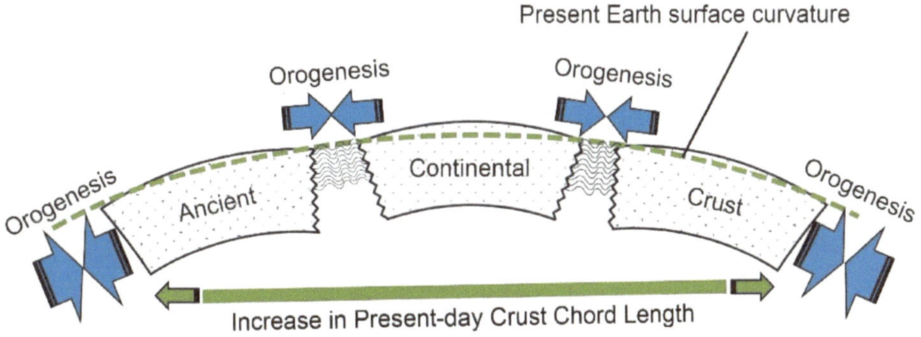

Figure 12.8 A schematic cross-section of fragments of an ancient supercontinental crust undergoing relief of surface curvature and compressional orogenesis within linear geosynclinal sedimentary basins. Crustal thickness not to scale.

Firstly, as the ancient crusts gravitationally collapse and re-equilibrate to the changing surface curvature, localised compression may occur between adjacent crustal fragments and within the surrounding network of geosynclinal sedimentary basins. Secondly, as the ancient crust re-equilibrates to the new surface curvature there is an increase in the overall chord length of the crust. This increase in chord length provides an outwardly-directed compression along the margins of the crust which may be localised within adjoining sedimentary basins. Both processes are then considered to result in orogenesis and uplift to form long linear fold mountain belts. Changing surface curvature also induces rotation and crustal jostling between the various ancient crustal fragments, which, in turn, is accompanied by localised folding, faulting, and shear movement.

This mechanism for orogenesis—the folding of rocks to form long linear fold mountain belts—on an Expansion Tectonic Earth differentiates between pre-break-up supercontinental crusts and modern post-break-up continental and seafloor crusts. It is important to take note of this distinction because orogenesis within the supercontinental crusts is not a random or isolated occurrence as portrayed in conventional tectonic studies; it is instead focused within the established global network of crustal weakness and sedimentary basins. This is because there were no large oceans, hence no mechanism or room for continental drift to promote continental collision. Similarly, break-up of the supercontinents and opening of the modern oceans is seen as a progressive increase in surface area involving an accumulation of new seafloor volcanic lava with again no provision or need for continental drift or collision.

The mechanism for orogenesis on an Expansion Tectonic Earth, shown schematically in Figure 12.8, shows that initially, as the Earth radius increases over time, the primitive crustal fragments gradually depress around their margins allowing eroded sediments to be deposited within newly formed or reactivated sedimentary basins. These sedimentary basins, in effect, are the *geosynclines* as discussed in Section 11.1.2. These eroded sediments were deposited at the surface curvature prevailing at the time of deposition. Deepening of the basin and vertical thickening of the accumulated sediments—a primary characteristic of geosynclines—is then seen as a direct result of changing surface curvature over time.

On Expansion Tectonic small Earth models, periods of orogenesis and mountain building occurring during Palaeozoic times were shown to be located within the global-scale network of sedimentary basins and crustal weakness. Within this network, localised compression occurred during periods of on-going or periodic crustal collapse as well as crustal jostling during changes to surface curvature. The resultant long linear inter-related fold mountain belts were, in turn, eventually broken-up and dismembered in conjunction with break-up of the supercontinents during formation of the modern continents and oceans. Fragments of these related ancient belts are now preserved within many of the modern continents throughout the world today.

It is significant to note that on an Expansion Tectonic Earth orogenesis, and its associated fold mountain building, is intimately related to the established global network of sedimentary basins and crustal weakness. As such, rather than being a random formation process as portrayed in conventional Plate Tectonics, orogenesis and fold mountain building is, in fact, an interconnected process extending over a wide range of geological times. It is only since supercontinental break-up and dispersal of the modern continents that we now see these various mountain belts as fragmented remains of formerly more extensive mountain systems.

For example, the Palaeozoic Mauritanides Mountains along the west coast of Africa once formed part of the Appalachian Mountains along eastern North America, which were, in turn, linked to the Hercynide Mountains of Western Europe. Similarly, the Cordilleran Mountains and associated Rocky and Sierra Nevada Mountains along the west coast of North America form an extension of the Andean Mountains along the west coast of South America. On Expansion Tectonic small Earth models, fragments of this same belt of mountains occurs in Eastern Australia, known as the New England fold belt, as well as the Alpine Mountains in New Zealand.

The Expansion Tectonic crustal compression processes involved in orogenesis and fold mountain building are precisely what are seen in all sedimentary basins exposed at the Earth's surface today. This crustal compression process, in effect, discredits the concerns of Brunnschweiler in 1983, who considered that "*Earth expansion was essentially a radial movement and therefore its tangential plate displacements are only apparent, not real.*" Sedimentary basins, along with their complex tectonic and geologic histories are real, and certainly can be accommodated on an Earth that is undergoing an increasing radius through time by the relatively simple process of changing surface curvature and relief of surface curvature.

Changing surface curvature is unique to an Expansion Tectonic Earth and cannot be comprehended on a Plate Tectonic model. Once supercontinental break-up was initiated during late-Permian times, true compressional orogenesis within the supercontinental crusts then reverted to vertical uplift around the margins of the modern continents. This formed the mountain escarpments and plateaux of Ollier and Pain as well as related crustal interaction between the various continental and seafloor crusts. Localised zones of compression within the modern continents also continued to occur during on-going changes in surface curvature, with the complex Himalaya and Alpine Mountain belts, as well as the Kunlun, Tian Shan, Baika, Carpathians, Dinarides, and Apennines being prime examples.

The Expansion Tectonic small Earth modelling studies, while not of a sufficient scale to model orogenesis and mountain building in detail, shows that orogenesis and mountain building is more complex than previously depicted in the models of Rickard and Carey, with every gradation arising from compressional to translational and torsional stresses involved. In reality, the small Earth models show that Palaeozoic orogenesis and fold mountain building results primarily from internal crustal interaction localised within the network of supercontinental sedimentary basins. This is in contrast to continent to continent collision as insisted on in Plate Tectonic models. The compressional and tangential directed forces involved in this Expansion Tectonic process therefore provide the means of transmitting the continuous stresses required for geosynclinal formation and also the means for discontinuous or intermittent periods of compression required for orogeny and mountain building.

12.5 Expansion Tectonic Crustal Model

The crustal mechanisms introduced in the previous sections are, out of necessity, schematic and simplistic. In reality, the Earths crusts are far more complex than depicted in either of the Plate Tectonic or Expansion Tectonic models. The simplistic *"conveyor belt"* process of subduction and mantle convection implied by some conventional Plate Tectonic figures, as introduced by Hess in the 1960s, bears little resemblance to the actual processes operating in the real world of geology. This next section will now attempt to present a more holistic crustal model for relief of surface curvature, orogenesis, fold mountain formation, and crustal interaction on an Expansion Tectonic Earth starting at the early-Archaean and extending through to the present-day.

The schematic models of the Earth presented in this section are based on an extension of ideas and mechanisms previously discussed throughout this and previous chapters. These models are based on ideas originally presented by Rickard, Carey, and others during the 1960s to 1980s and summarised elsewhere by Cwojdzinski in 2003. As discussed earlier, from an Expansion Tectonic perspective, when these historical proposals were first presented the ancient radius of the Earth was only thought to have increased from a Pangaea supercontinent of around 55 percent of the present Earth radius. Expansion was then thought to have only occurred during the past 160 million years or so. There was then very little consideration in the literature for changes to the continental crust or for any increase in Earth radius prior to that time. The primitive Earth was considered by these researchers to have been fully developed, or at least mostly developed, by 200 million years ago implying that it had also been fully developed since early-Archaean times.

Expansion Tectonic small Earth modelling studies presented here now show that Earth radius has been increasing exponentially throughout the entire geological history of the Earth, extending from early-Archaean times through to the present-day. From these small Earth modelling studies, crustal extension, as well as relief of surface curvature within the ancient supercontinental crusts, was shown to have occurred at a very much slower rate and over a far more extended period of time than that currently shown by the modern seafloor crustal spreading rates. The models and ideas of Rickard and Carey, in particular, are here modified to extend the Expansion Tectonic crustal model back to the known beginning of geological time.

Within the context of Expansion Tectonics, the primordial Archaean granite and volcanic supercontinental crust was initially formed and stabilised as a primitive Earth with an ancient radius of about 27 percent of the present Earth radius. By the time the ancient Pangaea supercontinental crust had started to rupture during the late-Permian, Earth radius had increased to about 50 percent of the present Earth radius. Since that time the Pangaea supercontinental crust then broke up to form the modern continents and modern oceans. What the crustal evidence from the Expansion Tectonic small Earth models then show is that there have been four distinctly different, but inter-related, crustal forming episodes during Earth history. These episodes are distinguished primarily by the differing styles of relief of surface curvature caused by the exponential increase in Earth radius.

The schematic model shown in Figure 12.9 shows three stacked cross-sections of an inner primitive Archaean and a Permian small Earth model located within a present-day Earth model.

Note, the crustal thicknesses shown are not to scale and in reality would be represented by a very thin skin on the surface of each model. The innermost model summarises the crustal development and tectonic processes operating during initial primordial crustal formation some 4,000 million years ago. It then steadily increased in radius throughout the Proterozoic and Palaeozoic supercontinental stages to the late-Permian model some 250 million years ago. This was then followed by a rapidly accelerating increase in Earth radius commencing with the late-Permian crustal break-up phase to form the modern continents and rifting of the crusts to form the modern oceans.

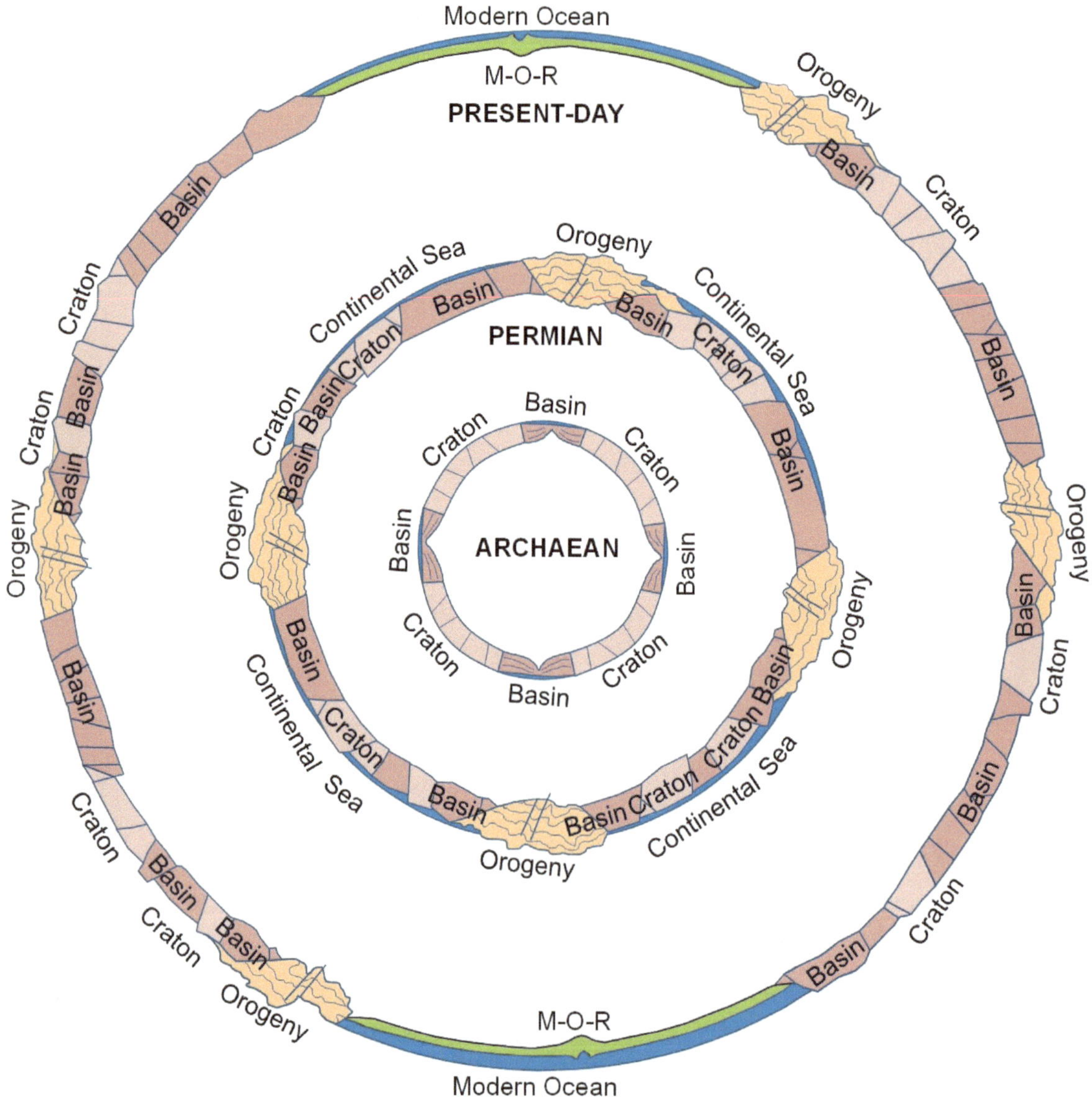

Figure 12.9 Schematic cross-sectional models of an idealised Archaean, Permian, and present-day Earth showing the Expansion Tectonic crustal model for development of the continental and seafloor crusts. Crustal thicknesses are not to scale.

In this figure, the inner-most primordial Archaean Earth crust has cooled and stabilised. The mainly crystalline granite and intruded volcanic rocks remain as a rigid supercontinental crust, covering the entire Earth. Onset of Earth expansion was accompanied by simple, global-scale fracturing of this brittle supercontinental crust. Fracturing gave rise to further intrusion of granite magma and volcanic lava along faults and fractures within a newly established network of crustal weakness. This network of crustal weakness also formed the precursor to a primitive network of low-lying sedimentary basins. The extremely small rate of change in both Earth radius and surface curvature occurring during this ancient time was enough to maintain a very limited super-elevation of the central parts of the fragmented crust, and to establish an elevation contrast between the locally elevated lands and low-lying sedimentary basins.

Between the inner Archaean and the middle Permian small Earth models, Earth radius increased from 27 percent to approximately 50 percent of the present Earth radius. The time interval shown by these two models is represented by the Archaean and Proterozoic Eons as well as the Palaeozoic Era—around 3,750 million years of Earth history. The Proterozoic Eon is marked by retention of the fragmented primordial crusts, along with on-going crustal extension and deposition of sediments localised between each of the ancient crustal fragments. This crustal extension process further formalised the global network of crustal weakness and accompanying platform sedimentary basins, which are now common to the ancient orogenic belts. Relief of surface curvature during these times was accompanied by localised super-elevation and on-going fragmentation of the primitive crusts, as well as depression of the land surface within the low-lying sedimentary basins. The extremely long period of time involved enabled these elevated lands to be continually eroded to a mainly flat and featureless landscape.

Because of the extended period of time available during the ancient Archaean and Proterozoic Eons, downward relief of surface curvature of the fragmented crusts also subjected localised areas to prolonged and sustained compression and folding. This initiated geosynclinal down-warping within the network of sedimentary basins. The amount of time available to compress these rocks also meant that erosion of the folded rocks exposed at the land surface continued to maintain an on-going essentially flat landscape, while still preserving the complexly folded geology below.

During the Palaeozoic Era—extending from around 540 to 250 million years ago—Earth radius and surface area began to steadily increase. Relief of surface curvature during that time was not able to keep pace with the accelerating changes to the land surfaces, leading to periods of orogenesis and formation of high elevated mountain ranges focused within the established network of sedimentary basins. As a result of these crustal changes, drainage systems became increasingly active and erosion of the elevated landscape became prominent. Older sediments deposited during the more ancient, smaller Earth radius times were also subject to periods of renewed folding, faulting, erosion, and orogenesis. All of this was accompanied by an increasingly elevated and disrupted landscape prior to eventual crustal rupture and failure to form the modern continents and oceans.

The post-Permian period of crustal break-up and opening of the modern oceans is represented in Figure 12.9 by the change from the middle Permian model to the outer present-day model, representing the remaining 250 million years of Earth history. The unprecedented change in Earth radius, surface area, and surface curvature occurring during this interval of time is characterised by intrusion and preservation of the modern seafloor volcanic crustal rocks. Relief of surface curvature during this post-Permian time then shifted from extension-style mechanisms involving sedimentary basin formation and erosion to completely new continental and seafloor crustal styles. These new styles mainly involved crustal uplift and erosion along the margins of the fragmented modern continents to form mountain escarpments and plateaux plus on-going mid-ocean-ridge extension and transform faulting within the oceans.

The process of crustal adjustment during gravity-induced change in surface curvature, while simplistic, provides a mechanism for both orogenesis and mountain building during an increase in Earth radius. On an Expansion Tectonic Earth this mechanism does not require, or rely on, continental collision or crustal subduction to promote horizontal or vertical tectonism. These changing surface curvature mechanisms also adequately explain the distribution and inter-relationship of both ancient and modern orogenic zones as preserved on Earth today, as well as the distribution of past and present mountain belts.

Throughout Earth history, multiple phases of orogenic activity have been recorded in the rock-record and these suggest a crustal extension-orogenesis-extension cyclicity. The radial and tangential vector components of this complex orogenic process, as shown in Figures 12.7 and 12.8, result in a continuum of potential orogenic models during on-going relief of surface curvature and crustal interaction. This continuum of potential orogenic models, along with observations of orogenic development on an Expansion Tectonic Earth, are consistent with my fundamental definition for orogenesis whereby orogenesis involves "*a sustained compression of rocks between mutual crusts giving rise to folding of the rocks to form long linear fold mountain belts.*" This also incorporates the extended definition of Burg and Ford whereby orogenesis is further "*characterised by intense internal rock deformation along long linear mountain belts, and, at places, by associated regional synkinematic metamorphism.*"

13 Palaeomagnetic Evidence

"The many geophysical and geological paradoxes that have accumulated during the past two or three decades are apparently the consequences of forcing observational data into an inadequate tectonic model." Storetvedt, 1992

Modelling studies have now reached a stage where I can start to introduce additional palaeomagnetic evidence in order to physically locate the ancient magnetic poles and corresponding equators on each small Earth model. Once the magnetic poles and equator are located, this information will then be used to establish a geographical latitude and longitude grid for each model which will be used in later chapters to model climate zones and to quantify plant and animal distributions on an Expansion Tectonic Earth.

Palaeomagnetics has long been considered the cornerstone of Plate Tectonic studies and has supplied a large amount of data about past locations of continents and plates. The data has also provided evidence about motion histories of the various continental crusts, continental growth, mountain belt formation, as well as measurements of the Earth's ancient radius. It was concluded in 1975 from the palaeomagnetic studies of McElhinny and Brock that "*...within the limits of confidence, theses of exponential Earth expansion, or even moderate expansion of the Earth are contradicted by the palaeomagnetic evidence.*" This therefore led McElhinny and Brock to conclude, "*...there has been no significant change in the ancient radius of the Earth with time.*" With this is mind, we will now look at palaeomagnetics in more detail to try and resolve this dilemma.

Palaeomagnetics is the study of remnant magnetism preserved in rocks and is credited with leading to the revival of the theory of Continental Drift and to its subsequent transformation into Plate Tectonics. The primary role of palaeomagnetic studies carried out on samples of magnetised rock is to provide measurements to determine the ancient latitude of a sample site and to measure a direction and distance to the ancient magnetic pole. From these measurements, the location of the ancient pole can then be calculated. From measurements taken from widely separated sample sites, the determination of apparent-polar-wander-paths–a Plate Tectonic concept referring to the apparent movement of an ancient magnetic pole location–for each ancient continental fragment can also be made. It is this evidence that provided the first clear geophysical evidence for Continental Drift. Palaeomagnetism is therefore used extensively in Plate Tectonic studies and, in particular, to extend the history of Plate Tectonics back in time and to aid in crustal plate assemblage.

As early as the 18th century it was noticed that compass needles were deflected near strongly magnetized rocky outcrops. In 1797, von Humboldt attributed this magnetization to lightning strikes as it had been noted that lightning strikes often magnetize rocks that outcrop at the surface. However, it is now known that this is not the sole cause. In the 19th century, early studies of the direction of magnetization in rocks showed that some recent lava flows were magnetized parallel to the Earth's magnetic field. Early in the 20th century further work also showed that many ancient rocks, in particular for the mid-Quaternary Period—around one million years ago—were magnetized in the opposite direction to the present magnetic field.

In 1956, the British physicist Blackett provided a major impetus to the study of palaeomagnetism by developing an extremely sensitive magnetometer that he used to measure the presence of remnant magnetism in samples of rock. His intent was to test his theory that the Earth's magnetic field was related to the Earth's rotation, a theory that he ultimately rejected. The magnetometer did, however, become the basic tool for palaeomagnetic studies leading to the revival of the theory of Continental Drift and showed that seafloor magnetic anomalies provided strong evidence for seafloor spreading.

Prior to the 1950s, before Continental Drift was accepted, it was assumed that seafloor crusts were simply oceanised or drowned continental crusts. It is now known that the vast majority of the seafloor crust comprises volcanic lava derived from the mantle, all of which has been intruded since at least early-Jurassic times. This is an important point to note, because it is only through the intensive magnetic mapping of the oceans that this information has become both available and evident.

A side-line of palaeomagnetic studies has been to use the various magnetic pole location measurements as a means to determine an ancient Earth radius. The overwhelming conclusion, based on these determinations, was that Earth radius has not changed significantly with time. It is therefore very important at this stage to interrogate palaeomagnetics before moving on to further detailed geological and geographical studies. If palaeomagnetics does indeed negate Expansion Tectonics, then there is a need to understand why. There is also need to interrogate the methods used by palaeomagneticians to see how they came up with these conclusions. If there are inconsistencies in the use of this primary geophysical data, then we also need to know what happens to this same data when it is applied to an Expansion Tectonic Earth model.

13.1 Palaeomagnetics

Palaeomagnetics is the study of Earth's remnant magnetism and this remnant magnetism is preserved in rocks that contain magnetic minerals, in particular the iron-rich mineral called magnetite. Over time these rocks acquire a weak but permanent magnetism during crystallisation, deposition, or cementation of the minerals. The orientation and polarity of this preserved remnant magnetism has been shown to be aligned parallel to the Earth's magnetic field that existed at the time of preservation.

Creer and others first appreciated the application of palaeomagnetics to tectonic studies in 1954. At that time they observed that remnant magnetism, measured in a series of rock samples from the same general area but of increasing age, gave a progressive change in the remnant magnetic vector and hence the position of the ancient magnetic pole derived from this information. This change in magnetic vector suggested that the ancient magnetic pole had moved either because of changes in the position of the pole or because of changes in the position of the continent containing the rock sample. A distinction between these two alternative suggestions was made possible by comparing the apparent-polar-wander-paths determined from sample sites located on different continents. On a Plate Tectonic Earth these apparent-polar-wander paths simply represent the remains of ancient magnetic vectors determined from the ancient continental crusts. Early tests confirmed that the ancient continents had indeed moved relative to each other, rather than the poles moving.

It is hard to imagine now, but at that time the observation that continents could move was a revelation to the Earth sciences. It was contrary to all that had previously been taught about our supposed *terra firma* Earth. Instead of the continents being fixed they were shown to be mobile and seemingly free to move about the surface of the Earth. A vast amount of research has since gone into understanding why and how this could possibly happen, as well as recreating past continental assemblages from the palaeomagnetic data.

The collection of palaeomagnetic data from sample sites located on all continents is now extensive and the measurement and statistical treatment of this information has reached a very high degree of precision. It should be noted that the fundamental basis of palaeomagnetics is not contradicted by Expansion Tectonics. A number of limitations to the use of palaeomagnetic data though must be clarified prior to interrogating palaeomagnetics and establishing a basis for palaeomagnetic modelling on an Expansion Tectonic Earth.

The limitations to the use of palaeomagnetic data include:

1. Measurement of remnant magnetism in a sample of rock can only be used to determine the precise ancient latitude–called *palaeolatitude*–of a sample site, as well as a precise direction–called *declination*–to the ancient magnetic pole–*palaeopole*.

2. Palaeomagnetics cannot determine, nor can it constrain the ancient longitude–*palaeolongitude*–of a sample site.

3. Determining an ancient Earth radius using palaeomagnetic data is based on an outdated assumption that continents have maintained a constant or near constant surface area over time.

4. Ancient magnetic pole locations, determined from palaeomagnetic data, are derived locations which are based on conventional formulae that are in turn based on the fundamental assumption that the Earth's radius is constant.

These limitations to the use of palaeomagnetics represent significant constraints during conventional tectonic reconstructions of the continents, as well as plate motion studies on a static radius Earth. The lack of any constraint for longitude during plate reconstructions is of particular concern. This lack of longitudinal constraint is the primary reason why Plate Tectonic reconstructions of past continental assemblages are constantly plagued by a multitude of plate-fit options. What this means is that the number of plate-fit options available is generally limited only by the number of researchers working on a particular reconstruction. For a conventional Plate Tectonic static radius Earth model there is no physical parameter that will definitively say that a particular plate can only be assembled at any one unique location.

To limit the number of plate-fit options during conventional reconstruction of the continents, heavy reliance is placed on establishing separate apparent-polar-wander-paths for each continent (Figure 13.1). In Figure13.1 the calculated ancient magnetic pole locations for various geological times are shown plotted for each continent. By joining the various pole locations for each continent, the resultant polar-wander paths from adjoining continents can then be compared to adjacent continents. These relative apparent-polar-wander paths are, in turn, merged with adjoining continents in an attempt to compare and constrain their plate assemblage options. Additional geological, geographical, and climatic information is also used to further limit the longitudinal plate fit-options presented by this method.

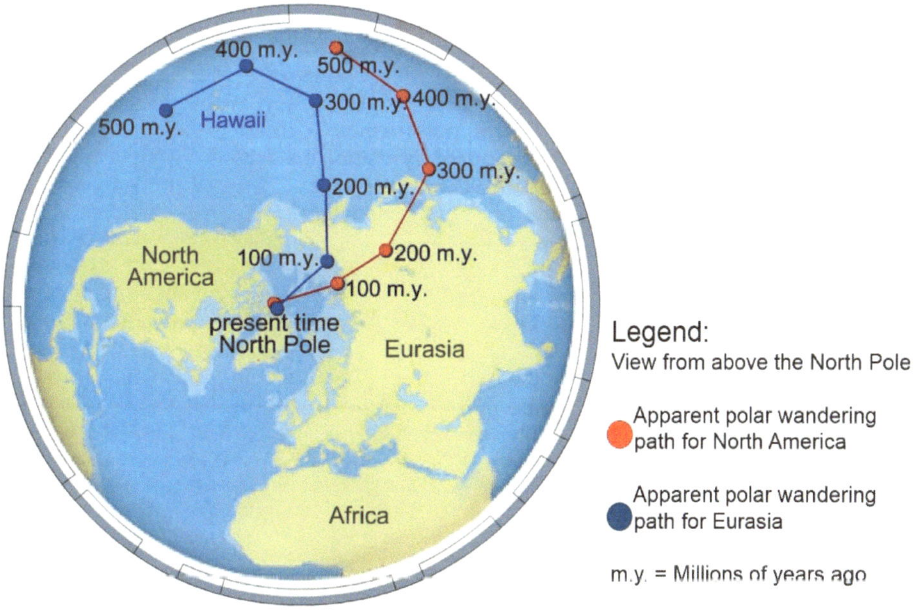

Figure 13.1 Conventional Apparent-Polar-Wander-Paths for North America shown in red and Eurasia shown in blue.

While seemingly adequate to constrain plate assemblage, I cannot stress too heavily that palaeomagnetics can only determine the ancient latitude of a sample site, plus indicate a direction to the ancient magnetic pole relative to the present geographical grid system. All other derived parameters are based on preconceived geophysical premises about Earth radius and rigidity of the crusts. If these premises, such as Earth radius, are changed, the locations of all derived ancient magnetic poles and hence continental assemblages are also changed. By changing Earth radius, as done in Expansion Tectonics, this then eliminates the need to use or rely on apparent-polar-wander curves to constrain plate assemblage.

13.2 Palaeomagnetic Dipole Formula

Before delving into the intricacies of palaeomagnetic formulae we need to know what it is that palaeomagneticians actually measure to determine the ancient latitude of a sample site, as well as to locate the ancient magnetic pole. These measurements are based on weak remnant magnetism preserved in rocks that contain small amounts of iron-rich magnetic minerals. The basis of all measurements is that the orientation and polarity of this remnant magnetism was originally aligned parallel to the ancient Earth's magnetic field that existed at the time and location of where these minerals were first preserved. What this means can be visualised by using a modern hand-held magnetic compass as an example, where the needle of the compass is analogous to the iron-rich magnetic minerals preserved in the rocks (Figure 13.2).

Figure 13.2 A simple hand-held magnetic compass commonly used for hiking and direction finding.

A magnetic compass is a navigational instrument that measures directions relative to the Earth's magnetic North Pole (Figure 13.2). The face of the compass shown may also highlight the directions relative to north, south, east or west. Simple compasses of this type consist of a magnetized needle—usually marked on the north end—which is free to rotate and align itself with the Earth's magnetic field. When the compass is held level, the magnetic field exerts a force on the needle which aligns one end toward the north magnetic pole and the other towards the south magnetic pole. The north end of the needle will always point to the prevailing magnetic North Pole and using that as a reference an angular bearing to the direction we intend taking, or conversely from where we have been, can be measured.

As well as pointing to the North Pole the compass needle will also be slightly inclined vertically, depending on the location on the surface of the Earth (shown by the black arrows in Figure 13.3). At the equator, the compass needle will always be horizontal because there is an equal magnetic force acting on the needle from both the north and south magnetic pole directions. If we were to travel to either the North or South Pole the north end of the needle will then point directly down or directly up respectively. This is because, at these locations, the respective north or south magnetic field is the dominant force acting on the needle at the expense of the other magnetic pole direction.

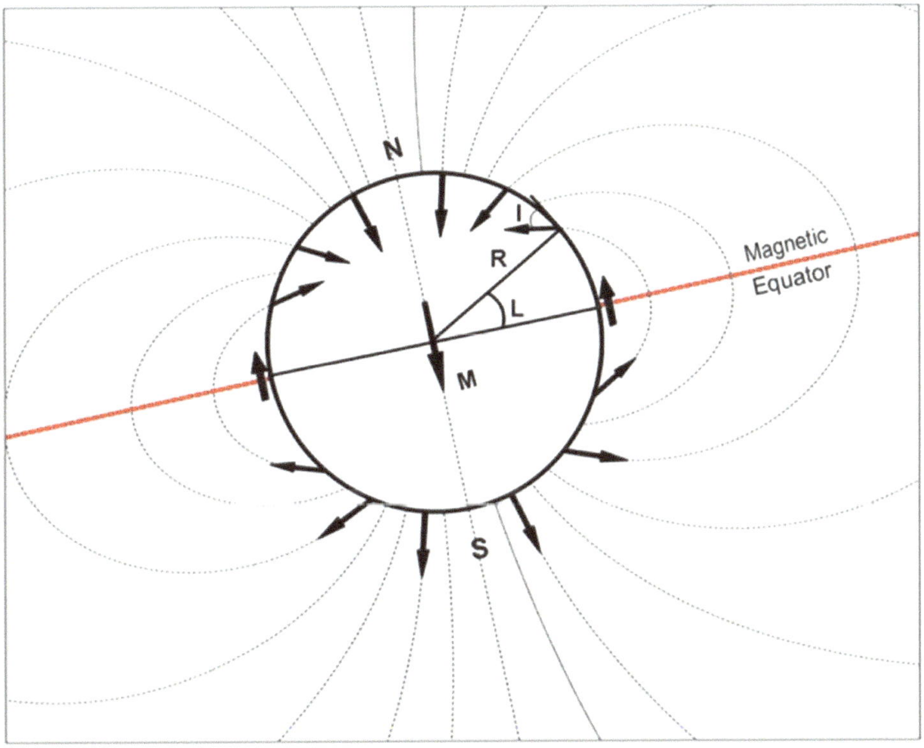

Figure 13.3 Conventional geocentric axial dipole model of the Earth. A magnetic dipole **M** is located at the centre of the Earth and is aligned with the rotation axis. The geographic latitude is **L**, and the mean Earth radius is **R**. The magnetic field directions at the Earth's surface, produced by the geocentric axial dipole, are schematically shown. Inclination **I** is shown for one location and **N** is the North magnetic pole.

As an example of how this relates to palaeomagnetism, if we were to go back in time to an exotic but known location on the Earth and freeze the compass needle solid, then years later we could pick the frozen compass up and take it to a palaeomagnetic laboratory. There, while still in its frozen state, we could measure both the direction the magnetic needle is pointing to, relative to the present-day North Pole, plus the inclination of the needle. From this information we could then calculate firstly, the latitude of the ancient site location as determined from the inclination of the needle and secondly, a direction to the ancient magnetic pole. By knowing the present-day geographic coordinates of the ancient site location we could then determine the location of the ancient magnetic pole by using simple spherical trigonometry. In reality, it is more complex than this because there may have been some tectonic disturbance to the site location as well as shifting crustal plates, but the analogy is basically correct.

To determine both the ancient latitude and ancient pole location from a palaeomagnetic rock sample there are a set of well-established, albeit conventional, formulae based on known magnetic properties. Derivation of these conventional palaeomagnetic formulae is shown in Appendix B. It must be reiterated that these formulae are based on the unwavering assumption that the Earth's ancient radius is equal to, or approximately equal to, the present Earth radius. Because of this assumption, these conventional palaeomagnetic formulae cannot acknowledge any physical variation in the Earth's ancient radius over geological time. This assumption was adopted by early researchers primarily because, at the time these formulae were first developed, there was no way of determining conclusively if Earth's radius was increasing or, if so, by how much. At that time palaeomagneticians then had only one option, which was to assume a constant radius Earth premise.

These conventional palaeomagnetic formulae simply use measurements of the natural remnant magnetic vector preserved in a rock sample to calculate ancient latitude or ancient colatitude. These relationships are shown in Figure 13.4 where latitude is the angular distance of a sample site north or south of the ancient equator and colatitude is the angular distance from a sample location to the ancient North or South Pole. From this information, the location of an ancient magnetic pole can then be calculated for each sample site measured.

In Plate Tectonic studies, from these derived ancient magnetic pole locations, plus the age-dating information of the rock samples used, an apparent-polar-wander-path is then established for the various continents. Similarly, for more detailed studies, the displacement history of small parts of the continental crusts may also be determined. These determinations are used to constrain the past latitudinal positions of the various continents as well as to limit the number of plate-fit options available for Plate Tectonic plate reconstructions on a static radius Earth model.

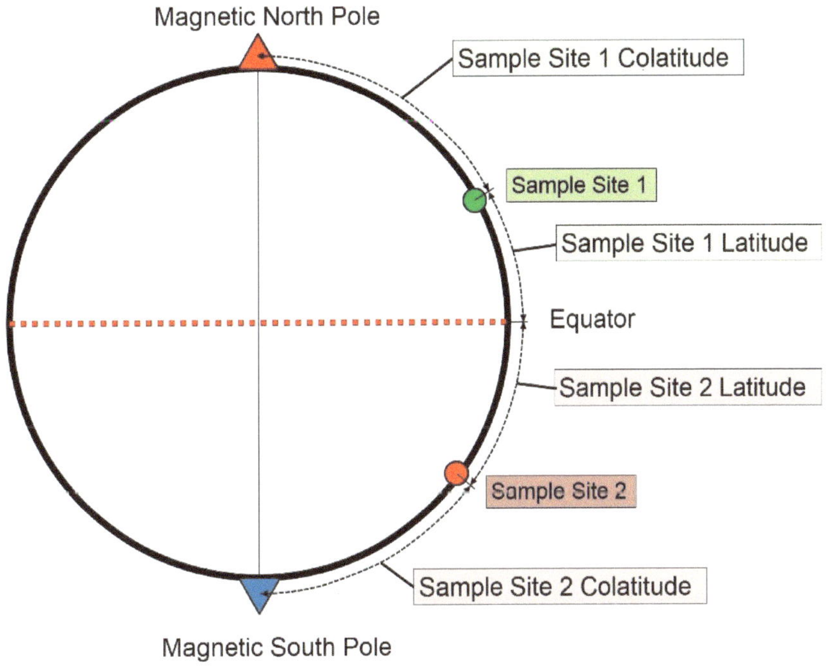

Figure 13.4 Cross section of the Earth showing latitude and colatitude relationships between two sample sites, the magnetic North and South Pole, and the equator.

These conventional palaeomagnetic formulae are based on what is referred to as the *geocentric axial dipole model* of the Earth. This model represents the basis for all palaeomagnetic studies and is equally applicable to a magnetised sphere or a planet of any radius. In this model, shown schematically in Figure 13.3, the geographical pole and rotational axis of the Earth are coincident with the mean magnetic north and south poles. A single magnetic north and south dipole vector, shown as M, is located at the centre of the Earth and is aligned with the geographical rotation axis. This single dipole produces the entire Earth's magnetic field, shown in the figure as grey dotted lines, and its configuration is exactly the same for any radius planet or magnetised object.

The magnetic field shown in this figure is also identical to childhood studies of basic physics, where iron filings are sprinkled over a bar magnet to highlight the magnetic lines of force. The formula used to determine the ancient latitude or ancient colatitude of a sample site was defined by Butler in 1992 as:

$$\tan I = 2 \tan L = 2 \cot P \qquad \text{Equation 1}$$

Rearranging the dipole equation gives colatitude P as:

$$\mathbf{P = cot^{-1} (tan\ I/2) = tan^{-1} (2/tan\ I)} \qquad \text{Equation 2}$$

Where:

- **I** is the mean angle of inclination of the remnant magnetic field, which is preserved in rocks located at a sample site. This angle increases from minus 90 degrees at the magnetic North Pole to plus 90 degrees at the magnetic South Pole.

- **L** is the ancient geographic latitude, calculated from the measured angle of inclination **I**, and it represents the angular distance along a meridian extending from the equator to the ancient sample site.

- **P** is the ancient geographic colatitude, calculated from the measured angle of inclination, and represents the angular distance along a meridian extending from the sample site to the ancient magnetic North or South Pole, depending on which hemisphere the sample site is located (Figure 13.4).

To minimise errors during sampling of the magnetised rock under study, and to give a more statistically representative measurement, a series of samples are generally taken from each sample site. These samples are structurally corrected to compensate for any Earth movements that may have occurred to the site since deposition of the rock sample. They are also magnetically screened for any overprinting magnetic vectors that may have occurred since initial preservation.

A statistical mean value for the angle of inclination is then determined from the corrected samples and this mean value is assumed to represent the ancient time-averaged magnetic field present at the time of preservation. This sampling procedure, in effect, compensates for any small annual variations in the magnetic field which may have been caused by wandering of the magnetic axis around the geographical axis as well as helping to mean out any errors in the determinations.

For the geocentric axial dipole field previously shown in Figure 13.3, the time-averaged and structurally corrected angle of inclination for the particular sample site is then used to calculate the ancient latitude or colatitude of the sample site for when it was first preserved in the rock-record.

Similarly, the time-averaged angle of declination—not shown in Figure 13.3 but it is simply a measured angular compass direction—measured from the same site sample is also used to determine the angular direction towards the ancient magnetic North Pole.

The primary purpose of this long and involved sampling and statistical procedure is to locate the position of the ancient magnetic North Pole, as determined by the site sample on the present-day Earth. The location of the ancient magnetic South Pole can also be determined from sample sites located in the southern hemisphere. Using conventional palaeomagnetic formulae, the angular distance to the ancient magnetic North Pole location is equal to the ancient northern colatitude, measured from the sample site along an ancient meridian in the direction shown by the measured angle of declination.

This procedure sounds very simple and in theory it is. In reality, however, the measured angle of declination, which may be taken from a sample that is say 10 centimetres long, is simply projected from the sample site for maybe thousands of kilometres to the ancient magnetic North Pole location. This pole location is inclusive of all errors in measurement, structural corrections, and statistical treatment of the sample data and hence has a very large built-in error factor to consider. It also does not—and in fact cannot—take into consideration the potential for an increasing Earth radius. Palaeomagneticians simply assume a constant Earth radius.

Using spherical trigonometry, additional formulae, as outlined in Appendix B, are then used to calculate the geographical coordinate location of the ancient magnetic North Pole on the present-day Earth. By assuming a constant Earth radius it is also assumed that the ancient geographical coordinate system was identical to the present-day coordinate system. By using these additional formulae, the colatitude and declination vectors of the ancient geographical system are then simply added, using spherical trigonometry, to the present geographical system in order to locate the ancient magnetic pole.

This calculation again sounds relatively simple, which it is because geographical grid systems do indeed use angular measurements to determine latitude and longitude. By using angular measurements and a constant radius Earth, both geographical grid systems can be readily added together. In practice however, while the angular measurements are in fact identical, the actual physical dimensions of the angular measurements on an Expansion Tectonic Earth—that is the length of a degree of latitude or longitude—are not identical and definitely cannot be simply added. For an Expansion Tectonic Earth, new formulae must be derived based on a varying Earth radius.

13.3 Understanding Apparent-Polar-Wander

Understanding apparent-polar-wander is crucial in understanding the difference between conventional palaeomagnetics and palaeomagnetics on an Expansion Tectonic Earth. The fundamental difference being that apparent-polar-wander is only applicable to a static radius Earth model and does not apply to an Expansion Tectonic Earth. Apparent-polar-wander on a static radius Plate Tectonic Earth model is simply a means of using the ancient magnetic pole locations from a number of continents in order to assist in reassembling the ancient continents. On an Expansion Tectonic Earth ancient pole locations coincide at a single ancient pole location and do not form apparent-polar-wander paths.

Prior to the 1950s, most scientists assumed that the seafloors were simply drowned continental crust and the continents were the exposed parts of this same rigid continental crust. The Earth's crust was also assumed to be permanently fixed at the same location for all time. In this situation, if palaeomagnetic measurements were taken on rocks from near say Brisbane in Australia and Little Rock in America, and, in fact, from anywhere in the world, the directions to the ancient magnetic pole measured from these sites would always be identical to the present-day measurements. In all cases these measurements would point to the same north or south magnetic pole and any variation in location would simply indicate a migration of the magnetic pole. If measured accurately, each of these magnetic measurements would intersect at a single point at the North or South Pole. In this ideal world there would be no apparent-polar-wander, nor a need to consider an Expansion Tectonic Earth.

After the 1950s though, it was discovered that the continents do in fact move—which was the basis of Continental Drift—and from this it was realised that the continents are not stationary at the same location. Because of this movement, the same magnetic measurements taken from the ancient rocks in the example above will not intersect at the present-day magnetic pole. Measurements from these different countries will in fact plot at completely different locations around the globe. As well as this, it is now known that pole measurements from rocks of different ages for each country also plot at different locations. When palaeomagneticians finally understood the meaning of this evidence it provided the first clear geophysical proof for Continental Drift and similarly, evidence from seafloor magnetic mapping also provided clear proof of seafloor spreading.

Conventional palaeomagnetic studies now use this measured palaeomagnetic data to plot the locations of the ancient poles for a range of ages for each country, as reproduced in Figure 13.5. In this figure the apparent-polar-wander-paths shown for North America and Eurasia also include the measured magnetic declination lines as red and blue dotted lines. These lines simply join each of the sample sites to their respective pole locations for each continent. The reader may recall that declination is measured at the sample site and is then projected from the sample site towards the ancient magnetic pole location.

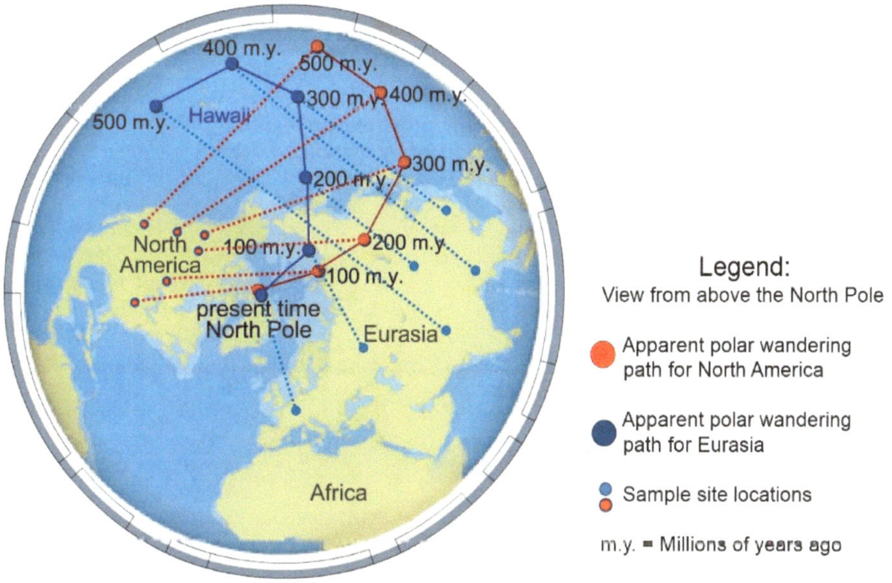

Figure 13.5 Conventional Apparent-Polar-Wander-Paths for North America and Eurasia showing hypothetical sample sites and declination lines to apparent-polar-wander poles.

In Figure 13.5 the blue European apparent-polar-wander path for the various ages is shown in relation to the red North American apparent-polar-wander path. By rotating the North American continent with respect to Europe, for instance, until its red apparent-polar-wander-path coincides with the blue European path, the past locations of each of the ancient poles will then coincide. While apparent-polar-wander on a global scale is far more complex than is shown here, this fundamental technique has since become the primary means of assembling and constraining the past locations of continental plates on a Plate Tectonic Earth.

In order to fully understand what apparent-polar-wander is actually telling us, Figure 13.6 is a schematic cross-section of the same ancient and present-day Earth crust extending vertically through the ancient magnetic pole.

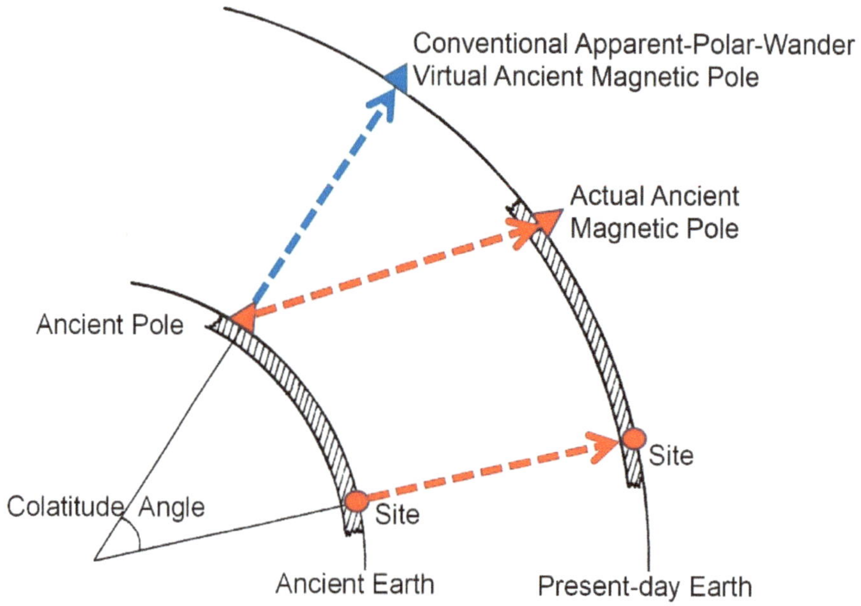

Figure 13.6 Schematic cross-section of an ancient and present-day Earth showing an ancient sample site plus its actual and virtual magnetic pole locations on the present-day Earth.

The figure shows an ancient sample site—the red dot shown on both the ancient and present-day Earth—located on the surface of the ancient crust. By taking palaeomagnetic declination and colatitude measurements from this sample site, the geographic location of the ancient magnetic pole can then be determined—shown as a red triangle on the ancient Earth. This red triangle represents the actual physical location of the magnetic pole on the ancient Earth as determined from the site sample data. When the Earth increases in radius from the ancient Earth to the present-day Earth radius, the same sample site and ancient magnetic pole locations are shown as a red dot plus red triangle on the present Earth. These symbols represent the actual magnetic pole and sample site locations on the present-day Earth.

In order to determine an apparent-polar-wander pole location for this sample site, conventional palaeomagnetic formulae adopt the premise that Earth radius has remained constant throughout time. In other words, Plate Tectonics does not consider that there has been an intermediate smaller radius Earth stage so what I am showing here has no relevance to conventional apparent-polar-wander studies. Instead, by adopting a constant radius Earth premise the ancient angular sample site measurements use the same present-day geographic coordinate system to determine the location of the ancient pole on a constant radius Earth.

What this means is that the colatitude angle measured from the ancient site will locate the ancient magnetic pole at the blue triangle shown in Figure 13.6 where the same colatitude angle is simply projected radially onto the present-day Earth surface. If there were two or more different ancient sites of the same age located on the ancient Earth crust, their ancient magnetic poles would then plot as a scatter of locations on the present-day Earth. This will become more evident in examples discussed in the next section. By joining each of these pole locations together they would then generate a separate apparent-polar-wander-path for each continent. This is equivalent to the blue and red apparent-polar-wander paths shown in Figure 13.5 for North America and Eurasia.

In contrast, on an Expansion Tectonic Earth the ancient magnetic poles from two or more different sample sites of the same age located on the ancient Earth will plot and coincide at the same Ancient Pole site. Again, this will become more evident in examples discussed in the next section. This coincident pole location will, in turn, transfer to the same Actual Ancient Pole site as shown on the present-day Earth. In the example above, the main point to understand is that conventional pole locations are what are called *virtual pole locations* and these can only be regarded as *actual pole locations* if they represent present-day sample sites on a present-day Earth. This discrepancy in actual pole and virtual pole locations is the fundamental reason why there is no need to consider an apparent-polar-wander on an Expansion Tectonic Earth. It is also why palaeomagnetic formulae cannot be used in their current format to locate ancient magnetic poles or to constrain crustal assemblage on an Expansion Tectonic Earth—***it is the seafloor crustal mapping that constrains these past locations of the continents, not palaeomagnetics***.

In the example above, the actual colatitude distance—the actual physical distance measured from the ancient sample site to the ancient magnetic pole—is what is used in Expansion Tectonics to locate the ancient magnetic pole position on a present-day Earth model. Only by using the actual pole locations can all poles from each sample site cluster as a single magnetic pole location. What must be appreciated from this introduction though is that, because Earth radius is constantly changing and with it the crust is constantly shifting position, the location of the magnetic poles at specific points on the crust will also constantly change with time. This is why both ancient Earth radius and time must be included in palaeomagnetic formulae.

13.4 Palaeomagnetics on an Expansion Tectonic Earth

The application of palaeomagnetic formulae to an Expansion Tectonic Earth is crucial in understanding both the limitations of conventional palaeomagnetics on a static radius Earth model and its usefulness on Expansion Tectonic models. In both cases, measurements of the remnant magnetism preserved in rocks will determine a direction to the ancient magnetic pole plus an estimate of the ancient colatitude of the sample site. The difference between conventional palaeomagnetics and palaeomagnetics on an Expansion Tectonic Earth is in the mathematical treatment of the raw field observational data plus the interpretation of the measured data.

The application of palaeomagnetic formulae to an Expansion Tectonic Earth is detailed in Appendix B where the conventional formulae have been modified by reconsidering the palaeomagnetic dipole formula in conjunction with both an exponential increase in Earth radius and geological time. Modified palaeomagnetic formulae used to determine the geographic location of an actual ancient magnetic pole on the present Earth are also presented in this appendix. It must be appreciated, however, that modifications to the Earth's crust during relief of surface curvature over time will influence these formulae in determining actual ancient pole locations.

The mathematical relationship for an exponential increase in the Earth's ancient radius, derived from empirical measurements of seafloor and continental surface area data, was previously found to be (derived in detail in Appendix A):

$$R_a = (R_0 - R_p)\, e^{kt} + R_p \qquad \text{Equation 3}$$

Where: R_a = ancient radius of the Earth, R_0 = present mean radius of the Earth = 6370.8km, R_p = primordial Earth radius = approx. 1700 km, e = exponential, t = time before Present (negative), k = a constant = 4.5366×10^{-9}/yr

By incorporating this formula into the conventional magnetic dipole formula introduced in the previous section, the actual ancient colatitude, extending from the sample site to the ancient palaeopole position on an Earth of present radius, is then equal to:

$$P = ((R_0 - R_p)\, e^{kt} + R_p))\, (\tan^{-1} (2/\tan I))/R_0 \qquad \text{Equation 4}$$

Where **I** is the mean inclination of the ancient magnetic field determined from the magnetic site data and **P** is the actual ancient colatitude determined from **I** and expressed in degrees. **I** and **P** were defined earlier in this chapter.

The application of this modified dipole formula to palaeomagnetic site sample data simply enables the ancient colatitude to be converted from the ancient geographical grid system to the present-day geographical grid system. It is important to keep in mind that a degree of ancient latitude does not have the same physical dimension as a degree of latitude on the present-day Earth. This formula converts and will then correctly locate the ancient magnetic pole position on the present Earth surface.

Understanding this most basic principle of spherical trigonometry is fundamental to understanding palaeomagnetics on an Expansion Tectonic Earth. As previously shown in Figure 13.6, on the present-day Earth an actual ancient magnetic pole is always located closer to the sample site than one determined using conventional palaeomagnetic formula. It is also important to appreciate that the location of an actual ancient magnetic pole on the present-day Earth cannot be calculated using conventional palaeomagnetic formulae. Conventional formulae simply give a dimensionless angular value for its geographical location, which is correct only if the ancient Earth radius equals the present Earth radius. For an Expansion Tectonic Earth, of course, it never does.

The unforeseen limitations placed on magnetic pole locations using conventional palaeomagnetic formulae, and similarly ancient radius determinations outlined later in this chapter, are shown schematically in cross-sections of an ancient and present-day Earth in Figures 13.7 and 13.8.

In Figure 13.7, a cross-section of a fragment of continental crust is shown as pink crust on an ancient Earth. This fragment contains three ancient sample sites, labelled Sample Site 1, 2, and 3—red, green and yellow dots respectively. The ancient magnetic pole, as determined from these sample sites, is shown by the blue triangle on the ancient crust and this pole location is coincident for each of the ancient sample sites.

As the ancient Earth radius increases radially to the present radius, the ancient continent is shown to fragment into three cratons—pink crusts—which are separated by younger sedimentary basins—green crusts. In this figure, the ancient Earth undergoes a simple idealised radial increase in radius to the present-day Earth radius and the distances between each sample site also increase.

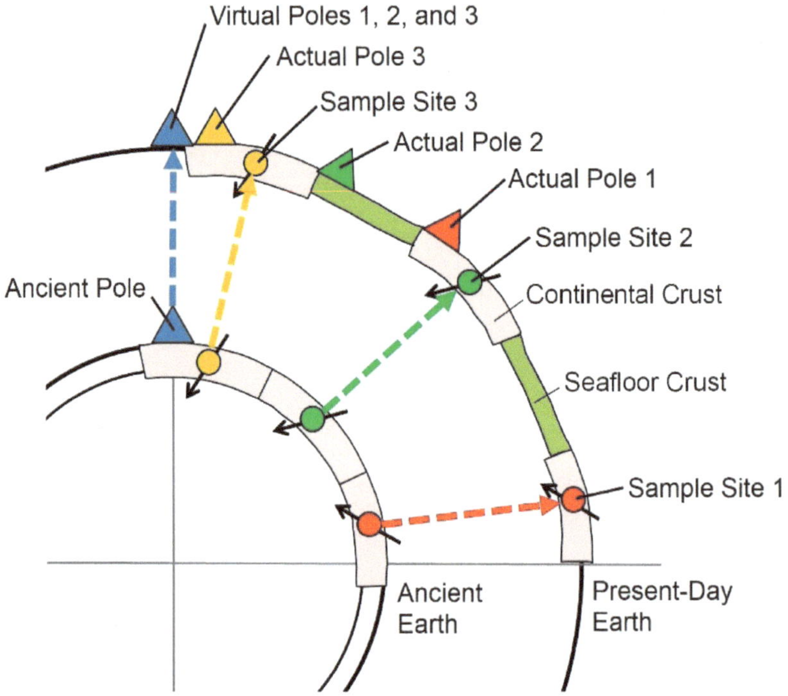

Figure 13.7 Schematic cross-sections of an ancient and present-day Earth showing sample site and pole locations determined from conventional palaeomagnetic formulae and modified Expansion Tectonic formulae.

On the present-day Earth the ancient magnetic pole locations, as determined using conventional palaeomagnetic formulae, are shown to remain coincident at the blue triangle labelled Virtual Poles 1, 2, and 3, which also coincides with the present-day magnetic pole location. This coincidence occurs because there has been an increase in separation of each of the sample sites on the present-day crusts during formation of the two green sedimentary basins.

A Plate Tectonic interpretation of this scenario would insist that, because each of the ancient magnetic pole locations are shown to be coincident, the original ancient continent must have remained intact, inclusive of the sedimentary basins, during the interval of time involved. This example would insist that there was no apparent-polar-wander of the continental fragments and there is no indication of any increase in Earth radius.

In contrast, the actual magnetic pole locations for each site shown in Figure 13.7, calculated from the modified Expansion Tectonic palaeomagnetic formulae, are shown dispersed at the Actual Poles 1, 2 and 3—red, green and yellow triangles on the present-day Earth. Because there is prior knowledge of the increase in Earth radius involved in this hypothetical scenario, it can be appreciated that the ancient continent has, in fact, physically fragmented and these crustal fragments have drifted apart and formed intervening sedimentary basins, but only if an expansion of the Earth is acknowledged.

In Figure 13.8, the same hypothetical ancient continental fragment, containing ancient sample sites 1, 2, and 3, again breaks-up and undergoes an idealised radial increase in Earth radius from the ancient radius to the present-day radius. In this example, one craton is shown to break-up and separate during mid-ocean-rifting, while the remaining two cratons remain intact.

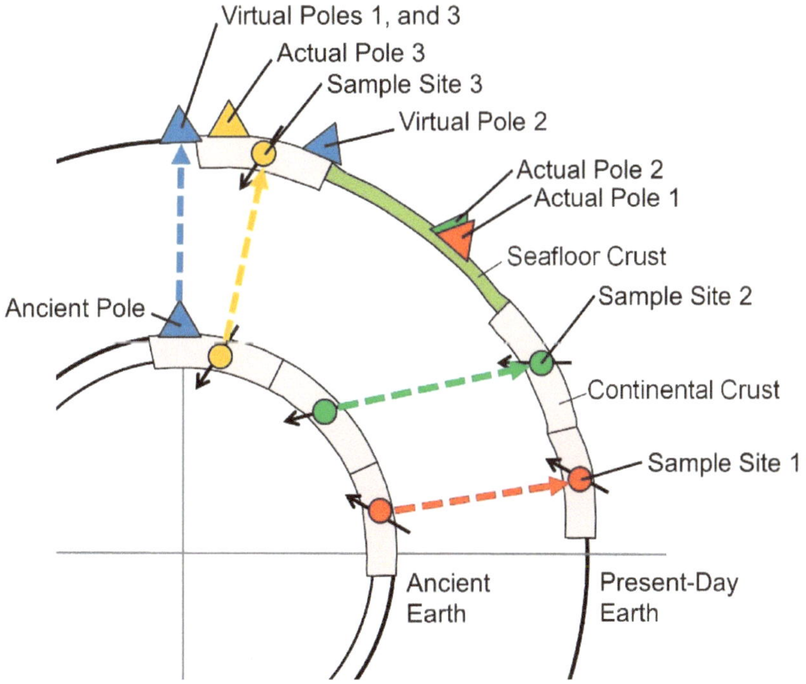

Figure 13.8 Schematic cross-sections of an ancient and present-day Earth_showing sample site and pole locations determined from conventional palaeomagnetic formulae and modified Expansion Tectonic formulae.

The locations of the conventional magnetic poles on the present-day Earth are shown as blue triangle Virtual Poles. The virtual magnetic poles for sample sites 1 and 3 are shown to be coincident, whereas the virtual pole for sample site 2 is shown to be separated from the others. For a Plate Tectonic constant radius Earth, the locations of the ancient magnetic poles would suggest that the two separate cratons have either undergone apparent-polar-wander relative to sample site 2, or they have remained intact relative to sample sites 1 and 3.

Plate Tectonic theory is now less rigid in its interpretation of continental crust than those of the 1960s when van Hilten first calculated Earth radius. Plate Tectonic theory would now suggest that the ancient virtual poles 1 and 2, shown in Figure 13.8, originated from the same continent and there has been either a displacement of the cratons relative to each sample site, or again, an apparent-polar-wander of poles has occurred relative to site 2.

In contrast, on an Expansion Tectonic Earth the positions of the actual ancient magnetic poles 1 and 2 on the present-day Earth coincide—red and green triangles on the present-day Earth. These require no displacement of the cratons and the isolation of the actual magnetic pole 3 confirms that there has indeed been rifting and continental break-up of the ancient continent. All three ancient poles again cluster as coincident magnetic poles on the ancient Earth.

While complex, the hypothetical examples shown in these figures serve to illustrate the potentially unforeseen and unimagined limitations imposed on Plate Tectonic interpretations of palaeomagnetic pole data. They also serve to illustrate the ramifications incurred in assuming continental crustal stability on a constant radius Earth when attempting to calculate an ancient Earth radius. More will be presented on this later.

The global palaeomagnetic database has increased considerably since early attempts at determining apparent-polar-wander-paths and ancient Earth radii were first made during the 1960s. Because of this large dataset, what conventional palaeomagnetic studies now show is a marked increase in the complexity of apparent-polar-wander between continents, as well as an increasingly complex crustal displacement history between individual fragments of continental crust. These increasingly complex apparent-polar-wander and crustal displacement histories are inherent and unique to Plate Tectonic reconstructions of continents on a constant radius Earth.

In contrast, these complexities cannot and indeed do not need to be reconciled on Expansion Tectonic small Earth models. Because of these inherent palaeomagnetic limitations, it is now important to see how palaeomagnetic data relate to an Expansion Tectonic Earth and, in particular, to see what happens when the same ancient magnetic pole data are plotted on Expansion Tectonic small Earth models.

13.5 Present-day Palaeomagnetic Poles

The international global palaeomagnetic database now available has an extensive range of data entries from all over the world for all geological time intervals. These entries range in age from the Archaean to the present-day, and this data will be used here to plot and investigate the distribution of present-day and ancient magnetic poles on each of the Expansion Tectonic small Earth models. From this, an equator and geographic grid system will also be established for each small Earth model.

In palaeomagnetic studies, the measured and structurally corrected angle of declination determined from an ancient site sample is simply projected, using spherical trigonometry, from the sample site—for maybe thousands of kilometres—to the ancient pole location on the present-day Earth (Figure 13.9).

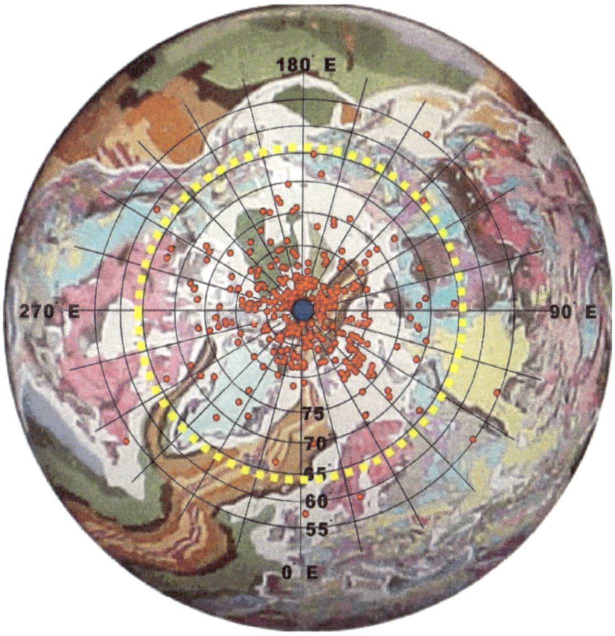

Figure 13.9 Present-day North Pole and magnetic pole locations. The magnetic pole locations are shown as red dots and a 25 degree general scatter of poles is highlighted in yellow. The poles are plotted from the published international database of McElhinny & Lock, 1996

In Figure 13.9, the locations of the present-day magnetic North Pole locations for each sample site recorded in the international database are shown as red dots. Each of these point locations also has an error factor ascribed to them but is not shown in this figure. This error factor is generally plotted as an ellipse around each pole, with a radius dependant on the accuracy of the determination—generally from a few to tens of degrees radius.

In this figure, the distribution of present-day magnetic pole locations determined from the various sample sites is shown to have an overall scatter of between 25 to 35 degrees radius centred on the present-day geographic North Pole. This scatter is referred to as *secular variation* of the magnetic pole and the scatter is largely due to errors introduced during measurement and structural correction of the site samples as well as to the presence of annual variations in the magnetic field.

This distribution of present-day magnetic pole data, where structural correction and magnetic interference is presumed to be at a minimum, is shown clustered around the geographic North Pole, yet the data still has an extensive scatter. It should be obvious from the distribution of the present-day magnetic poles shown in this figure that determining a true or mean pole position from a limited magnetic data set is potentially a very dubious undertaking. It can only be imagined what would happen to the distribution of older data, where structural correction and magnetic interference is far greater.

In conventional palaeomagnetic studies, in order to determine a pole location from this present-day magnetic data, the data is statistically treated to establish a mean, time-averaged, *geocentric magnetic pole*. If there is enough data, this magnetic pole should coincide precisely with the Earth's geographical pole, as shown by the blue dot in Figure 13.9.

For magnetic poles other than the present-day, conventional palaeomagnetic pole data from each continent, or selected portions of a continent, is statistically meaned to establish separate magnetic pole locations for each continent. These pole locations are then used in conventional studies to establish an apparent-polar-wander path for each continent. Similarly, statistically treated poles from selected sample sites within a continent are also used to investigate potential small-scale crustal displacements, referred to as *displaced terranes*, between the various sample site locations.

To overcome the problem of excessive scatter of ancient magnetic poles, in particular where data are limited, the present-day data, shown as red dots in Figure 13.9, can be re-plotted by drawing each pole location as an arc (Figure 13.10).

Figure 13.10 Swaths of present-day north magnetic pole locations plotted as small circle arcs on a present-day Earth model. The swaths represent clusters of arcs plotted for each continent highlighted.

In other words, the colatitude determined from the palaeomagnetic dipole formula—in this case using the conventional formula because it is true present-day data—can be simply projected from the known sample site and manually drawn as an arc of the same radius as the colatitude.

Because of the complexity of the hundreds of overlying arcs involved in this drawing method, in Figure 13.10 the full spectrum of arcs plotted from the published data is shown summarised as coloured swaths. Each swath represents the lateral distribution of the individual arcs plotted from each labelled continent shown on the figure. The radius of each arc plotted on the original figure is equal to the site colatitude—the distance from the sample site to the ancient magnetic pole—and this radius is measured from the site location in the general direction of the measured declination and then drawn as an arc. The swath of colatitude arcs in this figure typically range from 20 to 30 degrees wide. This width is also consistent with the error factors ascribed to each conventional pole location. From these swaths, the mean magnetic pole location can then be located at the intersection of each of the coloured swaths.

This non-conventional method, using colatitude arcs instead of single points, is presented as an alternative method for displaying the variation in pole locations for each continent. Each swath, or multiple swaths, defines the location of each continent relative to the magnetic pole and it can be seen that each swath straddles the coincident magnetic pole location. In this figure the intersection of each swath collectively determines a mean magnetic North Pole independent of statistical treatment, which is coincident with the geographical North Pole. In addition, the equator and magnetic South Pole can also be confidently plotted by measuring through 90 and 180 degrees of latitude respectively. As will be shown later, this is the only method available to plot ancient pole locations on a variable radius Expansion Tectonic Earth. This is because most of the more ancient data sets are limited in size and conventional use of this data may introduce unforeseen errors in locating the ancient magnetic poles.

13.6 Magnetic Poles on Small Earth Models

This colatitude arc method for plotting pole locations will now be adopted and used to establish mean magnetic pole locations and derive an ancient equator for each of the Archaean to present-day Expansion Tectonic small Earth models. This information will be further used in later chapters to establish a geographical grid for each model to enable additional geological and geographical information to be plotted. Based on the near-coincidence of the present-day mean magnetic pole with the present-day geographical pole in Figure 13.9, it is assumed that the ancient magnetic and geographical poles will also be near-coincident for each small Earth model.

To locate ancient magnetic poles on each of the small Earth models, the known published sample site locations are manually transferred from the present-day model to each small Earth model in turn using the detailed bedrock geology as a guide to locate the sample sites. The actual colatitude, as calculated for each site sample using the modified palaeomagnetic formula, is then manually drawn as an arc on each small Earth model in the direction of the measured declination. An example of the distribution of North and South Pole colatitude arcs, as well as the derived mean pole locations for the primordial Archaean small Earth model, is shown in Figure 13.11.

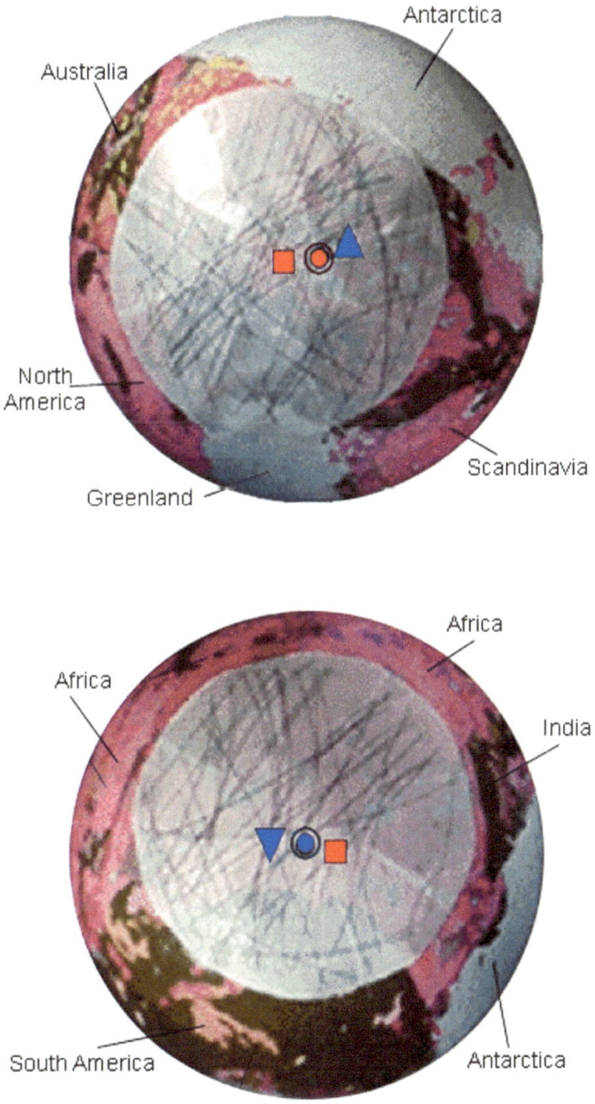

Figure 13.11 Archaean to mid-Proterozoic north and south magnetic poles. Red squares are the mean and projected north poles established from the arcs. Blue triangles are the mean and projected south poles. Red bulls-eye is the location of the meaned North Pole and the blue bulls-eye is the meaned South Pole.

It should be noted that the Archaean small Earth model is an extreme example of this method of plotting ancient poles and has never been done before, for any tectonic theory. This Archaean example has the dual role of showing how relatively easy it is to locate the ancient magnetic poles on small

Earth models as well as quantifying assemblage of the ancient crusts on the primordial small Earth.

For each of the small Earth models it was consistently shown that approximately 95 percent of the colatitude arcs plot within twenty-five degrees radius of each magnetic pole location. This compares favourably with the distribution of magnetic poles shown for the present-day North Pole palaeomagnetic data in Figure 13.9. The remaining arcs—less than five percent—are assumed to contain sporadic, site-related errors, so they were discarded. For each small Earth model, the magnetic pole positions were then located by visually averaging the clusters of North and South Pole arcs. Each pole location was transferred through 180 degrees to the opposite pole and again averaged to determine a mean north and south magnetic pole location. From these meaned pole locations, an equator was established for each model by scaling through 90 degrees of latitude, measured from each of the established poles. Longitude was then established by adopting the ancient location of Greenwich, located in England, as zero degrees longitude and scaling east and west from Greenwich along the ancient equator as per conventional geographic usage.

The locations of ancient north and south magnetic poles established for each of the Expansion Tectonic small Earth models are shown in Figures 13.12 and 13.13 respectively. These models extend from the Archaean to the present-day, plus one model extended to 5 million years into the future. The small Earth models shown in each of these figures are centred on the North and South Poles respectively. These images show the crustal development of each of the continents as they migrate, relative to the ancient magnetic poles, during crustal extension, break-up, and opening of the modern oceans to the present-day. It is important to remember when looking at these figures that: ***It is the continents that migrate relative to the poles, not the poles relative to the continents.***

In Figure 13.12, the Precambrian through to the Palaeozoic magnetic North Pole is shown located in eastern Mongolia. As the Pangaea supercontinent ruptured during the late-Permian and the various modern continents slowly migrated south, the distribution of poles shows there was an apparent northward migration of the magnetic pole through Siberia to its present location within the present Arctic Ocean. Similarly, in Figure 13.13, the Precambrian and Palaeozoic magnetic South Pole is shown located in west central Africa. As the Pangaea supercontinent ruptured and the various modern continents slowly migrated north, the distribution of poles shows there was an apparent southward migration of the pole along the South American and West African coastlines to its present location in Antarctica.

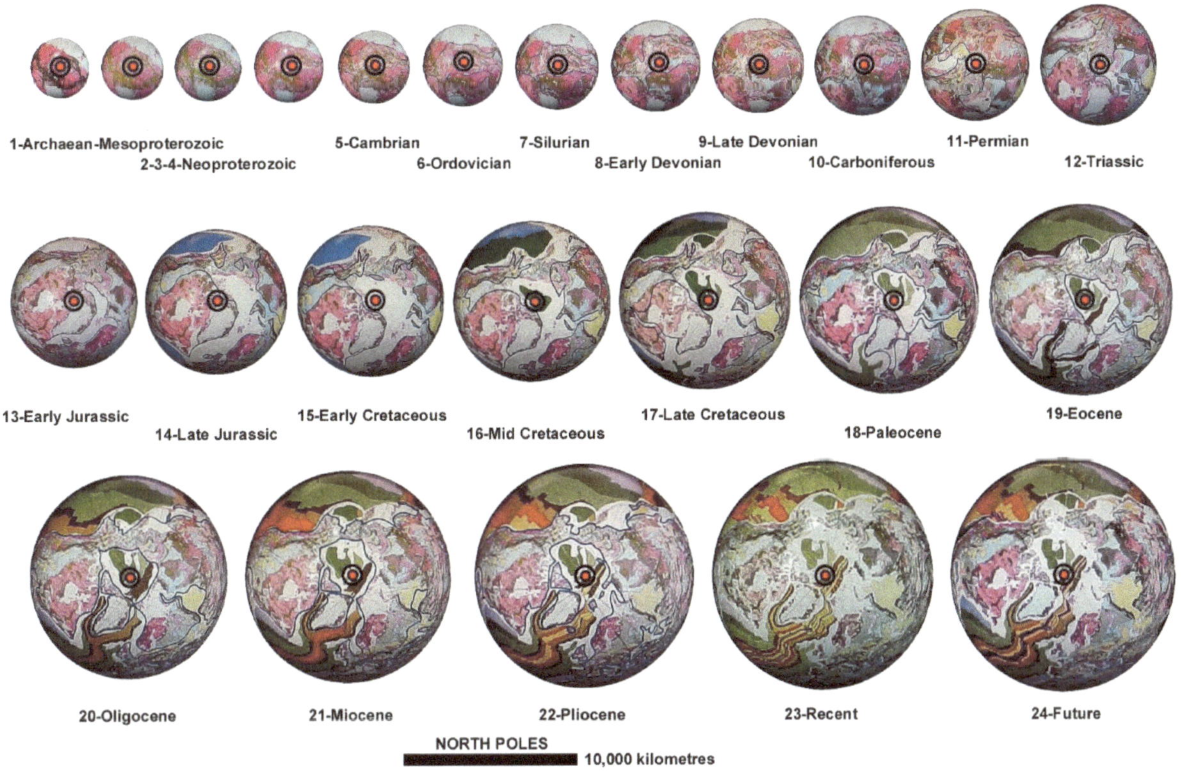

Figure 13.12 Expansion Tectonic small Earth Archaean to Future magnetic North Poles (red dots). During the Precambrian and Palaeozoic the North Pole was located within eastern Mongolia, prior to continental break-up and dispersal during the Mesozoic to present-day times.

The significance of both Figures 13.12 and 13.13 is that the distribution of North and South Poles, plotted independently on each Expansion Tectonic small Earth model, confirms that: ***Each pole remains stationary throughout Earth history***. And more significantly: ***Each pole plots as diametrically opposed North and South Poles***, precisely as they should. I cannot stress too heavily that both of these findings—stationary and diametrically opposed poles—are unique to Expansion Tectonic small Earth models and cannot be replicated on Plate Tectonic assemblages. These findings demonstrate conclusively that palaeomagnetic pole data are far better suited to an Expansion Tectonic Earth and therefore independently verify an Expansion Tectonic Earth process.

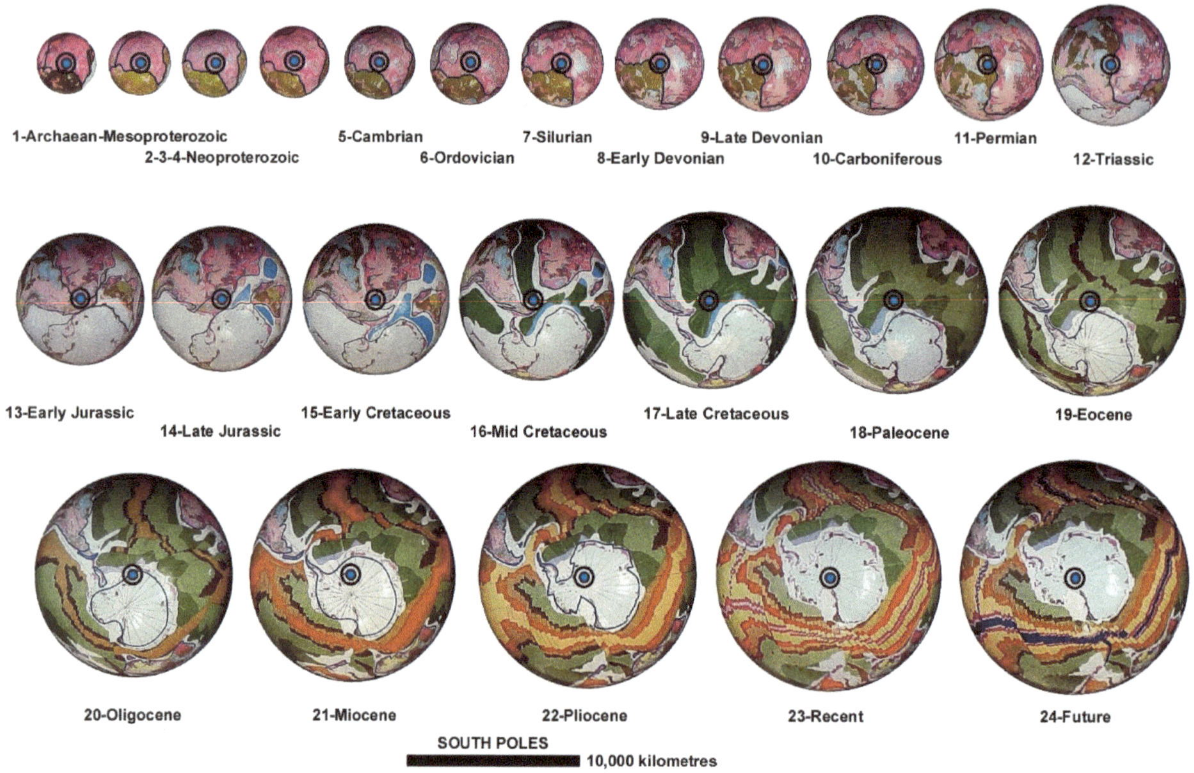

Figure 13.13 Expansion Tectonic Earth Archaean to Future magnetic South Poles (blue dots). During the Precambrian and Palaeozoic the South Pole was located within west central Africa, prior to continental break-up and dispersal during the Mesozoic to Recent.

13.7 Ancient Earth Radius Using Palaeomagnetics

Palaeomagnetic measurements were first used during the 1960s and early 1970s in an attempt to determine a potential ancient Earth radius. At that time there was considerable debate amongst scientists as to whether the Earth was or was not expanding, and, if so, by how much. Earth radius calculated from palaeomagnetic measurements was then used in an attempt to resolve debate once and for all. Conclusions from this research were that Earth radius was not increasing and this has since swayed popular scientific opinion towards Plate Tectonics.

Palaeomagnetic site data from Europe, Siberia, North America, and Africa was used in various studies during that time, with mixed and generally inconclusive results. Published estimates of ancient Earth radius made from these studies varied from large changes in Earth radius, comparable to expansion rates derived from expanding Earth modelling studies available at that time, to negligible changes in radius. Based on results from the African evidence in particular, McElhinny and Brock in 1975 concluded that, "*...for the past 400 million years the amount of potential Earth expansion is limited to less than 0.8 percent of the present Earth radius.*" No further estimates have been made since then and this explicit conclusion continues to be used to deny Earth expansion regardless of subsequent advances made to continental plate studies or completion of seafloor mapping.

During the 1960s and early 1970s, determining an ancient radius of the Earth was carried out using three different methods:

1. **Palaeomeridian method**: this was used for ancient radius calculations based on palaeomagnetic data situated approximately on the same ancient meridian.

2. **Triangulation method**: this was used for ancient radius calculations based on palaeomagnetic data situated on substantially different meridians.

3. **Method of minimum scatter of virtual magnetic poles**: this was calculated from a number of site observations.

It should be appreciated that all of this original work was carried out prior to completion of the seafloor geological mapping, at a time when continents were still considered to be rigid crusts. At no time since these ancient Earth radius calculations were first made have the palaeomagnetic methods been revised, nor have the original radius determinations been recalculated to accord with evidence from seafloor mapping.

In addition, for each of the ancient radius calculations used during that time, the distances between the various locations of the palaeomagnetic samples were typically in excess of 5,000 kilometres. The Expansion tectonic small Earth studies now show that these distances represent a significant proportion of an ancient Earth circumference, an observation that previous scientists were certainly not privy to. The significance of this observation is shown in Figure 13.14, where a 5,000 kilometre wide crustal fragment located on an ancient Pangaean Earth represents approximately 25 percent of the prevailing Earth circumference. This contrasts with the same crust on the present Earth which represents only 10 to 12 percent of the present-day circumference.

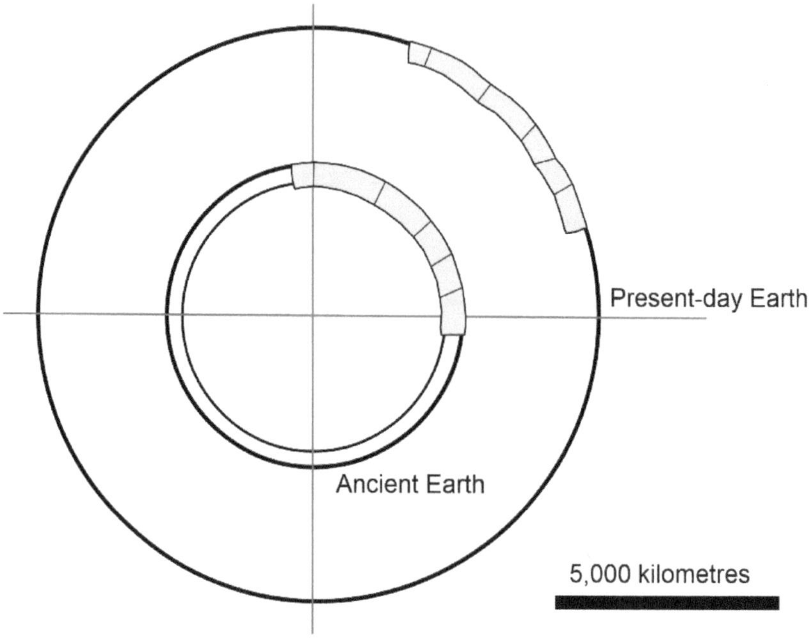

Figure 13.14 Schematic cross section of a 5,000 kilometre wide crustal fragment (shown as pink) on a Permian and present-day Earth.

The considerable amount of vertical and horizontal crustal distortion involved during an increase in Earth radius, extending in this example from the Permian to the present-day, is now known to be excessive. Any changes to surface curvature during an increase in Earth radius would impose extensive crustal distortions as well as changes to surface area during an increase in Earth radius to the present-day radius. None of this was accounted for in the conventional calculation methods previously developed to determine an ancient radius of the Earth from palaeomagnetic data.

Modern geological studies also show that, over this 5,000 kilometre distance, continents have indeed undergone considerable amounts of both crustal extension to form basins, as well as contraction during periods of orogenesis. This crustal evidence is contrary to the premises placed on the European and Asian ancient radius data of Egyed in 1960, and similarly Cox and Doell in 1961 and hence negates both their determinations and conclusions. The mechanics behind Plate Tectonic continental break-up and dispersal is now also well established, such as the extensive work of Scotese in 2001, as demonstrated by both seafloor mapping and conventional apparent-polar-wander studies. This seafloor mapping is contrary to the premises imposed on the European and North American ancient radius data of van Hilten in 1963, and again negates both their determinations and conclusions. Rotation and displacement of parts of continental crust are also well recognised within continents, and these displacements are contrary, in part, to the premises imposed on the African palaeomagnetic and ancient radius determinations of McElhinny and Brock in 1975.

13.8 African Palaeoradius Determinations

Because of the significance of McElhinny and Brock's damning published conclusions, I am going to delve a bit deeper into the methodology behind their calculations. In 1975, McElhinny and Brock published a paper outlining new palaeomagnetic results from Triassic and Cretaceous sample sites located in north, east, and South Africa. These results were used to estimate an ancient radius of the Earth for the Mesozoic Era in order to resolve criticism of previous, inconclusive estimates of radius determined from sample sites located in Europe, Siberia, and North America.

Africa was considered by McElhinny and Brock to be ideally suited for determining an ancient Earth radius because of its implied crustal stability. These implied stable African crustal conditions fitted the palaeomagnetic requirement prevailing at the time that the surface area of continents must remain near constant, and continents must act as rigid plates over time. In their paper, McElhinny and Brock went to great lengths to emphasise the quality of both the palaeomagnetic results and the age dating of samples taken from the various sample sites in Africa. In addition, their sample sites were shown to be located on approximately the same ancient meridian. This then enabled McElhinny and Brock to use the palaeomeridian calculation method to determine an ancient Earth radius.

McElhinny and Brocks' published sample site locations are shown as black squares and their calculated magnetic pole positions are shown plotted as yellow stars on a present-day Earth in Figure 13.15. In this figure the locations of the ancient magnetic South Poles, as calculated using conventional palaeomagnetic formulae, form a broad cluster for each age grouping. This clustering of ancient pole positions suggested to McElhinny and Brock that the ancient magnetic pole, with respect to Africa, had remained in essentially the same position throughout the Triassic to Cretaceous Periods. This conclusion is based on the similar scatter of poles as previously shown and discussed in Figure 13.9 for the present-day magnetic poles.

The distribution of these poles are shown highlighted by the dashed ellipse in the present southern Indian Ocean at the lower right of the figure.

Because the ancient magnetic poles were shown to cluster together, McElhinny and Brock then assumed that the ancient equator passed between the northern and southern African sample sites—shown as a dashed yellow line in Figure 13.15—similar to what it does today. To determine a past geographic separation of the sample sites they then simply added the ancient latitudes from the various sample sites located north and south of the ancient equator.

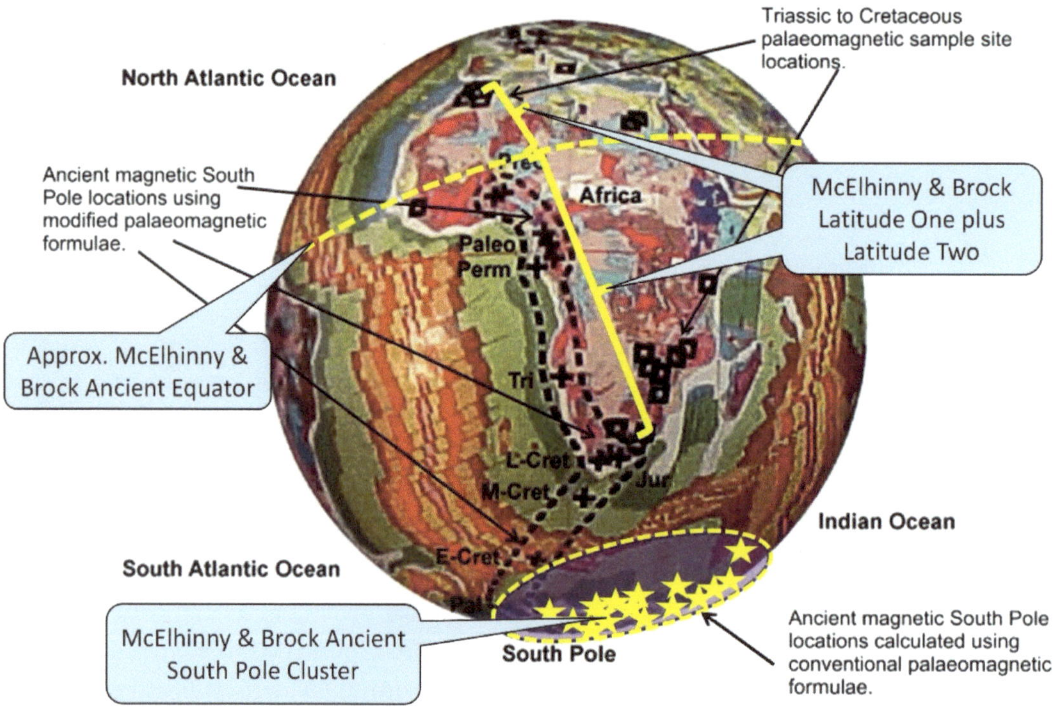

Figure 13.15 Published palaeomagnetic sample site locations for Africa shown as black squares, and calculated conventional magnetic pole locations shown as yellow stars and derived equator of McElhinny & Brock 1975.

Ancient radius was then calculated by McElhinny and Brock by comparing the ancient geographic separation with the present-day geographic separation between combinations of any two sample sites located on the same ancient meridian. In each calculation, and for each age group measured, the calculated ancient Earth radius was found to be very similar to the present Earth radius. From this, McElhinny and Brock were then able to conclude that Earth radius has not changed by more than 0.8 percent during the past 400 million years of Earth history. They further concluded that "*...hypotheses of Earth expansion were therefore very difficult to sustain.*"

These damning conclusions are still used to negate Earth expansion today and the African evidence presented by these authors has previously been very difficult to refute to the satisfaction of subscribers to Plate Tectonic theory.

In Figure 13.16, in addition to McElhinny and Brocks' sample site and magnetic pole data, the locations of each of the Expansion Tectonic magnetic South Poles are shown plotted as red crosses along the west coast of Africa on the same present-day Earth model. The locations of these poles were independently determined and transferred directly from each of the small Earth models as previously shown in Figure 13.13. As the reader may recall, each of these pole locations was established from published palaeomagnetic pole data where the poles were shown to plot as diametrically opposed North and South Poles for each small Earth model.

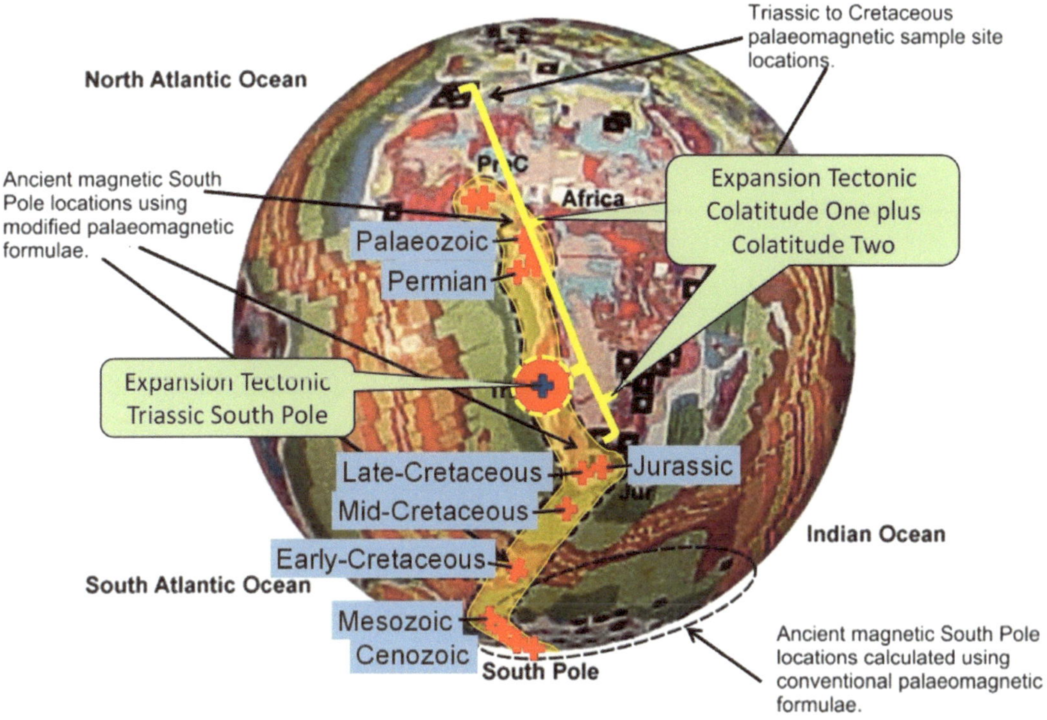

Figure 13.16 Published palaeomagnetic sample site locations of McElhinny & Brock 1975 for Africa (black squares) shown in relation to actual Expansion Tectonic Earth magnetic pole locations (red crosses and highlighted polar migration path).

The distribution of the Archaean to present-day south magnetic pole locations shown in Figure 13.16—the long narrow area highlighted in yellow along the west coast of Africa—shows that the ancient Expansion Tectonic

South Pole was originally located in central West Africa throughout the Precambrian and Palaeozoic Eras. This was followed by an apparent migration of the South Pole along the then joined coastlines of South Africa and South America prior to migration across the opening Atlantic Ocean to its present location within East Antarctica as each of the continents broke up and progressively migrated north. Once again it must be appreciated that it is the continents that migrate relative to the poles, not the poles relative to the continents.

It can be seen in this figure that the Expansion Tectonic magnetic poles are located much closer to the sample sites than those McElhinny and Brock determined using conventional palaeomagnetic formulae. Also, as it turns out, it is only the Triassic sample data, located in north and South Africa that lie along a common ancient meridian. As such, only these data can be used to calculate an ancient Earth radius on an Expansion Tectonic Earth.

In Figure 13.16 the Triassic Expansion Tectonic magnetic South Pole highlighted in red was located between the north and South African sample sites. This means that the ancient equator was positioned around the perimeter of the Earth model shown, well beyond all of the sample sites. To determine the past angular separation of the Triassic sample sites the ancient colatitude from each site must then be added, not the ancient latitude as McElhinny and Brock had done. It is also important to appreciate that it is the actual ancient colatitudes that are added, not the virtual colatitudes.

When compared to the present angular separation of the sites, the ancient radius of the Expansion Tectonic Earth existing during the Triassic Period then calculates as 52 percent of the present Earth radius. Needless to say this value is exactly the same as the Triassic small Earth radius established from measurements of seafloor surface area data. The outcome of this investigation is significant enough to reiterate in bold letters:

The radius of the Triassic Expansion Tectonic small Earth model calculates as 52 percent of the present Earth radius, exactly the same as the Triassic small Earth radius established from measurements of seafloor surface area data.

While McElhinny and Brock went to great lengths to present quality data and sound methodology, it should be appreciated that, at that time, there was very little agreement as to what a potential Earth expansion may or may not have been. Previous ancient Earth radius determinations from

Europe and North America were variable and inconclusive and there were many conflicting arguments for and against Earth expansion in the literature.

What McElhinny and Brock failed to comprehend was the significance of the magnetic pole locations as determined from their conventional palaeomagnetic formulae. The poles that McElhinny and Brock plotted were virtual pole locations and they used present-day geographic angular measurements in their calculation, not actual pole locations or ancient geographic measurements. Because of this oversight they mistakenly concluded that the ancient equator passed between the various sample sites and therefore added latitude, rather than colatitude, to determine an ancient Earth radius.

When the Expansion Tectonic magnetic pole locations are used for the African palaeomagnetic data, these locations—in contrast to McElhinny and Brocks' published conclusions—conclusively quantify a Triassic Expansion Tectonic Earth. This, in conjunction with the diametrically opposed North and South Pole plots, represents further considerable definitive palaeomagnetic proof in support of an Expansion Tectonic Earth.

14 Geographical Evidence

"Ultimately world reconstructions must be congruent not only with the data from geology and geophysics, but also with palaeobiogeography, palaeoclimatology, and palaeogeography." Shields, 1997

Being an Australian geologist, the conventional geographic history of Australia has long been a major concern for me. Conventional Plate Tectonic theory insists that Australia was originally located in southern polar latitudes and has then migrated north into its present position since the break-up of Pangaea. This would imply that Australia should comprise rocks and physical processes—such as weathering and erosion—indicative of low temperatures and even glacial conditions, much like Canada and northern Europe are today. Yet, what can be seen preserved throughout western, eastern, and northern Australia is a very deep, chemically weathered soil profile—several hundred metres deep in many places. This is also supported by the survival of remnant tropical plant and animal species still present in isolated areas throughout these regions. This style of weathering and remnant plant and animal species suggests warm tropical climates existed in the past, which, for Australia, was then followed by an increasing dryness marked by formation of extensive desert dune sands. A complete contrast to what is inferred from Plate Tectonic studies.

As the header above implies, it is all very well to have a set of geological and geophysical models as well as a convincing tectonic theory to explain the observed phenomena. But, if this modelling and tectonic theory does not fit with physical observations preserved in the rocks of the continents, it is either the model or the theory that is wrong, not the physical evidence.

With this in mind, additional physical evidence will now be sourced from the related fields of geography, climatology, and biogeography to plot on each of the established Expansion Tectonic small Earth models. This evidence will be used to further quantify the validity of the small Earth models and geophysical studies. Displaying this ancient geographic information on Expansion Tectonic models will then provide a valuable means to understand these natural phenomena and, more importantly, the distributions and inter-relationships existing between the various now widely scattered ancient supercontinents.

Geography—from the Greek geographia—is a discipline of science that studies the lands, the features, the inhabitants, and the phenomena of the Earth. A literal translation of its meaning is *to describe or write about the Earth.*

The first person to use the word geography is said to have been Eratosthenes (276-194 BC). Modern geography is now seen as an all-encompassing discipline that seeks to understand the Earth and all of its human and natural complexities, not merely where objects are but also how they came into existence and how they have changed over time.

Geography has also been called *the world discipline* and *the bridge between the human and the physical sciences.* This modern geography is divided into two main branches: human geography, and physical geography. In the context of this chapter, however, I will be focussing specifically on aspects of physical geography and, in particular, the ancient physical geography of the Earth in relation to the distribution of the ancient lands and seas.

Ancient geography, referred to in the geosciences as *palaeogeography*, is a study of the Earth's past geography—where *palaeo* simply means ancient—and, in particular, that of the ancient land surfaces. It includes studies of the distribution and configuration of the ancient continental landmasses, their topographic relief, the distribution of ancient seas and shorelines, as well as changing climates. Palaeogeography is also important in the study and distribution of plant and animal species where it is referred to as *palaeobiogeography*.

Evidence for this ancient geography is preserved in many of the rocks exposed on Earth today, especially the sedimentary rocks. Most people pay little attention to what rocks are, and may be unaware of the vast amount of geological and geographical information preserved within these rocks. Even the most mundane of rocks can tell a lot about what it is, where it came from, where it was deposited or intruded, as well as its subsequent weathering, erosional, and transport history—everyday stuff for most geologists.

Sedimentary rocks, for example, are made of rock and mineral particles that have previously been eroded and redeposited from much older rocks. These particles generally consist of quartz sand, mica, and clay, but may also consist of rock particles or volcanic detritus weathered from the exposed rocks that formed the ancient lands. Once eroded, these particles were transported and deposited elsewhere by water, wind, or ice, and in the process are further worn and broken down to form finer-grained particles. The types of sediments that are deposited and preserved in the rock-record will ultimately be dictated by the original rocks, the prevailing climate, and the environment where they are finally deposited.

In contrast, glacial deposits typically comprise mixed course cobbles and boulders deposited relatively close to their original source and set in a finer-grained matrix of powdered rock particles. These glacial deposits will, in turn, be markedly different from sediments that have travelled long distances along river systems and deposited in the deep seas or oceans. Similarly, limestone rocks generally only occur within about twenty five degrees north and south of the equator, coal prefers temperate climates and quite specific depositional environments, and so on. Information from these, and various other rock types, then gives us clear indications of the ancient geography and climates prevailing at the time they were eroded, transported, and deposited. However, while the entire history of the Earth has at some stage been preserved in the rocks, because of on-going geological processes such as folding, faulting, weathering, and erosion of these rocks, at best only remnants of this historical information are preserved today.

Now that the precise locations of the ancient magnetic poles have been plotted on the Expansion Tectonic small Earth models, this information can be further used to establish a geographical grid for each of the models. Even though each small Earth model has a different ancient radius, the angular latitude and longitude characteristics of these geographic grids remain identical to the present-day latitude and longitude grid system. This geographic system is based on degrees of latitude and longitude so it is independent of physical surface measurements and, more importantly, is independent of Earth radius.

Using this established geographic grid system for each small Earth model, new geographic information from around the world will be progressively introduced to investigate its global distribution and inter-relationships on each model. It is this evidence that holds the key to further quantifying the concept of Expansion Tectonics. Ideally, the distribution of this geographic evidence must fit the distribution of continents and seafloor crusts, as assembled on each of the small Earth models, as well as the established poles, equator, and climatic zones as dictated by the geographical grids.

Geographic evidence from the ancient continents supplies crucial information towards understanding the distribution of past exposed lands, ancient seas, rivers, shorelines, and mountain belts. This geographic distribution is constantly changing throughout Earth history and the processes involved are often progressive and, to a certain extent, evolutionary, while punctuated by brief periods of rapid change. Each process contributes to and influences the next within the particular timeframe and these processes may be further overprinted by a succession of new events, such as periods of orogenesis, metamorphism, and erosion during on-going crustal evolution.

Pre-Triassic models of an Expansion Tectonic Earth were previously shown to be covered by a single, all-encompassing supercontinental crust. At that time there were no modern oceans, only a network of continental seas covering low-lying parts of the continental crusts. Throughout this pre-Triassic time the distribution and areal changes to these seas was shown to define the presence, position, and outlines of each of the ancient supercontinents. As well as showing the inter-relationships between exposed supercontinental lands and intervening seas, the geographic information preserved in the rocks also enabled the ancient Pangaea, Gondwana, Laurentia, Baltica, Laurussia, and Rodinia supercontinents to be quantified on an Expansion Tectonic Earth.

Within this chapter, the locations and distributions of these ancient lands and seas will be quantified by plotting the locations of the ancient shorelines on each small Earth model. It should be noted, however, that documented shoreline records in the published literature are only available from the Cambrian to present-day time periods. This is not because the information doesn't exist in the Precambrian rock-record, but because Plate Tectonic assemblages cannot be confidently reconstructed for these ancient times. Because of the limited amount of published information available for the Precambrian times, no attempt is made here to reconstruct the shorelines prior to the Cambrian Period.

The application of this geographic information to an Expansion Tectonic Earth will also be used in later chapters to investigate the distribution of ancient climate zones in relation to the established geographic grids, as well as the distribution of selected examples of ancient animal species.

14.1 Ancient Coastal Shorelines

On an Expansion Tectonic Earth it has been established that prior to about 250 million years ago there were no modern oceans, only ancient continental seas. Similarly, before that time, ancient supercontinents existed as a complete continental crustal shell encompassing the entire ancient Earth. The outlines and configurations of these supercontinents were dictated by changes to the ancient sea levels, primarily as a result of changes to the surface areas of each of the ancient continental sedimentary basins. The transition from ancient seas to modern oceans only came about when the Pangaea supercontinent started to rupture and break-up during the late-Permian—about 250 million years ago. It was this break-up of Pangaea that finally established the modern continents and intervening modern oceans, as well as initiating draining of the waters from the ancient continental seas into the newly opening oceans.

The location and distribution of ancient coastal shorelines on small Earth models is used here to mark the transition from exposed land surfaces to seas or oceans—in other words, they represent the locations of the ancient beaches. The ancient shorelines have been established and published elsewhere by using geological features preserved in the rocks. These preserved geological features may include the presence of subtle sedimentary structures, such as ripple marks suggestive of a near-shore environment, to fossils that only lived in specific shoreline environments, to the presence of characteristic heavy mineral beach sands, near-shore dune sands, and so on.

Evidence from widely separated localities on the present-day continents will then build up a broad picture of the geographical environment for the particular sedimentary basin or ancient sea being studied. It may also enable the researcher to extend their investigation by predicting where a shoreline may have once been. Of course, this ancient shoreline will change over geological time as well as being subjected to on-going burial, erosion, or removal of much of the earlier evidence.

The aim of the following small Earth modelling exercise will be to establish, based on the published evidence, the location of the ancient shorelines for a particular time period or epoch under investigation. This, out of necessity, will be at a global scale without the detailed information shown by local geological mapping.

The distributions of ancient shorelines on each of the Cambrian to present-day small Earth models are shown plotted as blue lines in Figure 14.1. The ancient seas and modern oceans are also shown as shaded blue areas. The plotted shorelines, while only showing a generalised outline, were transferred directly from published shoreline data. On these small Earth models the published shoreline information from various sources, originally gathered to highlight the distributions on Plate Tectonic crustal assemblages, was then merged to give a complete global coverage. In some instances it was found necessary to simplify or modify the published outlines in areas where the information was more schematic or overlapping.

It is emphasised that the primary geological, geographical, and palaeontological—fossil-based—evidence used to define these ancient shorelines in the original publications is exactly the same data for both Plate Tectonic and Expansion Tectonic reconstructions. The only difference is that Expansion Tectonic reconstructions do not require arbitrary fragmentation of supercontinents, subduction of vast expanses of inferred seafloor crust, or random closure of oceans during assemblage of previously fragmented supercontinents.

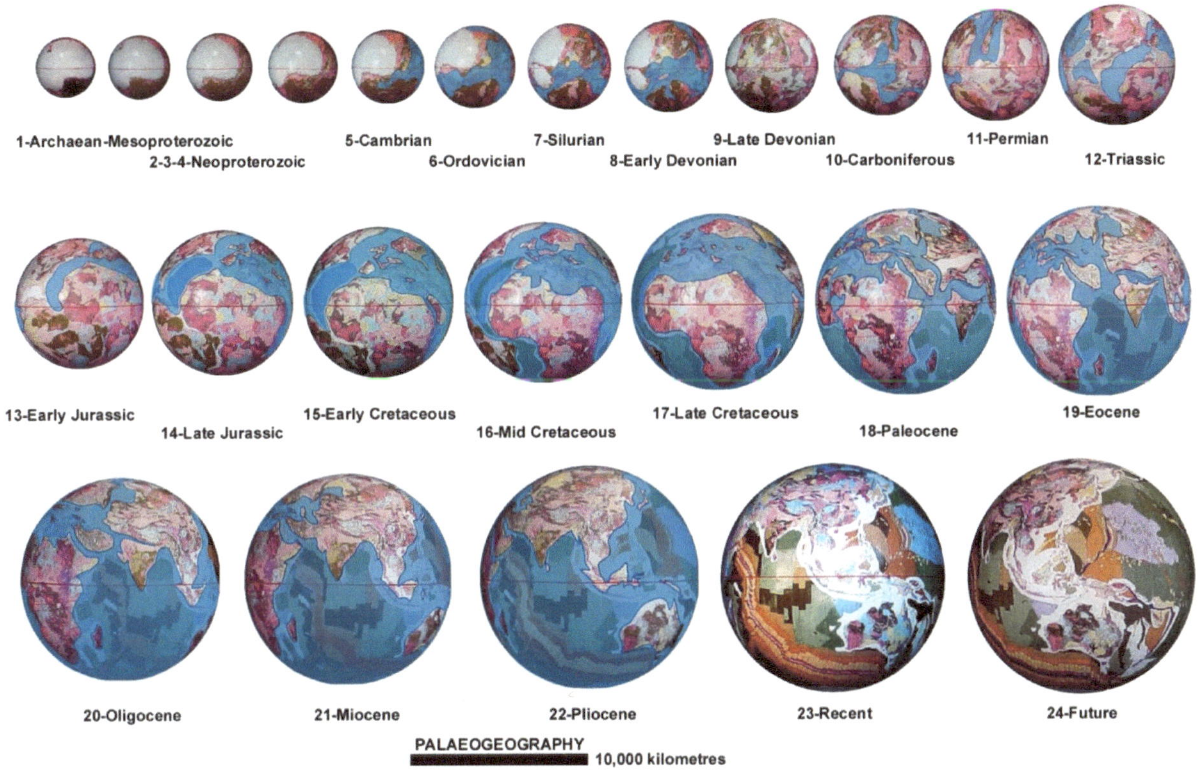

Figure 14.1 Shoreline geography on Archaean to present-day small Earth models. The ancient shorelines are shown as blue lines and the ancient seas and modern oceans are shaded blue. Each image advances 15 degrees longitude throughout the sequence to show a broad coverage of geographical development. Note: there was no published data available for the late-Devonian model.

By plotting the coastal outlines on each of the Expansion Tectonic small Earth models shown in Figure 14.1, it can be seen that the published shorelines merge to show consistent and interconnected shorelines and seas—as they should do. In the context of this chapter, these merged shorelines can then be confidently used to define and quantify the distribution of exposed ancient supercontinental land surfaces, as well as the distribution of continental seas and the break-up and opening history of the modern continents and oceans.

The distribution and variation of the ancient shorelines for the Cambrian to present-day small Earth models is shown in Figure 14.1 to be an evolving and progressive process. This distribution and variation is interpreted to be intimately related to changes in sea-level, which is, in turn, intimately related to changes in surface curvature during on-going increase in Earth radius and break-up of the supercontinents. As well as these global-scale sea-level changes, the distribution and variation of ancient sea levels may also reflect more regional-scale changes. These regional changes occur within a particular basin or sea during breaching and merging of any previously isolated seas or oceans.

The Precambrian small Earth models in Figure 14.1—models older than 560 million years ago—show that the distribution of progressively older sedimentary basins continued to form a global network surrounding the most ancient exposed lands. Without any published coastline information to fully confirm this, these sedimentary basins are inferred to further represent the distribution of ancient Precambrian continental seas. This inference is substantiated by an abundance of geological evidence from all continents showing that sedimentary basins, and hence continental seas, existed as far back as early-Archaean times.

It should be noted that the outlines of the emergent ancient land surfaces, as defined by both the distribution of ancient sedimentary basins and shorelines, are not the same as the outlines of the modern continents. Instead, these ancient emergent land surfaces define the ancient supercontinental assemblages on an Expansion Tectonic Earth. On an Expansion Tectonic Earth these supercontinental assemblages existed prior to when the ancient supercontinental crusts eventually broke-up and the modern oceans opened.

Figure 14.2 shows in more detail that the exposed supercontinental land surfaces, as shown by the late-Jurassic coastal shorelines, coincide with the distribution of the ancient continental cratons and orogens—shown as areas of mainly pink and red coloured rocks. Similarly, the ancient continental seas—highlighted in blue shading—coincide with the distribution of the underlying ancient sedimentary basins. By definition, these sedimentary basins represent low-lying regions where transported sediments were deposited and preserved, so it is to be expected that the margins of these sedimentary basins should coincide with the shorelines of the ancient seas.

Of significance to note, the distribution of coastal geography on Expansion Tectonic small Earth models shows that large conventional Panthalassa, Iapetus, and Tethys Oceans are not required. Instead, these oceans are replaced by continental Panthalassa, Iapetus, and Tethys Seas—see earlier Chapter 6—which represent precursors to the modern Pacific and Atlantic Oceans, as well as precursors to the ancient seas previously covering continental Europe and Asia respectively. Similarly, distributions of the Rodinia, Gondwana, and Pangaea supercontinents and smaller sub-continents on small Earth models confirms that their assemblages are evolutionary with no requirement for random conventional dispersion-amalgamation assemblage and break-up cycles. As shown here, the supercontinent configuration is instead defined by progressive changes to the surface areas of sedimentary basins, the distribution of ancient shorelines, and changes to sea levels.

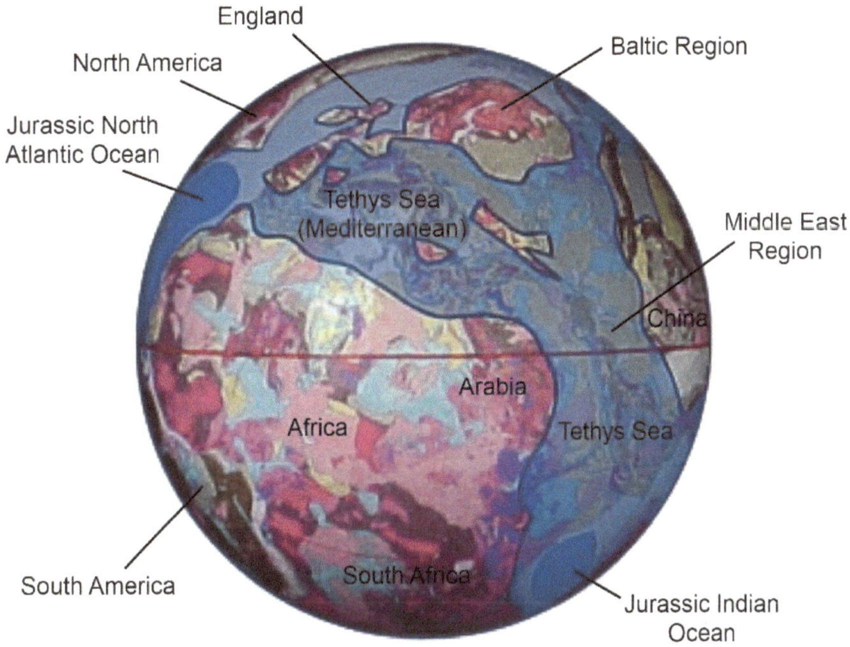

Figure 14.2 Late-Jurassic shorelines highlighted in blue shown in relation to the shallow continental seas, shaded in blue, as well as the distribution of exposed lands and the newly opening North Atlantic and Indian Oceans.

The ability for the supercontinental crusts to continue to increase in surface area, by stretching and extending within each of the sedimentary basins, was finally exceeded during late-Permian times. This process then initiated rupturing and break-up of the supercontinental crusts to form the modern continents and oceans. During late-Permian times, draining of the ancient continental seas was also initiated, which resulted in a progressive exposure of the ancient, relatively shallow seafloors. Once exposed, these ancient seafloors were then subjected to the ravages of erosion and vegetation colonisation. Eroded seafloor sediments were, in turn, redeposited elsewhere into new low-lying regions, in particular around the newly formed continental margins.

The relatively rapid changes in the distribution of shorelines for the younger small Earth models, as shown in Figure 14.1, then document a history of rapid rise and fall of sea levels from each of the emerging modern continents. These changes in shorelines and rapid rise and fall in sea levels are considered to be a direct result of supercontinental break-up and draining of the ancient seas from each of the ancient supercontinents.

14.2 Ancient Continental Sedimentary Basins

On an Expansion Tectonic Earth, the coastal outlines shown in Figures 14.1 and 14.2 are a reflection of the prolonged history of changing Earth surface area and surface curvature. These changes were also accompanied by depression of the lands around the margins of the continental crusts to form sedimentary basins. On each of the small Earth models, this sedimentary basin development is shown to form a global, inter-connected network, which also defines the distribution of ancient continental seas. These ancient basins and seas were centred mainly on what are now the European and Asian regions, as well as in the Australian and adjoining North American and South American regions. Each of these regions, in turn, show complex sea-level histories related to on-going increases in surface area and surface curvature, including climate induced sea-level changes, on-going erosion of the exposed lands, and ultimately to continental crustal break-up, opening of the modern oceans, and draining of the ancient seas.

On each small Earth model, the ancient European, Asian, Australian, and American sedimentary basins represent regions where there has been prolonged continental crustal extension. These low-lying sedimentary basin regions, in turn, are where sediments were ultimately deposited, after being eroded from surrounding elevated lands, throughout the Precambrian and Palaeozoic times. Over time, these sediments slowly filled each basin and in turn changed the surface area of the surrounding seas and supercontinents.

During late-Permian times, the previously long and extended history of supercontinental basin sedimentation began to rapidly decline. As the ancient supercontinental crusts began to rupture and crustal break-up was initiated, seawater from the previously existing continental seas then began to drain into the newly opening modern oceans. This draining further exposed the ancient seafloor sediments to the ravages of erosion and deposition elsewhere in new marine sedimentary basins. As the supercontinents continued to break-up, new marine sedimentary basins began to form and dominate over the older continental basins. These marine sediments were initially deposited in basins located in low-lying areas between the emerging modern continents, followed later by deposition around the margins of the modern continents.

Continental crustal rupture and break-up continued uninterrupted during Triassic to early-Jurassic times with new sediments being deposited within the newly formed Pacific Ocean, followed by deposition within the opening South Atlantic and Indian Oceans. The Southern Ocean later opened during Palaeocene times—crosscutting and displacing earlier formed ocean basins in these areas. Each of the marine sedimentary basins formed during these events are shown in Figure 14.1 as white continental shelf and marine plateau areas surrounding each of the modern continents. Remnants of these ancient marine sedimentary basins are now located adjacent to the continental shelves, located between the lands and the deep abyssal plains of the modern oceans, as well as preserved as plateaux's within older regions of the modern oceans.

14.3 Rise and Fall of Sea Levels

The variation in coastal outlines on Expansion Tectonic small Earth models show that the total volume of ocean water in the past was very much less than what it is now. This is an important point to note because many researchers still insist that the total volume of ocean water has remained constant throughout time, or has been periodically added to during meteor impacts or from passing comets. In the past, this inferred constant volume has also been used as a strong objection by others to refute Earth expansion. It was implicated by these researchers that on an assembled Pangaean Expansion Tectonic Earth model the entire lands would have been covered by over six kilometres of water, which, as can be seen in Figures 14.1 and 14.2, is not consistent with the geological evidence.

On an Expansion Tectonic Earth, small Earth modelling evidence shows the volume of ocean water has been increasing steadily since Archaean times and most prominently since the post-Permian opening of the modern oceans. The post-Permian increase in volume of new water has occurred in conjunction with intrusion of the huge volume of entirely new volcanic seafloor crust along the global network of mid-ocean-ridge spreading zones. This new water, plus accompanying atmospheric gases, represents the escaped volatile elements which occur naturally within the crystal lattices of all molten volcanic rocks. Petrological studies—the study and observations of rocks under the microscope—show that up to fifteen to twenty percent of the weight of a new volcanic rock may comprise fluid and gaseous elements. Graphic evidence of the release of these volatile elements can be seen from both volcanic eruptions on land and from *black smokers* pouring out huge volumes of new superheated water and gases from deep within the mid-ocean-ridges. Once the volcanic rocks are intruded as lava near the Earth's surface, these volatile elements are then expelled, or boiled off, during formation of new surface lava or seafloor volcanic crust.

On each of the Expansion Tectonic small Earth models, a record of this increase in new water is preserved as changes in the coastal outlines of the supercontinents. The full significance of this record can only be appreciated from an Expansion Tectonic Earth perspective, where changes in sea-level occur during changes in both Earth radius and associated surface curvature. Complicating this process is that major changes in sea levels also occur during merging of ancient continental seas, break-up of the ancient supercontinents, and opening of each of the modern oceans. Break-up of the supercontinents may have also destabilised and melted any ancient polar ice-caps existing at the time, which, in turn, may have affected global climate and sea levels. As will be shown later, these major changes in sea levels represent a primary mechanism for periods of mass extinction of ancient plant and animal species.

Global-scale changes in sea-level, where sea levels rise or fall the same amount in all oceans, are broadly referred to in geology as *eustatic* sea-level changes. Relatively small, local, or regional variations are in turn referred to as *transgressive-regressive* sea-level changes. These sea-level changes may be caused by a rise or fall of the land, where a quantity of water is displaced by the land or continental margins, by climatic change, by a variation in the volume of ocean water, by a variation in the surface area of a particular sea or ocean, or a combination of any of these causes.

Here, the term transgressive refers to when the sea-level rises and encroaches on the land for a period of time and regressive refers to when the sea-level falls and recedes from the land. Both of these are relatively common events in geology and leave behind characteristic features preserved in the sedimentary rock-record. On an Expansion Tectonic Earth, both transgressive and regressive changes in sea-level are primarily related to changes in surface curvature through time. These changes in surface curvature involve periods of both depression, or down warping of the outer margins of the supercontinental or continental crusts, as well as uplift as the crusts rise and re-equilibrate to on-going changes in surface curvature.

Depending on the severity of any change in surface curvature, these sea-level changes may also affect regional or even global climate, as well as ocean water circulation patterns, the distributions of marine species, and the type and location of sedimentary deposition. All of these ultimately affect the distribution, evolution, or extinction of the various marine and terrestrial animal and plant species as well as disrupting the various established species migration routes.

In conventional Plate Tectonic studies, these preserved changes to both sea-level and the various rock types preserved in sedimentary basins are interpreted as representing three basic kinds of sedimentary cycle. These sedimentary cycles include:

- Very long term cycles: These are caused by the inferred assemblage and subsequent break-up, rifting, and dispersal of supercontinents by seafloor spreading—referred to as *Wilson cycles*. These cycles take 400 to 500 million years to complete and the process is thought to have been operating for at least 2,000 million years.

- Convective overturn in the mantle: This convective overturn is inferred to lead to continental plate collision and rifting events during the assemblage and subsequent dispersal of supercontinents. This cycle is thought to be characterised by rapid global changes to seafloor spreading rates and accompanying changes to sea levels. The timing of these collision or rifting events is random, but may occur over periods ranging from about 10 to 100 million years.

- Climate induced global sea-level cycles—referred to as *Milankovitch cycles*: These cycles are inferred to result from mechanical variations in the Earth's orbit. The cycles are considered to occur over periods of tens to hundreds of thousands of years.

In contrast, on an Expansion Tectonic Earth the small Earth model studies show that there is no requirement for multiple cycles of random continental crustal assemblage, break-up, rifting, or dispersal to form supercontinents. Instead, the ancient supercontinents remained assembled as a complete supercontinental crust throughout the first 3,750 million years of Earth's history.

The small Earth model studies further show that this global supercontinental crust was subjected to a steady to accelerating process of crustal extension and basin opening, prior to a single phase of crustal rupture followed by continental break-up and opening of the modern oceans. Climate induced changes to the rock-record, emanating from these crustal extension and sea-level changes, may have then occurred as a result of disruption to the pre-existing climatic zones. It will be shown that these disruptions occurred during apparent migration of the modern continents into or out of the various climate zones during opening of the modern oceans.

Published changes to sea levels for the past 500 million years of Earth history are shown in Figure 14.3. The changes shown in this figure are readily recognisable in the rock-record, in particular by the relative severity of these changes above normal changes to sedimentary deposition or normal species extinction rates. Often, more recent changes to sea-level can be observed along coastlines in the form of raised wave-cut platforms or erosional features seen in rocky ledges and cliffs, as stabilised sand dunes, or as inter-dune swamps and lakes.

The distribution of first-order global changes in sea-level, relative to the present sea-level, is best seen in the smoothed red curve on Figure 14.3. This cycle is referred to in the literature as part of the very long term Wilson cycle. In contrast, on an Expansion Tectonic Earth this cycle is the most basic sea-level cycle operating during the past 500 million years of Earth history. Small Earth model studies show that this cycle records the rupturing and break-up of the Pangaea supercontinent and formation of the modern continents, including draining of the ancient seas into the newly opening modern oceans.

Sea levels are shown in Figure 14.3 to steadily fall from a peak level during Cambrian (Cm) and Ordovician (O) times to a minimum level during the Triassic (Tr). A steady rise during Triassic to late-Cretaceous (K) times was followed by a steady fall in sea levels to the present-day (N). Also superimposed on this first-order global cycle are more detailed second-order sea-level cycles as shown in the blue curve in Figure 14.3. These second-order cycles approximate the major first-order changes in sea levels, but show greater detail. The effects of these second-order changes were first recognised by naturalists and geologists well before geology became a recognisable science. The recognition of these secondary cycles has since traditionally been used in geology to define the major time periods and epochs of Earth history, as well as to correlate sea-level changes and extinction events across each of the modern continents.

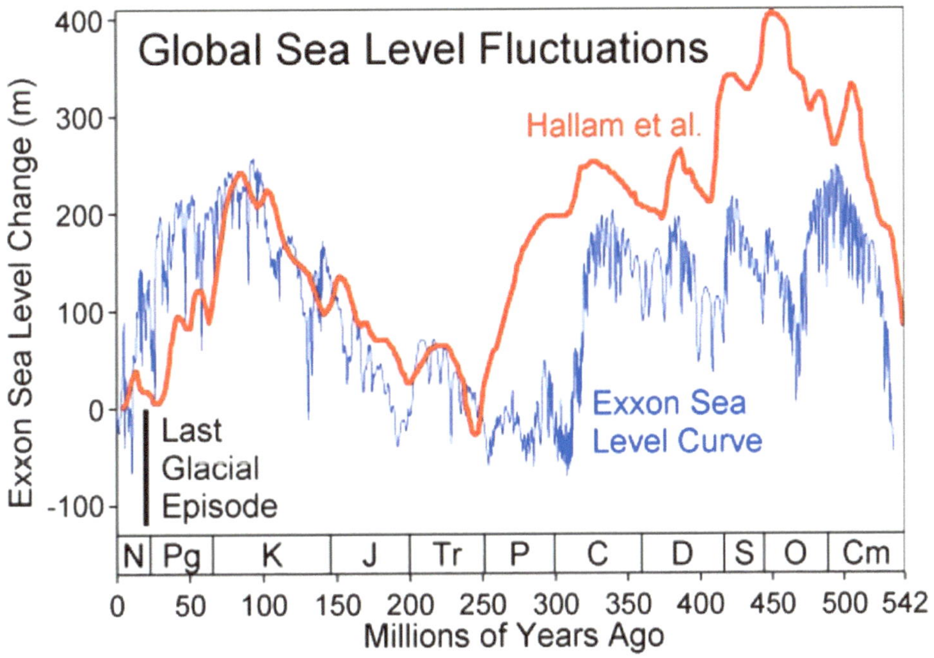

Figure 14.3 Changes in both sea-level and glacial record during the past 542 million years showing the Hallam et al. (1983) and Exxon global sea-level reconstructions for the Phanerozoic eon. The Exxon curve is a composite from several reconstructions published by the Exxon Corporation (after Hallam et al., 1983).

On an Expansion Tectonic Earth, these first- and second-order sea-level cycles reflect the variation in both coastal outlines and the volume and distribution of available ocean water existing during the past 500 million years. The second-order cycles, in turn, reflect more local variations in sea-level during the merging of previously isolated, smaller continental basins and seas. These local variations again coincide with local to global-scale extinction events caused by either rapid draining or flooding of land areas or seas.

These cycles, in effect, are a reflection of the progressive changes in surface area and changes in surface curvature operating throughout geological history. It is significant that the steady lowering of sea levels in the central part of Figure 14.3 coincides precisely with draining of the ancient continental seas during rupture and break-up of the ancient Pangaea supercontinent. The steady rise in sea levels during the latter part of this cycle coincides with both increases in new water from the mid-ocean-ridge spreading centres plus disruption and melting of any pre-existing ancient polar ice-caps.

In addition to the major changes to sea levels occurring during break-up of the ancient supercontinents, each of the second-order variations shown in Figure 14.3 also reflect changes to climate, to a shift in the distribution of the continental seas, to periods of relative crustal mobility, mountain building, erosion, as well as in response to progressive opening of the modern oceans. This is somewhat like water in a puddle, where the water level will always respond to any external influence placed on it such as more water coming in, evaporation, or someone stepping in it.

On an Expansion Tectonic Earth, the whole continental crustal and ocean system represented by these sea-level cycles can be viewed as cause-effect-cause feedback cycles. These cycles are intimately and directly related to the progressive increase in surface area and crustal extension during changes in surface curvature and ultimately to break-up of the ancient supercontinents, opening of the modern oceans, and merging of previously separate oceans. The fossil, ancient climate, and stratigraphic records now preserved in the rock-record represent inter-related physical evidence, which, in turn, reflects and preserves the changes occurring throughout geological time.

For example, the sea-level changes occurring on an Expansion Tectonic Earth, in what later became continental Europe and Asia, define and mirror the changes in surface area of the ancient continental Tethys Sea. This change in surface area accords precisely with the changes to sea levels shown previously in Figure 14.1 and these changes are now faithfully preserved in the rock-record as a succession of second order sea-level events. Throughout Palaeozoic times and extending into the Triassic Period, this Tethys Sea is shown to steadily reduce in surface area during initial draining during opening of the modern oceans. The areal extent of the Tethys Sea then increased to a maximum during mid- to late-Cretaceous times which mirrored a shift in sedimentary deposition, disruption of existing polar ice-caps, and associated climatic change. This maximum was followed by a rapid opening of the modern oceans plus a rapid increase in new water being expelled along the mid-ocean-ridges. The area of the Tethys Sea then slowly decreased to a minimum during more recent Cenozoic times continuing to the present-day. This rapid opening also coincided with rupturing of the ancient Tethys seafloor crust and opening of the modern Mediterranean Sea, which then exposed the ancient Tethys seafloor as an elevated continental land surface and now makes up the present-day European and Asian land masses.

As will be discussed in later chapters, the preserved fossil evidence throughout the world also shows that these rapid sea-level changes have produced unprecedented disruption to all marine and terrestrial plant and animal systems. The most dramatic changes to sea-level coincide with recorded extinction events which are shown in Figure 14.3 to coincide with periods of dramatic rise or fall of sea levels. The record of much of this fossil evidence is now preserved in sedimentary basins, remnants of which are located on each of the modern continents.

15 Climate Evidence

"Climatic features of the Mesozoic and Cenozoic have been shown to depart significantly from latitudinal predictions based on palaeomagnetism." Barron, 1983

When talking about climate evidence, in particular about ancient climate evidence, I must stress that it is not the day-to-day or seasonal climate changes that will be discussed here, but rather the very long-term evidence preserved in the rock-record. This long-term evidence is used to identify the major climate zones present on the ancient Earth and will be used here to quantify their location on each of the small Earth models. This information will also be used to compare the various climate zones with the locations of the magnetic poles and equators, as previously established from palaeomagnetic evidence and tentatively confirmed by geography.

Throughout Earth history, variations in climate are mainly affected by the amount of incoming solar radiation. The major variables over geological time include the rate of *insolation* and *albedo*, as well as the chemical composition of the atmosphere, and atmospheric humidity. Insolation is defined as a measure of the amount of solar radiation energy received from the Sun on a given surface area, and albedo is defined as a measure of the reflecting power of the Earth surface. The absorption and reflectance of the oceans and land may also change with time, for example by the presence of ice sheets or deserts, by modifications to the biosphere, and particularly by the proportion of greenhouse gases present in the atmosphere.

Insolation is also sometimes called *solar irradiation* and the amount of insolation received on the Earth's surface is largest when the Earth's surface directly faces the Sun. As the angle between the surface and the direction of the rays of sunlight changes, the insolation is reduced in proportion. This projection effect is the main reason why the polar-regions are much colder than equatorial regions on Earth and why the poles receive less insolation than does the equator.

All assumptions about ancient climate also involve the study of three complex systems: ocean circulation, atmosphere, and the past distribution of continents. Ocean circulation, in particular the development of deep ocean currents during opening of the modern oceans, is strongly influenced by the presence of land barriers. Similarly, topographic relief on the continents strongly influences the amount of rainfall received from the atmosphere as well as the accumulation of snow and ice in mountainous regions.

Correlation of coal swamps, thick bedded sandstone sequences, and glacial rocks are all considered to be excellent indicators of wet climates, while dry climates are indicated by evaporate rocks such as salt deposits, and equatorial regions by the presence of coal, carbonate-bearing sandstones, and carbonate reef strata.

Plate Tectonic studies suggest that the climate during late-Palaeozoic times—around 300 million years ago—was variable and complex. A period of glaciation also occurred at that time and lasted for about 60 million years. This was considered by Klein in 1994 to be related to the geography of the ancient Gondwana supercontinent, as well as a subsequent reduction in atmospheric carbon dioxide during assemblage of Pangaea, based on reconstructions on a Plate Tectonic Earth. The Permian climate was also considered by Erwin in 1993 to be related to an almost complete Pangaean continental crustal assemblage. The associated pattern of climate and glaciation was further considered to be related to an alternation of greenhouse and icehouse states.

In 1994 Crowley also showed that the Pangaea climate involved a late-Palaeozoic period of glaciation, with a subsequent transition to arid and nearly ice-free conditions during Triassic and Jurassic times. Geological studies indicated to Crowley that aridity during Triassic and Jurassic times was extensive over much of Pangaea, where global evaporite formation and distribution—mainly deposition of salt beds—was at a peak. Other evidence, such as shifts in vegetation patterns, pointed to periods of warming and drying during the late-Jurassic. Some changes were also considered by Crowley to be associated with a rise in sea-level, which was said to have accompanied the break-up of Pangaea.

Cretaceous climates were described by Barron in 1983 as warm and equable. The polar-regions in particular were considered to have been much warmer than today, with no evidence for polar ice. During that time, warmth-loving plants and animals were able to spread into high latitudes. In contrast, Francis and Frakes in 1993 showed there were distinct variations and trends in warming and cooling during the Cretaceous and Barron considered features of the Mesozoic and Cenozoic climates suggested that these ancient climates departed significantly from latitudinal predictions based on conventional palaeomagnetism.

In addition to changes in climate and the associated distribution of plant and animal species, Crowley also showed there were significant variations in the presence of carbon—in the form of coal—during Pangaean times. Organic, carbon-rich layers in the rock-record were found to be coincident with peaks in high sea-levels. The largest period of deposition of carbon in the rock-record reflected a time of major coal formation during the Permian and Carboniferous times. The high carbon burial rates during that time coincided with when land plants first became established on the continents and suggested there was a general absence of organisms capable of consuming decaying organic matter. These carbon changes also contributed to the presence of low atmospheric carbon dioxide levels during the Carboniferous glaciation times, with levels estimated to be comparable to the present-day levels, increasing to levels of up to five times the present value during the late-Jurassic.

The presence of coal—including hydrocarbons such as oil and gas—and limestone throughout much of the Palaeozoic Era represent important mechanisms for locking up carbon in the rock-record. Coal locks up carbon in the form of primary carbon derived from vegetative matter as well as complex carbon compounds in the form of hydrocarbons. Similarly, limestone is essentially calcium carbonate—calcium, carbon and oxygen—deposited in the rock-record by chemical precipitation of calcium carbonate or by the action of marine organisms. During early-Proterozoic times this locking up of carbon, in particular carbonate rocks, is also seen as an important mechanism for observed changes from a prior reducing to a present oxidising atmosphere.

In discussing and establishing ancient climate zones on an Expansion tectonic Earth I will be mainly considering the global distribution of available latitude-dependant climate indicators—that is, rocks that are specific to the various climate zones. This will be done by superimposing the distribution of various rock types on each of the small Earth models. These climate indicators will include a variety of sedimentary rock types that form, or were deposited within specific ancient climate zones, such as the carbonate sediments and glacial rocks.

15.1 Ancient Climate Zones

The idea of geographical and climate zoning was first hypothesized by the ancient Greek scholar Aristotle. He considered that the Earth was divided into three types of climate zones based on their distance from the equator. Thinking that the area near the equator was too hot for habitation, Aristotle called this region the "*Torrid Zone*"–now called the Tropics. He reasoned that all lands near the poles were permanently frozen and he called this uninhabitable zone the "*Frigid Zone*"–now called the Polar Region. The only area that Aristotle believed was liveable was the "*Temperate*

Zone," located between the Frigid Zone and the Torrid Zone. Aristotle's reasoning was vastly over simplistic, although the general idea was correct. Since then, humans and many animals have been found to inhabit almost all climate zones present on Earth including inside both the Arctic and Antarctic polar-regions.

The Earth's surface is now known to comprise five main geographical and climatic zones which are essentially defined by the major circles of latitude. The main differences between each of these zones relate to climate distribution as well as the attitude of the Sun (Figure 15.1). On this figure a circle of latitude on the Earth's surface is defined as an imaginary east-west circle connecting all locations on the surface with the same north or south latitude. The position of the Equator is fixed at 90 degrees from the Earth's axis of rotation, while the latitudes of the other circles depend on the tilt of the axis of rotation, relative to the plane of the Earth's orbit around the Sun.

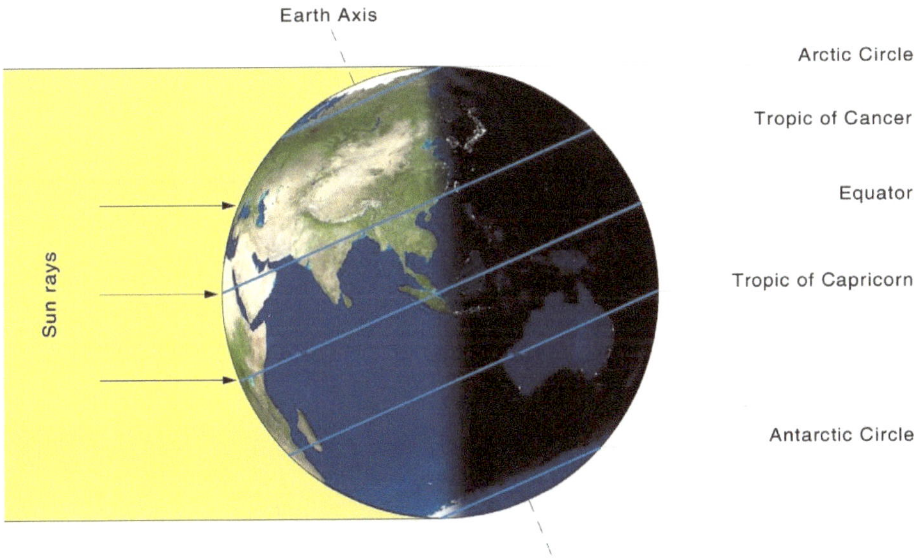

Figure 15.1 Diagram showing the location and derivation of the major circles of latitude on the Earth. (http://en.wikipedia.org/wiki/File:Earth-lighting-summer-solstice_EN.png)

The latitudinal values for these established circles of latitude vary slightly over time and each of the circles shown in Figure 15.1 are presently located at:

- Arctic Circle: 66° 33′ 44″ north latitude.
- Tropic of Cancer: 23° 26′ 16″ north latitude.

- Equator: 0° latitude.

- Tropic of Capricorn: 23° 26′ 16″ south latitude.

- Antarctic Circle: 66° 33′ 44″ south latitude.

By definition, the positions of the Tropic of Cancer, Tropic of Capricorn, Arctic Circle, and Antarctic Circle all depend on the tilt of the Earth's rotation axis relative to the plane of its orbit around the Sun—called the obliquity of the ecliptic. If the Earth were upright, that is, if its axis was at right angles to the solar orbital plane, there would be no Arctic, Antarctic, or Tropic circles. If this were the case, at the poles the Sun would always circle the horizon and at the equator the Sun would always rise due east, pass directly overhead, and set due west. In reality, the Earth's rotation axis is tilted and this is the reason why there are relatively well defined geographical and climate zones. The Earth's rotational axis is not exactly fixed though. Measurements have shown that the axis undergoes small fluctuations over time. This has a very small effect on the location of the Tropic and Polar circles and also on the location of the Equator. In year 2000, the mean value of the tilt of the Earth's rotational axis was 23° 26′ 21″.

The circles of latitude, excluding the equator, mark the divisions between the five main geographical zones. These zones are defined as follows:

- The North Polar Region, located north of the Arctic Circle.

- The North Temperate Zone, located between the Arctic Circle and the Tropic of Cancer.

- The Tropics, located between the north and south Tropical Circles.

- The South Temperate Zone, located between the Tropic of Capricorn and the Antarctic Circle.

- The South Polar Region, located south of the Antarctic Circle.

The Tropics straddle the equator and in this zone the Sun is directly overhead at least part of the year. In the two adjacent Temperate Zones the

Sun is never directly overhead and the climate is mild, generally ranging from warm to cool temperatures. The two Polar Regions experience a midnight sun and a polar night for part of the year. The polar-regions are the coldest parts of the Earth and are generally covered with semi-permanent to permanent ice and snow.

The locations of each of these present-day and ancient circles of latitude and climate zones on Expansion Tectonic small Earth models are shown plotted in Figure 15.2. The ancient circles of latitude, while appearing to be non-parallel in this figure, are in fact parallel and their geographical locations have been adopted as being the same as the present day latitudes. The tapered, non-parallel representation in this figure is meant only to simulate the progressive evolution of these circles and zones between each of the small Earth models. In reality, this representation may not be strictly correct as the location of the circles and zones is dependent on the tilt of the Earth's axis. If the tilt of the ancient Earth axis varied significantly from what it presently is, then it is the width and location of the zones that would be affected.

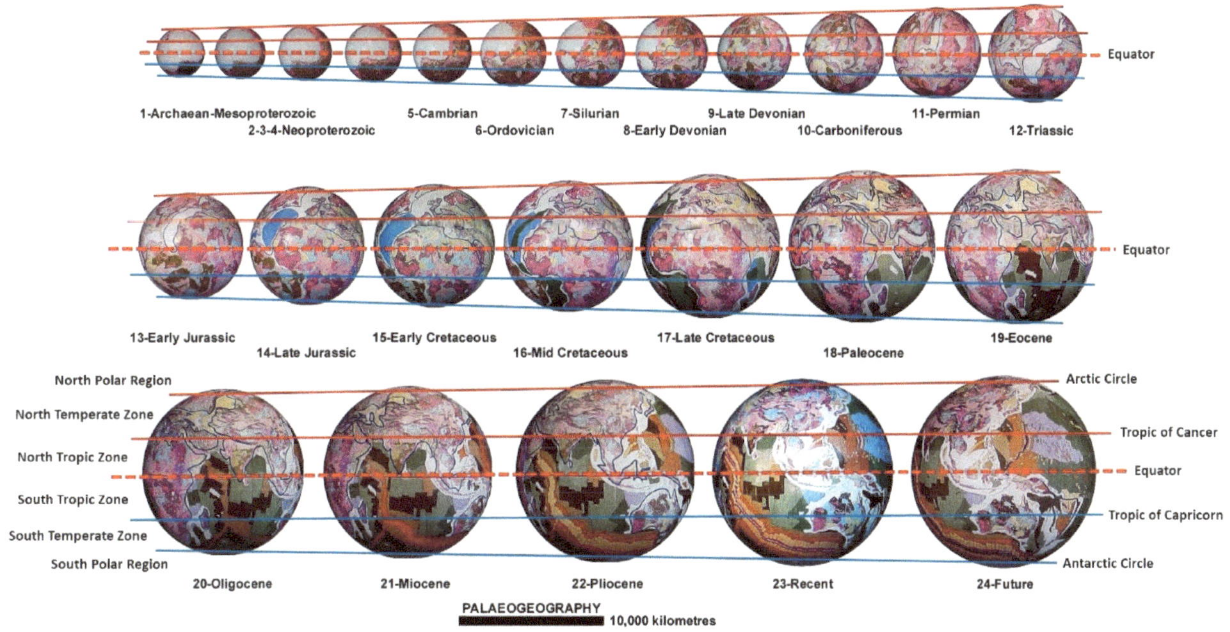

Figure 15.2 The location of climate and geographical zones on Expansion Tectonic small Earth models

Because of the limited amount of available data and research effort, absolute confirmation of the Earth's ancient tilt is very difficult to determine. Modern understanding of the Earth considers that the Earth's interior is molten to semi-molten, hence the angular momentum of the Earth would likely tend to resist any changes to its axis of tilt similar to the motion of a gyroscope. Because of this, for the purpose of this investigation, at this stage the Earth's ancient tilt is here considered to be fixed, or near fixed, over time.

While largely hypothetical at this stage, it should be appreciated that for the small Earth models to be valid, the distribution of climate-dependant rock types—such as reefs and carbonate sediments, glacial rocks, coal, and evaporite deposits—must all plot within their respective climate zones. If the location of a Cambrian coral reef, for example, plots either north or south of the Tropic of Cancer or Tropic of Capricorn respectively, then it is either the assemblage of the crusts on the Cambrian model or the inferred tilt of the Earth's axis that may be wrong, not the data itself.

Complicating this formal climate and geographical zonation is the break-up of the ancient supercontinents and opening of the modern oceans during post-Permian times. Once the Pangaea supercontinent broke-up and the modern continents dispersed to their present locations, evidence for each of these ancient zones will remain preserved in the rock-record but won't necessarily coincide with the present-day climate zones. This can be likened to coloured patterns on a jigsaw puzzle. Spread out randomly on a flat surface the patterns appear meaningless, but when the pieces are fitted together the patterns coincide to form a picture. For the ancient supercontinental Earth, this scenario is reversed. The climate and geographical zones—representing the picture—coincide with their respective ancient locations, and during supercontinental break-up the modern continents and fragmented ancient climate zones are dispersed to their present locations.

For the first 3,750 million years of Earth history, small Earth modelling studies have shown that the configuration of the ancient supercontinents remained intact. During that time the continental crusts were subjected to a steady and protracted process of crustal stretching focussed primarily within a global network of sedimentary basins. Under these conditions, established climate zones remained relatively intact throughout this extended period of time. As seen in Figure 15.2, any variation of these zones during the progression of time would have been limited to increasing their widths as a result of the increase in Earth radius. The more rigid ancient crusts may have also undergone localised jostling across, or within, the various established circles and zones during formation of the intervening sedimentary basins.

Once rupture and break-up of the supercontinental crusts commenced during late-Permian times, the rapid increase in Earth surface area gave rise to unprecedented disruption, fragmentation, and dislocation of the established ancient climate and geographical zones. In addition, new water, atmospheric gases, and volcanic lava were continually being expelled or intruded along the newly forming mid-ocean-rift zones in sympathy with opening of the modern oceans. Previously established polar ice-sheets may have also been severely disrupted during break-up of the supercontinents and opening of the modern oceans. In addition, where present, these ice-sheets would have then been subjected to periodic melting and refreezing during migration of the underlying continents.

This increase in new seawater and atmospheric gases may have also had a profound effect on the amount of absorption or reflectance of incoming solar radiation from both the newly formed oceans and the lands. It is interesting to note that these atmospheric and crustal changes coincide with the period of rapid plant and animal evolution on Earth. This evolution history may have also contributed significantly to modifications to the biosphere as well as the presence and quantity of atmospheric greenhouse gases.

15.2 Ancient Limestone and Coral Reefs

The global distribution of ancient carbonate limestone and coral reefs represents a unique means of locating and quantifying the position of ancient tropical zones on Expansion Tectonic small Earth models. Based on present day distributions, limestone and coral reefs only occur within a broad zone located plus or minus 25 degrees of latitude north and south of the equator. Within this zone, warm ocean waters and currents enable corals and marine creatures to thrive along the continental shelves of islands and continents. Plotting the distribution of ancient coral reefs on the small Earth models will then enable the location of the ancient equator, previously established from palaeomagnetic studies, as well as the established climate zones to be independently verified.

When talking about coral reefs we generally think of modern corals in context with tropical islands and warm equatorial climates. These reefs are underwater structures made from calcium carbonate secreted by marine organisms plus accumulated shelly fragments from deceased marine creatures. Coral reefs represent colonies made up of tiny living animals called *polyps* and these reefs are found in marine waters that contain few nutrients. Most coral reefs are built over a long period of time from what are referred to as stony corals, which, in turn, represent the habitats of the coral polyps. These polyps are like tiny sea anemones to which they are closely related.

Unlike sea anemones, coral polyps secrete hard external carbonate skeletons which support and protect their bodies. These coral reefs grow best in warm, shallow, clear, sunny, and agitated waters.

Throughout the latter part of Earth history, ever since hard skeletons were first developed by marine organisms around 500 plus million years ago, there have almost always been coral reefs. The origin of these external carbonate skeletons may simply have developed from the utilisation of a lime-rich waste product, which, no doubt, set like cement around the particular organism. During the Cambrian and following periods, the various marine shelly and coral organisms utilised this material and rapidly evolved and diversified in the process. The times of maximum development of these marine organisms were in the middle-Cambrian, Devonian, and Carboniferous periods, as well as the late-Cretaceous to more recent times.

Not all reefs in the past were formed by corals though. Stromatolites provide some of the most ancient records of fossil life on Earth, dating back to more than 3,500 million years ago. *Stromatolites* are microorganisms, commonly known as blue-green algae, which form layered structures in shallow water by trapping, binding, and cementing fine sand grains. Similarly, reefs in the early-Cambrian—around 540 million years ago—were first formed from calcareous algae and small marine creatures related to sponges. In the late-Cretaceous—around 100 to 65 million years ago—reefs were also formed by an evolving group of bivalve shelly creatures called *rudists*.

In the rock-record, these reef structures accumulated over large areas and formed what are referred to in geology as limestone reefs, or when mixed with sand or silt are simply called carbonate sediments. These carbonate limestone rocks are mainly composed of secreted carbonate and fragmented skeletal remains of marine organisms such as algae, coral, shells, and sponges. Limestone is a sedimentary rock composed largely of the minerals calcite and aragonite which are different crystal forms of calcium carbonate. The metamorphosed—heated and recrystallized—equivalent of limestone is called marble. Over time, deposition of these limestone rocks may extend along an ancient coastline for hundreds of kilometres, and, in turn, accumulate in thicknesses varying from tens to hundreds of metres.

On Plate Tectonic reconstructions, the assemblage, crustal accretion, and eventual break-up and dispersal of Pangaea during Permian to Triassic times resulted in substantial continental changes. These changes, in turn, are said to have controlled the spread or decline of the established carbonate reefs. These changes are also said to have greatly influenced the dispersal and distribution patterns of all marine species as well as the development of carbonate reef structures. In 1994 Flügel considered that the setting and location of Pangaean reefs on Plate Tectonic reconstructions is unusual, compared to the distribution of modern reefs, because most Pangaean reefs were shown to be formed in and around continental basins rather than along exposed continental margins flanking large oceans.

This observation of Flügel contradicts Plate Tectonic reconstructions of Pangaea where the growth of Permian and Triassic reefs supposedly took place along the margins of, or in basins adjacent to, a large inferred east-west trending Tethys Ocean. This inferred ocean was, in turn, located at the western end of an even larger inferred Panthalassa Ocean. Additionally, the Pangaean reefs were also interpreted to be located on, and associated with, small continental fragments—such as small continents or islands. These small fragments were inferred from the presence of exotic continental rocks of unknown origin, such as presently seen in rocks from western North America, South East Asia, and Japan. On conventional reconstructions, the plate motions of these small continents were then thought to have formed marginal seas located around the margins of these continental fragments. These continental fragments may have, in turn, favoured growth of coral reefs. Evolutionary crises affecting the marine species from each of these sites were considered by Flügel to correlate with a major end-Triassic reef extinction event.

In contrast, on an Expansion Tectonic Earth the presence of a large, inferred, Tethys Ocean, and similarly an even larger Panthalassa Ocean, is not required. Instead, these hypothetical oceans were shown to form a network of ancient continental seas located on a smaller radius supercontinental Earth. These ancient seas, in turn, formed part of a global network of sedimentary basins separating and defining the supercontinents and were the precursors to break-up of and opening of the modern oceans.

The distribution of published occurrences of mid- to late-Palaeozoic carbonate reef deposits—around 300 to 250 million years ago—is shown on the Permian small Earth model in Figure 15.3, along with the location of the established tropical climate zones and circles of latitude. Also shown is the distribution of the various ancient continental seas, previously established from the distribution of published coastal outlines on small Earth models. It should be noted that this distribution of carbonate reefs is based on published records and does not necessarily represent the entire global distribution of reefs in lesser studied regions. Note also, out of interest, the location of Antarctica straddling the ancient Permian equator in the third model from the left.

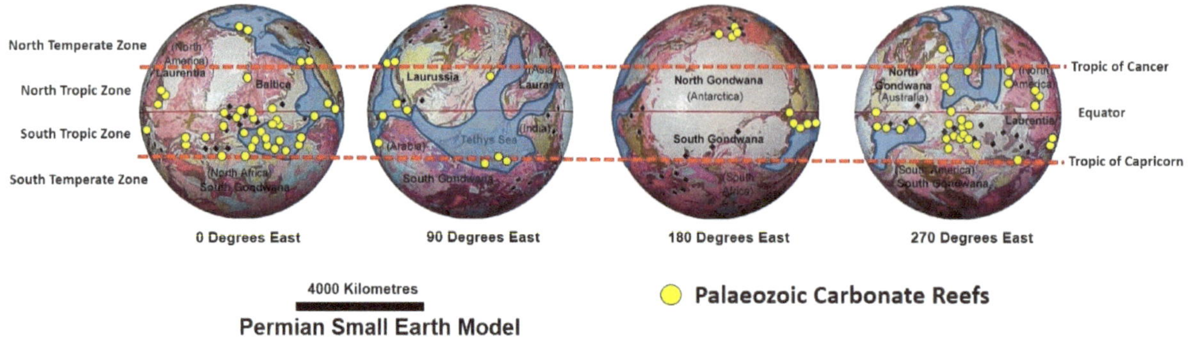

Figure 15.3 Distribution of Palaeozoic carbonate reefs (yellow dots) plotted on a Permian small Earth model, as well as ancient Permian coastlines and continental seas.

In this figure the reader will note that there is some minor disparity between the plotted reef locations and the published ancient shorelines. This is only apparent because the reef data extend back in time a further 100 to 200 million years to times when these coastlines were different to those shown on this Permian Pangaean small Earth model—representing coastlines extending back to Gondwanan times. On this figure, the distribution of carbonate reefs is shown to neatly straddle the equator and shows a good correlation with the adopted equatorial Tropics, as well as the Tropic of Cancer and Tropic of Capricorn circles of latitude. Outlying reefs shown within the northern Temperate Zone may either suggest the presence of warm tropical currents extending into these regions or because of poorly dated reef locations.

It can be seen that the width of the ancient Topic Zone, as shown by the distribution of Pangaean reefs, is also very close to the anticipated width shown by the dashed red lines. This suggests that the axis of tilt of the ancient Earth may have been the same or very similar to what it is now. The distribution of carbonate reefs within this ancient Pangaean Tropic zone then quantifies the location of the Pangaean equator on an Expansion Tectonic Earth, as previously established from palaeomagnetic studies, as well as quantifying the location of the ancient tropical climate zones.

During break-up of the Pangaea supercontinent during the late-Permian, it should be noted that these ancient climate zones remained as fixed zones of latitude as defined by the tilt of the Earth's axis of rotation. Similarly, the presence of living marine organisms is totally dependent on the presence and location of their preferred tropical climate zone and over time these living organisms must remain within this preferred zone. During crustal break-up the marine organisms were then forced, over tens to hundreds of millions of years, to continuously relocate and re-establish new reefs in order to remain within the ever changing climate zone. This change would be very subtle and would occur over a very long period of time without the organisms being aware of the changes. From this it should be appreciated that migration of the reef and marine species is then only apparent; it is the continent and its established ancient reef that moves, not the climate zones or marine species.

It should also be appreciated that if an ancient continent and its attached reef migrated out of a preferred tropical climate zone, the marine organisms living on this reef would gradually die out. This migration would then leave behind the abandoned carbonate sediments or ancient coral reef. In contrast, if a continent migrated into an equatorial tropical climate zone the marine organisms would begin to establish and colonise a new reef. As the continents continued to migrate, any new reef colonies established in these new regions would then be subject to further on-going relocation or extinction of the reefs over time. After supercontinental break-up, the rock-record will then show the modern continents and fragmented remains of the ancient reefs migrating relative to their previous ancient tropical climate zones.

In order to emphasise this migration, the distributions of Palaeozoic through to Jurassic carbonate reef deposits are shown as various coloured dots plotted on an early-Jurassic Expansion Tectonic small Earth model in Figure 15.4. These reefs are again shown in conjunction with the anticipated Tropic Zone, the Tropic of Cancer, and the Tropic of Capricorn circles. Also shown in this figure is the early opening of the modern Pacific and Atlantic Oceans, shaded in blue, as well as the locations of the early-Jurassic coastlines shown as blue lines.

On an Expansion Tectonic Earth, the Palaeozoic to Jurassic times represent a transitional period of Earth history leading to the eventual opening and rapid development of the early North and South Pacific, North Atlantic, and Arctic Oceans. Because of this, earlier formed carbonate reefs were subject to fragmentation, disruption, and migration away from their previously established ancient tropical zones and into new climatic zones during migration of the continents.

The various reef deposits shown in Figure 15.4 highlight this period of disruption, where most of the abandoned ancient reefs are shown to have migrated well north into temperate climate zones. This is highlighted by the distribution of ancient Palaeozoic reefs, shown as yellow dots, which can be compared to their same distributions shown in Figure 15.3.

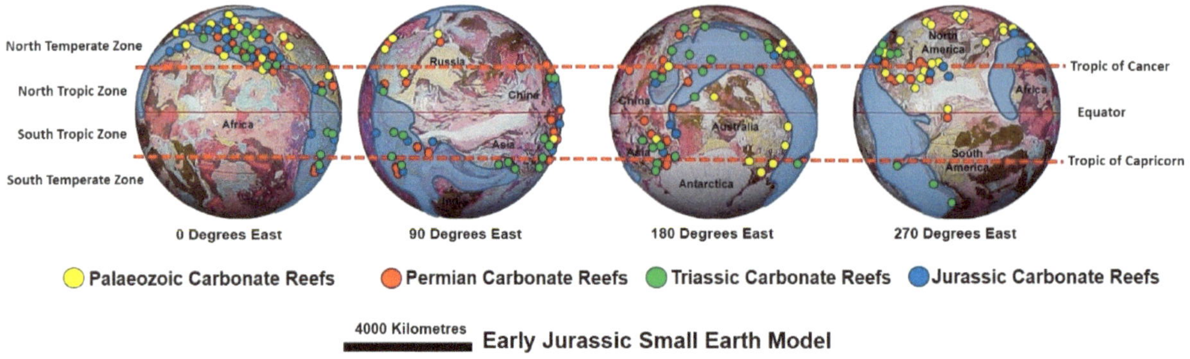

Figure 15.4 Distribution of Palaeozoic to Jurassic carbonate reefs plotted on an early-Jurassic small Earth model, as well as ancient Jurassic coastlines and opening of the modern oceans.

The distribution of the youngest Jurassic reefs—shown as blue dots in Figure 15.4—is reasonably well constrained within the Jurassic Tropic Zone, but these also extend north into the newly opening Mediterranean Sea north of Africa. This may either represent a hold-over effect of marine species in this warm sea—where particular species didn't die out—or a variance of the axis of tilt of the Jurassic Earth. It may also represent incorrect dating of the reefs themselves, as well as forced migration associated with opening of the modern oceans.

In addition, during the late-Triassic Period—the time interval between Figures 15.3 and 15.4—a pre-existing land connection between California in North America and Queensland in Australia was breached (Figure 15.5)—more on this in Chapter 16. This breaching resulted in the ancient Panthalassa Sea, located in the north, merging with the newly opening South Pacific Ocean to form a single modern Pacific Ocean during the Jurassic. This breaching globally disrupted the previously separate sea levels existing at the time and coincides with the end-Triassic reef extinction event as noted by Flügel in 1994. This Pacific Ocean breaching event, plus similar breaching events in the Atlantic Ocean, also opened up new avenues for species migration and development into other oceans. Marine organisms remaining after the extinction event were then able to migrate into, or away from, areas previously not favourable or accessible to new colonisation.

By the early-Jurassic Period, the ancient continental seas were also continuing to drain from the ancient Pangaea supercontinent. This draining of the seas further exposed the older Palaeozoic reef deposits, forcing the marine organisms responsible for reef development to migrate into new locations, to evolve, or to perish over an extended period of time lasting many tens of millions of years. The continued presence of Triassic and Jurassic reefs suggests that these marine organisms had indeed migrated and were able to keep pace with the migrating continents and lowering sea-levels. Figure 15.4 also shows that, by the early-Jurassic, these marine species had successfully migrated into and colonised each of the modern oceans within their preferred equatorial tropical climate zones.

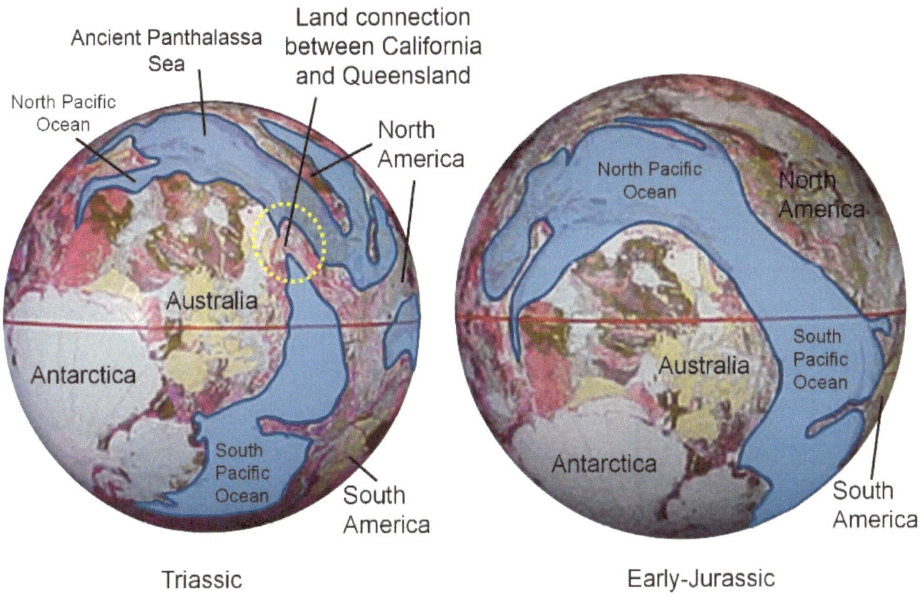

Figure 15.5 Triassic to early-Jurassic breaching of the land connection between the ancient Panthalassa Sea and the opening South Pacific Ocean to form the modern Pacific Ocean.

15.3 Ancient Glacial Record

There have been a number of major ice-ages over the course of Earth history where ice-sheets have periodically covered the higher latitudes of the ancient and modern continents and seas. Evidence for ancient glacial occurrences in the rock-record comes from various sources. These include the presence of striated rock surfaces and rock formations created by the passing of a glacier, large foreign rocks embedded within sedimentary strata, and the accumulation of characteristic rock debris. This rock debris accumulates where a glacier or ice-sheet melts and includes deposits that accumulate adjacent to streams flowing from a melting glacier, through to debris rafted well out to sea by icebergs where melting later deposits the debris on the seafloor. Because of the wide variation in types of glacial debris, the generic term *glacigenic* will be used here to include all glacial-related rocks. In this context glacigenic simply means rock, formations, or rock debris that have a glacial origin and includes debris that was deposited both on the lands and on the sea floors.

The geologic rock-record shows that, rather than the continuous presence of polar ice-sheets on Earth throughout time, there have been five major well defined periods of glaciation during Earth history. These periods include a number of prominent events in early- and late-Proterozoic times, the early- and late-Palaeozoic, and more recently during the late-Cenozoic, each of which lasted from tens to hundreds of millions of years. These glacial events were often cyclical, with individual cycles lasting for millions of years, interspersed with relatively warm interglacial periods. The Earth is currently passing through an ice-age, known as the *Quaternary glaciation,* and is presently in the warmer *Holocene interglacial period*—hence the global warming that we are experiencing.

Evidence from glacial studies suggested to Eyles and Young in 1994 that during the late-Archaean and early-Proterozoic, Earth surface temperatures may have been 30 degrees Celsius less than now and the Sun's luminosity was about 70 percent of the present value. Low Archaean temperatures were considered by Eyles and Young to be the result of an insulating effect caused by development of a primitive crust in conjunction with declining geothermal temperatures in the mantle and crust. It will be shown later that this may also be an artefact of modelling the data on a Plate Tectonic Earth model.

Conventional studies further suggest that glacial conditions during that time may have existed as a *Snowball Earth,* with frigid conditions inferred to have extended throughout the entire Earth's surface. This Snowball Earth hypothesis was first presented by Hoffman and others in 1998 who considered that the Earth's surface became entirely or nearly entirely frozen at least once during Precambrian times. One of the problems of a Snowball Earth, though, is that if the ancient Earth were covered in ice there would have been no open water to evaporate to keep the snow falling. Another problem is that if the whole Earth were covered in ice the albedo—the reflectivity of the Earth's surface—may have been so high that it may have been difficult to melt the ice again.

Proponents of the Snowball Earth hypothesis argue that it best explains the global distribution of ancient glacigenic rocks, in particular those known to occur within ancient tropical latitudes. Opponents of this hypothesis contest the presence of a global glaciation, with objections based on the implications surrounding an ice- or slush-covered ocean, as well as the perceived difficulty of the Earth to escape from an all-frozen condition. There are also a number of unanswered questions proposed by these opponents, including whether the Earth was a full snowball, or a slushball with a thin equatorial band of open, or seasonally open water.

I will dwell on this point for a while longer because this hypothesis is now becoming entrenched into conventional Plate Tectonic thinking, to the point where it is almost considered a meme—a paradigm—a foregone conclusion. The Snowball Earth hypothesis was proposed to explain the global distribution of ancient glacigenic rock deposits; in particular deposits located in ancient equatorial regions where the distributions are based on a Plate Tectonic Earth model. On Plate Tectonic crustal assemblages the distributions of ancient continental fragments and supercontinents are random and hence the presence of glacigenic rocks in ancient equatorial regions is also random and very difficult to explain other than from a Snowball Earth perspective. I must point out, though, that there are no definitive Plate Tectonic crustal assemblages available for Precambrian times hence a Snowball Earth scenario remains largely hypothetical.

The abundance of normal, water-lain sedimentary rocks preserved within Archaean and Proterozoic terranes throughout the world today, including warm-climate carbonate rocks, is testament to the presence of ancient rivers, streams, and seas as well as variable climate conditions during these Precambrian times. The presence of sedimentary and warm-temperature rocks demonstrates that some form of latitudinal distribution of climate zones must have also been present during Precambrian times, which may have also been separated into distinct zones as they are today.

I must further point out that the presence of glacigenic rocks in the rock-record does not automatically dictate their location within either of the polar-regions. Most glaciated regions on the Earth today, such as the Himalaya, the Alps, the Andes in South America, Australian Eastern Highlands, the High Sierras in North America, Mount Kilimanjaro in Africa, and so on, are well away from the poles and these occurrences simply represent high mountainous regions, not polar-regions. This mountainous distribution differs markedly from the presence of extensive ice-sheets within the polar-regions. Forcing observational data into a Snowball Earth concept is therefore potentially very misleading.

In contrast to the random distribution of glacigenic rocks on a Plate Tectonic Earth, these same rocks on an Expansion Tectonic Earth will now be investigated to identify the presence of ancient high latitude climate zones. It is anticipated that the perceived reduced luminosity of Eyles and Young as well as the Snowball Earth of Hoffman on an Expansion Tectonic Earth may then simply be a reflection of the much smaller surface area of the primitive

Precambrian Earth. If this is the case, then the widely scattered distribution of glacigenic debris on the present-day Earth may definitely be generating a misguided climatic hypothesis.

15.3.1 Distribution of Ancient Polar Regions

The locations of the ancient magnetic North and South Poles were previously established from small Earth modelling of available palaeomagnetic pole data in Chapter 13. From these pole locations it is possible to show the hypothetical locations of each of the north and south polar-regions. Plotting these polar-regions is based on locating the polar circles of latitude, similar to how they are located on the present-day Earth. For the purpose of this exercise it is assumed that the tilt of the Earth's ancient rotation axis was the same or very similar to what it is now. This, of course, is conjectural but is a reasonable starting point.

Figure 15.6 shows the location of the hypothetical North Polar Region, centred over the ancient North Pole on each of the Expansion Tectonic small Earth models. Similarly, Figure 15.7 shows the location of the South Polar Region, centred over the ancient South Pole. Within each of these polar-regions, if climate conditions were favourable we would also expect to find polar ice-sheets. The increase in size and surface area of the polar-regions shown in these figures is directly related to the increase in radius and surface area of the Earth. In other words the circle of latitude remains constant for each model, but the physical dimension of a degree of latitude, and hence size of the polar region, increases as the Earth radius increases.

The ancient Precambrian to Palaeozoic magnetic North Pole was previously shown to be located within what is now eastern Mongolia. This pole location is shown in Figure 15.6 as red dots centred within the North Polar Region. As the northern part of the Pangaea supercontinent ruptured during the late-Permian, the ancient Arctic Ocean opened and the various modern continents began to slowly migrate south.

The location of the ancient North Pole on each small Earth model in Figure 15.6 shows there was then an apparent northward migration of the pole through Siberia to its present location within the modern Arctic Ocean. It can be seen that rupturing of the northern Pangaea supercontinent began during late-Permian to early-Triassic times. This event then destabilised the ancient North Polar ice-sheet within the ancient North Polar Region. Since then, the modern Arctic polar ice-sheet has ceased being a continental ice-sheet and has remained as a marine ice-sheet located over the Arctic Ocean, which is now subject to seasonal melting and freezing.

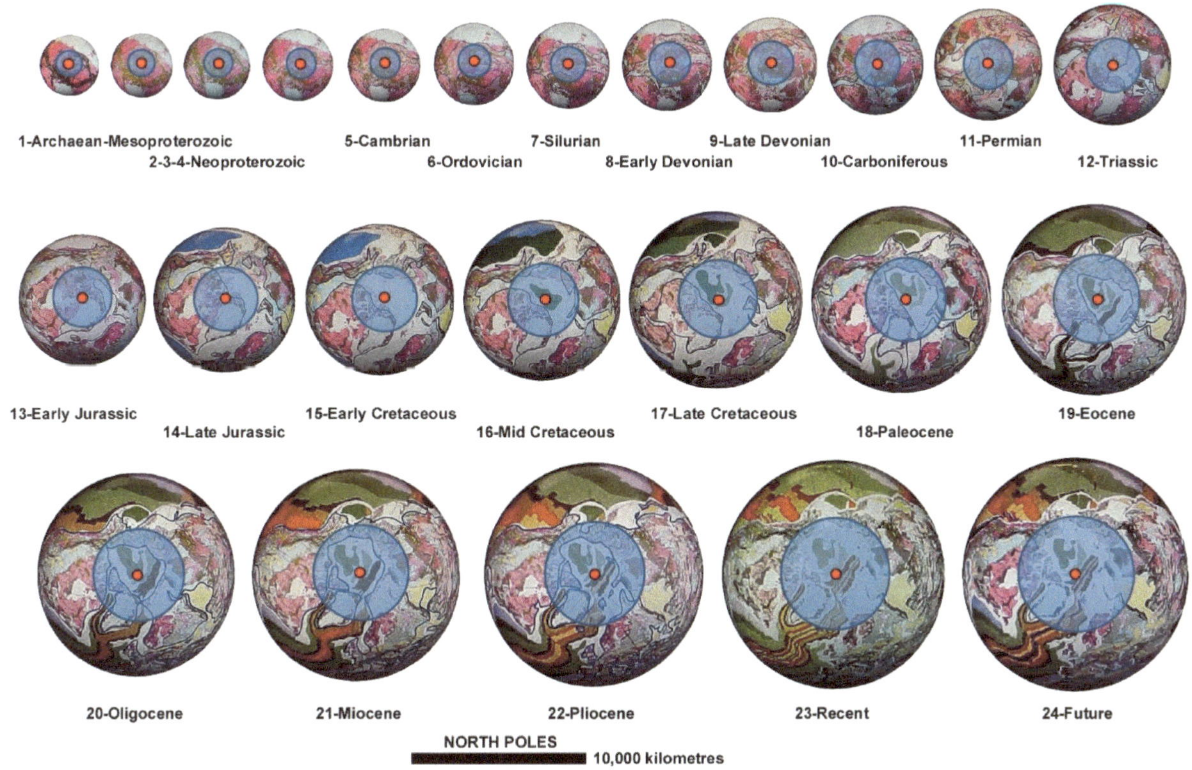

Figure 15.6 Locations of the hypothetical ancient North Polar Region shaded in blue on Expansion Tectonic small Earth models.

The ancient Precambrian and Palaeozoic magnetic South Pole was shown to be located in what is now west Central Africa. This is shown in Figure 15.7 as blue dots centred within the ancient South Polar Region. As the southern part of the Pangaea supercontinent commenced rupturing and the South Atlantic Ocean began opening during the Jurassic, the various modern continents slowly migrated north relative to the ancient South Pole. The location of the ancient South Pole on each of the small Earth models shows there was then an apparent southward migration of the ancient pole along the South American and West African coastlines to its present location in Antarctica.

As the ancient South Polar Region crossed from South Africa onto the opening South Atlantic Ocean, any ancient ice-sheet, if present, would have been disrupted and fragmented. As the polar region crossed onto the

Atlantic Ocean, the continental ice-sheet and prevailing climate conditions would have then changed to a seasonal marine ice-sheet similar to the present-day Arctic ice-sheet. It is interesting to note that this migration of the polar region onto the Atlantic Ocean coincides precisely with the duration of the Mesozoic Era, which was accompanied by changes to sea levels and global climate.

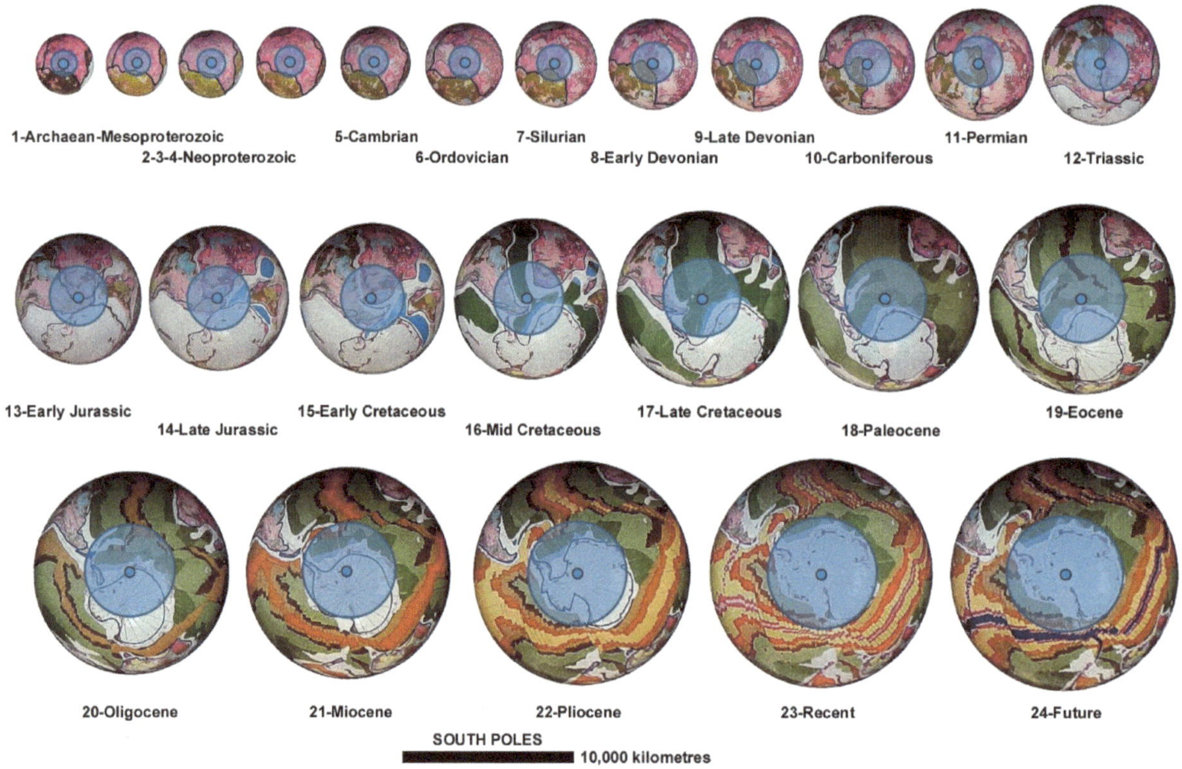

Figure 15.7 Locations of the hypothetical ancient South Polar Region shaded in blue on Expansion Tectonic small Earth models.

It is also shown on the small Earth models that during Cretaceous to present-day times, Antarctica migrated from equatorial regions into the South Polar Region before crossing onto the South Pole. During this migration, a South Polar ice-sheet was progressively established on Antarctica as a permanent continental ice-sheet. This continental migration and establishment of the ice-sheet coincides with the duration of the Cenozoic Era.

From this study what must be appreciated is, by definition, the North and South Polar Regions must remain centred over each of the ancient poles. This means that, as the continents migrate—relative to the ancient poles—the evidence presented by the distribution of ancient glacigenic rocks, which are firmly attached to the continents within each of these polar-regions, also migrate along with the respective migrating continents. In reality, what would be expected is that during migration of a continent through a polar region, the leading edge of an established ice-sheet would melt and break-up as it moves out of the respective polar region. Similarly, the trailing edge of the ice-sheet would then freeze and increase in area as it moves into the respective polar region.

A further important point to note is that opening of the modern oceans beneath an ancient ice-sheet exposes the ice to the influence of circulating ocean currents. This then changes the ice-sheet from a permanent continental sheet—such as the modern Antarctic ice-sheet—to a seasonal marine sheet—such as the modern Arctic ice-sheet. This will then affect the presence, size, and extent of ice cover within these zones and will directly affect global climate, sea levels, and the distribution or decline of various plant and animal species.

The known glacigenic occurrences shown in the following figures throughout the rest of this chapter include all rocks of recognisable glacial origin, as well as marine debris and non-glacial rocks. These non-glacial rocks include sedimentary rocks that consist of a wide range of consolidated, non-sorted, to poorly sorted, sediment of any glacigenic origin ranging from sand to larger size particles that are suspended in a mud matrix. Similarly, marine glacigenic rocks represent glacial debris that was dropped on the seafloor from melting icebergs or sea-ice. This marine glacigenic debris is a very important and characteristic rock type, which can be distributed well away from the original source areas depending on sea currents and rate of melting of the icebergs at the time.

15.3.2 Early-Archaean and Proterozoic Glaciation

Apart from small glacigenic occurrences from South Africa and western North America, evidence for glaciation during the Archaean Era is scarce. This suggests that the Archaean record, in contrast to what Eyles and Young inferred from low surface temperature evidence, may have been essentially barren of glacial deposits. On a primordial Archaean Earth this lack of glacigenic occurrences may have also simply been due to a shortage of surface water required to form ice-sheets. Similarly, a low elevation contrast between the lands and seas would have limited the ability to form mountain glaciers and hence glacial erosion to form glacigenic debris. The small, known glacial occurrences in South Africa and western North America may then simply represent all that is left after subsequent erosion of the ancient Archaean terranes.

In contrast to Archaean times, small Earth models show the Proterozoic to be characterised by an extensive network of stable sedimentary basins and shallow seas. There is abundant evidence in the rock-record for widespread glaciation occurring during early-Proterozoic times from several regions of North America, Finland, South Africa, Australia, India, and Scandinavia. Other Proterozoic glacigenic deposits occur elsewhere in the Baltic region as well as the east-central and northern Russian regions. The glacigenic deposits from each of these regions range in age from about 2,560 to 1,620 million years. In some areas there is evidence for multiple glaciations separated by warmer interglacial periods with associated sediments and chemical weathering of the exposed land surfaces. As previously mentioned, chemical weathering is indicative of warm surface temperature conditions, in particular within the equatorial region.

In contrast, the mid-Proterozoic Era is shown from the rock-record to be a non-glacial period. The only reported occurrence of glacigenic deposits is from Russia, but this has an uncertain age. This lack of glacial activity may therefore indicate the presence of warmer climates persisting during that time.

15.3.3 Late-Proterozoic Glaciation

During the late-Proterozoic Era there is again abundant evidence from the rock-record for widespread and long-ranging periods of glaciation during that time. These periods of glaciation lasted from around 1,000 million years ago into the early-Cambrian Period—around 500 million years ago.

Most regions where late-Proterozoic glacigenic rocks have been found show evidence of several periods of glaciation during this major event. These glacial periods occurred on a scale of tens of millions of years, and are similar to glaciations previously observed for the early-Proterozoic Era. Each of these individual glacial events were also preceded, separated, and followed by periods of relatively warm inter-glacial climates.

In 1989, Chumakov and Elston noted that late-Proterozoic glaciers recorded from Russia descended to sea-level in low latitudes close to the ancient equator. In 1994 Eyles and Young also showed that most late-Proterozoic glacial deposits accumulated as marine sediments along the ancient continental margins. In contrast, carbonate rocks, interbedded with many of the preserved glacigenic deposits, show that many of these glacigenic rocks were deposited in restricted warm-water seas prior to onset of continental uplift along the basin margins.

The presence of carbonate rocks interbedded with marine glacigenic deposits represents a major and very contentious dilemma for Plate Tectonic reconstructions. How can icebergs, or ice-sheets, drop debris within restricted warm-water seas near the equator without having access to high continental topography along the basin margins? This is especially applicable in areas where warm-water carbonate rocks were then deposited during interglacial periods lasting for millions of years.

15.3.4 Precambrian Expansion Tectonic Glaciation

The known Archaean and Proterozoic glacigenic occurrences are plotted on the late-Proterozoic Expansion Tectonic small Earth model in Figure 15.8. These occurrences include all deposits from each of the early- and late-Proterozoic glacial events as well as the few Archaean glacigenic occurrences.

The Precambrian glacigenic deposits are plotted together because palaeomagnetic evidence shows that, on an Expansion Tectonic Earth, the North and South Poles were relatively stationary throughout these ancient times. During these times a south-polar ice-sheet was located in what is now west Central Africa and a north-polar ice-sheet was located in what is now Mongolia-China. The ancient poles remained at these locations throughout the Precambrian Eras and their continued presence and locations extended well into the early-Palaeozoic Era.

Figure 15.8 shows that the global distribution of Precambrian glacigenic deposits extended throughout each of the ancient Archaean and Proterozoic terranes. As noted by Eyles and Young in 1994, most of these Precambrian glacigenic occurrences were deposited as marine sediments and in most cases this is confirmed in Figure 15.8 by their close association with the network of sedimentary basins and shallow continental seas shown as a khaki colour.

As previously discussed in Part 1 of this book, Earth expansion during Archaean and Proterozoic times amounted to about 60 kilometres increase in Earth radius, which was in turn spread over nearly 2,500 million years of Earth history. The very long period of erosion involved in this history, along with the presence of an extensive global network of stable sedimentary basins, suggests that the elevation contrast between the lands and seas may have been quite low during these times.

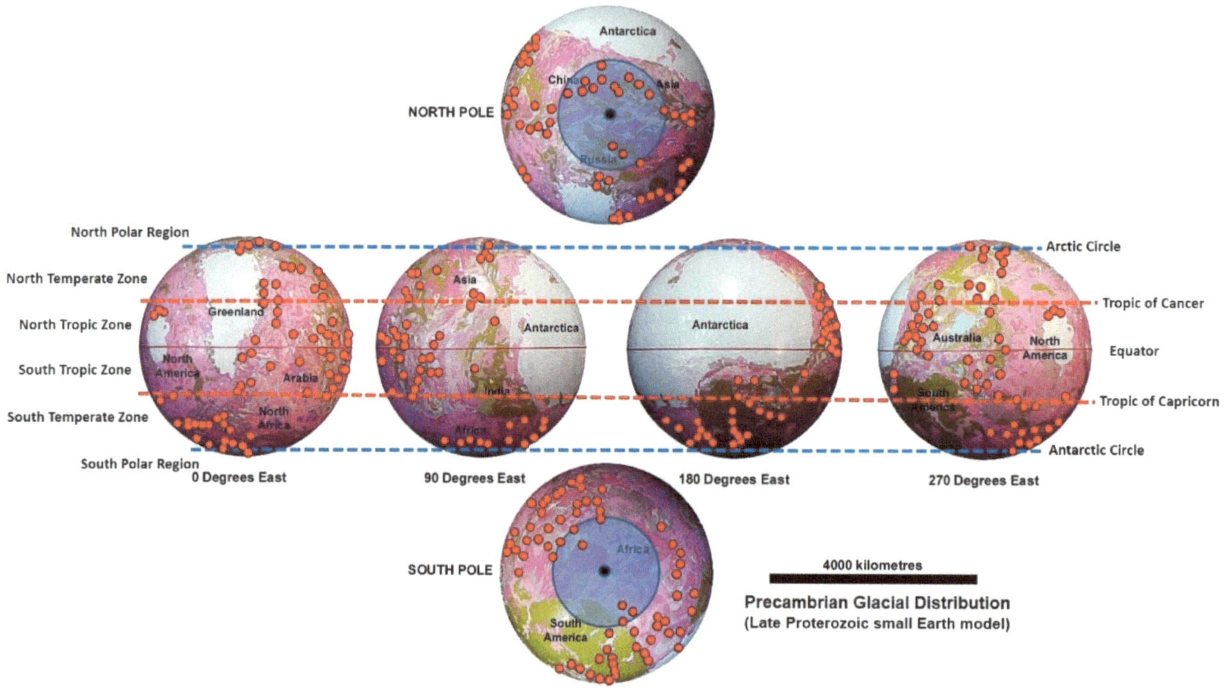

Figure 15.8 Precambrian glacigenic deposits plotted on a late-Proterozoic Expansion Tectonic small Earth model. Known Precambrian glacigenic occurrences are shown as red dots.

This low elevation contrast then limits the potential for elevated topography suitable for forming mountain glaciers. At first glance this observation may seem to be contradicted by the widespread distribution of glacigenic rocks shown in Figure 15.8, which extend from near-polar to equatorial regions. This distribution also shows a general absence of glacigenic rocks within the polar-regions. This absence from the polar-regions is normal, as glacigenic material from glacial ice-flow would have been transported away from these regions.

To understand the distribution of Precambrian glacigenic rocks on an Expansion Tectonic Earth, the distribution must be considered in relation to the much reduced primordial Earth radius. During the Archaean to late-Proterozoic Eras, the surface distances extending from the ancient poles to the ancient equator varied from about 2,600 to 3,200 kilometres respectively, compared to about 10,000 kilometres now. These are maximum pole to equator distances and the actual distances would be less than this. When the radius of the ancient polar region is subtracted from these distances—say 1,000 kilometres—this reduces the maximum distances to between 1,600 to 2,200 kilometres, which are well within the realms of iceberg migration into the equatorial regions.

These dimensions suggest that transport and dispersal of debris by floating icebergs fragmented from the polar or floating ice-sheets may have been a viable means of dispersing marine glacigenic debris during Precambrian times. The network of continental seas and circulation currents on the ancient Earth would then enable icebergs to readily disperse and migrate into warmer equatorial regions, which was accompanied by melting and dropping of large amounts of rock debris on the seafloor. Similarly, the erosion and transport of glacigenic debris by melt-water streams on the lands may also explain the presence of glacigenic rocks deposited on and around the margins of many of the ancient continents.

Precambrian warm-water sediments interbedded with marine glacigenic rocks are reported from Australia, North America, Africa, Greenland, Scotland, and from Spitsbergen in Norway. On an Expansion Tectonic Earth each of these continents were in close proximity to the ancient equator during Precambrian times and each was surrounded in part by ancient continental seas. At these locations, the distribution of glacial debris by both sedimentary transport and floating ice mechanisms provides a plausible explanation for the presence of glacial marine deposits recorded from sediments located near the ancient equator. The presence of interbedded warm-water carbonate rocks, including iron-silica-rich banded iron and chert sediments, would then be perfectly justifiable during warm interglacial times.

The distribution of Precambrian glacigenic rocks shown in Figure 15.8 adequately confirms the location of the ancient polar-regions. It also subtly confirms the location of the Precambrian equator on Expansion Tectonic small Earth models, without any need for an arbitrary fragmentation of continents. This glacigenic evidence is further supported by documented palaeomagnetic pole data, which confirms low-latitude equatorial locations in many of these known glacigenic regions.

15.3.5 Early-Palaeozoic Glaciation

A significant early-Palaeozoic glacial episode, peaking at about 440 million years ago, is recorded from late-Ordovician rocks in West Africa, Morocco, and Saudi Arabia. Less well defined deposits occur in Scotland, Ireland, Normandy, South Africa, Spain, Ethiopia, and Portugal, and small mountainous ice-centres also occur in eastern North America. In some areas the late-Ordovician glacial episode lasted into the following early-Silurian Period, in particular those recorded from South America and possibly Europe.

Geographic small Earth model studies have shown that during the Palaeozoic Era the ancient Earth was undergoing a steady to rapid increase in both radius and surface area, along with relatively rapid changes to surface curvature. The Palaeozoic Era is noted for its onset of complex geosynclinal and tectonic activity within sedimentary basins, along with an increase in deposition of eroded sediments. This implies that, on an Expansion Tectonic Earth, there was an increasing elevation contrast between the lands and sedimentary basins along with an increasing potential for elevated topography suitable for forming mountain glaciers.

The distribution of early-Palaeozoic glacigenic occurrences is shown plotted on the Ordovician small Earth geographic model in Figure 15.9, along with published coastal outlines and ancient seas. This distribution of glacigenic rocks agrees in principle with early-Palaeozoic glaciation on Plate Tectonic reconstructions of Gondwana and in each case the location of the South Polar ice-sheet coincides with the location of the ancient magnetic South Pole, which, at that time, was located in west central Africa.

The distribution of early-Palaeozoic glacigenic deposits on an Expansion Tectonic Earth is shown in Figure 15.9 to be restricted to the southern hemisphere during this glacial period. No significant deposits are recorded from the northern hemisphere. On an Expansion Tectonic Earth this distribution coincides with an exposed South Gondwana supercontinent, which was centred over the South Pole, and an extensive Tethys Sea centred over the North Pole. This would imply that during these times there was a continental ice-sheet located over the South Pole and a seasonal marine ice-sheet located over the North Polar Arctic Sea.

The distribution of glacigenic rocks in this figure, shown in conjunction with the labelled climate zones, also highlights preferential deposition of these rocks within the South Temperate climate zone. This is again understandable as glacigenic material derived from the South Polar Region would have been transported and deposited away from this region by movement of the ice-sheet. Similarly, there is also some evidence shown in this figure for on-going deposition of marine glacigenic rocks within the equatorial regions via melting icebergs.

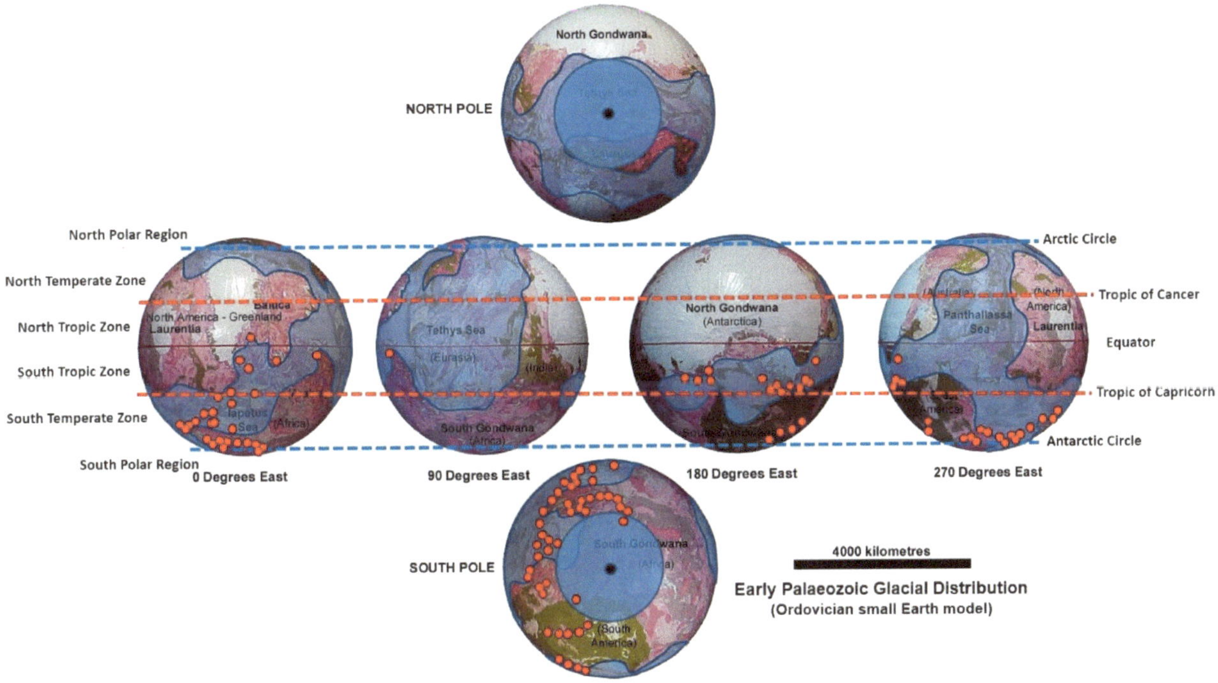

Figure 15.9 Early-Palaeozoic glacigenic deposits on an Ordovician Expansion Tectonic small Earth model, shown in relation to climate zones. Known glacigenic occurrences are shown as red dots and continental seas are shaded blue.

On the early-Palaeozoic small Earth model shown in Figure 15.9, the ancient Tethys Sea is shown to be open to equatorial and polar latitudes, which suggests that both the northern sea temperatures and climates may have been moderated by warm equatorial currents. A seasonal North Polar ice-sheet, similar to the ice-sheet within the present Arctic Ocean, may have existed during that time, which would have then limited the ability to form, or deposit, glacigenic rocks and may explain their general absence from the northern hemisphere.

On an Expansion Tectonic Earth, a slow southern apparent migration of the ancient South Pole commenced during mid- to late-Palaeozoic times prior to eventual rupture and break-up of Pangaea and opening of the Atlantic Ocean. The presence of early-Silurian glacigenic rocks persisting into South America is testament to a slow waning of the African ice-sheet during its migration into Brazil. Low-latitude glacigenic deposits in Europe, persisting to the Early Silurian, may also suggest the presence of small highland ice-centres in these regions.

15.3.6 Late-Palaeozoic Glaciation

A late-Devonian period of glaciation is recorded from Brazil and Bolivia, and by the following early-Carboniferous Period this glaciation had extended from Bolivia into Argentina and Paraguay. By the mid-Carboniferous Period, glaciation had spread even further and extended into parts of Antarctica, Australia, central and southern Africa, the Indian sub-continent, northern Asia, and the Arabian Peninsula. By the late-Carboniferous Period a very large area of the ancient Gondwana supercontinent was also experiencing glacial conditions that persisted into the following late-Permian times.

Magnetic pole locations on an Expansion Tectonic Earth show that the ancient South Pole remained within central West Africa during early-Palaeozoic to Devonian times. By the late-Palaeozoic Era the South Polar Region and accompanying ice-sheet had shifted to east Brazil before migrating south along the west coast of Africa and into South Africa during Jurassic times. These apparent shifts in location of the pole and ice-sheet are consistent with a northward migration of the African and South American continents during the late-Palaeozoic Era prior to continental break-up and opening of the South Atlantic Ocean. Fluctuations in glaciation, noted by others, during that time are considered to be due, in part, to a destabilisation and disruption of the previously established South Polar ice-sheet. This destabilisation resulted in periodic melting and refreezing of the ice-sheet as the underlying continents progressively migrated north relative to the ancient South Pole.

The distribution of Carboniferous to Permian glacigenic deposits is shown on the Permian small Earth model in Figure 15.10. These deposits are shown in relation to the predicted climate zones plus established continental seas and shorelines. The distribution of these glacigenic deposits on the small Earth model extends from high-southern to high-northern latitudes. This distribution, in particular the extensive distributions within the temperate and tropical zones, suggests the presence of mountain glaciation with lesser iceberg activity within the continental seas. This is consistent with geographical studies where the Palaeozoic Era is noted for its increasing elevation contrast between the lands and sedimentary basins, along with an increasing elevated topography suitable for forming mountain glaciers.

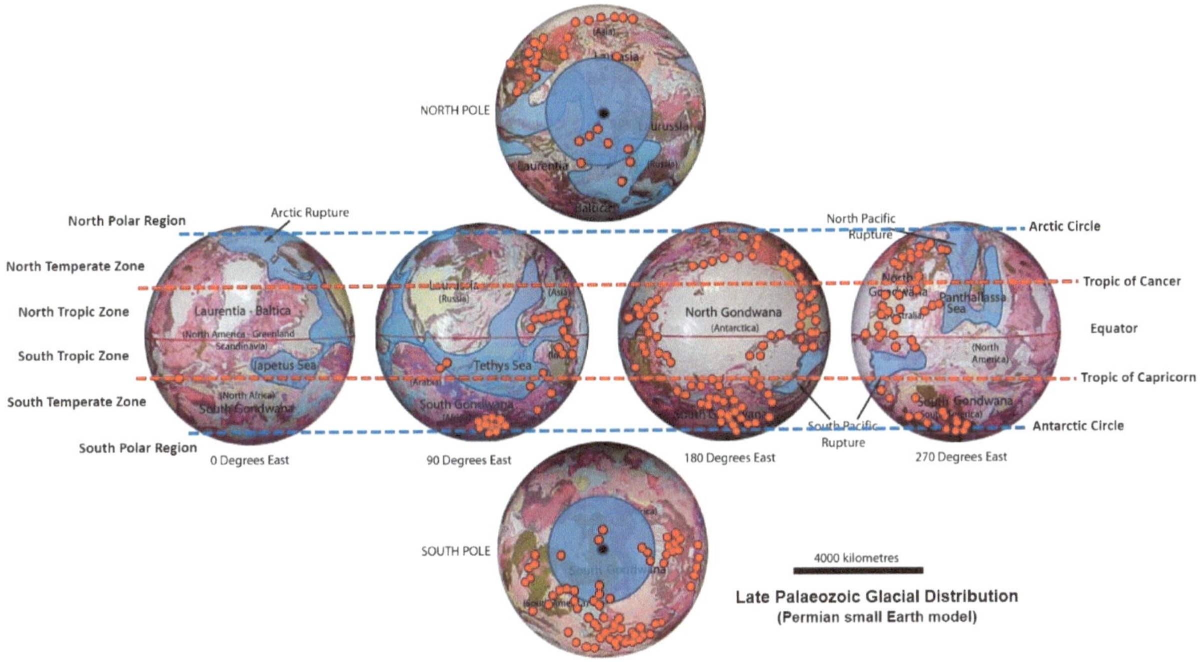

Figure 15.10 Late-Palaeozoic glacigenic deposits on a Permian Expansion Tectonic small Earth model shown in relation to climate zones and continental seas. Known glacigenic occurrences are shown as red dots and continental seas are shaded blue.

The distribution of southern hemisphere glacigenic deposits coincides with an exposed South Gondwana supercontinent, which included a permanent continental ice-sheet. Deposits in the north-polar region partly straddle the ancient emergent Laurussian and Laurentian supercontinents, as well as northern extensions of the ancient Tethys and Panthalassa continental seas. The presence of continental seas in the north-polar region may have again moderated climatic conditions at the North Pole giving rise to the presence of a seasonal marine ice-sheet.

No direct geologic evidence for glaciation exists for the following Triassic, Jurassic, or Cretaceous Periods, which, on an Expansion Tectonic Earth, coincides with break-up and opening of the modern oceans. During that time the South Pole migrated from South Africa across the opening South Atlantic Ocean. This resulted in melting of the south-polar ice-sheet and destabilisation of climate and sea levels. Eyles and Young, as well as Frakes and others in 1992 and 1993, also considered that ice-rafted material from Jurassic and Cretaceous rocks in Siberia and Cretaceous rocks in Australia may have indicated seasonally cold mountainous conditions in these areas.

15.3.7 Late-Cenozoic Glaciation

The late-Cenozoic glacial era covers the time interval from around 3.5 million years ago to the present-day. Glacial events during this time are therefore geologically relatively recent and well-documented phenomena. The glacial distributions during this time can be considered to coincide approximately with the present-day climate zones. Glaciation within high mountainous regions also occur throughout the world and the Earth is at present experiencing a relatively warm interglacial period. In 1994 Eyles and Young considered these late-Cenozoic glacial events to be the result of cyclical climatic changes, which began about 60 million years ago. During this time it was noted that a north-polar ice-sheet may have extended south into North America and Europe and a south-polar ice-sheet may have extended well into the Southern Ocean.

On an Expansion Tectonic Earth, the Cenozoic Era—commencing around 65 million years ago—coincided with the establishment of symmetrical seafloor spreading and on-going opening of each of the modern oceans. More importantly, the climatic changes occurring during that time also coincided with the Antarctican continent migrating from an equatorial position into the south-polar region. A permanent continental ice-sheet was then established on Antarctica about 35 million years ago and global sea-levels began to lower.

This migration of Antarctica is shown in Figure 15.11 where, during Oligocene times—around 60 million years ago—the Antarctican continent is shown to have commenced migrating into the South Polar Region. Migration then continued steadily to its present location centred over the South Pole. It should be noted that, due to the sheer size of the Antarctican continent, prior to the Oligocene Period parts of Antarctica extended from equatorial to polar-regions. Evidence from the rock record will then preserve all variations in climate zonation, which should not be taken as representative of any one zone.

It is also interesting to note that Antarctica is currently continuing to migrate across the South Polar Region, beneath the existing polar ice-sheet. On an Expansion Tectonic Earth Antarctica will continue to migrate across this region into the future, dragging the ice-sheet with it. As a result of this migration, the ice-sheet will be subject to melting and refreezing along the outer edges as the Antarctic continent attempts to displace the ice-sheet away from the South Pole. This phenomenon is considered to be continuing today which may have a direct influence on the presently observed melting of the South Polar ice-sheet as well as its long term influence on climate.

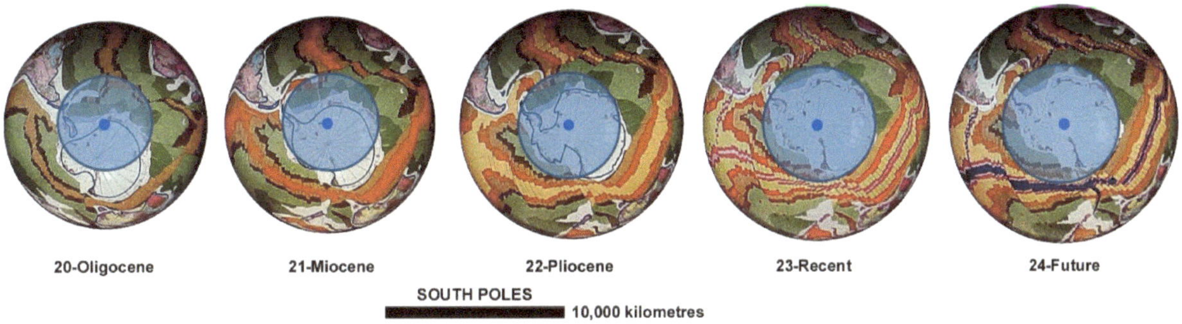

Figure 15.11 A series of late-Cenozoic small Earth models centred over the South Polar Region showing migration of the Antarctican continent onto and across the South Pole.

In summary, in this chapter the distribution of sediments characteristic of the various latitudinal climate zones, including carbonate and glacigenic rocks, were shown on a range of Expansion Tectonic small Earth models. In all cases these distributions are shown to coincide precisely with predicted latitudinal climate zones paralleling the ancient equator as well as magnetic poles established from palaeomagnetic data. In many cases the distribution of these sediments also coincides with, and confirms the distribution of, ancient coastlines established from geographical data.

In this context the presence of reef or glacigenic development within areas previously considered not suitable, such as carbonate sediments and reef development in inferred polar-regions or glacigenic rocks in equatorial regions, then seriously challenges conventional Plate Tectonic interpretations of this data.

16 Biogeographic Evidence

"Plate Tectonic reconstructions and proposed movements of continents do not correspond to the known or necessary migration routes and directions of biogeographical boundaries." Meyerhoff et al, 1996

Biogeography is defined as *a study of the distribution of plant and animal species and ecosystems in geographic space and through geological time*. The distributions of plant and animal species and biological communities on Earth are known to vary in a regular manner depending on their latitude, elevation, isolation, and habitat area. Biogeography is also seen as an integrative field of inquiry that unites concepts and information from the various fields of ecology, evolutionary biology, geology, and physical geography. Ancient biogeography—called *palaeobiogeography*—goes one step further to include ancient geographic information, in particular, in relation to Plate Tectonic studies. This information is used to help constrain hypotheses on the timing of biogeographic events, such as extinctions, and provides insights into the regional development of life forms.

It has long been recognised, but rarely acknowledged in the published literature, that there are significant inconsistencies in both biogeographic and geographic reconstructions in Plate Tectonic studies. In 1992, Smiley reported that the distributions of land-based plants and animals existing during late-Palaeozoic, Mesozoic, and Cenozoic times, as well as many marine creatures, just do not support continental drift-based Plate Tectonic reconstructions. The evidence instead suggested to Smiley that the relative positions of both the ancient continents and ancient poles have been relatively stable since late-Palaeozoic times. This continental and polar stability though contrasts strongly with the complex, random, amalgamation-dispersal-amalgamation plate motions required for Plate Tectonic reconstructions. From this, Smiley concluded that *"Plate Tectonic theories create more biogeographic problems than those already existing."*

In 1989, Noël recognised from his extensive studies into the global distribution of fruit and nut trees that all species that are related must have a common ancestry. The only way for the various species to have spread naturally is for the areas of population to have been in close contact in the past and to have since moved apart during *"Continental Drift"*. From this, Noël concluded that *"...distributions are all readily explicable on the assumption that the current land areas of the Earth were once all physically linked...* [and]*...the Earth has since expanded"*.

Just like Noël, in 1990, Cox showed that the distribution of various plant and animal species on the present Earth form a number of distinct east-west and north-south distributional patterns. He considered that these patterns "*...better substantiate the alternative theories, such as Earth expansion, which do not require any form of complex plate motions.*" This was further supported by the work of McCarthy in 2003 where he showed that faunal and floral species matches across the Pacific Ocean were good. He also showed that species that matched on opposite sides were absent in between, disproving any ad-hoc stepping stone or land bridge explanations for species migration and development.

Stepping stone, land bridge, and island hopping explanations are primarily Plate Tectonic—and before it Continental Drift—concepts conceived in an attempt to explain the global distribution and migration of species on a Plate Tectonic Earth. Each of these concepts requires ad-hoc land bridges or connections between continents at some time in the past in order to explain the past dispersal or migration of species and observed biogeographic distributions.

Geographic isolation must also be accounted for in these concepts where a species, or a population of species, becomes separated by a physical barrier such as an ocean, allowing each to diverge along separate evolutionary paths. A good example of geographic isolation is Australia, where a distinctive range of species has evolved. On Plate Tectonic reconstructions India should have also evolved the same as Australia as it supposedly fled Gondwana and drifted to a distant Asia. Instead, biogeographic evidence shows that the Indian species did not become distinctive, like Australia, and hence there are significant biological problems facing conventional reconstructions of both India and Asia.

Similarly, in 1990 Stevens, in discussing the various marine faunal links across the Pacific Ocean, showed that a number of distinctive Jurassic marine fossil species from New Zealand have very marked similarities to species from Mexico. On Plate Tectonic reconstructions, a direct link between these now remote continents is impossible because of the need to include the presence of vast Panthalassa and Tethys Oceans. To maintain any form of faunal link between New Zealand and Mexico on Plate Tectonic reconstructions additional links would be required through polar-regions between New Zealand, West Antarctica, and South America. These fossil species links just do not occur on any of these now remote continents.

Stevens suggested instead that tighter palaeobiogeographical assemblages could be achieved by assuming an Expansion Tectonic Earth thesis. By reducing the Earth radius, the need for a theoretical Panthalassa or Tethys Ocean is then drastically reduced. This reduction then makes direct linkages between New Zealand, and both Central America and South America, a physical possibility and, in turn, simplifies the required migration routes between Australia and Mexico.

In 1996 Meyerhoff and others carried out an extensive study of the Cambrian to present-day plant and animal regions of the Earth. They showed that the *"...assemblage and movements of the continents on Plate Tectonic reconstructions do not correspond to the known, or necessary, migration routes shown by these established biogeographical regions."* In most cases the discrepancies between the actual physical distributions of species and the Plate Tectonic continental assemblages were impossibly large—so large that not even an approximate match could be made with the Plate Tectonic reconstructions.

This observation of Meyerhoff was further substantiated by the continental reconstructions of Shields. In 1996, Shields considered that none of the existing Plate Tectonic or partial Earth expansion assemblages of continents—referring to the partial expanding Earth reconstructions of Owen in 1976—surrounding the Indian and Pacific Oceans appear to completely satisfy all of the constraints imposed by palaeobiogeography.

The detailed concerns and information presented by these, and many other authors, is based on extensive, well-documented field evidence and it is this evidence that must be fully accounted for on all past reconstructions of the Earth. The conflicting evidence confronting biogeography, with regard to the ancient geographical and biogeographical reconstructions, can be briefly summarised as:

- The present-day distribution of plant and animal species evidence is best explained on reconstructions that display relative continental and polar stability, rather than complex, random, continental fragmentation and dislocation.

- With the possible exception of Australia, the global distribution of many land-based animals confirms that a Pangaean assemblage of ancient continental land masses is feasible.

- There is sufficient evidence, based on the biotic patterns and regional distributions of ancient species on the present-day Earth, to refute various aspects of Plate Tectonic based palaeobiogeographical reconstructions.

- There is also strong opinion that alternatives to Plate Tectonic constrained biogeographic reconstructions, such as Expansion Tectonics, far better explains the distribution of the observed physical evidence.

The quantity and quality of data available from biogeographical studies is now extensive and in most cases comes from many specialised fields of research. In this chapter, the global distribution of ancient reptiles and dinosaurs will be used as an example to investigate their evolution and demise on Expansion Tectonic small Earth models. These examples will also be used to test their distribution in context with the established small Earth crustal assemblages, as well as to further test the locations of the poles and equator established from palaeomagnetic studies. But first, I will delve into the meaning of life and extinction to see how evolution in general fits with the various phases of Expansion Tectonic crustal development.

16.1 Evolutionary History of Life

Life is defined as *a characteristic distinguishing objects that have communication and self-sustaining processes from those that do not*; either because such objects have died, or because they lack such functions and are classified as inanimate. In biology, an *organism* is regarded as any living system—such as an animal, fungus, micro-organism, or plant. In at least some form, all types of organisms are capable of responses to stimuli, reproduction, growth, and development, as well as maintenance of *homeostasis*—a process by which the body's internal environment is kept stable.

The term organism—Greek *organismos* meaning *instrument, implement or tool; an organ of sense or apprehension*—first appeared in the English language in 1701 and, according to the Oxford English Dictionary, took on its current definition by 1834. It is directly related to the term *organization* whereby there is a long tradition of defining organisms as self-organizing beings. A diverse array of living organisms can be found on Earth today and the properties common to all of these organisms are carbon and water-based cellular forms with complex organization and heritable genetic information.

The first known attempt to classify organisms was said to have been conducted by the Greek philosopher Aristotle (384–322 BCE) who classified all living organisms known at that time as either plant or animal, based mainly on their ability to move. He also distinguished animals with blood from animals without blood and divided the blooded animals into five groups: mammals, reptiles and amphibians, birds, fishes, and whales. The bloodless animals were also divided into five groups: cephalopods, crustaceans, insects, shelled animals, and zoophytes—a type of animal that visually resembles a plant but is obsolete in modern usage. Aristotle's work was considered the grandest biological synthesis of the time and remained the ultimate authority for many Millennia after his death.

The preserved remains or traces of animals, plants, and other organisms from the remote past are called *fossils*—from Latin *fossus,* literally meaning *having been dug up.* The presence of fossils in rock formations and sedimentary layers is known as the *fossil record.* A study of fossils across geological time, how they were formed, and the evolutionary relationships between the various organisms are some of the most important functions of the science of *palaeontology.* Observations that certain fossils were associated with various rock strata led geologists in the 19th century to use this information to formalise a geological timescale. The development of radiometric dating techniques in the early 20th century have since allowed geologists to determine more exact ages for the various rock strata, as well as the ages of any included fossils.

Scientific evidence suggests that life began on Earth approximately 3 to 3.5 billion years ago. The mechanism by which life emerged is unknown and various hypotheses are still being formulated. Since then, life has evolved into a wide variety of forms which biologists have classified into a hierarchy of *taxa*—a group of one or more populations of organisms. For about the first 2,000 million years of Earth history, microbial mats—multi-layered colonies of different types of bacteria—were the dominant life on Earth. The evolution of *oxygenic photosynthesis* then enabled microbes to play a major role in the oxygenation of the atmosphere from about 2,400 million years ago. These changes in the atmosphere were considered to have increased the microbe's effectiveness as *nurseries of evolution.* While *eukaryotes*—cells with complex internal structures—may have been present earlier, their evolution increased markedly after they acquired the ability to transform oxygen from a poison to a powerful source of energy in their metabolism. This innovation is thought to have come from primitive eukaryotes capturing oxygen-powered bacteria and transforming them into *organelles*—a specialized subunit within a cell such as *mitochondria.*

Multi-cellular life during these ancient times was initially composed only of eukaryotic cells and the earliest evidence for this is from around 2,100 million years ago. The earliest evidence of complex eukaryotes with organelles dates from around 1,850 million years ago. Specialization of cells for different functions first appeared between 1,430 million years ago—thought to be a possible fungus—to 1,200 million years ago—a probable red alga. Sexual reproduction may have been a prerequisite for this specialization of cells, as an asexual multi-cellular organism was considered to have been at risk of being taken over by rogue cells if they had the ability to reproduce.

The earliest known animals are the *cnidarians*—a species of animal found mostly in marine environments from about 580 million years ago. These were considered to have been so modern-looking that even earlier animals may have been present before then. Fossils of earlier animals are rare because they had not developed mineralized hard parts until about 548 million years ago. The earliest modern-looking *bilaterian animals*—animals having a bilateral symmetry comprising a front and a back, as well as an upside and downside—appeared in the Early Cambrian, along with several species that bear little obvious resemblance to modern animals. From this early development, there is a long-running debate about whether the Cambrian *species explosion* was truly a rapid period of evolutionary experimentation. Alternative views consider that modern-looking animals began evolving earlier than the Cambrian but fossils of their precursors have not yet been found.

The spread of life from water to land required organisms to resolve several problems, including protection against drying out and supporting themselves against gravity. The earliest evidence of land *invertebrates*—animal species that do not develop a vertebral column, including the insects, worms, snails, and related clams, crabs, octopus, and starfish—dates back to 490 million years ago. This was followed by the land plants, dating back to about 476 million years ago. These land plants were so successful during late-Devonian times that they initially caused an ecological crisis, which lasted until the evolution and spread of fungi that could digest dead wood. The lineage that produced land vertebrates—animals with backbones and spinal columns, including the jawless fishes, bony fishes, sharks and rays, amphibians, reptiles, mammals, and birds—evolved rapidly from between 370 million and 360 million years ago.

During the Permian Period, *synapsids*—terrestrial vertebrates, including the ancestors of mammals—are thought to have dominated the land environments, but the end-Permian extinction some 250 million years ago came very close to wiping out most of these existing complex life forms. During the slow recovery from this extinction event, a previously obscure group called the *archosaurs*—a group that includes the extinct dinosaurs and whose living representatives consist of birds and crocodiles—became the most abundant and diverse terrestrial vertebrates. The dinosaurs were the dominant land vertebrates for the rest of the Mesozoic Era and birds are thought to have later evolved from one group of the dinosaurs. During that time the ancestors of mammals survived only as small, mainly nocturnal insect-eating animals, although larger specialised mammals are known from the fossil record but only in small numbers. The end-Cretaceous extinction some 65 million years ago is a well-publicised event thought to have killed off the remaining dinosaurs, allowing the mammals to rapidly increase in size and diversity with some also taking to the air and the sea.

Fossil evidence indicates that flowering plants first appeared and rapidly diversified during the early-Cretaceous Period, between 130 million and 90 million years ago. Their rapid rise to dominance of the lands is thought to have been assisted by co-evolution with pollinating insects. Social insects—insects belonging to large colonies—appeared around the same time and, although they account for only small parts of the insect family, they now form over 50 percent of the total mass of all insects.

To complete the picture, humans are thought to have first evolved from a lineage of upright-walking apes whose earliest fossils date from over 6 million years ago. There is a long-running debate about whether modern humans are the descendants of a single small population located in Africa, who then migrated all over the world less than 200,000 years ago to replace previous hominine species, or they arose worldwide at the same time as a result of interbreeding.

This brief evolutionary history of life on Earth is shown schematically in relation to the Expansion Tectonic Earth in Figure 16.1. In this figure, shown in conjunction with the exponential Earth radius curve, it can be seen that the presence of ancient microbial life forms coincides with the extremely long Precambrian times—which includes the Archaean and Proterozoic Eras. Small Earth modelling has previously suggested that throughout this extended period of time the ancient Earth had a relatively flat featureless landscape, devoid of mountains, and included an extensive network of relatively shallow continental seas. Deposition of sediments within these shallow seas included *chemically-precipitated* siliceous chert, carbonate, and banded iron formation rocks. In other words, the presence of these sediments suggests that environmental conditions on the ancient Earth were very stable and remained virtually unchanged for billions of years.

By the beginning of the Cambrian Period, some 560 million years ago, rapidly increasing Earth radius and changes in surface curvature were starting to occur which had the effect of increasing the elevation contrast between the lands and seas. This elevation contrast, in turn, gave rise to elevated topography, which increased the rate of erosion of the exposed lands and also influenced changes to the distribution of seas and corresponding coastal outlines of the supercontinents.

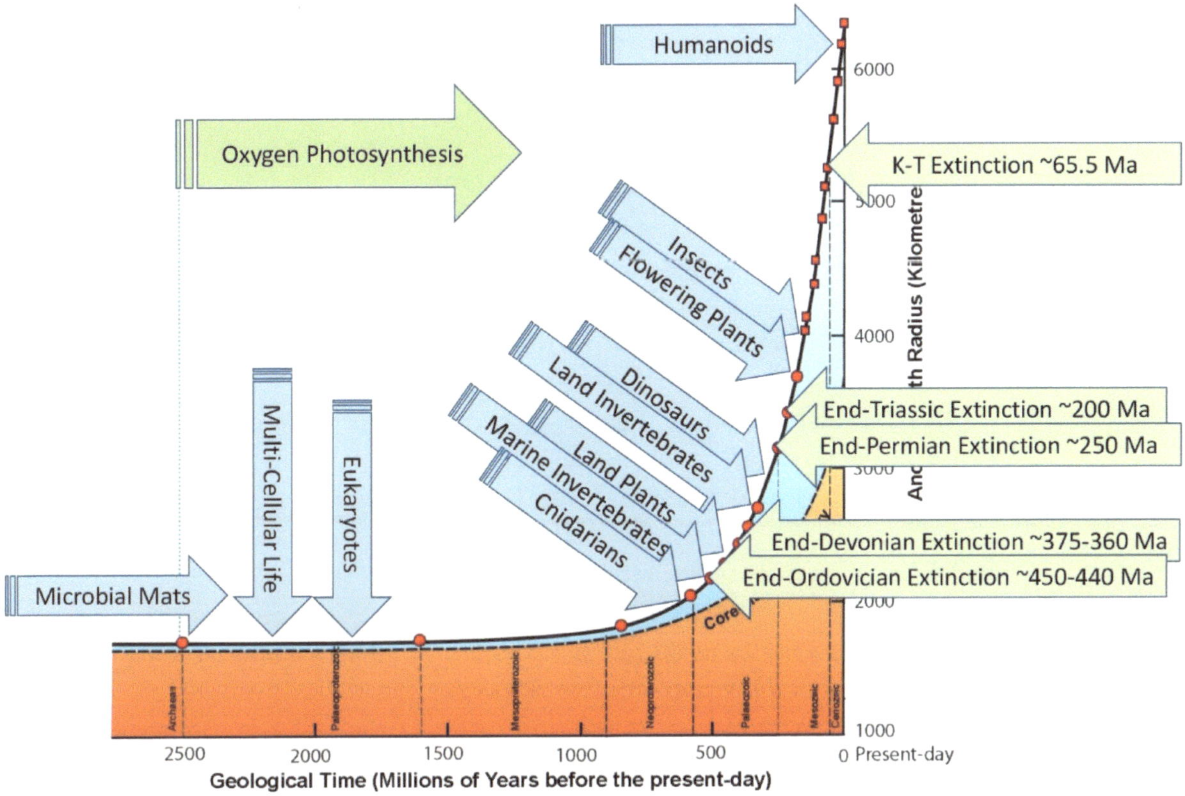

Figure 16.1 Summarised evolutionary history of life on an Expansion Tectonic Earth, shown in blue in relation to exponential increase in Earth radius and the five main extinction events.

As can be seen in Figure 16.1, the changing Palaeozoic times in particular coincided with the development and rapid evolution of all life forms on Earth. Environmental change during these times went from stable featureless land conditions, to elevated topography, to increased disruption and geographic isolation of the lands and continental seas. The degree of crustal and environmental change during these times then had the capacity to markedly influence evolutionary change in all life forms. Similarly, extinction of the species is seen as a by-product of both rapid sea-level change and the inability to keep pace with these environmental changes. It is considered that evolution of species during these times was then driven by the need to survive amongst these ever increasing changes to the environment.

16.2 Mass Extinction Events

An extinction event, also known as *mass extinction,* is defined as *a widespread and geologically rapid decrease in the amount of life on Earth*—where geologically rapid refers to events lasting for millions to tens of millions of years. Such an event is identified by a change in the diversity and variety of life forms occurring when the rate of extinction increases with respect to the rate of *speciation*—the evolutionary process by which new biological species arise. As the majority of organisms on Earth are *microbial*—microscopic organisms—and therefore difficult to detect in the rock record, extinction events are best observed and recognised in the biologically more complex organisms rather than from the total diversity and abundance of life on Earth.

Published literature suggests that over 98 percent of all recorded species of life forms are now extinct. What this means though is that not all of these life forms were in existence at the same time, and, in fact, most only existed because of the environmental changes and periodic extinction events that have driven past evolution. This statement should, however, be understood in context with the background rate of extinctions. Based on the fossil record, the background rate of extinctions on Earth has been estimated by others to be about two to five families of marine invertebrates and vertebrates every million years. Marine fossils are mostly used to measure these extinction rates because of their well-preserved fossil record compared to land organisms. It is also emphasised that this background rate of extinction is the natural rate of dying out as species either evolve to replace pre-existing species or new species emerge to take their place.

Since life began on Earth, several major mass extinction events have significantly exceeded this background extinction rate. It has been estimated that in the past 540 million years there have been five major events, where over 50 percent of animal species existing at the time have died out (see Figure 16.1). It has also been acknowledged in the literature that mass extinctions seem to be a phenomenon associated with the past 560 million years or so of Earth history, occurring after the large complex organisms came into existence. Before that time extinction rates were considered much lower. This may of course simply be a factor of incomplete preservation of soft-bodied fossil species in the rock record.

The most common causes of mass extinctions currently recorded in the literature are listed below.

- **Flood basalt events:** Where formation of large volcanic eruptions may have produced dust, sulphur oxides, and carbon dioxide, which may have inhibited photosynthesis and caused food chains to collapse on land and at sea.

- **Sea-level falls:** These falls in sea-level reduce the continental shelf area—the most productive part of the oceans—sufficiently to cause a marine mass extinction, as well as disrupting weather patterns enough to also cause extinctions on land. Variations in sea-level are known to be associated with most of the known mass extinction events.

- **Impact events:** The impact of a sufficiently large asteroid or comet is thought to cause food chains to collapse, both on land and at sea, by producing dust and particulate aerosols which may inhibit photosynthesis.

- **Sustained and significant global cooling:** This is thought to make the Earth's climate more arid by locking up much of the planet's water in ice and snow. This is considered to then kill many polar and temperate species and force others to migrate towards the equator, thus increasing competition for tropical species.

- **Sustained and significant global warming:** This is thought to make the Earth's climate wetter by melting ice and snow and thus increasing the activity of the water cycle. It may also expand the area available for tropical species, thus killing temperate species or forcing them to migrate towards the poles, possibly causing severe extinctions of polar species.

- **Anoxic events:** These events are said to occur when the middle and upper layers of the ocean become deficient, or totally lacking in oxygen. The causes are complex and controversial, but all known instances are associated with severe and sustained global warming, mostly caused by massive volcanism.

- **Oceanic overturn:** Overturn is a disruption of ocean circulation which lets surface saline-rich water sink to the bottom. This brings anoxic deep water to the surface and kills most of the oxygen-breathing organisms that inhabit the surface and middle depths. It is thought to occur either at the beginning or the end of a period of glaciation.

- **A nearby nova, supernova, or gamma ray burst:** A nearby gamma ray burst is thought to be potentially powerful enough to destroy the Earth's ozone layer, leaving organisms vulnerable to ultraviolet radiation from the Sun. These gamma ray bursts are fairly rare, occurring only a few times in a given galaxy per million years.

- **Plate tectonics:** Movement of the continents contributes to extinctions by initiating or ending ice ages, by changing ocean and wind currents and thus altering climate, or by opening seaways or land bridges which exposes previously isolated and poorly adapted species to competition.

- **Other hypotheses:** Including the spread of a new disease, super-volcanic events, the eventual warming and expanding of the Sun and human activities—including changes in climate, overhunting, overfishing, invasive species, and habitat loss.

It should be noted that many of these causes are speculative and are based on conventional tectonic theses and assumptions. Irrespective of this long list of possible causes, current media hype has all but convinced us that extinctions are the result of random *impact events*. This hype is driven by the theory of *catastrophism* where the Earth is considered to have been affected in the past by a number of sudden, short-lived, violent events, possibly worldwide in scope. In this context, we are now led to believe that an impact event is caused by the collision of a large meteor, asteroid, comet, or other celestial object with the Earth which may reoccur on Earth at any time in the near future.

It is interesting to note that while many palaeontologists agree that an asteroid may have struck the Earth at about 65 to 65.5 million years ago, there is still an on-going dispute as to whether this impact was the actual cause of the much-publicised end-Cretaceous extinction event. Additional evidence has also been presented to suggest that there was an interval of about 300 thousand years between the impact and the actual mass extinction, and in 1997 Chatterjee also drew attention to the possibility of a multiple-impact scenario.

It should be noted that this catastrophism-driven media hype is actually very misleading as biogeographic evidence shows that extinction events generally occur over many millions of years, not as singular moments in time. Contrary to what scientists, or the media, tell us, the classic Chicxulub end-Cretaceous impact crater in Mexico has been drilled for oil and has been shown to contain many metres of Cretaceous sediment. This sediment overlies the impact site and age-dating shows that, being Cretaceous, the sediments are actually older than the impact event. An end-Cretaceous asteroid impact of this magnitude would have surely destroyed this Cretaceous sediment, which must have clearly been formed before the extinction event and should in reality either lie under the impact site or have been totally destroyed.

Rather than catastrophic extinction, on an Expansion Tectonic Earth the small Earth modelling has shown that, during early-Palaeozoic to present-day times, there have been a number of drastic and prolonged changes to sea-levels. On these models, major changes in sea-levels have been shown to occur as a result of merging of previously separate ancient continental seas, during break-up of the ancient supercontinents, opening of each of the modern oceans, and draining of the ancient continental seas. Break-up of the supercontinents was also shown to have destabilised climate zones and influenced melting of existing ancient polar ice-caps. Depending on the severity of these events, it is considered that sea-level changes of this magnitude may also affect regional to global-scale climate, as well as ocean-water circulation patterns, species habitats, and the type and location of sedimentary deposition.

In this context it is further considered that, on an Expansion Tectonic Earth, it is the prominence of drastic changes in sea-levels that led to mass extinction of the various species. It was these changes in sea-levels that ultimately affected the distribution, evolution, or extinction of the affected marine and terrestrial animal and plant species as well as disrupting the established habitats and species migration routes.

16.2.1 End-Ordovician Extinction Event

The end-Ordovician extinction event occurred over a period of 10 million years at the Ordovician to Silurian transition, extending from 450 to 440 million years ago. Within this time frame two events have been documented which showed that 27 percent of all families, 57 percent of all genera, and 60 to 70 percent of all species were killed off. Together, these events are ranked by many scientists as the second largest of the five major extinctions in Earth's history in terms of percentage of genera that became extinct. On an Expansion Tectonic Earth, the timing of this event places it within the early part of the early-Palaeozoic Gondwana supercontinental assemblage (Figure 16.2).

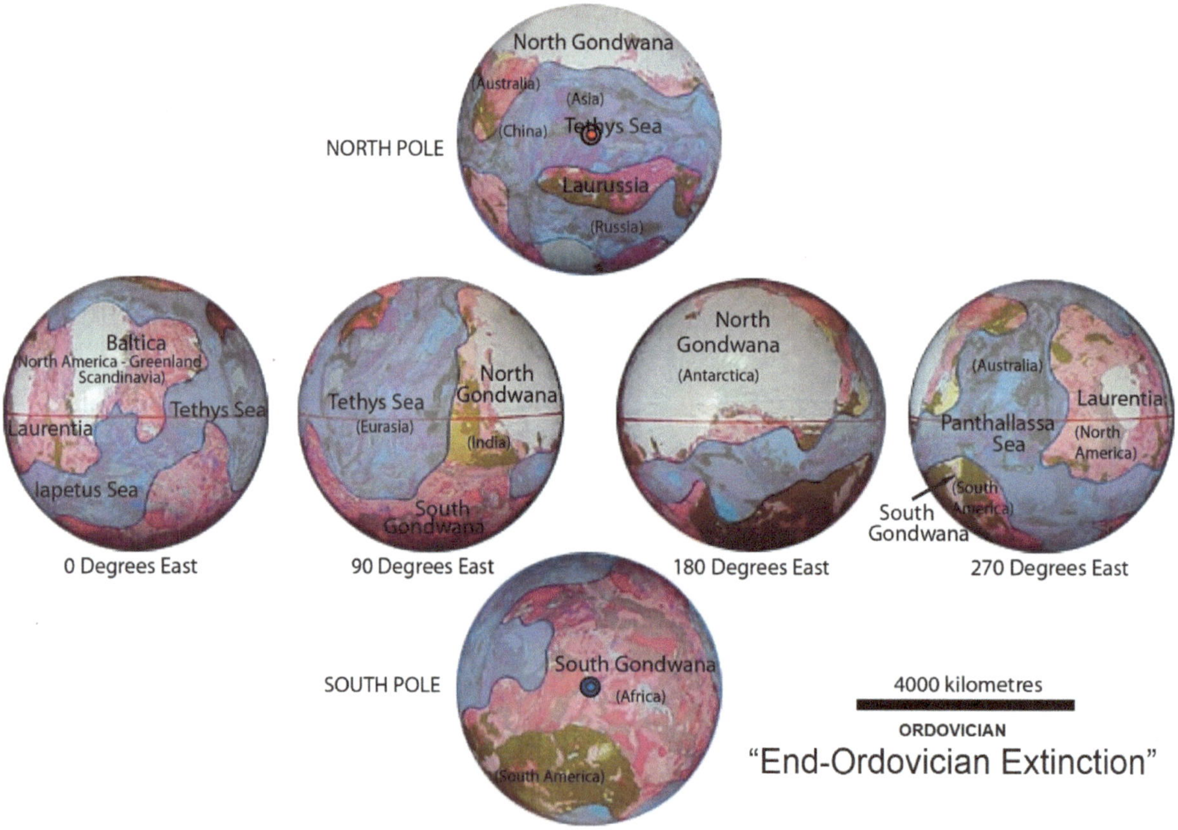

Figure 16.2 The Ordovician Gondwana Expansion Tectonic small Earth assemblage showing the distribution of ancient coastlines and seas. The ancient Tethys, Iapetus, and Panthalassa Seas form part of a global network of continental seas surrounding each of the exposed supercontinents.

Small Earth modelling shows that during the early-Palaeozoic times there was an on-going and well-established network of interconnected continental seas which separated each of the known supercontinents. At the scale of the modelling shown in Figure 16.2 there is no evidence to suggest that there were any obvious, separate, or isolated seas—although this is not to say there weren't any small localised seas.

Extinction during the end-Ordovician, in particular over the 10 million year time-frame involved, is considered to have occurred as a result of increasing changes to surface curvature and elevation of the lands during geosynclinal and orogenic activity. In effect, these changes disrupted the configuration of the previously existing coastlines and continental seas. Any crustal changes in one area during changing surface curvature would have globally affected coastlines elsewhere by changing the overall sea-levels. This change in sea-levels would then disrupt favoured continental shelf habitats as well as any established migration routes.

16.2.2 Late-Devonian Extinction Event

The late-Devonian extinction event is known to have occurred over a period of 15 million years near the Devonian to Carboniferous transition, extending from 375 to 360 million years ago. During that time a prolonged series of extinctions eliminated about 19 percent of all families, 50 percent of all genera, and 70 percent of all species. This extinction event is thought by others to have lasted perhaps longer, to as much as 15 to 20 million years, and there is also some evidence for a series of extinction pulses within this time.

On an Expansion Tectonic Earth the timing of the late-Devonian extinction event places it within the latter part of the Gondwana supercontinental assemblage. During that time the previously well-established network of interconnected continental seas was being subjected to ever-increasing adjustments to Earth surface curvature and orogenesis. As can be seen in Figure 16.3, by the Devonian times the ancient continental seas had started to separate into a number of isolated seas and became more restricted in surface area. This restricted surface area suggests the seas were becoming deeper as a result of increased geosynclinal and orogenic activity. This activity also provides confirmation for the series of extinction pulses occurring during this event, as noted by others.

By comparing the network of Ordovician continental seas shown in Figure 16.2 with the early-Carboniferous seas in Figure 16.3—the configuration of seas just after the late-Devonian extinction—it can be seen that there is a considerable amount of difference in both coastal outlines and surface areas when compared to the older ancient seas. This phenomenon highlights the geographical observation that changing surface curvature and tectonic activity during that time was severely disrupting the land surface, giving rise to an increasing elevation contrast between the lands and seas. This elevation contrast in turn suggests there was an increase in orogenesis and mountain building, which in turn increased the rate of erosion of the exposed lands. These changes also influenced changes to the distribution of seas, depths of the seas, and corresponding coastal outlines of the supercontinents.

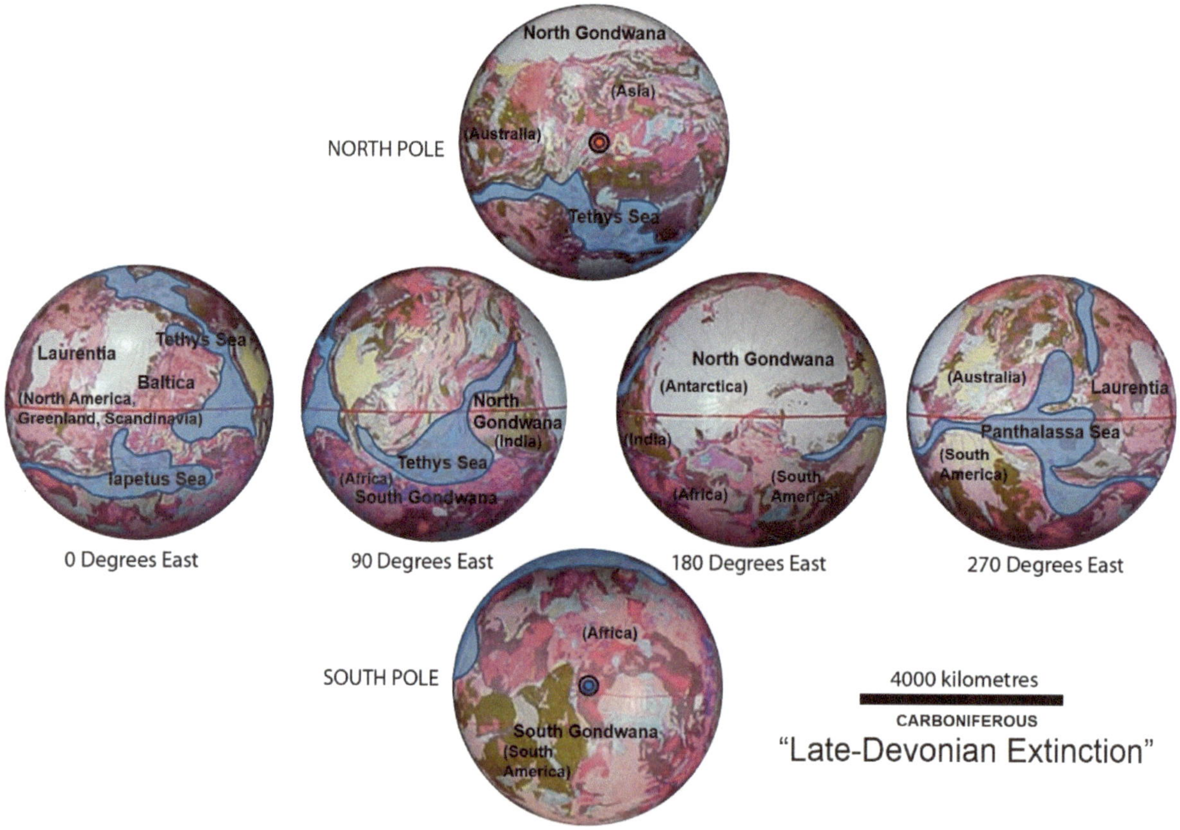

Figure 16.3 The early-Carboniferous Gondwana Expansion Tectonic small Earth assemblage showing the distribution of ancient coastlines and seas. The ancient Tethys, Iapetus, and Panthalassa Seas form part of a disrupted global network of continental seaways surrounding each of the exposed supercontinents.

It is considered here that the late-Devonian extinction event on an Expansion Tectonic Earth primarily occurred during these Gondwanan times as a result of increasing changes to surface curvature, severe disruption to the configuration of the coastlines, and isolation of continental seas into a number of relatively deep and restricted seaways. All of these changes adversely affected the well-being of established marine and terrestrial species by isolating the seas and locally changing sea levels. This then severely disrupted favoured continental shelf habitats and any established migration routes.

16.2.3 End-Permian Extinction Event

The end-Permian extinction event occurred around 250 million years ago at the Permian to Triassic transition. This is considered by most researchers to have been the largest extinction event, devastating life on Earth and estimated to have killed off 57 percent of all families, 83 percent of all genera, and 90 to 96 percent of all marine and land species. While the subsequent recovery of vertebrates took around 30 million years, the vacant niches left by the extinction event also created the opportunity for the *archosaurs*—the species that includes the extinct dinosaurs—to become the dominant species on Earth.

Uncertainty exists in the literature regarding the overall duration of the end-Permian extinction event. Recent research shows that different groups of species became extinct at different times. An older theory, still supported in the literature, shows that there were two to three major extinction pulses 5 million years apart separated by periods of extinctions well above the background level. The final extinction pulse killed off about 80% of all remaining marine species, while the other losses occurred during the first and intervening pulses. Research indicates that the long recovery time for this extinction event was due to successive waves of extinction as well as prolonged environmental stress to organisms, which inhibited recovery well into the following early-Triassic Period.

On an Expansion Tectonic Earth, the continental crusts were previously shown to encompass the entire Earth as a single Pangaea supercontinent up to the late-Permian. During this extended period of time there were no modern oceans, only a network of continental seas covering low-lying parts of the supercontinental crusts. The transition from ancient seas to modern oceans came about when the ancient Pangaea supercontinental crust first started to rupture and break-up some 250 million years ago. The end-Permian extinction event is therefore significant as it coincides with break-up of the Pangaea supercontinent to form the modern continents and intervening modern oceans (Figure 16.4).

Of more significance is that break-up of the Pangaea supercontinent initiated draining of the waters of the ancient continental seas into these newly opening modern oceans. This crustal break-up also destabilised the previously existing polar ice-caps and the subsequent migration of the modern continents in turn destabilised established climate zones. On an Expansion Tectonic Earth, draining of the continental seas then shifted the location of the sea waters, exposing the ancient sea-beds as new lands, which in turn resulted in rapid and unprecedented changes in sea-levels, erosion, and disruption to species habitats.

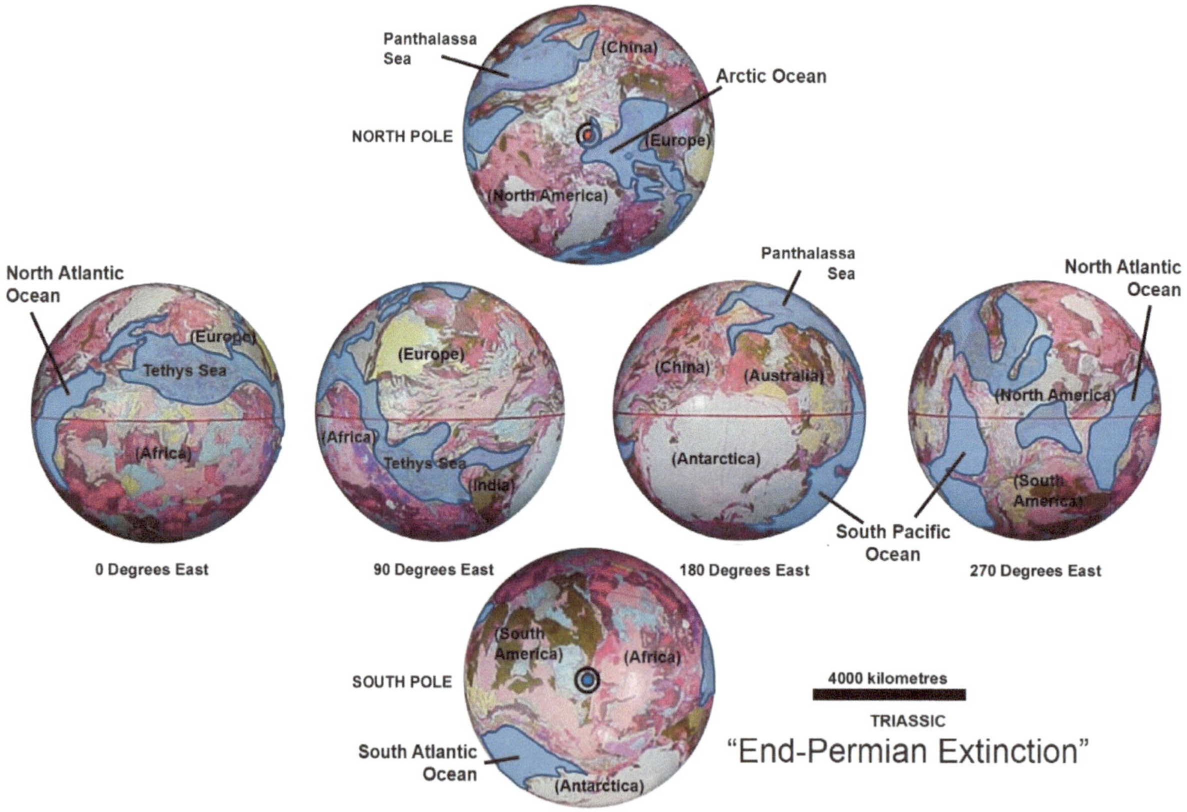

Figure 16.4 Triassic Expansion Tectonic small Earth crustal assemblage showing the ancient coastline distribution as well as remnants of the ancient Pangaea supercontinent. The figure also shows the locations of continental rupture which commenced in the north and south Pacific and Arctic Ocean regions to form the modern oceans.

Shown in Figure 16.4 are the outlines of the newly opening modern oceans and seas in relation to the emerging modern continents, along with remnants of the pre-existing supercontinents and seas. Also shown, and of significance to later extinction events, is the fragmentation of the previous network of supercontinental seas into a number of separate and isolated oceans and seas—the precursors to the modern oceans. During that time, each of these separate oceans and seas had different sea-levels, seawater salinities, currents, and seawater temperatures. Break-up of the continents and merging of these isolated seas over time is then seen as the prime cause of extinction during Permian and later post-Permian times.

Isolation of these ancient seas and modern oceans may have also promoted unique marine species evolution, as well as extinction of prior species during periods of environmental change, in particular during changes in sea-level and disruptions to climate zonation. On an Expansion Tectonic Earth, the end-Permian extinction event is therefore considered to have resulted from the prolonged period of break-up and disruptions to pre-existing continents and seas. This extinction was a direct result of break-up of the Pangaea supercontinent to form the modern continents and draining of the supercontinental seas to form the precursors to the modern oceans.

On an Expansion Tectonic Earth the post-Permian times are then seen as representing a shift in the style of extinction events occurring on Earth. These events shifted from simple coastline and sea-level changes occurring prior to the Permian Period, to more disruptive sea-level changes during catastrophic merging of previously isolated seas and oceans during post-Permian times.

16.2.4 End-Triassic Extinction Event

The end-Triassic extinction event occurred some 200 million years ago at the Triassic to Jurassic transition. About 23 percent of all families, 48 percent of all genera—20 percent of marine families and 55 percent of marine genera—and 70 to 75 percent of all species became extinct. During that time most of the non-dinosaurian archosaurs, as well as most *therapsids*—a group of the most advanced terrestrial vertebrates, including the ancestors of mammals—and most of the large amphibians became extinct, leaving the dinosaurs with little competition to contend with on land.

On an Expansion Tectonic Earth, by the end of the Triassic Period break-up of the Pangaea supercontinent and opening of the modern oceans was well established (Figure 16.5). Draining of the ancient supercontinental seas into the modern ocean basins was also well advanced, with only remnant continental seas remaining within the Mediterranean to Middle Eastern

Tethys Sea regions. Small Earth modelling also shows that a number of isolated, separate oceans and seas still remained.

It is significant to note that the end-Triassic represents the period of time when the previously separate North and South Pacific Oceans merged to form a single modern Pacific Ocean. This is highlighted by comparing Figures 16.4 and 16.5 where a land connection between North America and Australia is shown to have been breached, allowing the ancient North and South Pacific Oceans to merge as a single ocean. In doing so, the previously separate sea-levels within each of these early oceans then equilibrated—where sea-level in one ocean rose and sea-level in the other fell.

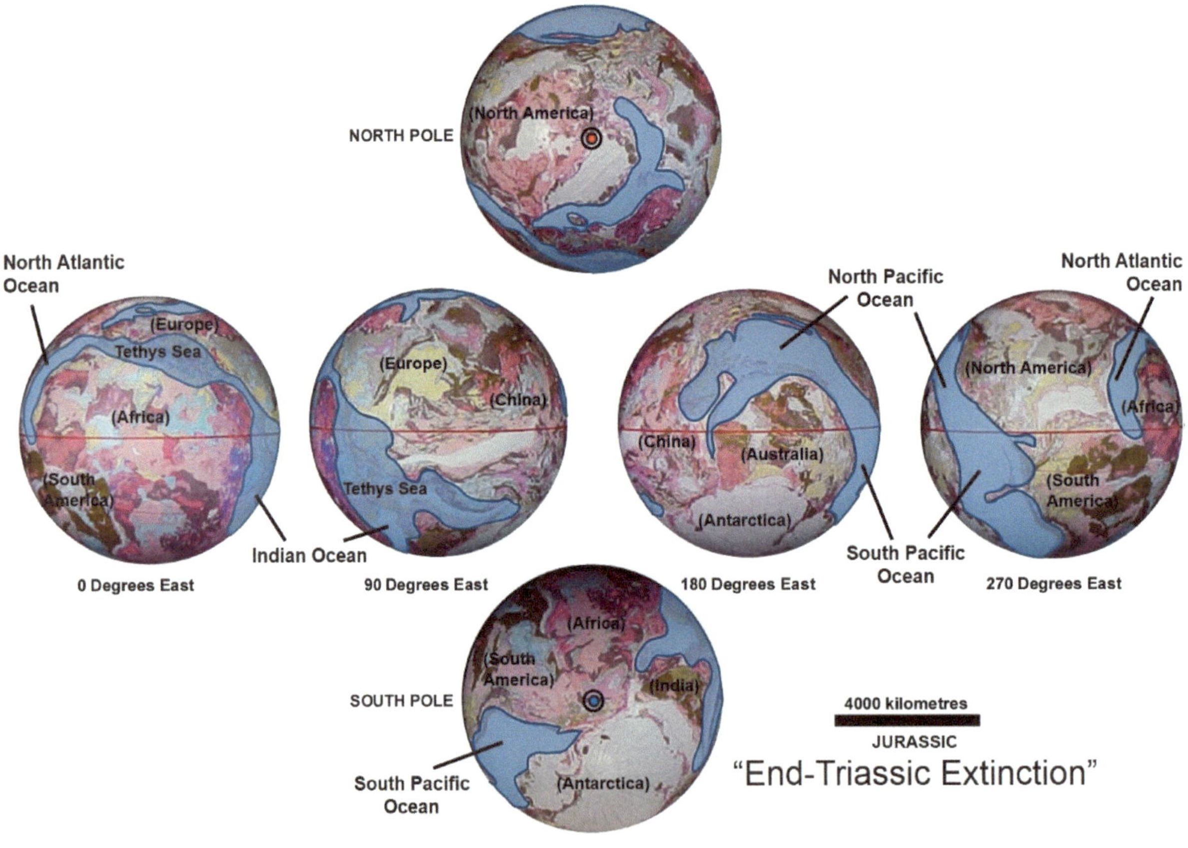

Figure 16.5 Jurassic Expansion Tectonic small Earth crustal assemblages showing the ancient coastline distribution as well as break-up of the ancient supercontinents to form the modern continents and oceans. The figure also highlights the location of continental breaching in the north and south Pacific to form the modern Pacific Oceans.

The end-Triassic extinction event is therefore considered to be the direct result of continental breaching between North America and Australia and merging of the previously separate North and South Pacific Oceans to form the modern Pacific Ocean. Breaching and merging of these two separate oceans then resulted in drastic changes to global sea-levels as the previously separate sea-levels equilibrated, disrupting the various habitats and established species migration routes. Merging of these oceans also allowed previously separate species to migrate and colonise new habitats, potentially displacing or eliminating less adaptive species.

16.2.5 End-Cretaceous Extinction Event

The end-Cretaceous extinction event—commonly referred to as the K-T event—is the most publicised and well known all of the extinction events, occurring around 65.5 million years ago at the Cretaceous to Paleocene transition. This extinction event was a large-scale mass extinction of animal and plant species, most notably the dinosaurs, in a geologically short period of time. During that time about 17 percent of all families, 50 percent of all genera, and 75 percent of all species became extinct. In the oceans it also reduced the number of marine species to about 33 percent.

Scientists currently hypothesize that the end-Cretaceous extinction was caused by one or more catastrophic events, including either an asteroid impact event or an increase in volcanic activity. In 1980, a team of researchers lead by Luis Alvarez discovered that concentrations of the element iridium, located at the Cretaceous–Paleocene time boundary, were higher than in surrounding sedimentary layers. This gave support for consideration of a collision of the Earth by a large celestial object—similar to the Shoemaker-Levy impact observed on Jupiter in 1994. This potential celestial object was later identified to be the Chicxulub crater located in Central America.

Several other impact craters, as well as massive volcanic activity such as that in the Deccan Traps of India, have also been dated to the approximate time of this extinction event. Each of these events was postulated to have released massive amounts of dust and ash into the atmosphere, reducing surface sunlight, hindering photosynthesis, and severely disrupting the Earth's biosphere.

Other researchers though believe the extinction was more gradual, resulting instead from sea-level and climate changes already occurring during the late-Cretaceous, which may or may not have been aggravated by impact events or increased volcanic activity. The end-Cretaceous dinosaur extinction was considered by Dodson and Tatarinov in 1990 to have been a highly selective world-wide phenomenon, which lasted for approximately 8 million years. From this, they concluded that too much is being attributed to a postulated single, or even multiple catastrophic impact events.

Biogeographic evidence shows that extinction events generally occur over many millions of years, not as singular moments in time. Once again, contrary to what scientists tell us, the classic Chicxulub end-Cretaceous impact crater in Mexico has been drilled for oil and has been shown to contain many metres of Cretaceous sediment. This Cretaceous sediment overlies the impact site and age-dating shows that the sediments are actually older than the impact event. An end-Cretaceous asteroid impact of this magnitude would have surely destroyed this Cretaceous sediment, which was clearly formed before the extinction event happened and should in reality either lie under the impact site or be totally destroyed.

While many researchers still believe that an impact event may have occurred during that time, its primary cause for the end-Cretaceous extinction event cannot be reconciled with conclusions based on Expansion Tectonic small Earth models. As was shown for the end-Triassic extinction event, on an Expansion Tectonic Earth extinction during these post-Permian times occurred when two or more separate oceans or seas merged, resulting in drastic changes to sea-levels and species habitats.

It is significant to note that on the Expansion Tectonic small Earth models, after the end-Triassic extinction and prior to the end-Cretaceous Period there were two very large, separate, and distinct oceans: a single Pacific Ocean and a combined Indian, Atlantic, and Arctic Ocean (Figure 16.6).

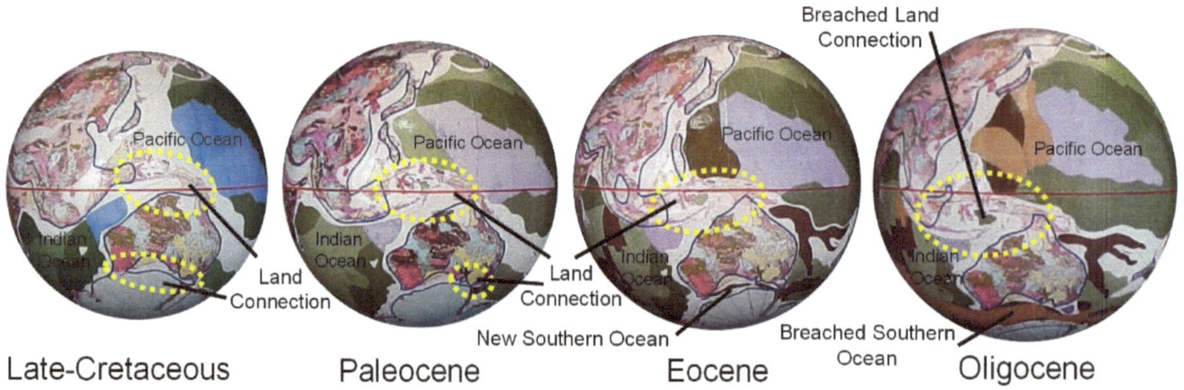

Figure 16.6 Late Cretaceous to Oligocene Expansion Tectonic small Earth globes showing land connections between Australia, Asia, and Antarctica, prior to breaching of the Australia to Antarctica connection during the Paleocene to form the Southern Ocean.

Each of these oceans may have had different sea-levels, different salinities, water temperatures, currents, and marine species, and they were separated by land connections between the various modern continents which, at that time, had not yet been breached.

The land connection existing between Australia and Asia appears to have been a series of densely packed island arcs with intervening shallow equatorial waters. I might also add that, contrary to conventional thinking, both Australia and Asia are currently separating and moving away from each other, not colliding.

Figure 16.6 also shows that the Australian to Antarctica assemblage had land connections adjacent to Western and Eastern Australia with a shallow basin adjacent to Southern Australia prior to the late-Cretaceous—known as the Eucla Basin. This land connection was then partly breached along the Western Australian coastline, causing flooding of the Eucla Basin, prior to complete breaching and opening of the Southern Ocean during Paleocene to Eocene times—around 65 million years ago.

It is considered that this breaching between Antarctica and Australia, giving rise to formation of the Southern Ocean, resulted in a period of unprecedented sea-level change—the combined Atlantic, Indian, and Arctic Ocean sea-level potentially going down and the Pacific Ocean sea-level going up—possibly of the order of a 100 metres or more. This continental breaching and sea-level change coincided precisely with the end-Cretaceous extinction event. It therefore supports the conclusions of others where the extinction event was considered to be more gradual, resulting from catastrophic sea-level changes already occurring during the late-Cretaceous, which extended over a greater period of time than an impact event would imply.

16.2.6 Mass Extinction Summary

What has been seen from this discourse into mass extinction events is that extinction on Expansion Tectonic small Earth models primarily occurred during periods of either local or global-scale changes in sea-levels. Prior to the Carboniferous Period—around 360 million years ago—sea-level changes were essentially global and were the outcome of inter-related changes in Earth radius, surface area, and surface curvature over time. The severity of these changes then increased relatively rapidly during the Carboniferous and Permian Periods as deep geosynclinal activity and orogenesis became increasingly active, forming prominent fold mountain belts and a number of relatively deep seaways. Rupture of the ancient supercontinental crust then came about during late-Permian Pangaean times, which initiated opening of the modern oceans.

Progressive draining of the ancient continental seaways during late-Permian times gave rise to formation of a number of isolated and discrete seas as well as birth of the early modern oceans. On-going break-up of the continental crusts and opening and enlarging of the modern oceans ultimately breached many of the prior land connections, resulting in merging of these separate oceans and seas to form the present-day modern oceans. Breaching and merging of the oceans and seas then resulted in periods of unprecedented sea-level changes occurring over a number of million years, resulting in catastrophic disruption to both marine and terrestrial species habitats.

With this mechanism for extinction on an Expansion Tectonic Earth in mind, we will now digress and use the distribution of dinosaurs as an example to investigate the evolution and demise of these animals on small Earth models. This will be used primarily to test their distribution in context with the established crustal assemblages, and to also investigate the environmental changes that the dinosaurs had to go through during their extended time on Earth.

16.3 Mesozoic Dinosaurs

Dinosaur fossils have been known for millennia, although their true nature was not recognized at the time of their discovery. The Chinese, whose modern word for dinosaur is *konglong,* or *terrible dragon,* considered them to be dragon bones. Villagers in central China have long unearthed these fossilized dragon bones for use in traditional medicines, a practice that continues today. In Europe, dinosaur fossils were originally believed to be the remains of giant lizards or other creatures killed during the *Great Biblical Flood.*

Descriptions of dinosaur bones first appeared in the late 17th century in England. Part of a bone, now known to have been the femur of a *Megalosaurus,* was discovered in 1676 in a limestone quarry at Cornwell, England. Between 1815 and 1824 the Rev William Buckland, a professor of geology at Oxford University, collected more fossilized bones of Megalosaurus and became the first person to scientifically describe the fossils. The study of these so called *great fossil lizards* soon became of great interest to European and American scientists and in 1842 the English palaeontologist Richard Owen conceived the term *"dinosaur"*, meaning terrible lizard. Owen went on to establish the Natural History Museum of London in order to display the national collection of dinosaur fossils and other biological and geological exhibits.

The fossil dinosaurs found today are now known to form a well-defined group of extinct reptiles. These reptiles first evolved from the ancient *Archosauria* group, characterised by limbs held erect beneath their bodies. The oldest known archosaur is called *Archosaurus*, which is known from fragmentary remains found in late-Permian rocks from Russia. From their earliest beginnings, archosaurs then evolved and radiated extensively during the following Triassic Period. After the end-Triassic extinction event, survivors of the dinosaurs became the dominant terrestrial vertebrates throughout the remaining Mesozoic Era. Other groups of animals existing at the same time were more restricted in size and habitat. Survivors of the end-Triassic and end-Cretaceous extinction events are now considered to be closely linked to the birds and reptiles of today.

Dinosaurs are a diverse group of animals belonging to the group *Dinosauria*. They first appeared during the Triassic period, approximately 230 million years ago, and became the dominant terrestrial vertebrates for 135 million years. The dinosaurs lasted until the end of the Cretaceous Period—approximately 66 million years ago—where the end-Cretaceous extinction event led to the extinction of most, if not all, dinosaur groups. It is interesting to note that Paleocene—the epoch following the Cretaceous—fossil dinosaur remains have since been found from above the Cretaceous—Paleocene extinction boundary in both New Mexico and China. These fossils provide evidence to support the suggestion that some dinosaur populations, in fact, survived the end-Cretaceous extinction and lasted at least a further half million years into the Paleocene Epoch.

Knowledge about dinosaurs is derived from a variety of fossil and non-fossil records, including fossilized bones, eggs, faeces, track-ways, stomach stones, and feathers, impressions of skin, internal organs, and soft tissues. Current evidence suggests that the average size of a dinosaur varied throughout the Triassic, Jurassic, and Cretaceous periods. The mainly carnivorous *Theropod* dinosaurs varied from 100 to 1,000 kilograms, whereas many other dinosaurs ranged up to ten metric tonnes or more. The lizard-hipped *sauropod* dinosaurs were the largest and heaviest of the dinosaurs. For much of the dinosaur era even the smallest sauropods were larger than anything else in their habitat, and the largest were an order of magnitude more massive than anything else that has since existed on land on the Earth. Sauropod remains are now mostly found in rock formations that suggest they existed in dry or seasonally dry climates, and their ability to eat large quantities of low-nutrient browse was considered to have been advantageous in such environments.

Current knowledge suggests that dinosaur evolution after the Triassic Period followed changes in vegetation, along with changes in the physical locations of the continents. *Gymnosperms*, a group of seed-producing plants represented mainly by the conifers, became well-established during the late-Triassic. The dominance of gymnosperms continued into middle- and late-Jurassic times and these plants provided an important food source for the dinosaurs. By the early-Cretaceous, as a result of on-going break-up of the Pangaea landmass, evolution of the dinosaurs was then becoming increasingly diversified. A major change in the early-Cretaceous was also marked by the evolution of flowering plants, along with the first grasses, which continued into the late-Cretaceous. This change in vegetation enabled several groups of herbivorous dinosaurs to evolve more sophisticated ways to process food. By the late-Cretaceous times a number of dinosaur clusters had become well established. These were mainly located in North America and Asia, in the southern continents that had previously made up Gondwana, and in Europe.

The end-Cretaceous extinction event has been extensively studied by others and has been shown to have resulted in the extinction of all but a few remaining dinosaur groups, including ancestors to the modern birds. Other groups, such as the crocodiles, turtles, lizards, snakes, and various related but now extinct species, also survived the extinction, as did a number of Paleocene dinosaurs. Though the consensus in conventional studies is that an impact event was the primary cause of dinosaur extinction, other scientists cite other possible causes or support the idea that a confluence of several factors was responsible for the disappearance of dinosaurs from the fossil record.

It should be noted from the fossil record that misguided confusion surrounds the real meaning of extinction of the dinosaurs. Dinosaurs have always been an extremely varied group of animals. According to a 2006 study, over 500 dinosaur genera have been identified with certainty so far and the total number of genera preserved in the fossil record has been estimated at around 1850, nearly 75 percent of which are said to still remain to be discovered. On average, throughout the reign of the dinosaurs, each dinosaur genus existed for only about five to six million years. As well as this, individual dinosaur species existed for only two to three million years prior to becoming extinct through natural background rates of extinction. This typical rate of dinosaur species turnover suggests that, rather than a single or even multiple catastrophic extinction events, natural species turnover and dying out was a common phenomenon throughout their entire 135 million years of existence.

Once again the end-Cretaceous dinosaur extinction was considered by Dodson and Tatarinov in 1990 to have been a highly selective world-wide phenomenon, which lasted for approximately 8 million years. What this means is that the disappearance of the dinosaurs at the end of the Cretaceous Period only represented the loss of a relatively few remaining species. This final demise was a small aspect of the entire population of dinosaurs that had previously already died out. What is now mistakenly portrayed in the fossil record is the decline of the entire 135 million year population of dinosaurs superimposed on top of each other, not the individual extinction events or the separation of a final catastrophic event.

The remains of fossil dinosaurs so far discovered are extensive. These fossils are known throughout the world from sedimentary deposits ranging in age from late-Triassic to late-Cretaceous times. The locations of currently known Triassic and Jurassic dinosaur sites, as well as Permian ancestral reptiles from around the world, are shown plotted on the Triassic small Earth model in Figure 16.7. These sites were plotted from the extensive world-wide listings in published literature, in particular Weishampel (1990), and these are plotted in conjunction with the distribution of Permian reptiles to emphasise their common ancestry. Also shown on this figure are the Triassic climate zones and distribution of continental seas, as well as locations of the ancient poles and equator previously established from geographic and palaeomagnetic studies.

On the Triassic Expansion Tectonic small Earth model shown in Figure 16.7, the Pangaea supercontinent had already started breaking up to form the modern continents. The intervening modern oceans were in turn starting to open and water was progressively draining from the ancient seas into these newly opening oceans. The newly formed marine sedimentary basins were also receiving new sediment eroded from the exposed lands, and, along with established climate zones, the polar ice-caps were being severely disrupted. The presence of separate, isolated seas and juvenile modern oceans suggests the presence of variable sea-levels, as also noted in conventional studies.

The apparent migration of the South Pole and south polar climatic region are shown in this figure to have commenced migrating south along the west coast of Africa. By the following Jurassic Period, the South Pole and accompanying climate zone were then located in South Africa prior to migrating across the newly opening South Atlantic Ocean. This break-up and opening of the South Atlantic Ocean caused significant disruption to the previously existing south polar climatic zone, which then resulted in seasonal melting and refreezing of any prevailing ice-cap with considerable changes to both climate and sea-levels throughout the world.

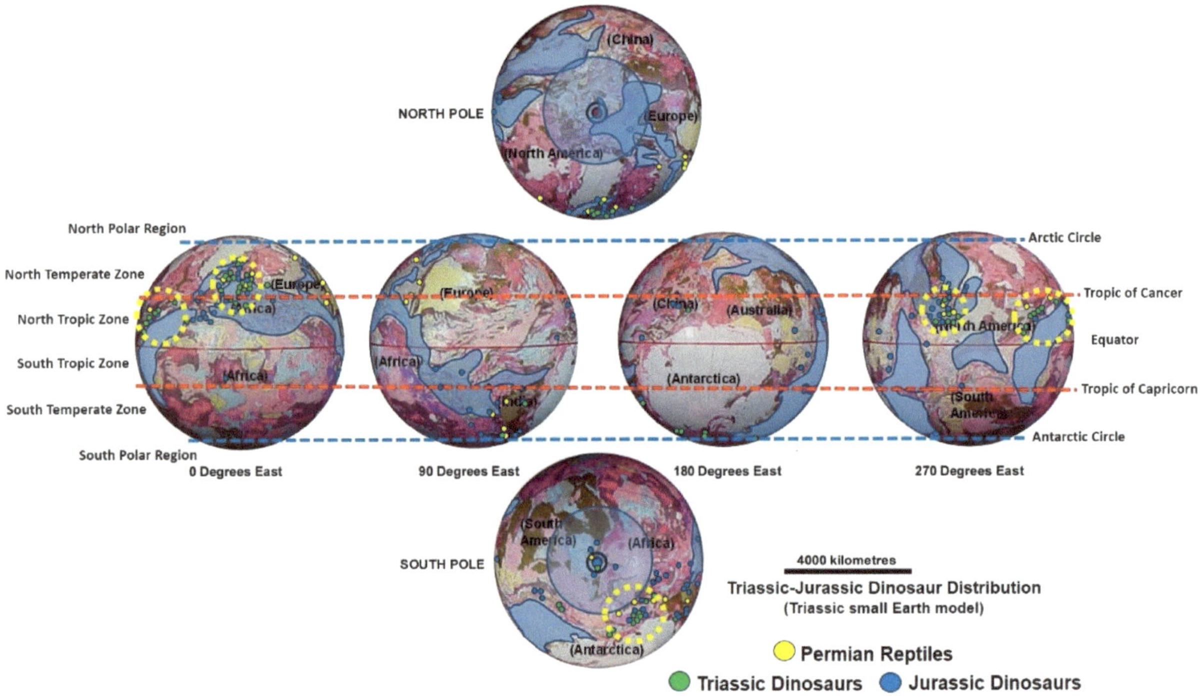

Figure 16.7 Permian reptile and Triassic and Jurassic dinosaur locations on the Triassic small Earth model. Their distribution shows clusters within North America, Europe, South Africa and southern South America. Isolated occurrences are also shown in eastern Australia, southern China, and western South America.

In addition to climate and crustal changes during that time, it is significant to note that on an Expansion Tectonic Earth, the 135 million year Permian to late-Cretaceous interval of time when the dinosaurs prevailed represents an increase in Earth radius from 50 percent of the present radius to approximately 80 percent. Prevailing surface gravity calculated from the radius data shows that, during that time, gravity then increased from around 50 percent of the present surface gravity during the Permian to approximately 75 percent during the late-Cretaceous. Considering the large size of many dinosaur species, the dinosaurs existing during these times would well have benefited from the much lower surface gravity prevailing during the early part of the Mesozoic Era.

Hurrell, in 1994, provided independent evidence for a reduced gravity during these times after analysing the mechanical aspects of dinosaur bones. He concluded that it would have been impossible for the larger dinosaurs to exist in environments of high surface gravity—such as exists now. From his mechanical studies of dinosaur bones Hurrell also independently calculated a surface gravity during the Permian of 50 percent of the present surface gravity. The subsequent increase in surface gravity during the remaining Mesozoic Era may then offer an additional explanation for the large turnover of dinosaur genera throughout the history of the dinosaurs, as noted by Dodson and Tatarinov in 1990.

The distribution, migration, and eventual demise of the dinosaurs on an Expansion Tectonic Earth should then be visualised in conjunction with this very involved period of crustal break-up, new ocean development, and change in surface gravity. The Triassic small Earth model shown in Figure 16.7 represents only one momentary aspect of this long and involved history of the dinosaurs. While the prolonged period of time involved during this crustal break-up allowed the dinosaur species to migrate to more equitable locations, it also presented new physical and environmental barriers. The draining of the ancient continental seas, for instance, influenced the distribution of the modern seas and oceans and hence adversely affected available dinosaur habitats, migration routes, and escape avenues. Migration away from these physical barriers may have then encouraged the dinosaurs to evolve into new species. In contrast, failing to evolve, migrate, or being land-locked may have also caused their localised demise.

In Figure 16.7 the distribution of Permian ancestral reptiles—displayed as yellow dots—and Triassic dinosaurs—displayed as green dots—are shown to cluster in a number of distinct provinces. These provinces are primarily located in Eastern Europe, western North America, and South Africa, with links to eastern North America and lesser links elsewhere. The Triassic assemblage of continents shown suggests that the South African cluster may have also formed a more extensive province extending into Antarctica, now covered by the Antarctican ice-sheet. From these primary clusters, depending on crustal break-up constraints, the distribution of Jurassic dinosaurs—shown as blue dots—can then be seen to radiate and extend outwards from these provinces.

During the remaining Cretaceous times, continental crustal break-up continued unabated as each of the modern continents became well established. By then all of the intervening modern oceans, except for the Southern Ocean, had opened up. Breaching of the North American and Australian land connection had already occurred, giving rise to merging of the north and South Pacific Oceans, which resulted in the end-Triassic extinction event. During this end-Triassic extinction event most of the non-dinosaurian archosaurs, as well as most of the terrestrial vertebrates, the ancestors of mammals, and most of the large amphibians became extinct. This extinction enabled the surviving dinosaurs to then migrate and extend their coverage well beyond their original ancestral provinces. Because this end-Triassic extinction event mainly affected coastal environments, those dinosaurs and other species adapted to more inland habitats may have been well suited to survival.

The distribution of Cretaceous dinosaurs is shown plotted as red dots on the late-Cretaceous small Earth model in Figure 16.8 in relation to Cretaceous climate zones.

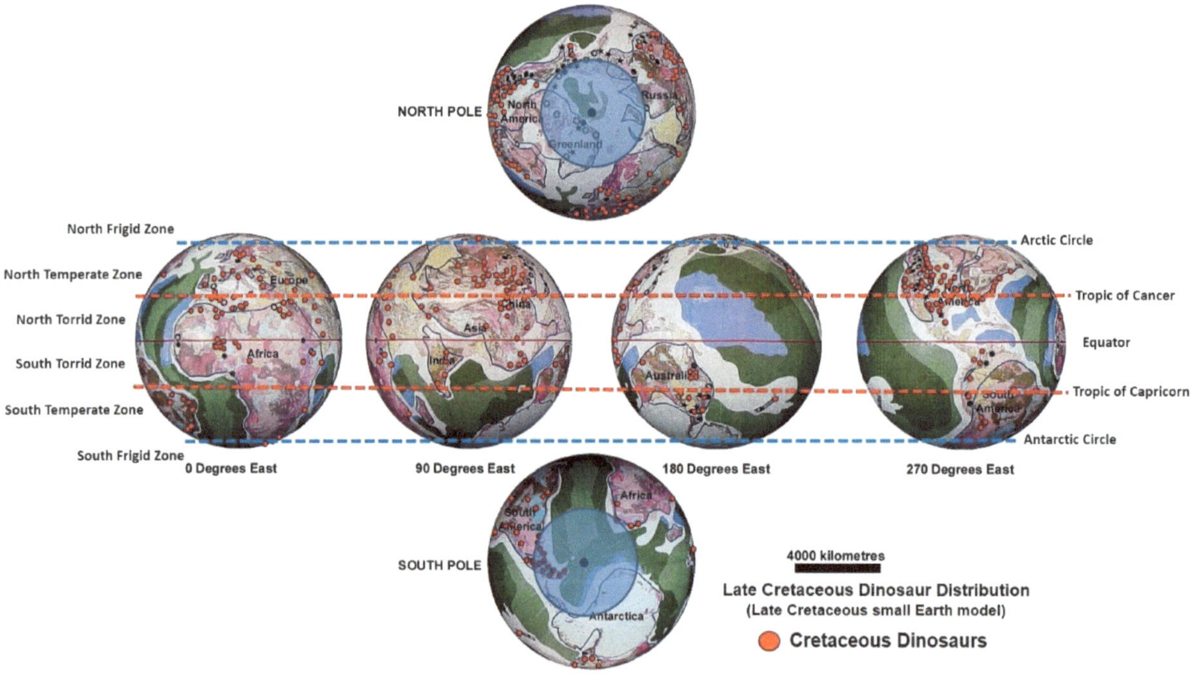

Figure 16.8 Cretaceous dinosaur distributions on the late-Cretaceous small Earth model. Their distribution shows an extension and isolation of previous provincial clusters into South America, Asia, China, and Australia.

The dinosaur distributions on this figure show that throughout Jurassic and Cretaceous times the dinosaurs progressively dispersed and radiated away from each of the three main Triassic provinces. Depending on the presence of physical and climatic barriers, the surviving dinosaurs were then able to extend their habitats into new regions in Australia, Asia, China, and South America.

On Figure 16.8 it can also be seen that during these times both the north and south polar-regions were located over oceans, suggesting that the presence of any polar ice-caps may have been seasonal and climates were moderated—as also noted in conventional literature. It is interesting to note the subtle preference of dinosaur habitats within the northern and southern Temperate Zones, with lesser occurrences within the equatorial and polar-regions.

From their beginnings, the distribution of European and Mediterranean dinosaurs extended their habitats during the Cretaceous Period into the northern Africa and eastern Mediterranean regions. Similarly, from their beginnings the western North American dinosaurs migrated north-east to include western Canada, south to include parts of Central America, and west to include East Australia—although this link was breached during the Jurassic. During that time an eastern North American dinosaur sub-province also became isolated from Europe during opening of the North Atlantic Ocean, with links remaining to western North America.

During the Jurassic Period the African and South American continents commenced breaking-up and the South Atlantic and Indian Oceans commenced opening. The common South African and South American dinosaur province was then separated and isolated during the late-Jurassic Period. The shift in the southern polar region, from West Africa to South Africa, then encroached on the existing South American and South African dinosaur province. The dramatic changes to climate, coupled with break-up of the two continents, then forced these dinosaurs to migrate away from the increasingly cold areas to more equitable climates in northern South America and India.

During the Cretaceous Period, the Australian dinosaurs had become separated from their North American ancestors and progressively dispersed throughout east and central Australia. Similarly, the dinosaur occurrences within the south China sub-province dispersed throughout the rest of China, as well as into Mongolia, Southeast Asia, and Eastern Asia.

This unique provincial distribution of the earlier Permian ancestral reptiles and Mesozoic dinosaurs on an Expansion Tectonic Earth demonstrates consistent evolutionary links between Permian, Triassic, Jurassic, and Cretaceous species. Many of these links were permanently disrupted during the Triassic and Cretaceous Periods as a result of crustal break-up and continental dispersal, draining of the ancient seas, variable sea-levels, and disruption to existing climate zones.

These environmental changes then forced the dinosaurs to migrate, or rapidly evolve, in order to keep pace with on-going environmental changes.

Failure to keep pace with these changes may explain the relatively rapid species turnover recorded in the fossil record, as well as the periodic extinctions seen. During these times, the European and Mediterranean provinces remained relatively intact, with common links remaining with Africa and possibly extending into Asia. Elsewhere, links between east Australia, New Zealand, and North and Central America were severed, as were links between South America and South Africa. Dinosaur species in each of these areas then evolved independently, with new species retaining a common ancestral heritage.

On Plate Tectonic reconstructions, the end-Triassic and end-Cretaceous extinction of the dinosaurs remains a contentious issue. In contrast, the Permian to Palaeocene Expansion tectonic small Earth models show that as each of the modern oceans opened and developed, there were a number of separate oceans existing throughout much of this Mesozoic Era. The presence of completely separate oceans suggests there may have been separate sea-levels, separate salinity's, separate temperatures, and separate marine species in each of these oceans, all co-existing at the same time.

The small Earth reconstructions then show that, rifting and break-up between the Australian and North American continents occurred during the Triassic Period which resulted in merging of the previously separate North and South Pacific Oceans. This relatively rapid change in sea-levels gave rise to the end-Triassic extinction event. During the end-Cretaceous to Palaeocene Periods, rifting between the Australian and Antarctican continents also occurred which resulted in merging of the Pacific, Indian, and Atlantic Oceans and opening of the Southern Ocean. This coincidence of mergence and disruptions to sea levels suggests that the primary cause of mass species extinctions during these times was ultimately linked to periods of rapid sea-level change, rather than to inferred catastrophic, asteroidal, or meteoric impact events.

17 Space Geodetics

"Proof of any theory comes through direct observation or by direct measurement."

Space geodetics is a relatively new and sophisticated physical science that basically uses a network of radio telescopes, satellites, and ground-based receiver and transmitter stations from around the world to routinely measure the precise dimensions and continental plate motions of the Earth. This technology also forms the framework for modern GPS technology that is more familiar today. Because of its implied sophistication and complexity, understanding space geodetics, including its limitations, is generally well beyond most people's comprehension. It is therefore very important at this stage to investigate space geodetics in the context of an Expansion Tectonic Earth.

Space geodetic measuring techniques developed to measure the dimensions of the Earth stem from the early 1970s and include Very Long Baseline Interferometry (VLBI), Satellite Laser Ranging (SLR), Lunar Laser Ranging (LLR), Global Positioning Systems (GPS), and Doppler Orbitography and Radiopositioning Integrated by Satellite (DORIS). Observational data are now routinely recorded from each of these measurement techniques and the mathematically treated data from all receiver stations are combined and used to calculate a solution to the *global geodetic network*—a three dimensional measurement framework of the Earth.

In 1993 when Robaudo and Harrison first combined SLR data, including all global geodetic data from 1976 to the beginning of 1991, as well as VLBI data containing data up to the end of 1990, they allowed all observational stations to have three independent *"X-Y-Z motion velocities"*—vector motions of the ground-based observational stations in three dimensional coordinates. When these motion velocities were used to establish a *"...global geodetic network"* they calculated *"...a Root Mean Squared value of up-down* [variation in Earth radius] *motions of over 18 mm/year."* In other words the Earth was found to be potentially expanding by up to 18 millimetres per year.

At that time, Robaudo and Harrison considered this up-down motion to be extremely large. They obviously did not consider Earth expansion when making this judgement, but instead compared it to values that were expected from areas undergoing *"crustal rebound"* during glacial melting—the relaxation and rise of a continent after glacial melting—estimated to be less than 10 millimetres per year. It is significant that Robaudo and Harrison *"...expected that most VLBI stations will have up-down motions of only a few mm/year,"* and they went on to recommend that the vertical motion *"...be restricted to zero, because* [they considered that] *this is closer to the true situation than an average motion of 18 mm/year."* Robaudo and Harrison were, in fact, faced with a daunting problem. When they calculated this variation in Earth radius from 15 years' worth of observational data they found, but failed to acknowledge, that the Earth was potentially expanding by 18 millimetres per year. It is significant to note that this value is very close to the value of 22 millimetres per year calculated here using published seafloor mapping, especially when Robaudo and Harrison's error margins are also taken into consideration.

Instead, this value of 18 millimetres per year was assumed by Robaudo and Harrison to be an accumulation of systematic errors during the collection and mathematical and statistical treatment of the raw observational data. Since then the mathematical formulae and applied correction parameters have been refined, which has now enabled this assumed accumulation of errors to be statistically meaned out to zero—precisely as Robaudo and Harrison recommended! Additional estimates of present-day Earth radius from space geodetic methods are shown in Table 17.1.

Author	**Year**	**Rate [cm/yr]**	**Method**
Blinov[1]	1987	2.43	Doppler Surveying (general uplift)
Carey[2]	1988	2.08 ± 0.8	SLR (chord analysis)
Maxlow[3]	2000	>1.8	VLBI (general uplift)
Koziar[4]	2011	>1.0	VLBI (apparent baselines contraction)
Koziar[5]	this paper	2.72	increase in the equatorial semiaxis of global geodesic ellipsoid

1) correct interpretation of the results published by Anderle & Malyevac (1983)
2) W.D. Parkinson's calculations
3) correct interpretation of the results obtained by Robaudo & Harrison (1993)
4) correct interpretation of the results obtained by Heki *et al.* (1989)
5) correct interpretation of the results published by McCarthy ed. (1992) and McCarthy & Petit eds (2003)

Table 17.1 Present-day rates of increase in Earth radius obtained by space geodetic methods. (From Koziar 2011).

The application of advanced space geodetic techniques to studies of the Earth has now progressed to the point where estimation of Earth radius and present-day plate motion is quoted to sub-millimetre accuracy. In other words, geodesists can now measure the dimensions of the Earth, along with the motions of the continents, extremely accurately. Results of Shen and others in 2011 now show that "*...both geodetic and gravimetric observations support the conclusions that the Earth is expanding at a rate of 0.2 millimetres per year in recent decades.*" This is all very well but this value is a factor of 100 too low when compared with the expansion rate based on seafloor mapping data used in this book, so we need to understand why?

17.1 Space Geodetic Data

Regardless of the supposed accuracy and sophistication of space geodetic measurement, the limitations of space geodetics must first be understood so as to fully appreciate the measurement techniques from an Expansion Tectonic point of view. It is far too easy to mathematically manipulate observational data to suit a particular need, in this case a static radius Earth model, and once this data is encoded into general practice it is often very difficult to change or later refute.

The primary limiting factor to the accuracy of measurements in all space geodetic measuring techniques is the systematic errors that come from seasonal atmospheric interference to the signals being measured. What this atmospheric interference does is to alter the timing of the optical or radio signals being measured due to refraction of the signal as it passes through the atmosphere. To minimise errors, correction factors are routinely introduced during mathematical treatment of the raw measured data. These correction factors make allowance for the variations in wet and dry atmospheric conditions throughout the year. If these factors are wrong then they can directly affect the final measurements.

For SLR, GPS, and DORIS measurements, additional limiting factors also include satellite tracking and modelling of the Earth's magnetic force field. Force field premises imposed on the mathematics are based on adopting a constant universal gravity G, a constant Earth mass M, and a constant product G·M. This product is then used to calculate Earth's surface gravity and to locate the physical centre of the Earth, which is used in both satellite altimetry control and as the X-Y-Z co-ordinate reference point. Satellite positioning and altimetry control are also known to be sensitive to both universal time and to the value of G·M.

For an Expansion Tectonic Earth, the universal gravity constant is also assumed to be constant, while Earth mass has been shown here to be potentially increasing with time. Current known errors in the positioning of the centre of the Earth, and hence the co-ordinate system used by space geodesists, have been quoted as plus or minus 50 to 100 millimetres, so constraining Earth mass to a constant value may well be introducing additional unforeseen errors during data processing.

Vertical crustal motion studies carried out by geodesists have shown that the systematic and random errors in space geodetic measurements are largest in the vertical component of a site measurement, which, of course, directly affects measurement of Earth radius. On the other hand, VLBI is shown to be insensitive to changes in the vertical. Similarly, for SLR the greatest discrepancies between estimates of plate motion and established plate model values are again in the vertical component and the largest systematic error source for absolute height determination is the value of G·M. Regardless of this, in 1997 Larson and others concluded from the observational data that there is "*...no persuasive evidence for any significant net vertical crustal motions*"—that is, increase in Earth radius. Similarly, Soudarin and others in 1999 concluded that "*...with the exception of a few stations, vertical plate motion rates are generally less than 5 millimetres per year.*"

More recently, Shen and others in 2011 gave a brief overview of how raw observational data is treated before using it to calculate a variation in Earth radius. It is worth quoting in detail the process they went through to tidy up the data before using it in their calculation. In their overview they used spacial geodetic data published in 2008. In this data "*...there are 1572 readings at present from various stations (including GPS, VLBI, SLR, and DORIS)...Due to discontinuities of many recordings, only 930 recordings...are available for present study.*" Also, "*...stations that are very close to each other...are merged to one station...Then, there are 841 stations left...in this study the stations located in active tectonic zones...have been removed from our calculations.*" Furthermore, "*Another concern is that the absolute values of the vertical velocities of some stations are beyond 0.02m/year, and so large vertical movements of such kinds of stations are not related to Earth expansion...Hence, such kinds of stations are not included in our calculations...After removing the stations located in the orogenic zones and the stations whose vertical velocities are greater than 0.02m/yr, there are 625 stations left...Our calculations show that the Earth is expanding at present at an expanding rate of 0.24±0.04mm/yr.*"

From this discourse it can be seen that 60 percent of the raw observational data are eliminated before calculating a rate of change in Earth radius simply because this data doesn't fit the geodesists' pre-conceived static radius Plate Tectonic model of the Earth. In other words, what they did was remove all data that might otherwise indicate Earth expansion. This, in effect, smoothed out the raw data before making their calculation. By doing this Shen and others then concluded "*...the Earth is expanding at present at an expanding rate of 0.24±0.04mm/yr.*" This smoothed and calculated data is now routinely published on the internet for all to see and use.

I am now going to digress slightly to use an analogy to explain where these space geodetic calculations and pre-conceived misunderstandings may have gone wrong. It is a relatively common misconception in science that the way the Earth may expand is like a balloon, in other words it is merely pumped up with a relatively uniform increase in radius each year. This can be visualised as a thin smear of damp clay coating the outside of a balloon to simulate the Pangaea supercontinental crust. By merely pumping up the balloon, the damp clay will then stretch and extend just like the Earth's supercontinental crusts did before break-up of Pangaea. Ultimately the clay will rupture and break apart simulating break-up of the ancient Pangaea supercontinent to form the modern continents and oceans. Measuring an increase in radius could then be done by simply taking a number of measurements from around the balloon, irrespective of how the data are treated, and calculating a mean value for change in radius. It is unfortunate that this analogy could not be further from reality if it tried.

In reality, I have shown in early chapters that the Earth's continental and seafloor crusts are not uniform; hence the Earth does not simply pump up like a balloon. The continents instead comprise a mixture of cratons, orogens, and sedimentary basins, each with their own definitional requirements, as well as the seafloor crusts which now make up around 70 percent of the surface area of the Earth and are predominantly intruded volcanic lava.

This crustal composition can be visualised by varying the balloon example to take into account these different types of crusts. The cratons, by definition, are crusts that have stabilised before about 2,400 million years ago. This can be simulated by partly drying the coating of clay on the surface of the balloon, while leaving the underlying clay still damp. By again pumping up the balloon, the dried clay outer crust will crack and fragment to accommodate for the increase in balloon radius. By continuing to pump up the balloon three things will happen: firstly, the underlying damp clay will stretch–simulating the opening and development of sedimentary basins; secondly, the dry outer crust will continue to fragment–simulating ongoing fragmentation of the cratons; and thirdly, the soft clay located around the margins of the dry fragments will wrinkle and crumple—simulating formation of the orogens.

In addition to fragmenting, because of its increased rigidity, the dry outer clay coating will retain a partial super-elevation at the centre of each fragment—in other words it will begin to rise in relation to the rest of the balloon surface due to the change in balloon radius. The outer edges of each dry fragment will also be forced down into the balloon along the respective margins. Ultimately, the exposed damp clay will then rupture and break apart as the balloon continues to be pumped up, again simulating break-up of the ancient Pangaea supercontinent to form the modern continents and oceans.

In this second example the pumped-up surface of the balloon is not smooth and uniform, as would be anticipated in the first damp clay coated example. The surface will instead reflect the variation in super-elevation within the dried crusts, the depressed margins around these same crusts, and a relatively even surface of the stretched and ruptured damp clay. In other words the crusts will be going both up and down depending on where the surface measurements are taken. This variable *pumping up* process will then reflect the variability of the *up-down motions* of the various crusts, identical to what is seen in modern space geodetic measurement of the Earth.

The point I am making with both of these examples is that, in the first example we could simply measure the increase in radius from a relatively few surface measurements—which is what Shen and others assumed they could do by eliminating 60 percent of the raw space geodetic observational data. But, in the second example, because the crust of the Earth is not smooth or well behaved, we definitely should not eliminate 60 percent of the space geodetic observational data.

In the second example the observational data measurements would in fact reflect everything from crusts going up in the centre of the orogens, to crusts going down along the margins, to others that have a relatively uniform increase in radius. To eliminate 60 percent of the measurements before calculating Earth radius, especially when the vertical motions are so small, is therefore misleading and erroneous and also forces the observational data into a preconceived static radius Earth model. All of the measured space geodetic measurements must therefore be considered relevant and must be used during calculation.

In addition to this observation, the choice, location, and amount of observational stations distributed around the world would also potentially bias the outcome of this space geodetic calculation. By "*... removing the stations located in the orogenic zones and the stations whose vertical velocities are greater than*

0.02*m/yr*" this will definitely bias the outcome of the radius calculation by removing the very measurements that should show an increase in Earth radius. Many more observational stations than currently available around the world would definitely be needed before this bias could be minimised and a meaningful increase in Earth radius be calculated from space geodetic data.

17.2 Continental Plate Motion

In contrast to measuring the vertical crustal motions of the Earth, the primary function of space geodetics is not to measure variation in Earth radius. The primary function is to measure the horizontal plate motion and plate motion directions of each of the continents in order to determine the movement of continents according to a static radius Plate Tectonic Earth model. Measuring vertical plate motion is of secondary importance to space geodetics because it is assumed that Earth radius is constant.

The horizontal plate motions of each of the continents are routinely published as horizontal velocity field vectors relative to the centre of the Earth, as shown in Figure 17.1, published from all ITRF data—the International Terrestrial Reference Frame—up to 2008.

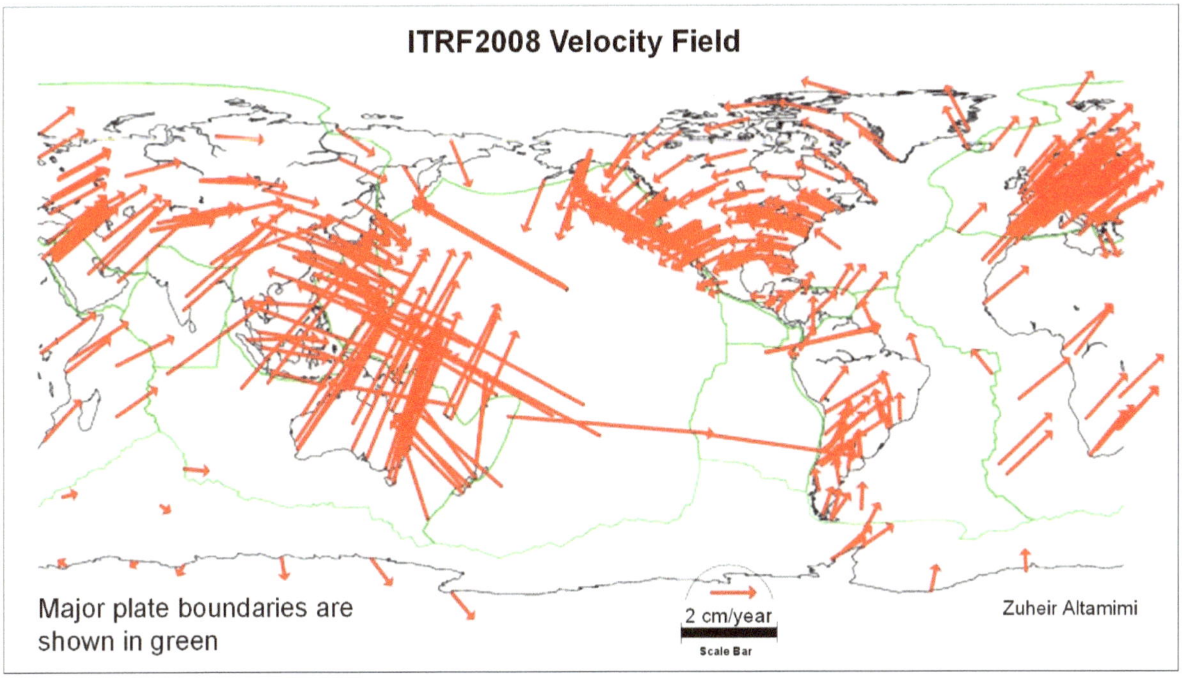

Figure 17.1 International Terrestrial Reference Frame 2008 horizontal velocity field vectors. A two centimetre per year vector scale bar is shown at lower centre.

Because the Earth is not flat, as depicted in this figure, earlier 1997 versions of this plate motion vector data from my original research are also shown in spherical format in Figure 17.2 along with vectors for the mid-ocean-ridge spreading zones.

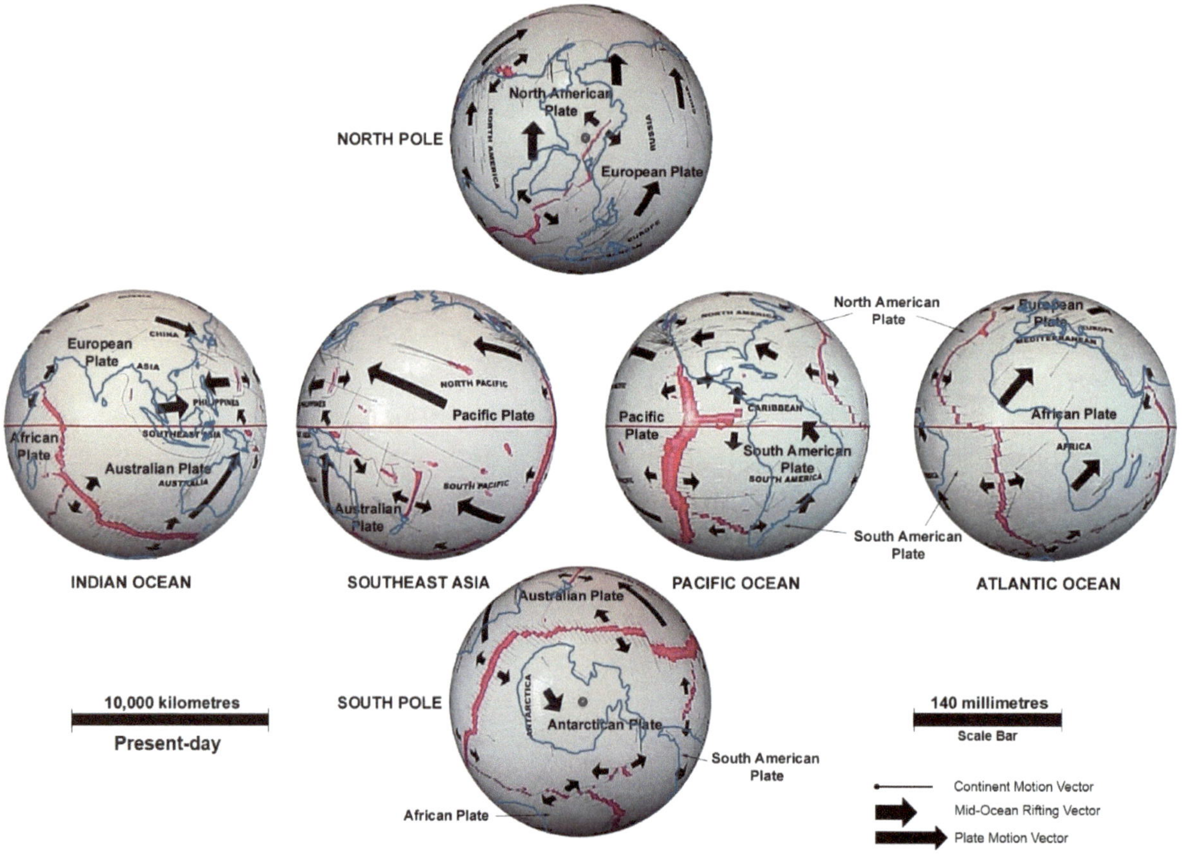

Figure 17.2 1997 VLBI, SLR, GPS, and DORIS horizontal plate motion vector plots. Images show continent outlines and Pliocene to Recent mid-ocean-rifting (pink and red) on the present-day small Earth model. Large arrows (excluding arrowhead) represent plate vectors to scale shown at lower left, small arrows represent mid-ocean-rifting, fine-line arrows represent motion vectors plotted from published solutions. Plate motion vector scale bar at lower right represents the circumferential increase per year required for an Expansion Tectonic Earth.

The plate motion vector arrows shown in both Figures 17.1 and 17.2 are scaled horizontal vectors with the arrow heads showing the directions of plate motion constrained to a Plate Tectonic Earth model. It should be noted that the International Terrestrial Reference Frame only measures and publishes the horizontal plate motion vectors. They do not, and cannot, take into account any significant vertical vector components that may be present. This is because a constant Earth radius is a built-in assumption. Space geodesists do not want, nor can they accommodate for an increase in Earth radius in their Plate Tectonic model.

In order to better understand what horizontal and vertical plate motion vectors mean on a Plate Tectonic and Expansion Tectonic Earth, consider the simple cases shown in Figure 17.3 and Figure 17.4 where an ocean basin, such as the Atlantic Ocean, opens. Both figures show a cross section of the Earth containing continental crusts and an ocean basin.

Figure 17.3 shows a Plate Tectonic cross section where, because Earth radius is assumed to be constant, as the ocean basin opens during mid-ocean-rifting only the blue horizontal plate motion vector is required. Note: these vectors are grossly exaggerated and are meant to only represent centimetres of crustal movement per year as distinct from hundreds of kilometres of ocean basin width. In this simple case, as the ocean widens both continents located to the left and right of the ocean must then *"drift"* away from the mid-ocean-spreading-ridge by Continental Drift. Furthermore, in order to maintain a constant Earth surface area, an equivalent amount of pre-existing crust must also be disposed of elsewhere.

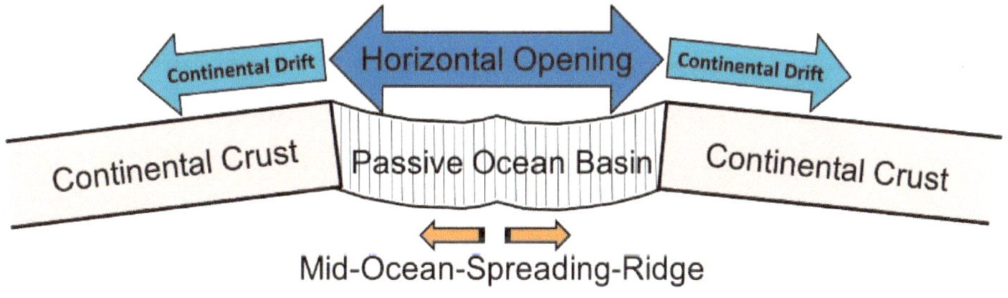

Figure 17.3 A cross-section of a Plate Tectonic Earth showing a horizontal plate opening vector arrow (grossly exaggerated) and associated Continental Drift for a passive ocean basin such as the Atlantic Ocean and two adjoining continents. Crustal thicknesses are not to scale.

In contrast, Figure 17.4 shows an Expansion Tectonic cross section of part of the Earth crust where, because Earth radius increases, horizontal, vertical, and actual plate motion vectors are shown. Again, the scale of these motion vectors are grossly exaggerated and represent centimetres of crustal movement per year. In this example Earth radius has increased from the lower (A) to the upper (B) crustal position simulating a simple radial expansion of the Earth of 22 millimetres per year.

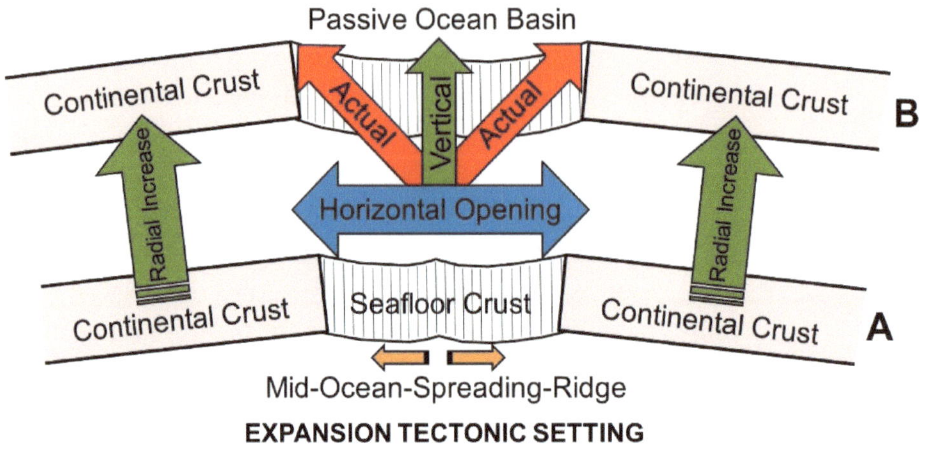

Figure 17.4 A cross-section of an Expansion Tectonic Earth showing horizontal, vertical, and actual plate opening vector arrows (grossly exaggerated) in relation to radial expansion for a passive ocean basin such as the Atlantic Ocean and two adjoining continents. Crustal thicknesses are not to scale.

In this case the continental crusts are shown to move out radially from A to B and, as such, there is no requirement for continental drift. The annual increase in Earth radius equals the green vertical motion vector, while opening of the ocean along the mid-ocean-ridge spreading centre equals the blue horizontal vector—again in centimetres per year. For an Expansion Tectonic Earth, the sum of half the blue horizontal vector plus the green vertical motion vector equals the red actual plate motion vector.

The simple case shown in Figure 17.4 is precisely what happens for all ocean basins on an Expansion Tectonic Earth. To demonstrate this, the example is extended in Figure 17.5 to show the same motion vectors in relation to opening of two separate oceans. This example can be further extrapolated to the rest of the Earth. The figure is again shown in cross-section where the Earth increases in radius from the lower crust A to the upper crust B position. On an Expansion Tectonic Earth, both of the oceans depicted simply open and spread apart at the mid-ocean-ridges in sympathy with a radial Earth expansion.

In this figure, the increase in Earth radius is again represented by the vertical green radial motion vector arrows and opening of each ocean is represented by the blue horizontal motion arrows—representing centimetre scales. The net opening of each of the oceans plus the increase in Earth radius is then depicted by the red inclined actual crustal motion vector arrows. Note, there is again no requirement for Continental Drift of any of these continents.

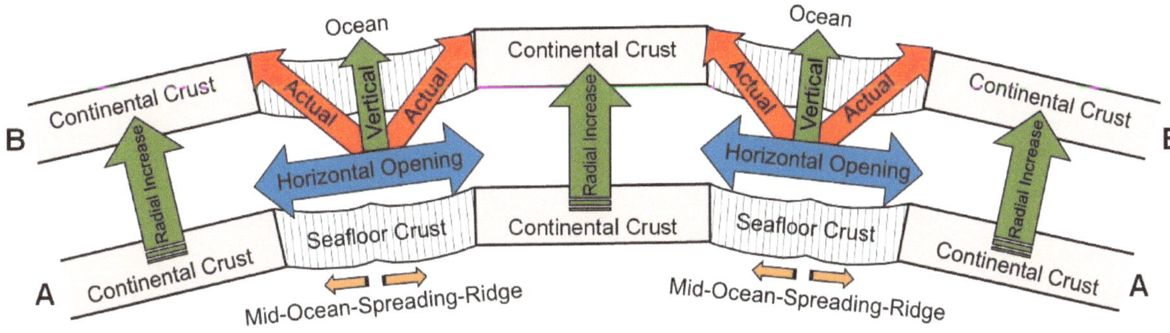

Figure 17.5 A cross-section of an Expansion Tectonic Earth increasing in radius from crustal position A to B, showing horizontal, vertical, and actual plate motion vector arrows (grossly exaggerated) for three continents plus two adjoining ocean basins. Crustal thickness not to scale.

In Figures 17.4 and 17.5 it should be noted that the positions of both the mid-ocean-ridge spreading axes and the continents, relative to the centre of the Earth, remain relatively stationary. Their primary displacement is radial. This situation is unique to an Expansion Tectonic Earth and the figures simulate the actual motions required for each of the continents during an expansion of the Earth. For all oceans, opening along each of the mid-ocean-spreading-centres and accumulation of new seafloor crust is then dictated by the increase in Earth surface area as a direct result of increase in Earth radius. In these Expansion Tectonic examples there is again no requirement for disposal of excess crusts by subduction, and certainly no requirement for continental drift.

In conventional space geodetic and Plate Tectonic studies only horizontal plate motion vectors are used because Earth radius is assumed to be constant and therefore the net vertical vector is zero. Because of this insistence it should be appreciated that no two adjacent ocean basins can open without continental drift because their horizontal motion vectors oppose each other. The word *drift* in this context is, however, somewhat misleading. Prior to the discovery of mid-ocean seafloor spreading Wegener likened this drift to "*...icebergs of low density granite floating on a sea of denser basalt*" in a somewhat aimless manner. Now, with the knowledge that seafloor is being continually created worldwide at the mid-ocean-ridge zones, the apparent motion of the continents relative to each other becomes quite orderly—dependent mainly on the increase of seafloor area—and hence the word "drift" is not an accurate description.

On an Earth that is expanding, the increase in total surface area does not represent a problem as it is being driven by the ever increasing Earth radius. However, on a fixed radius Earth some mechanism must be postulated to accommodate for the observed increase in surface area.

This can be seen in Figure 17.6 where, relative to the fixed continental crust B, located in the centre of the figure, both continental crust A and continental crust C must move away from crust B by the amount of horizontal crustal opening shown in each ocean. In this example both horizontal vectors are trying to move each adjacent continent, as well as their respective mid-ocean-ridge spreading zones, away from the fixed continent B. This is the fundamental basis of, and reason for, Continental Drift in Plate Tectonic studies.

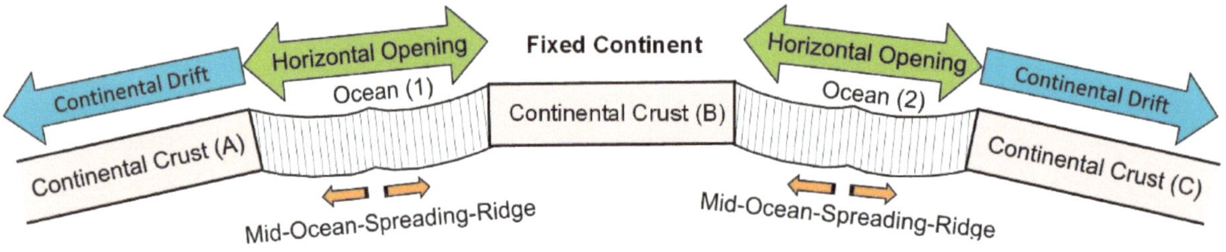

Figure 17.6 A cross-section of a Plate Tectonic Earth showing the required sideways drift motion of two continental crusts A and C relative to a fixed continent B during mid-ocean-spreading within two separate oceans (vector arrows are grossly exaggerated). Crustal thickness not to scale.

From Figure 17.6 it should be appreciated that sideways continental drift away from the fixed continental crust B is purely a local phenomenon. At some point around the globe continental drift will then interact and oppose each other. This is the mechanism predicted in Plate Tectonic studies for both collision to form mountains and subduction along trenches.

In order to compensate for opening the oceans, such as the Atlantic Ocean relative to the Pacific Ocean, a further option, as adopted during the 1960s, was to dispose of excess crusts by subduction. This proposal now forms the fundamental basis of Plate Tectonic theory, whereby excess crusts are conveniently disposed of within, for example, the Pacific Ocean via an inferred *conveyor-belt* motion of the continents driven by mantle convection. In the same east-west example replicated in Figure 17.7, both of the ocean basins open and excess seafloor crust is then consumed beneath the continents along subduction zones.

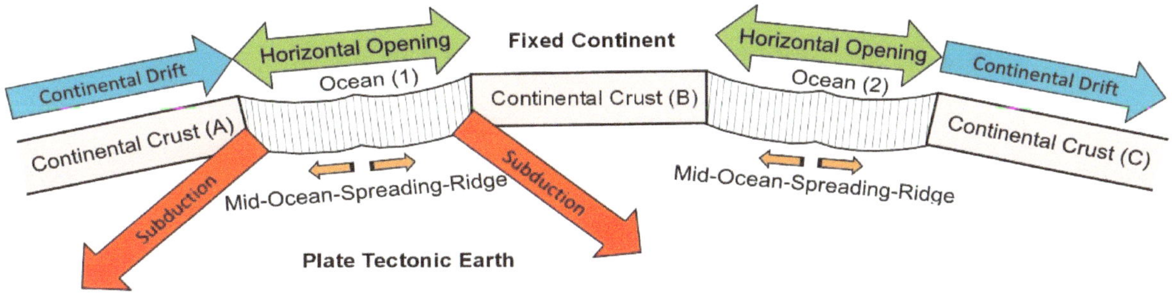

Figure 17.7 A cross section of a Plate Tectonic Earth showing horizontal plate motion vectors for a number of continental crusts plus two intervening ocean basins. In this example excess crusts must be disposed of by subduction. Crustal thickness not to scale.

Reflecting back to the published horizontal plate motion vectors shown in Figures 17.1 and 17.2, it can be seen that the two scenarios shown in Figures 17.6 and 17.7 accord with the published plate motion results. This is because calculating the horizontal motion vectors uses a Plate Tectonic crustal model in order to maintain a static radius Earth premise. The space geodetic derived vectors in each of these figures is therefore unique to a constant radius Plate Tectonic Earth model and cannot be reconciled on an Expansion Tectonic Earth. These same horizontal continental plate motion vectors on an Expansion Tectonic Earth are meaningless because they do not take into account vertical motion vectors relative to an increase in Earth radius.

On an Expansion Tectonic Earth the horizontal motion vectors are, in fact, the same as the seafloor spreading rates measured at each of the mid-ocean-ridge spreading centres. The issue with space geodetics then is not the actual mechanism behind the physical measurements, but the mathematical and statistical treatment of the raw data in order to constrain the data to a static radius Plate Tectonic Earth model.

18 Causal Mechanism

"...it may be fundamentally wrong to attempt to extrapolate the laws of physics as we know them today to times of the order of the age of the Earth and of the Universe." Creer, 1965

Discussion on the causal mechanism for Expansion Tectonics has been intentionally left until last, regardless of when this most fundamental of questions first arose. This was done to allow empirical evidence to be presented in a logical manner, without biasing outcomes of the evidence that needed to be presented. Unfortunately, most humans instinctively want to know or at least comprehend a cause well before acknowledging any physical evidence. This also touches on our innate and instinctive hesitation towards change where; *...it is better to err towards what we already know and are comfortable with rather than be seen to be different.*

Carey, on a number of occasions stated that he *"...may not necessarily be expected to know"* the cause of Earth expansion since the answer could only be expected to be known if all relevant fundamental physics were already known. I, myself, have been chastised at various conferences because I have stated quite honestly that, at the time, *"I did not know."* Similarly, as Cwojdzinski wrote to me in 2005, *"The insinuation that we still do not know a physical process responsible for an accelerated expansion of the Earth is not a scientific counter-argument."* He further maintained that, *"It is not a task of the geologist to explain problems beyond their discipline. Their task is to see and correctly explain all geological facts."*

It is also acknowledged here that any proposal for a causal mechanism of an increase in Earth radius will open all sorts of controversial criticism and intellectual backlash. This is acceptable, as it is consistent with what Arthur Schopenhauer said about science in general. *"All truth passes through three stages. First, it is ridiculed. Second, it is violently opposed. Third, it is accepted as being self-evident."*

This is all very well though, but at least an attempt must be made to assimilate the new ideas about Expansion Tectonics presented here into a working hypothesis, even if just to stimulate on-going research or intellectual backlash. Over time though *...true science insists that we must continually challenge new ideas until such time as the original concepts and ideas are indeed fully understood and they can be confidently accepted within the hypothesis as truth.* If time shows that these basic ideas are incorrect, true science also decrees that the particular reining hypothesis must be either modified, or rejected in favour of a hypothesis that better fits observation.

18.1 Historical Considerations

Concerning the physical cause of Earth expansion, in 1965 Creer maintained that *"...we should beware of rejecting the hypothesis of Earth expansion out of hand on grounds that no known sources of energy are adequate to explain the expansion process."* Creer further considered that *"...it may be fundamentally wrong to attempt to extrapolate the laws of physics as we know them today to times of the order of the age of the Earth and of the Universe."* This remains equally true today since historically, in the evolution of knowledge, empirical phenomena have often been recognized long before their cause or reason has been understood.

A classic example of this was 19th century spectroscopy. Today it is common knowledge that spectral lines are caused by electrons in an atom changing orbits. A photon is absorbed by an electron when the electron jumps to a higher orbit, and conversely, a photon is released when an electron jumps to a lower orbit. However, when spectral lines were first recognized at the beginning of the 1800s, the electron and the model of an atom was not known. Despite not knowing about electrons and atomic structure, spectroscopy flourished in the 1800s and was used for element determination and astronomical observations for a century before the electron was finally discovered in 1897 and the Bohr model of the atom was introduced in 1911.

During the 1950s and 1960s there was active debate amongst many scientists, in particular centred on the new discoveries in seafloor crustal geology and how they relate to global tectonics. In 1963, Egyed summarized hypotheses proffered for the cause of Earth expansion since the theory of Earth expansion first gained recognition in the 1800s, and in 1973 Wesson gave a comprehensive review of the cosmological implications of an Expansion Tectonic Earth. Carey in 1983 also provided a comprehensive list of authors who have contributed to the causes of Earth expansion.

It should be appreciated, however, that most, if not all, of the following historical considerations were proffered many decades ago. Furthermore, it is important to appreciate that these considerations were made well before completion of the bedrock geological map of the oceans and continents, where small Earth modelling studies presented here represent the first time that this mapping data has been used to accurately quantify and constrain an Earth expansion process.

Regardless of this, these early researchers recognized five main reoccurring themes for a causal mechanism of Earth expansion:

1. **A pulsating Earth:** This is where cyclical expansion of the Earth was considered to have opened the oceans and contractions formed the mountains. This proposal fails to satisfy an exponential increase in Earth expansion, and Carey considered the theme to have arisen from the *"false axiom"* that *"...orogenesis implies crustal contraction"* or collision to form mountains. Carey saw no compelling evidence for intermittent contractions of the Earth and this observation is now borne out by the symmetric seafloor spreading patterns as mapped in all of the modern oceans for the past 170 million years at least.

2. **Meteoric and asteroidal accretion:** This proposal was rejected by Carey as the primary cause of Earth expansion as the increase in Earth radius should decrease exponentially with time, not increase. Nor does this proposal explain the seafloor spreading patterns in each ocean. In order to maintain an increase in Earth radius of 22 millimetres per year this proposal would necessitate that the entire Earth, including both continents and oceans, be uniformly covered in meteoric dust and debris for many kilometres thick, which it clearly is not.

3. **Constant Earth mass, with phase changes to an originally super-dense core:** This was rejected by Carey as the main cause of Earth expansion because he considered the theme would imply too large a surface gravity throughout the Precambrian up until late in the Palaeozoic. Carey further considered that there is no evidence to suggest there was a large surface gravity in the past. Furthermore, the expansion of the Earth from an initial radius of 27 percent of its current radius to its current radius indicates a volume increase of the order 50 times the initial volume. Nowhere in the Solar System is a body with a mean density of this required magnitude, making it extremely unlikely there was ever a super-dense Earth.

4. **Secular reduction of the universal gravitation constant G:** Such a decline in G would cause expansion through the release of elastic compression energy throughout the entire Earth, resulting

in physical changes in the minerals and rocks to lower densities. Carey again rejected this proposal as the main cause of expansion for three reasons; that formerly surface gravity would have been unacceptably high during the Precambrian; that the magnitude of expansion from this proposal would have probably been too small; and the arguments for such reduction in G were considered not to indicate an exponential rate of increase in Earth radius.

5. **A cosmological cause involving a secular increase in the mass of the Earth:** This remains the most acceptable causal mechanism, however, where the additional new matter comes from to increase Earth mass has not been forthcoming until now.

Carey considered that each of these five proposals briefly summarised above are soundly based and may have contributed in part to some form of Earth expansion. He further considered that, "...because *of the limitations of surface gravity in the past there may be no alternative to an exponential increase of Earth mass with time.*" Because conventional science considers that the size of the Earth has remained essentially constant for all time, the question as to where any additional Earth mass may come from has never been required to be asked. Because it has never been required to be asked, none of these historical proposals have been taken seriously. Moving in full circle, this is also why Earth expansion continues to be shunned and nothing has been successfully proffered to encourage conventional science to change its stance.

A lot of new science and global evidence has since been gathered over the past decades to seriously challenge conventional Plate Tectonics and to offer a plausible cause for Earth expansion. In addition to the five main themes above, new themes for a causal mechanism have recently been proposed, including:

6. **Ionic mass expansion:** Where the atoms of various elements gain additional mass which then doubles the radius of the Earth in about 500 million years. This theory is based on the ionization potentials of various elements, with an initial ionization beginning the mass accumulation process. The theory primarily differs from other theories in that the new mass accumulates within existing elemental atoms and not by the addition of new atoms.

7. **Matter transfer from the Sun:** Where electrons and protons ejected from the Sun enter the Earth and recombine within the 200 to 300 kilometres thick D" region, which is located at the base of the mantle directly above the core-mantle boundary.

With all of this in mind, a causal mechanism presented here is based on a suggestion put forward by John Eichler in 2011. Eichler posed the question, "*Does plasma from the Sun cause the Earth to expand?*" and by presenting a new argument based on known physical phenomena, he suggested that this might indeed be the case. Eichler elaborated with his statement that:

"To assume that the Earth is gaining matter and that this may be due to nucleosynthesis within the Earth seems to fly in the face of conventional wisdom–and it does. Based on empirical geologic evidence which strongly indicates that Expansion Tectonics is indeed valid, the task confronted is to formulate a viable mechanism whereby this occurs. In a plasma universe, the Earth is under constant bombardment from space, with all the necessary components to reconstitute matter from its component parts deep within the Earth not requiring theoretical constructs which have never been experimentally observed. The Earth, having a magnetic field strong enough to interact with impinging particles, gathers more than sufficient fundamental particles, namely electrons and protons, to account for a slow increase in matter internally over hundreds of millions of years. There is therefore no lack of component particles to create new matter deep within the body of the Earth. The exact process by which this occurs is complex in nature and, like the interior of the Earth itself, involves speculation as to its dynamics. It is argued that the avenue of approach proposed here is plausible and warrants further serious scientific investigation. If new matter has been added to the interior of the Earth, there must be an answer to the riddle of the dynamics of the process."

This proposal also incorporates the observations of Kremp who, in 1992, suggested that new geophysical evidence indicates that the Earth has been growing rapidly in the past 200 million years. Kremp indicated that seismologists have located the existence of a zone, about 200 to 300 kilometres thick, located at the base of the mantle directly above the core-mantle boundary, designated the D" region. Yuen and Peltier in 1980, as well as Boss and Sacks in 1985, earlier postulated the existence of a substantial flow of heat across this core-mantle boundary and concluded that if whole-mantle convection–a Plate Tectonic requirement for convection within the mantle–were to occur in the Earth's mantle, this D" region should be the lowest thermal boundary layer of the whole-mantle system. With the temperature of the outer core of the core-mantle boundary estimated to be about 800 degrees higher than the D" layer of the mantle, or perhaps even 1500 degrees higher, Kremp concluded that this thermal increase in the outer core may be a fairly recent process forcing a rapid expansion of the Earth.

The following is a discourse on what is currently known about matter transfer from the Sun into the rest of the Solar System, and in particular its interaction with the Earth. My intent here is to present this discourse in order to stimulate awareness of a physical process impinging on the Earth for other, more qualified, researchers to then take it to the next level. That is, with so much interstellar material impinging on the Earth, why hasn't anyone taken the next logical step and asked, where is all this material going to and what is it doing to the Earth itself?

18.2 Solar System

The Sun is the Solar System's only star and is by far its main physical component. Its large mass, estimated to be 332,900 Earth masses, produces temperatures and densities in its core that are high enough to sustain nuclear fusion. This nuclear fusion releases enormous amounts of energy, mostly radiated into space as electromagnetic radiation peaking in the 400 to 700 nanometre band of visible light.

Along with light, the Sun radiates a continuous stream of plasma—known as the *solar wind*—which represents a stream of charged particles released from the upper atmosphere of the Sun consisting mostly of electrons, protons, and other ions. This stream of charged particles spreads outwards at roughly 1.5 million kilometres per hour, creating a tenuous atmosphere called the *heliosphere* that permeates the Solar System out to well beyond the orbit of Neptune. This atmosphere is also known as the *interplanetary medium.* Activity on the Sun's surface, including *solar flares* and *coronal mass ejections,* also disturb the heliosphere creating a form of space weather which in turn generates *geomagnetic storms.*

Coronal mass ejections and other similar events in turn produce a magnetic field and eject huge quantities of material from the surface of the Sun. The presence of the Earth's own magnetic field stops its atmosphere from being stripped away by the solar wind. This contrasts with Venus and Mars, for instance, that do not have well developed magnetic fields and as a result it is speculated that the solar wind has caused their atmospheres to gradually bleed away into space. The interaction of this magnetic field and ejected material with the Earth's magnetic field funnels charged particles into the Earth's upper atmosphere, where it forms *auroras*—a natural light display in the sky seen near both of the magnetic poles.

The presence of a continuous stream of particles flowing outward from the Sun was first suggested by British astronomer Richard C. Carrington. In 1859, Carrington, as well as Richard Hodgson, independently made the first observation of what would later be called a *solar flare*—a sudden outburst of energy from the Sun's atmosphere. On the day following this observed outburst, a geomagnetic storm was also observed on Earth and Carrington suspected that there might be a connection. George Fitzgerald later suggested that matter was being regularly accelerated away from the Sun which reached the Earth after several days.

In 1910 British astrophysicist Arthur Eddington suggested the existence of a solar wind, but the idea did not catch on at the time and it was generally ignored. He postulated that the ejected material consisted of electrons, while in his earlier studies he supposed them to be *ions*—an atom or molecule in which the total number of electrons is not equal to the total number of protons. The first person to suggest that they were both was Norwegian physicist Kristian Birkeland. Birkeland's geomagnetic surveys showed that auroral activity went on essentially uninterrupted. As these displays and other geomagnetic activity were being produced by particles from the Sun, he concluded that the Earth was being continually bombarded by *"...rays of electric corpuscles emitted by the Sun."* In 1916, Birkeland proposed that *"From a physical point of view it is most probable that solar rays are neither exclusively negative nor positive rays, but of both kinds."* In other words, the solar wind consists of both negative electrons and positive ions. Three years later in 1919, Frederick Lindemann suggested that these particles are both protons and electrons emitted from the Sun.

The total number of particles carried away from the Sun by the solar wind is now known to be about 1.3×10^{36} per second. From this, the total mass loss of the Sun each year is estimated to be about 4 to 6 billion tonnes per hour, or 35 to 53 trillion tonnes per year. This is equivalent to losing a mass equal to the Earth every 150 million years, or conversely providing a mechanism to double Earth radius in the past 250 million years.

18.2.1 Coronal Mass Ejection

The solar wind can be interrupted by large, fast-moving bursts of plasma called *interplanetary coronal mass ejections*. These interplanetary coronal mass ejections are the interplanetary manifestations of *solar coronal mass ejections*, which are caused by the release of magnetic energy at the Sun and are often simply called *solar storms*. These are sometimes, but not always associated with solar flares which are another manifestation of magnetic energy release at the Sun. These interplanetary coronal mass ejections cause shock waves in the thin plasma of the heliosphere, launching electromagnetic waves and accelerating particles—mostly protons and electrons—to form showers of ionizing radiation that precede the coronal mass ejections.

When a coronal mass ejection impacts on the Earth's magnetosphere (Figure 18.1) it deforms the Earth's magnetic field, changing the direction of compass needles and inducing large electrical ground currents in Earth itself. This is the result of a *geomagnetic storm* and is a global phenomenon. Coronal mass ejection impacts can induce magnetic reconnection in parts of the Earth's *magnetosphere*—the area of space around the Earth where charged particles are controlled by the magnetic field. This interaction launches protons and electrons downward toward the Earth's atmosphere where they form an aurora at both magnetic poles.

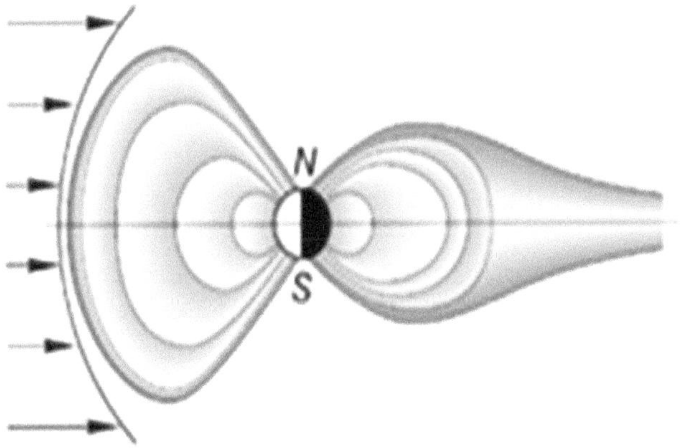

Figure 18.1 A typical cross-section through the Earth, shown at the centre, and the surrounding magnetosphere.

As the solar wind approaches a planet that has a well-developed magnetic field, such as Earth, Jupiter, and Saturn, the charged particles are now known to be deflected within the magnetosphere. The magnetosphere is shaped roughly like a hemisphere on the side facing the Sun and is drawn out in a long wake on the opposite side. The boundary of this region is called the *magnetopause* and some of the charged particles are able to penetrate the magnetosphere through this region by partial reconnection of the magnetic field lines.

The solar wind is responsible for the overall shape of the Earth's magnetosphere and fluctuations in its speed, density, direction, and entrained magnetic field strongly affect Earth's local space environment. For example, the levels of ionizing radiation and radio interference can vary by factors of hundreds to thousands. The shape and location of the magnetopause and the bow shock wave upstream of it can also change by several Earth radii exposing many satellites to the direct influence of the solar wind. These phenomena are collectively called space weather.

In 2000, four identical Cluster II satellites were launched to study the impact of the Sun's activity on the Earth's space environment by flying in formation around Earth. For the first time in space history, this mission was able to collect three-dimensional information on how the solar wind interacts with the magnetosphere and how it affects near-Earth space and its atmosphere, including auroras. This study was undertaken by the European Space Agency and results have shown that it is much easier for the solar wind to penetrate the magnetosphere than had previously been thought.

A group of these scientists directly observed the existence of certain waves in the solar wind that were not expected. These waves enabled incoming charged particles of solar wind to breach the magnetopause, suggesting that the magnetosphere responds more as a filter rather than a continuous barrier. These discoveries were considered by the European Space Agency's project scientists to be of great importance because it showed how the Earth's magnetosphere can be penetrated by solar particles under specific interplanetary magnetic field circumstances. The study also suggested that these waves may be a lot more common and possibly represents a means for the constant penetration of solar wind into terrestrial magnetospheres.

18.2.2 Plasmasphere

The *plasmasphere*, or inner magnetosphere, is a region of the Earth's magnetosphere consisting of cool, low-energy plasma and is located above the Earth's ionosphere. The outer boundary of the plasmasphere is known as the *plasmapause* which is defined by an order of magnitude drop in plasma density (Figure 18.2). The plasmasphere was discovered in 1963 by Don Carpenter. Traditionally, the plasmasphere was regarded as *well-behaved cold plasma* with particle motion dominated entirely by the geomagnetic field of the Earth and hence it generally co-rotates with the Earth.

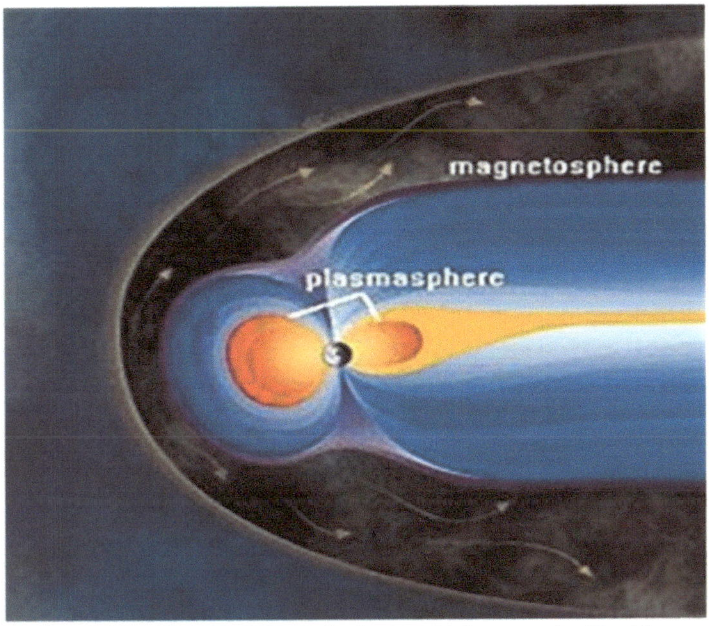

Figure 18.2 A typical cross-section through the Earth showing the Earth's magnetosphere and inner plasmasphere.

The most familiar portions of this region are the Van Allen radiation belts (Figure 18.3) which contain vast amounts of protons, ions, and electrons. The Van Allen belts circle the entire Earth equatorially at distances of 13,000 out to 60,000 kilometres from the surface of the Earth and hence greatly increase the particle cross-sectional capture area, making the Earth appear much larger to the influx of charged particles than the Earth's actual physical size.

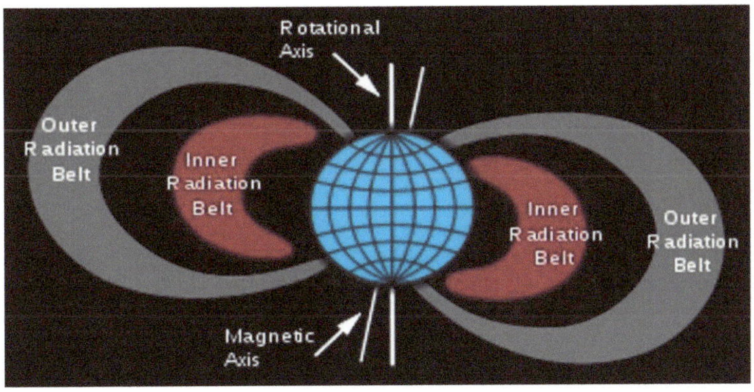

Figure 18.3 The inner and outer Van Allen radiation belts extending from altitudes of 13,000 to 60,000 kilometres above the Earth's surface. (https://en.wikipedia.org/wiki/Van_Allen_radiation_belt)

18.2.3 Ionosphere

The ionosphere is a shell of electrons and electrically charged atoms and molecules that surrounds the Earth closer than the Van Allen belts, stretching out from a height of about 50 kilometres to more than 1000 kilometres above the Earth. It owes its existence primarily to ultraviolet radiation from the Sun and contains plasma, whereby negative free electrons and positive ions are attracted to each other by an electrostatic force but are too energetic to stay fixed together as electrically neutral molecules.

Ultraviolet, X-Ray, and shorter wavelengths of solar radiation are ionizing, whereby photons at these frequencies contain sufficient energy to dislodge an electron from a neutral gas atom or molecule upon absorption. The reverse process to ionization is recombination, in which a free electron is captured by a positive ion and occurs spontaneously. Ionization depends primarily on the Sun and its activity. The amount of ionization in the ionosphere varies greatly with the amount of radiation received from the Sun. Thus, there is a diurnal effect and a seasonal effect.

The activity of the Sun is associated with the sunspot cycle, with more radiation occurring with more sunspots. Radiation received also varies with geographical location. There are also mechanisms that disturb the ionosphere and decrease the amount of ionization, such as solar flares and the associated release of charged particles into the solar wind which reach the Earth and interact with its geomagnetic field.

18.2.4 Plasma

In physics and chemistry, *plasma* is a state of matter similar to a gas in which portions of the particles are both ionized and moving in a stream analogous to an electric current. Heating a gas ionizes its molecules and atoms, thus turning it into a plasma containing charged particles—which are positive and negative ions and electrons. Ionization can also be induced by other means, such as a strong electromagnetic field applied with a laser or microwave generator and may also be accompanied by the dissociation of molecular bonds.

Plasmas are electrically conductive, so they respond strongly to electromagnetic fields. Plasma has properties quite unlike those of solids, liquids, or gases and is considered a distinct state of matter. Like a gas, plasma does not have a definite shape or a definite volume unless enclosed in a container. Unlike gas, under the influence of a magnetic field it may also form structures such as filaments, beams, and layers. Much understanding of plasmas has come from the pursuit of controlled nuclear fusion and fusion power, more commonly known as plasma physics.

Plasma is now known to be the most common state of ordinary matter, making up an estimated 99.99 percent of all matter in the universe. Plasma occurs mostly as rarefied *intergalactic plasma* as well as occurring within the Sun and stars. In the solar system, interplanetary space is filled with plasma from the solar wind which extends from the Sun out to the heliopause. Lightning is also an example of plasma present at the Earth's surface. Typically, a lightning discharge may release 30,000 amperes of energy at up to 100 million volts, and emits light, radio waves, X-rays, and even gamma rays.

Over the past half century, increased attention has been focused on the role of plasma in the formation of the universe. Magnetic fields generated by plasmas are millions of times more powerful than the force of gravity and are considered by many scientists to shape not only galaxies but also solar systems, including stars, planets, and satellites.

This new field is referred to as *plasma cosmology*. Because it is now known how widespread and important plasma is, based on satellite and space probe measurements since the start of the space-age, plasma may in fact be playing a major unrecognized role in the elusive search for a mechanism accounting for matter increase within the Earth. Furthermore the characteristics of plasma, as well as electron and proton conduction within the Earth itself, is based on empirical laboratory investigations rather than being reliant on theoretical constructs, many of which defy direct experimental verification. As Alfvén has pointed out, as noted by Bruce from PlamaCosmology.net, time after time the underlying assumptions of cosmologists today *"...are developed with the most sophisticated mathematical methods and it is only the plasma itself which does not 'understand' how beautiful the theories are and absolutely refuses to obey them."*

18.2.5 Aurora

An aurora—from the Latin word *aurora,* meaning dawn—is a natural light display in the sky particularly in the high latitude Arctic and Antarctic regions. This display is caused by the collision of energetic charged particles with atoms in the high altitude atmosphere. These charged particles originate in the magnetosphere from solar wind and are directed by the Earth's magnetic field into the atmosphere. Most auroras occur in a band known as the auroral zone. The auroral zone typically extends from 10° to 20° around the Earth's magnetic pole. During a geomagnetic storm, the auroral zone may also increase in area to lower latitudes.

Auroras are associated with the solar wind, a flow of electrons, protons, and ions continuously flowing outward from the Sun. The Earth's magnetic field traps these particles, many of which travel toward the poles where they are accelerated toward Earth. Collisions between these particles and atmospheric atoms and molecules cause energy releases in the form of auroras, appearing as large zones around the poles. Auroras are also more frequent and brighter during the intense phase of the Sun's solar cycle where coronal mass ejections increase the intensity of the solar wind.

A *diffuse aurora* is a featureless glow in the sky which may not be visible to the naked eye even on a dark night and defines the extent of the auroral zone. The more *discrete auroras* are sharply defined features within the diffuse aurora. These vary in brightness from just barely visible to the naked eye to bright enough to read a newspaper at night. Discrete auroras are usually observed only in the night sky because they are not as bright as the sunlit sky. Auroras may occasionally be seen pole-ward of the auroral zone as diffuse patches or arcs, which are generally invisible to the naked eye.

In northern latitudes, this effect is known as the *aurora borealis*—or northern lights—named after the Roman goddess of dawn, *Aurora*, and the Greek name for the north wind, *Boreas*. Auroras seen near the magnetic pole may be high overhead, but further away they illuminate the northern horizon as a greenish glow or sometimes a faint red. Discrete auroras often display magnetic field lines or curtain-like structures and can change within seconds or glow unchanging for hours, most often in fluorescent green.

Its southern counterpart, the *aurora australis*—or southern lights—has almost identical features to the aurora borealis and changes simultaneously with changes in the northern auroral zone. It is visible from high southern latitudes in Antarctica, South America, New Zealand, and Australia. Auroras also occur on other planets where astronomers have observed them close to the planet's magnetic poles.

18.3 Proposed Causal Mechanism

A proposed causal mechanism for an increase in Earth mass causing an exponential increase in Earth radius through time remains speculative, but is presented here in order to stimulate further discussion and research. What should be appreciated is that the Earth is now known to be constantly immersed in a solar wind, a rarefied flow of hot plasma emitted by the Sun in all directions, which is associated with the two-million-degree heat of the Sun's outermost layer. This solar wind usually reaches Earth with a velocity around 400 kilometres per second, with a density of around 5 ions per cubic centimetre. During magnetic storms, the flow of ions can be several times faster and the interplanetary magnetic field may also be much stronger.

The Earth's magnetosphere has been shown to be full of trapped plasma as the solar wind passes the Earth. This flow of plasma into the magnetosphere increases with increase in solar wind density and speed, as well as with increases in turbulence in the solar wind. The flow pattern of magnetospheric plasma is from the magnetotail toward the Earth, around the Earth and back into the solar wind through the magnetopause. In addition to moving perpendicular to the Earth's magnetic field, some magnetospheric plasma travels down along the Earth's magnetic field lines where it loses energy to the atmosphere within the auroral zones. New research now suggests that this process may be more common and possibly a constant means for the penetration of solar wind into the terrestrial magnetosphere.

Beyond the known observations about coronal mass ejection related plasma flow into the Earth, all else remains speculative once it enters the Earth. Again, as Eichler states, "*There is therefore no lack of component particles to create new matter deep within the body of the Earth. The exact process by which this occurs is complex in nature and, like the interior of the Earth itself, involves speculation as to its dynamics. It is argued that the avenue of approach proposed here is plausible and warrants further serious scientific investigation. If new matter has been added to the interior of the Earth, there must be an answer to the riddle of the dynamics of the process.*"

At this stage it is envisaged that electrons and protons enter the Earth and recombine within the 200 to 300 kilometres thick D" region, located at the base of the mantle directly above the core-mantle boundary. Kremp concluded in 1992 that, "*...this thermal increase in the outer core may be a fairly recent process forcing a rapid expansion of the Earth.*" New mass formation requires not only pure energy but the presence of both electrons and protons and science now knows that these are plentiful from the Sun. The recombination of charged electrons and protons within this D" region then provides a mechanism to form new matter. This matter in turn becomes the building blocks of all elements and mineral species present on and in the Earth including the new lava, water and gases now being expelled from the mid-ocean-ridge spreading centres in all of the ocean basins.

From an Expansion Tectonic perspective, this new matter accumulates within the D" region, located at the core-mantle interface. This increase in new matter results in an increase in Earth mass and volume which manifests itself as a swelling of the mantle. The increase in Earth volume and associated mantle swell is then transferred to the outer crust where it results in crustal extension, which is currently occurring as extension along the full length of the mid-ocean-rift zones. Extension within the mid-ocean-rift zones is accompanied by intrusion of new basaltic mantle-derived lava, along with expulsion of new water and gases. These in turn increase the surface area of all of the modern oceans in strict accordance with the seafloor mapping shown in the Geological Map of the World.

19 What the Earth has to Say

"...the past history of our globe must be explained by what can be seen to be happening now." Hutton, 1788

We have now finished looking at a range of available global tectonic data, albeit a brief overview, and it is time to put everything into perspective to see how this relates to what the Earth has to say in relation to current scientific thinking. In modern science it has become far too easy for scientists to become over complacent in their chosen field of research by focusing on the sophistication and complexity of their particular field and ignoring extended field observation. This over specialisation of modern science also prevents many scientists from venturing outside their chosen field of study, preferring instead to leave related fields of study to those deemed better informed.

A renowned professor once said to me that "*...in order to convince him that the Earth is expanding I must provide demonstrable evidence to support my claim.*" I have long since come to the conclusion that it doesn't really matter what this renowned professor, or anyone else says because ultimately it is what the Earth has to say that really counts.

In this book I have simply looked at modern, freely available geological, geophysical, geographical, and biogeographical evidence from a different perspective, a perspective that requires a reassessment of the fundamental premise that the size of the Earth has remained near-constant throughout time. I have done this to not only see what has happened to the Earth since it first formed, but to also see if science has, in fact, got the interpretation of this evidence wrong. If science has incorrectly interpreted this evidence, it may then suggest that conventional science has now reached *paradigm paralysis*—a stage where the inability or refusal to see beyond the current assumptions and models of thinking has set in.

To me, the evidence presented in this book does indeed show that the concept of an Expansion Tectonic Earth, whereby Earth radius increases exponentially with time, is viable and represents a demonstrable global tectonic process. Similarly, it has been shown that all geological, geophysical, geographical, and biogeographical evidence, when displayed on models of an Expansion Tectonic Earth, substantiate an Archaean-to-future Expansion

Tectonic process and far better explains this readily available physical phenomena.

Understanding these well documented phenomena on an Expansion Tectonic Earth model therefore presents a new tectonic perspective and hence represents a *paradigm shift* in scientific thinking. This new understanding also presents an opportunity to speculate—a process of thinking or meditating on a subject allowing a judgement or conclusion to be reached in order to stimulate and promote further alternative research opportunities.

The small amount of speculation presented in the following section on the Pre-Archaean Earth is generally beyond the scope of this book, but as Cwojdzinski said to me in 2005, *"It is not a task of the geologist to explain problems beyond their discipline. Their task is to see and correctly explain all geological facts."* With all due respect to all scientists though, it is acknowledged that in addition to what I have to say there are many other and potentially more authoritative views in existence today, in particular views regarding speculation of the pre-Archaean and Earth-Moon formation.

19.1 Pre-Archaean Earth-Moon

The 24 small Earth models presented in this book extend from the early-Archaean, some 4,000 million years ago, through to the present-day with one model extended to 5 million years into the future. Age dating studies suggest that during early-Archaean times the ancient Earth crusts were just starting to stabilise, implying that the first minerals were cooling and crystallising to form recognisable rocks. Similarly, age dating from Moon rocks has returned ages extending to approximately 4,500 million years old, which potentially extends knowledge about the Earth-Moon system back a further 500 million years. Because of the age constraints on Earth, prior to 4,000 million years ago there is no physical evidence in the Earth's rock-record to say what the Earth was like during these pre-Archaean times. There is enough subtle evidence preserved in the early Earth and Moon rock-record though to make some speculation, and this first section will now offer an alternative viewpoint as to how this evidence may fit with an Expansion tectonic Earth model.

Conventional cosmological studies consider that formation of the Earth was by accretion of interstellar debris during pre-Archaean times. Accretion was considered to be complete by the early-Archaean and, once stabilised, the size of the ancient Earth was assumed to be the same or very similar to the size of the present-day Earth. In order to account for the presence of continental crusts on a constant radius Earth, the ancient crusts were considered to have either fully covered the Earth by about 4,000 million years ago and have since been recycled through the mantle throughout geological time with no net change in crustal mass or, that the earliest continental crust nucleated as small localised patches and this juvenile continental crust has not been recycled back into the mantle.

Likewise, the prevailing hypothesis for formation of the Moon is that the Earth–Moon system formed as a result of a giant impact whereby a Mars-sized body hit the newly formed proto-Earth and blasted material into orbit around it. This ejected material subsequently accreted to form the Moon. Simulations of this event are said to show that most of the Moon material may have come from the impactor, not from the ancient Earth. However, more recent simulations suggest that more of the Moon may have come from the Earth rather than the impactor. It must be appreciated that both of these conventional Earth and Moon formation considerations are highly speculative and, as such, are open to further interpretation. It should also be appreciated that these conventional speculations are not necessarily relevant to an Expansion Tectonic Earth.

As far back as 1982, Shuldiner presented evidence from ancient *high-grade*—high temperature and pressure—metamorphic rocks to suggest that Earth surface temperatures during the pre- to early-Archaean were 300 ± 100 degrees Celsius hotter than what it is now. This implies there was no liquid water, no oceans, and no rain but probably some semblance of a primitive atmosphere. This is conceivable, because the bulk of the early-Archaean rocks surviving through to the present-day are granite and volcanic rocks, along with high-grade metamorphic rocks, which all require high temperatures to form. By inference, this may also extend to include the igneous and volcanic rocks on the Moon.

Also in 1982, Kröner estimated that the upper mantle temperatures on Earth during the late-Archaean to early-Proterozoic times were still up to 150 degrees Celsius hotter than the present. High surface and mantle temperatures during early-Archaean to Proterozoic times suggested that these temperatures were attributed to a higher mantle geothermal gradient still prevailing during these ancient times.

Evidence from glacial studies suggested to Eyles and Young in 1994 that by late-Archaean to early-Proterozoic times the Earth surface temperatures had also fallen significantly to maybe 30 degrees less than now. This is inferred to have enabled atmospheric gases to condense to form surface water and ice. These low Archaean temperatures were considered to be the result of an insulating effect caused by the development and stabilisation of a primitive crust in conjunction with a declining mantle geothermal gradient.

In contrast, small Earth modelling studies presented here have shown that the size of the early-Archaean Expansion Tectonic Earth was around 27 percent of the present Earth radius—a similar size to the present Moon. It is important to remember that, at that time the entire primordial Earth crust comprised granite and volcanic rocks with very little sedimentary rocks and no large oceans. It is further speculated from the crustal and mantle temperature studies that the pre-Archaean Expansion Tectonic Earth—that is times prior to 4,000 million years ago—may have been incandescent, in other words, hot enough to remain molten without crystallisation to form stable dateable rocks.

In order for the primordial Earth, and by inference the Moon, to be incandescent during pre-Archaean times a possible suggestion is that the Earth-Moon system was located much closer to the primitive young Sun than it is now. The very much reduced Earth-Moon size and mass existing at that time would insist that centrifugal forces and angular momentum would be vastly different to what they are now and hence this suggestion may be plausible. By Archaean to early-Proterozoic times the Earth-Moon system may have then moved sufficiently far away from the influence of the Sun's surface temperature and gravity to cool and form stable primitive crusts. Sufficiently far to also lower Earth-Moon surface temperatures enough to retain liquid water and ice on the primitive Earth surface while still retaining a high residual geothermal gradient in the mantle.

During the distant pre-Archaean times—an indeterminate time span of maybe billions of years—it is considered that the Earth-Moon system may have also been originally combined as a single molten planet. By combining the volume of the present Moon and the volume of the early-Archaean Earth, the pre-Archaean Earth-Moon system is readily calculated to have been approximately 2,100 kilometres radius prior to separation. At, or sometime prior to 4,500 million years ago, the Earth and Moon may have then separated. It is envisaged that this separation occurred as a result of gravitational instability in the more basaltic molten surface layer of the primitive incandescent Earth-Moon, possibly as a result of a high rotational velocity and centrifugal forces, forming a double planet. This mechanism is described in conventional literature as *fission of the Moon from the Earth's primitive basaltic crust through centrifugal forces occurring during a period of high angular momentum of the Earth–Moon system.* A high angular momentum and rotational velocity may also add support to my suggestion that the Earth-Moon system was originally located much closer to the primitive young Sun.

It is further suggested that this double planet scenario would have originally comprised two bodies, the Earth and the Moon, of equal size and in close proximity co-rotating around each other as they do today. Currently, the Earth and Moon are continuing to move further apart and are now separating at a measured rate of 38 millimetres per year. This more passive separation process, in contrast to a speculative impact event, may also go a long way to explain why the Moon is currently in synchronous, albeit retrograde rotation around the Earth, always showing the same face to the Earth.

This suggestion is further supported by research in 2001, where a team at the Carnegie Institute of Washington reported precise measurement of the *isotopic signatures*—variants of a particular chemical element—of lunar rocks. To their surprise, the team found that the lunar rocks gathered during the Apollo program carried an isotopic signature that was identical with rocks from Earth and were similarly different from almost all other bodies in the Solar System. In 2012, further analysis of titanium isotopes from surface lunar samples also indicated that the Moon has the same composition as the Earth. This conflicts strongly with the current speculation that the precursor to the Moon originated far from Earth's orbit and was formed as a result of a giant impact between Earth and a Mars-sized body.

This is a significant point to take note of because not only is it telling us that the Moon did not originate from debris left over after a giant impact between Earth and a Mars-sized body, it is also saying that the Moon and Earth may not have been involved in later heavy bombardment from asteroids or comets colliding with the lunar surface. Debris from foreign asteroids or comets would have been readily detected in the research carried out by the Carnegie Institute by showing a distinctly separate isotopic signature, which it clearly hasn't.

It is here envisaged that the original molten surface basaltic material that separated from the Earth-Moon body to form the primitive Moon was somewhat cooler than the remnant core-mantle material left over to form the Earth. The Moon is now known to have a distinct crust, mantle, and core, and Moon rock samples collected from the surface confirm the composition is iron rich, similar to the seafloor crust and ancient volcanic rocks on Earth. The inner core of the present Moon is known to be small, about 20 percent the size of the Moon, in contrast to about 50 percent estimated for most other terrestrial bodies. This small core and mantle would also offer an explanation as to why the Moon has a limited surface gravity and magnetic field and why there is no atmosphere or hydrosphere.

During the Earth-Moon separation process it is suggested that the hotter primitive core-mantle, as well as underlying remnants of the surface layer of the former Earth-Moon body, remained intact, forming the primordial Earth (Figure 19.1). After gradual separation of the Moon, it is envisaged that there would have been a period of time when remnant spatter of fragments left over from the separation process would have bombarded both bodies, forming the now familiar cratering on the near side of the Moon's surface. Much cooler temperature conditions on the Moon may also explain why the primitive Moon cooled and crystallised 500 million years earlier than the primitive Earth, thus preserving this period of surface bombardment and cratering.

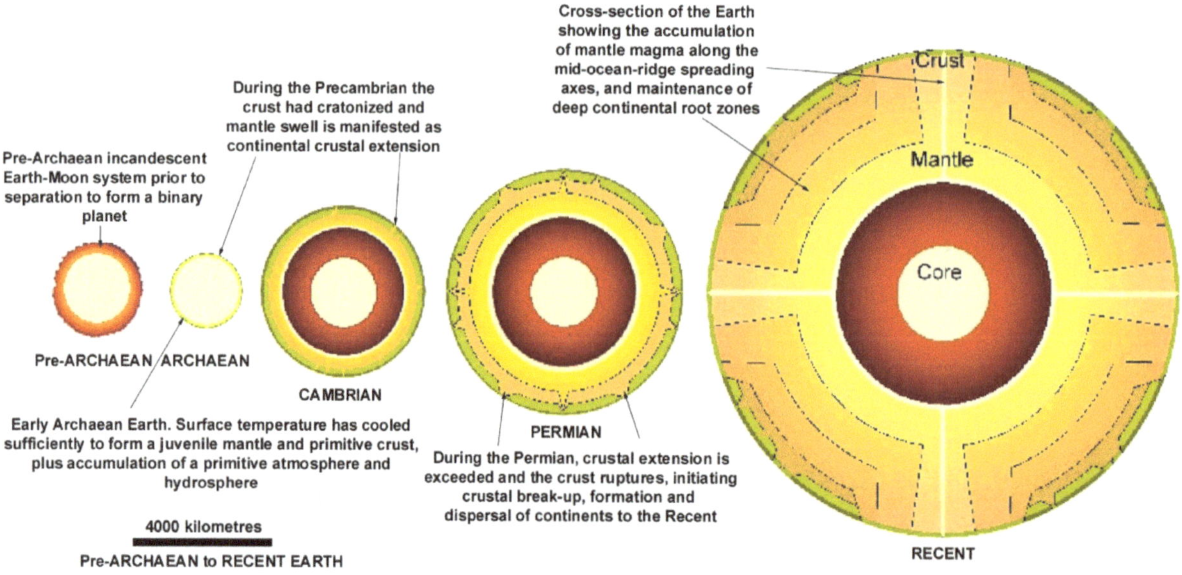

Figure 19.1 Proposed development of the Earth's core, mantle, and crust followed by continental break-up and dispersal throughout Earth history, extending in time from the pre-Archaean to the present-day.

Once the Earth-Moon body separated, it is further envisaged that the remaining molten core-mantle material forming the primitive Earth began to differentiate, whereby the molten elements and minerals separated into a distinct core, mantle, and crust. Over time, surface temperatures then gradually cooled sufficiently to form a new primitive silica-rich outer crust. On Earth, the expulsion of fluids and gases from the mantle similarly began to accumulate to form shallow seas and a primitive atmosphere with an accompanying reduction in geothermal temperatures over time.

The schematic representation of cross sections of the Earth extending from the pre-Archaean to present-day (Figure 19.1) shows the proposed development of the Earth's core, mantle, and crust followed by continental break-up and dispersal throughout the later part of Earth history. Once again it is reiterated that this section on the pre-Archaean Earth-Moon system is purely speculative and scientists don't really know for sure what happened during that time. It is beyond the scope of this book to delve deeper into speculation on this matter and is simply offered here as an alternative consideration.

19.2 Primitive Atmosphere and Hydrosphere

Fundamental to the concept of Expansion Tectonics is the premise that the ocean waters and atmospheric gases have been accumulating throughout much of geologic time in sympathy with the formation of ancient supercontinental crusts and new seafloor volcanic crusts. It was considered by Bailey and Stewart in 1983 that "*...for an Earth undergoing expansion with time, the bulk of the oceans would have to be outgassed since the Palaeozoic, requiring fundamental changes in atmosphere, climate, biology, sedimentology and volcanology.*" It was also considered by Carey in 1988 that, "*as the generation of the ocean floors depends fundamentally on the out gassing of juvenile water, it would therefore be expected that the volume of seawater* [and atmospheric gases] *and capacity of the ocean basins both increased, but not necessarily precisely in phase, in a related way.*"

The primitive atmosphere and *hydrosphere*—the combined mass of water found on Earth—was considered by Lambert in 1982 to have been formed largely from elements and molecules degassed from the Earth's interior and subsequently modified by physical, chemical, and biological processes. Rubey proposed as early as 1975 that *degassing*—the removal of dissolved gases from liquids [inclusive of molten magma]—has been a continuous or recurrent process, which is still occurring today. Rubey further suggested that "*the whole of the waters of the oceans have been exhaled from the interior of the Earth, not as a primordial process, but slowly, progressively and continuously throughout geological time.*" This degassing process is analogous to opening a bottle or can of soft drink or carbonated water. At cool temperatures or elevated pressures, carbon dioxide is dissolved in the water and is invisible, but when the container is opened the carbon dioxide bubbles out of solution and degasses.

Studies of melted igneous rocks carried out since the 1970s and 1980s have shown that the solubility of water in melted rocks increases with increasing pressure and temperature until a maximum value is reached in the mantle. Quoted examples range from 14 to 21 percent by weight of water dissolved in volcanic rocks at temperatures varying between 1,000 to 1,200 degrees Celsius and high pressures. For silica-rich magmas, carbon dioxide was also shown to be readily dissolved, in particular under high pressures. It was concluded from these studies that if water and carbon dioxide were available, they would both be highly soluble in the magmas normally generated in the upper mantle.

Eggler in 1987 further considered that the elements carbon, oxygen, hydrogen, and sulphur would all exist in the Earth's mantle within *volatile-bearing* minerals—minerals capable of retaining these elements in their crystal lattice. It was also considered possible that solution of elements in crystalline minerals represents a significant repository for each of these elements within the Earth's mantle and crust. These volatile-bearing minerals occur in most volcanic and metamorphic rocks, as well as in carbonate rocks and sulphide ores. Similarly, each of the volatile minerals such as water, carbon dioxide, carbon monoxide, methane, hydrogen, sulphur dioxide, and hydrogen sulphide are soluble in silica-rich melted rocks, with water and methane being more soluble than the rest.

The main controls on the chemistry of the Archaean hydrosphere were considered by Lambert to be reactions with hot igneous rocks, seafloor weathering, and additions of new water degassed from the crust and mantle, with minor input from river systems. Oxygen isotope studies from Archaean sediments suggest that fluids were in existence at that time which had a similar composition to modern seawater, with prevailing seawater temperatures of up to 70 degrees Celsius.

One of the most important events occurring during the Proterozoic Era was the accumulation of oxygen in the Earth's atmosphere. Though oxygen may have been released by photosynthesis well back in Archaean times, it was unable to build up to any significant degree until *chemical sinks*—the presence of unoxidized sulphur and iron—had been filled. Until about 2,300 million years ago, it has been estimated by others that oxygen was probably only 1 percent to 2 percent of its current level today. Banded iron formation rocks, which now provide most of the world's iron ore, were also a prominent chemical sink where most iron accumulation ceased after 1,900 million years ago. Red beds, which are sediments coloured by the iron oxide mineral hematite, indicate an increase in atmospheric oxygen after 2,000 million years ago as they are not found in older rocks. This oxygen build-up is considered in the literature to be due to two main factors: a filling of the chemical sinks, and an increase in carbon burial, which sequestered organic compounds that may have otherwise been oxidized by the atmosphere.

On an Expansion Tectonic Earth, it is suggested that degassing of the dissolved volatile elements and minerals from a primitive mantle and crust commenced during early-Archaean times. This commenced once the primitive crust had cooled sufficiently to start crystallising and expelling excess volatile elements—a process known as *devolatilization,* or simply *degassing.* It is envisaged that, once degassed, these elements would have been retained as water-rich gases in the atmosphere until such time as temperatures had cooled sufficiently to retain liquids on the surface of the Earth. Evidence suggests that a reduced atmosphere was well established during Archaean times comprising mainly water and carbon dioxide, with lesser carbon monoxide, nitrogen, sulphuric acid, sulphur dioxide, hydrochloric acid, and small amounts of nitric acid and carbonic acid. The absence of an oxygen source implied to Lambert that oxygen was not a stable component of the early atmosphere but instead was generated later by secondary processes, such as photosynthesis and microbial action.

It is worth considering that at an average modern ocean depth of 3,800 metres, the volume of the ocean waters represents about 40 percent of the estimated total volume of the seafloor crusts. When water from the underlying mantle and the volume of the ancient continental seas are also taken into consideration, this volume of ocean water is then well within the realms of degassing the 14 to 21 percent by weight of water dissolved in volcanic rocks from primary mantle melts during formation of the seafloor crusts, as required on an Expansion Tectonic Earth.

19.3 Archaean Crust-Mantle

The Archaean extends in time from around 4,000 to 2,500 million years ago and commenced after the Earth-Moon separation. On an Expansion Tectonic Earth (Figure 19.2) this represents an era where surface temperatures were becoming sufficiently low to allow crystallisation and solidification of a thin crust over a mantle that still retained a relatively high temperature gradient. As cooling progressed, the early crust thickened and extended in area as primitive volcanism became widespread. Volcanic activity covered the entire primordial Earth surface and this activity was associated with a complex system of mantle plumes, where hot volcanic rock was rising through the Earth's mantle to form the Earth's crust.

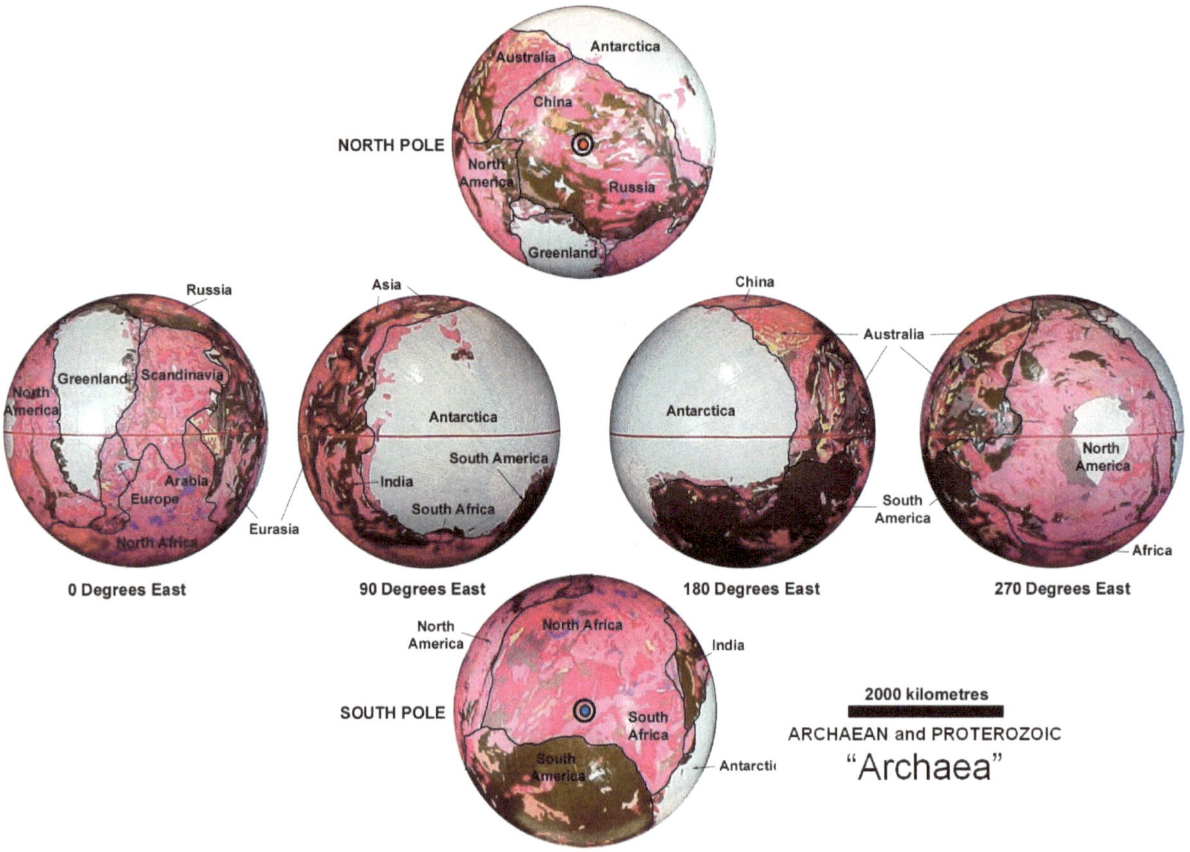

Figure 19.2 Primordial Archaean to Proterozoic Expansion Tectonic small Earth model. Cratons are shown as pink and red, Proterozoic basement rocks are shown as khaki. Outlines of the remnant present-day ancient crusts are shown as black lines. (Present-day Antarctic and Greenland ice-sheets are shown as pale blue areas).

During this process, hot gases began degassing from the mantle to form a primitive atmosphere while later, once surface temperatures had lowered sufficiently, condensed fluids started to pool in low-lying depressions.

The primitive early-Archaean crust shown in Figure 19.2 enclosed the entire ancient Earth at a radius of approximately 1,700 kilometres. It is suggested that *differentiation*—a chemical processes whereby molten magmas undergo bulk chemical change—of a molten pre-Archaean Earth magma separated to form a distinct core and a mantle prior to crystallisation to form a primitive surface crust. The presently known Archaean geological history commenced around 4,000 million years ago with the stabilisation and preservation of a thin outer primitive crust prior to onset of Earth expansion and fragmentation of the crust during mid- to late-Archaean times.

The formation of localised stable patches of juvenile crusts, as suggested by Moorbath in 1982, is consistent with an early-Archaean Expansion Tectonic Earth model. During the early-Archaean, as the primitive crust progressively cooled and crystallised, these patches of juvenile crusts would have then eventually merged and enclosed the entire primordial Earth with stabilised crust. Because of the small size of the primordial Earth, this event was probably *globally synchronous*—that is, it was occurring at the same time all over the primordial Earth. With this in mind the following proposal for crust-mantle evolution and development of volcanic, granite, and metamorphic crusts during the Archaean is based on a similar, albeit conventional proposal put forward by Kröner in 1982.

After the Earth-Moon separation and prior to the early-Archaean cooling and crystallisation phase, the primordial Earth was molten and volcanically turbulent. At that time patches of primitive juvenile crusts may not have survived the extensive volcanic *plume activity*—upwelling of hot rock within the Earth's mantle—and may have been, in part, re-absorbed back into the mantle, further reducing mantle temperatures in the process. By about 4,000 million years ago, mantle temperatures and heat flow had decreased sufficiently to maintain a volcanic lava-dominated crust, allowing the patches of juvenile volcanic rocks to rapidly increase in height, area, and thickness. Growth of the volcanic crust may have also occurred through chemical changes and crystallisation of the underlying upper mantle, giving rise to further cooling, thickening, and stabilisation of the overlying crusts. In contrast, regions away from the main volcanic centres may have maintained a relatively thin upper crust, which may have ultimately formed the observed loci for on-going crustal weakness, crustal extension, and basin formation during later phases of Earth expansion.

During early-Archaean times, the primitive crust was made up of predominantly silica and alumina-rich rocks comprising a mixture of intruded granites and an assortment of iron and magnesia-rich volcanic rocks. Once finally cooled and stabilised, the surface of the primordial Earth was then exposed to erosion by wind and waters, in particular erosion by hot juvenile waters from volcanic eruptions. This erosion process initially formed what are referred to as *dirty sediments* which were washed away and deposited in low lying sedimentary basins. Initially, these dirty sediments may have comprised volcanic ash and rock debris and later included chemically weathered and eroded sediment derived from the action of the waters on the exposed rocks. The early sedimentary basins were initially small and isolated, becoming more extensive by about 3,800 million years ago. This early crust now forms the oldest preserved sedimentary crustal remains known on Earth today.

The formation of silica and alumina-rich granite and related volcanic rocks from a dense iron-magnesia-rich upper mantle source suggested to Lambert that the primordial molten crustal material may have initially formed as a surface layer of not more than two to three kilometres thick. Lambert considered that the high thermal gradient existing during pre- to early-Archaean times suggested that the crustal material would have continued to segregate to also form a lower region of iron-magnesia-rich rocks extending from depths of 30 to 100 kilometres thick. The present-day survival of a 30 to 40 kilometre thick Archaean crust further implied the presence of a thick stable sub-crustal zone, which may have ultimately extended still further to greater than 200 kilometres thick.

Kröner envisaged that by about 3,600 million years ago continental crust had attained sufficient thickness and rigidity to support large sedimentary basins. On an Expansion Tectonic Earth, once cooled, this continental crust was subjected to the initial phases of crustal extension and fracturing during early Earth expansion. The magnitude of the increase in Earth radius and crustal extension during these times has been shown to be of the order of microns per year. Over a very long period of time—hundreds of millions of years—this small crustal extension process ultimately gave rise to crustal fracture, fissuring, and rifting, generating large sedimentary basins or sharply bounded, fault controlled landscapes. These low-lying basin areas started to collect water, shallow-water sediments, and extruded volcanic lava originating from the upper mantle. A small proportion of this volcanic lava reached the surface as magnesia-rich lava, which are now shown to be unique to the high mantle temperature conditions existing during Archaean times. A much larger volume of magma was also retained near the base of the crust to become the source for later volcanic eruptions and granite intrusives.

It is envisaged that on-going crustal extension of the stabilised Archaean crust during onset of Earth expansion resulted in early crustal fracturing and initiation of a global network of crustal weakness. Faulting and rifting—crustal extension—was focused within this network and was accompanied by deposition and accumulation of sequences of volcanic lava and eroded sedimentary rocks. The Archaean is also renowned for the presence of *high-grade metamorphic terranes*—high temperature and pressure metamorphic rocks—as preserved throughout the ancient crusts of today. These terranes represent largely pre-volcanic silica and alumina-rich rocks, together with early sedimentary deposits and granites produced during early stages of crustal fissuring and basin formation. Crustal extension and sub-crustal mantle movements during on-going increases in Earth radius and surface area then provides a mechanism for stretching and extension of these rocks within the Archaean crust. Crustal extension, accompanied by high mantle temperatures also provided the necessary heat-flow for recrystallization of these high-grade gneissic terranes.

The Archaean Eon extended over an extremely long period of time and it has been shown that expansion of the Earth during this time was less than the thickness of a human hair per year. It is envisaged that over this 1,500 million year history the crustal rocks were fractured, intruded by volcanic lava, eroded during weathering, and sediments were deposited in low-lying basins, all of which were complexly contorted and folded as a result of the steadily changing surface curvature. Because of the extremely long time involved, as well as the minute rate of expansion per year, it is further envisaged that on-going erosion of the surface of the entire Earth maintained a very flat landscape throughout this eon. In other words, erosion kept pace with changing surface curvature. This landscape may have been barren, rocky, and inhospitable by present-day standards.

The latter phase of Archaean basin development eventually gave rise to formation of very large, stable sedimentary platforms and shallow seas with a low elevation contrast between the continents and seas. Continental sedimentary input during this time was limited to mainly chemically precipitated sediments, such as calcium and silica-rich chert and banded iron formation rocks, with lesser fine-grained sediments. These rocks were common during mid- to late-Archaean times and formed laterally extensive deposits that are now known throughout most of the ancient crust on the present-day Earth.

During this Archaean Eon the ancient magnetic poles were located in what are now Mongolia-China and west Central Africa, and the ancient equator passed through North America, the Baltic region, Antarctica, and Australia. There was also a well-developed climate zonation established, which persisted for the entire duration of the eon. It is significant that primitive microbial life forms first evolved in the seas during this eon, possibly evolving from within carbon-rich sediments formed during the early anoxic atmosphere and seawater phase, prior to onset of an oxygen-rich atmosphere.

19.4 Proterozoic Earth

The Proterozoic Eon extended in time from around 2,500 to 541 million years ago (Figure 19.3). This time span represents another extremely long eon. The eon was characterised by comparative crustal stability dominated by stable exposed continental lands, extensive sedimentary basins, and shallow seas with slowly increasing changes in surface curvature.

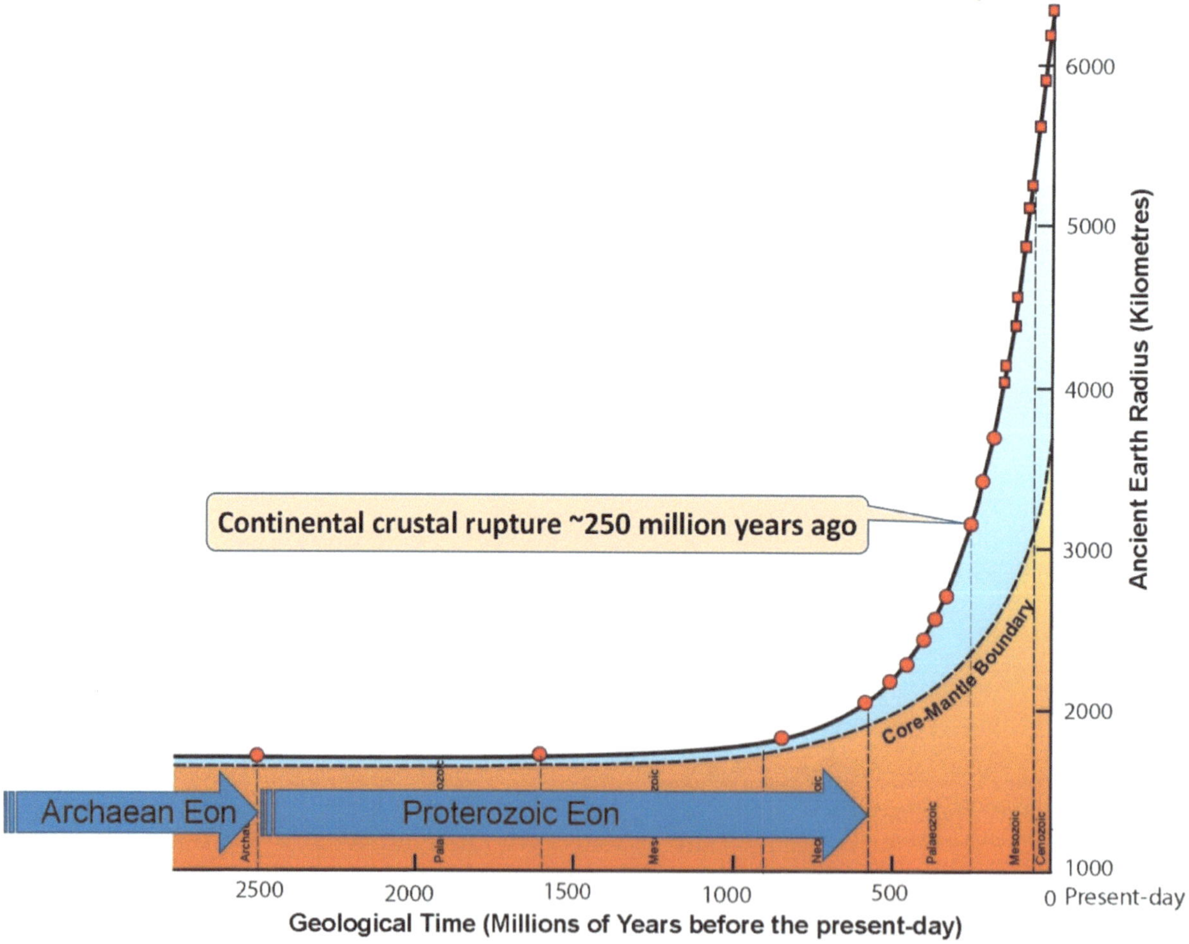

Figure 19.3 Expansion Tectonic Earth eons and eras in relation to increase in Earth radius.

The change in Earth radius throughout the Archaean to mid-Proterozoic times amounted to approximately 60 kilometres increase in radius spread over a 3,000 million year history, and around 400 kilometres increase in Earth circumference. The increase in circumference during that time was reflected by the steadily increasing surface areas of the sedimentary basins surrounding each of the ancient cratons.

On a Proterozoic Expansion Tectonic Earth there was not a lot of difference between late-Archaean and Proterozoic times. The location of the poles and equator were essentially the same, and the distribution of the ancient exposed supercontinental lands was also much the same. The distinguishing feature of the Proterozoic history was that increasing changes in surface curvature were slowly starting to take effect by increasingly elevating the exposed lands. Like the Archaean, the rate of change in surface curvature during the Proterozoic was still small enough for erosion to continually maintain a relatively flat landscape. In time this increasing elevation contrast began to slowly increase the rate of erosion of the exposed lands, which also influenced changes to the distribution of seas and corresponding coastal outlines of the established supercontinental lands. These changes to the coastal outlines then formalised the existence and configuration of both the ancient seas and the ancient Rodinia supercontinent (Figure 19.4).

After the mid-Proterozoic—about 1,700 million years ago—the steadily increasing Earth radius and crustal conditions was reflected in the gradual onset of increasing surface curvature and increasing elevation contrast between the lands and seas. This increasing elevation contrast gave rise to increasing changes to and distribution of sediments deposited in the established network of shallow seas during these times, changing from predominantly chemically precipitated sediments to predominantly courser grained sands and silts. The courser grain-size of these sediments, in turn, was a reflection of the increased erosional energy available to both erode and move these sediments, which, in turn, is confirmation of the increased elevation contrast of the land surface during these times.

Another distinguishing feature of Proterozoic times was the changing atmospheric conditions. These conditions gradually changed from chemically acidic atmospheric conditions throughout much of the earlier Archaean, through to the accumulation of atmospheric oxygen during the Proterozoic. This transition was marked in the rock-record by the increasing accumulation of iron oxide-rich banded iron formation, calcium and magnesium carbonate rocks, and haematite-rich red beds during this time. It is significant to note that these rock types are now found within most of the ancient terranes located on the present-day Earth and they signify globally synchronous conditions existing throughout Archaean and Proterozoic times. This transition also coincided with development and preservation of the earliest life forms, soon to explode in diversity during the following Phanerozoic Eon.

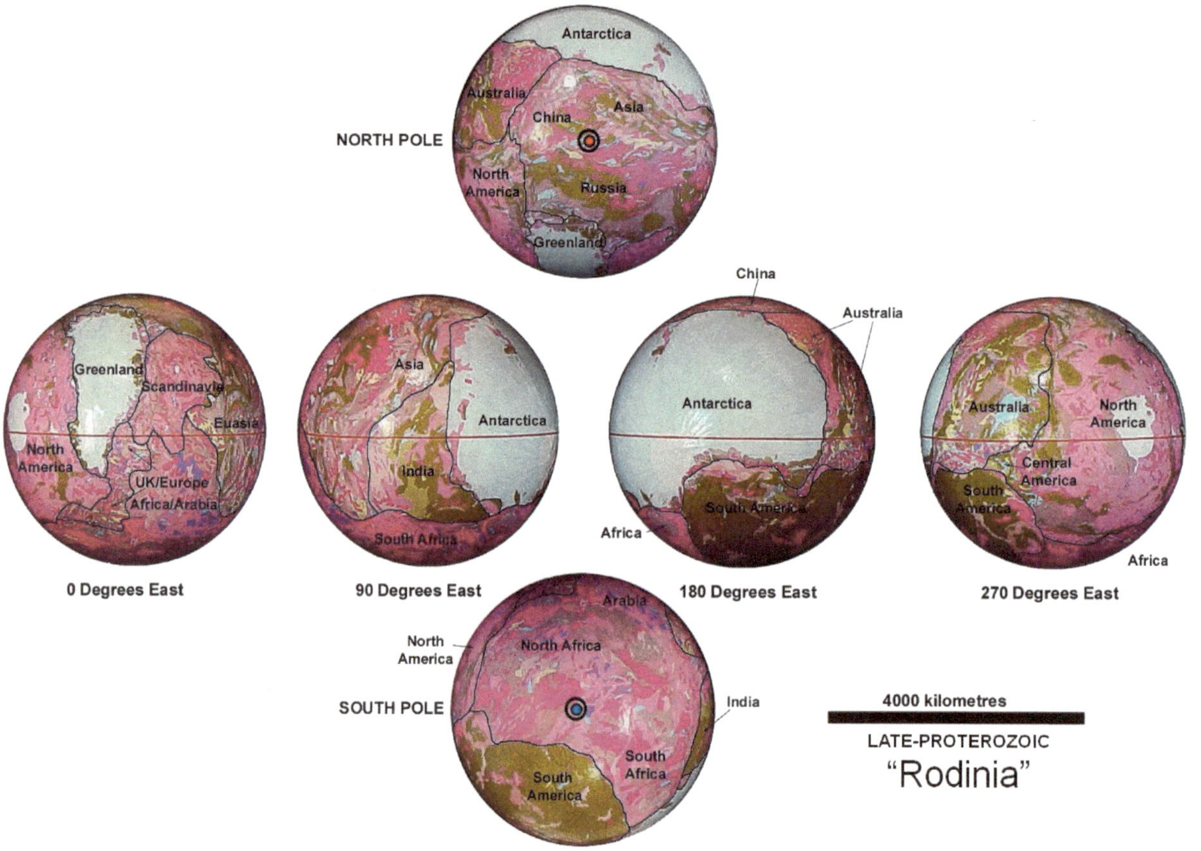

Figure 19.4 late-Proterozoic Rodinia Expansion Tectonic supercontinent. The model shows the distribution of Precambrian cratons (pink and red) inferred to represent exposed land, and a network of basins (khaki and brown) inferred to represent continental seas. The black lines represent outlines of remnant present-day continents.

Throughout the Proterozoic Eon, on an Expansion Tectonic Earth the ancient North and South Poles continued to be located within what is now northern China-Mongolia and West Africa respectively. The ancient equator continued to pass through what is now North America, East Antarctica, Australia, Greenland, and Scandinavia, as well as Europe, Asia, and India, essentially the same as during Archaean times. This continuity of magnetic pole locations, as well as continuity of the equator and associated climate zonation is further testament to the extremely long period of crustal stability during the Rodinia supercontinental configuration as reflected in the Expansion Tectonic small earth models.

On subsequent, younger Expansion Tectonic small Earth models this early Rodinia supercontinental assemblage represents a precursor to the better known Gondwana supercontinent and ultimately to the Pangaea supercontinental configurations. This transition from one supercontinent to the next was intimately related to the variation in coastal outlines during change in sea levels and the passive nature of this transition is unique to an Expansion Tectonic Earth. On an Expansion Tectonic Earth it was only Pangaea that eventually broke up and dispersed to form the modern continents and modern oceans at around 250 million years ago.

Tectonism during the Proterozoic was essentially confined to the established global network of crustal weakness, which was coincident with the network of sedimentary basins. This increasing tectonism was intimately related to increases in surface curvature and was marked by increased jostling of the ancient cratons. Tectonism was also accompanied by on-going crustal extension within the sedimentary basins. This tectonism generated long, narrow, elongate orogenic zones and may have also been accompanied by renewed volcanic activity and granite intrusion. Each of these Proterozoic orogenic zones coincided with the established network of sedimentary basins and in turn represent a global network of crustal weakness.

Proterozoic crustal extension during Earth expansion was confined to this network of sedimentary basins and was also accompanied by high heat flow from the mantle. Multiple phases of crustal extension-mobility-extension are a prime feature of Expansion Tectonic Earth models, continuing through to Palaeozoic times, prior to continental crustal rupture during the early-Permian and continental break-up and dispersal during the Mesozoic and Cenozoic to the present-day.

19.5 Palaeozoic Earth

The Palaeozoic Era represents the early part of the Phanerozoic Eon and extends in time from about 541 to 252.6 million years ago. The boundary between the Proterozoic Eon and the Palaeozoic Era is located at the base of the Cambrian Period, coinciding with when fossils of hard body marine animal species became noticeably abundant. Compared to the preceding Precambrian times, the Palaeozoic was a time of dramatic geological, climatic, and species evolutionary change. The Cambrian Period in particular witnessed the most rapid and widespread diversification of life in Earth's history in which most modern life forms, including fish, arthropods, amphibians, and reptiles first appeared and evolved. Life during that time began in the ocean but eventually extended onto land, and, by the late-Palaeozoic, great forests of primitive plants covered the continents, many of which formed the coal beds of Europe, Australia, and North America. Towards the end of this era, large reptiles became dominant and modern plants also appeared.

On an Expansion Tectonic Earth the transition from late-Proterozoic to the Palaeozoic represents an accelerating phase of increasing Earth radius along with accelerating increases in surface area and surface curvature. During the Palaeozoic distinct elevated ancient land surfaces were becoming prevalent and rivers were actively carving valleys and modifying landscapes. Formation of soils during these times was still the result of physical and chemical weathering of the rocks, rather than soils with organic matter that we are now more familiar with. These land surfaces were, in turn, surrounded by a network of equally distinct, relatively shallow continental seas, which, in turn, defined the outline of the ancient supercontinents.

Gondwana (Figure 19.5) at that time was subdivided into a North and South Gondwana, separated in part by an early Panthalassa Sea—the precursor to the modern Pacific Ocean. North Gondwana was made up of Australia, East and West Antarctica, and India, and also possibly included Tibet and Afghanistan. South Gondwana, joined at Madagascar, comprised Africa, Arabia, and South America. The ancient Laurentia and Baltica sub-continents were made up of North America, Greenland, and Scandinavia, as well as smaller islands making up England and Ireland. Similarly, the ancient Laurussia sub-continent was centred on the Precambrian regions of Mongolia and northern Russia.

These exposed Gondwana land surfaces were surrounded by the ancient continental Tethys, Iapetus, and Panthalassa Seas. Remnants of these seas are now preserved and represented by many of the ancient sedimentary basins that are located in Eastern Australia, North and South America, Europe, Asia, and Africa. These continental seas continued to maintain a global network and their outline and distribution were dictated by the changing coastal outlines and sea-levels. Accelerating changes to surface curvature gave rise to a marked increase in erosion of the lands and deposition of new sediments within the network of sedimentary basins. The increasing changes in surface curvature and super-elevation of the lands also initiated localised compression of the sedimentary basins to form long linear fold mountain belts, further disrupting established seaways. The presence of mountain glaciers and ice sheets during a number of Palaeozoic glacial events is testament to the marked increase in land surface elevation contrast. The presence of marine glacial debris deposited within equatorial regions is also testament to the presence of continental seas extending into polar-regions.

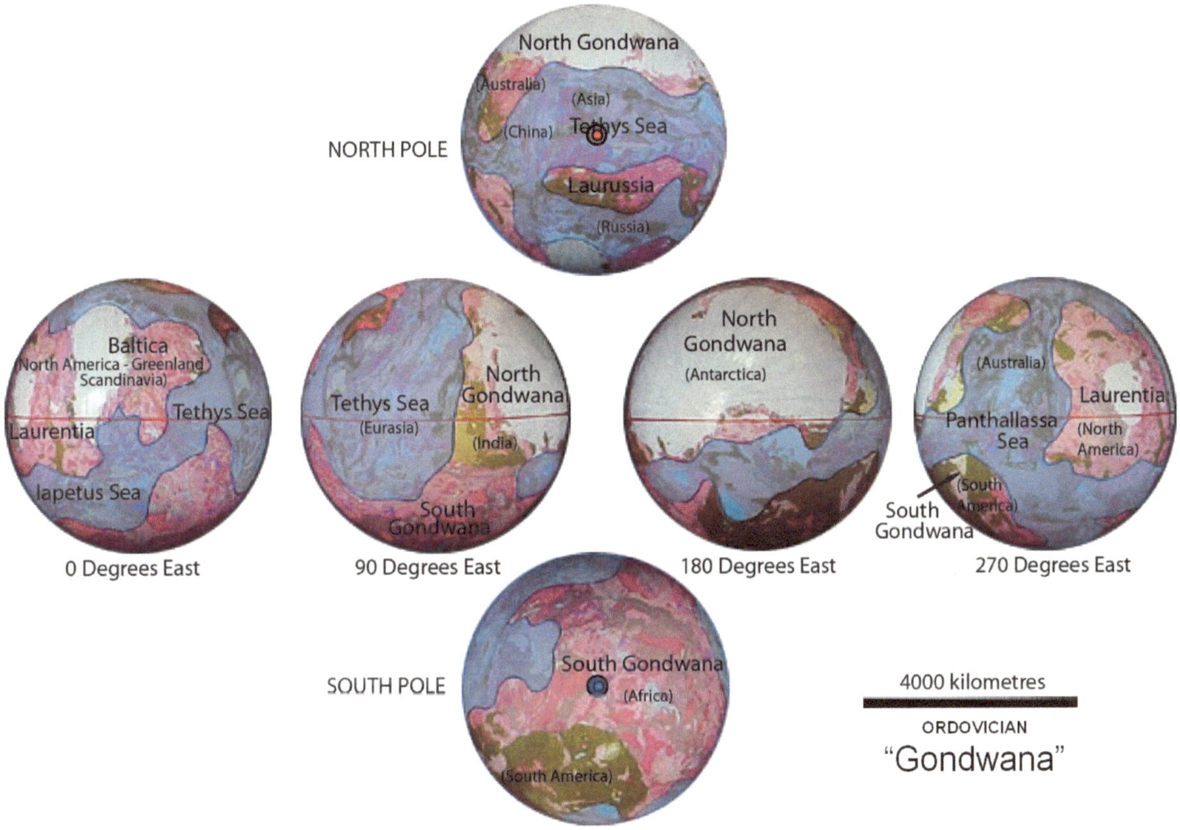

Figure 19.5 The Ordovician Gondwana Expansion Tectonic small Earth assemblage. This model shows the ancient coastline distribution (blue lines) defining North and South Gondwana in relation to Laurentia, Baltica, and Laurussia. The ancient Tethys, Iapetus, and Panthalassa Seas form part of a global network of continental seas (shaded areas) surrounding each of the exposed supercontinents.

During Palaeozoic times, the South Pole continued to be located within central West Africa in what was then South Gondwana. Similarly, the North Pole was located within Northern China-Mongolia in what was part of the northern Tethys Sea. The ancient equator passed through East Antarctica, central Australia, North America, central Eurasia, and India through what was then North Gondwana. This crustal configuration approximates Plate Tectonic reconstructions in part, but differs substantially in the South Pacific region because of the need for an expanse of inferred Panthalassa and Tethys Oceans.

By late-Permian times crustal rupture of the Pangaea supercontinent (Figure 19.6) had commenced, which resulted in break-up of Pangaea and formation of the modern continents and modern oceans during the following Mesozoic and Cenozoic Eras.

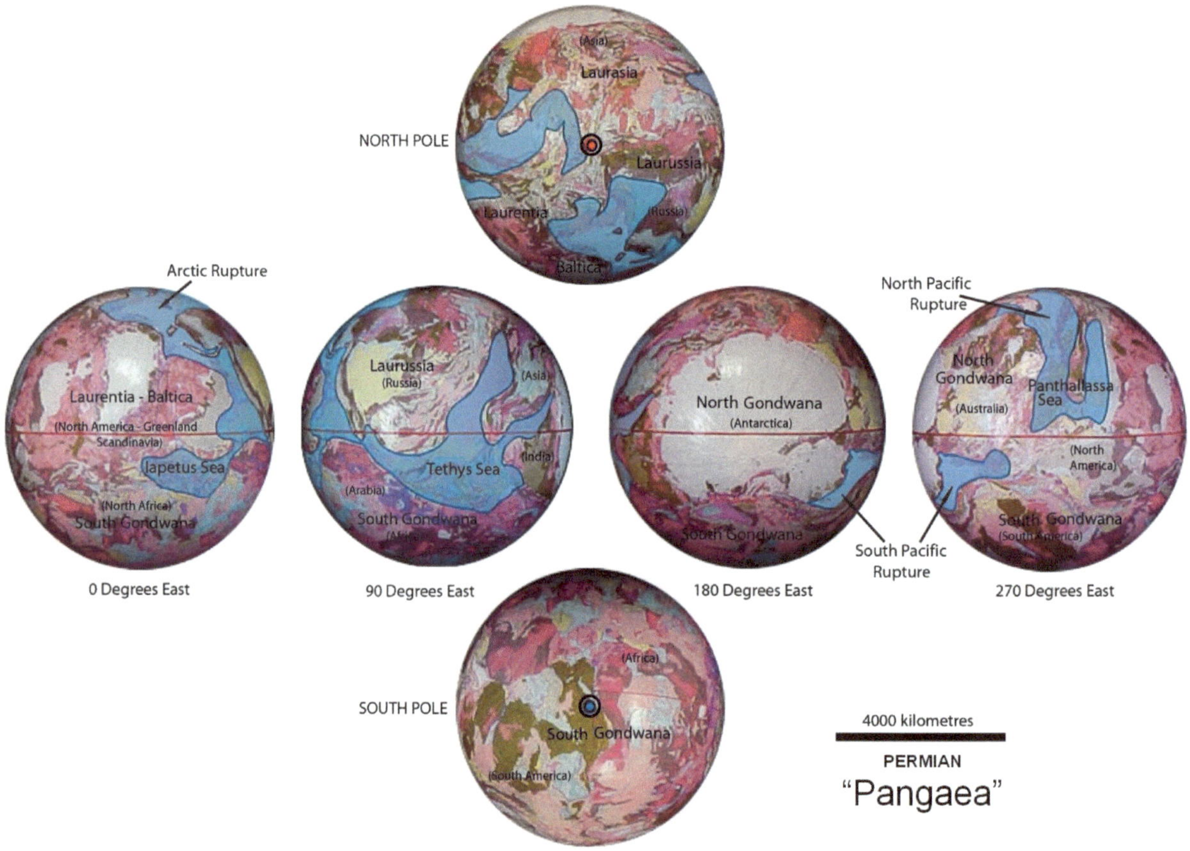

Figure 19.6 Permian Pangaea Expansion Tectonic small Earth crustal assemblages. The model shows the ancient coastline distribution (blue lines) as well as the ancient Tethys, Iapetus, and Panthalassa Seas (blue shaded areas) forming part of a global network of continental seas. The figure also shows the locations of continental rupture commencing in the north and south Pacific and Arctic Ocean regions to form the modern oceans.

This rupturing also initiated draining of the continental seas into the newly emerging marine sedimentary basins–the forerunners of the modern oceans–and formation of new waters from mid-ocean-rift zones. Disruptions to the continental seas initiated exposure of the former sea-floors to the ravages of erosion, which resulted in an unprecedented accumulation of sediment and organic matter in the form of coal beds in newly emerging coastal environments.

These Palaeozoic times coincided with the rapid development and evolution of all modern life forms on Earth. On an Expansion Tectonic Earth this evolution of life forms was driven by the need to keep pace with the rapidly changing environmental conditions and, similarly, extinction is seen as a by-product of not being able to keep pace with environmental change. The degree of crustal change during this interval of time certainly had the capacity to markedly influence evolutionary change in all life forms. The Palaeozoic Era then ended with rupture and break-up of the Pangaea supercontinental crust and draining of the continental seas into the newly opening modern oceans–the end-Permian extinction event. The effects of this extinction event were so devastating that it took remaining life on land a further 30 million years to recover.

19.6 Mesozoic Earth

The Mesozoic Era was an unprecedented interval of geological time extending from 252.6 million years ago to about 66 million years ago. The era began in the wake of the end-Permian extinction event, the largest mass extinction in Earth's history, and ended with the end-Cretaceous extinction event, which is best known for the demise of the dinosaurs along with many plant and animal species. The Mesozoic is recognised as a time of significant tectonic, climate, and evolutionary change, and characterised by vast Triassic salt deposits in rifts and isolated seas that appeared at that time. The climate of the Mesozoic was varied, alternating between warming and cooling periods. Dinosaurs appeared in the mid-Triassic and became the dominant terrestrial vertebrates early in the Jurassic, occupying this position for about 135 million years until their demise at the end of the Cretaceous. Birds first appeared in the Jurassic, having evolved from a branch of Theropod dinosaurs. The first mammals also appeared during the Mesozoic, but would remain relatively small to about one metre in size until the Cenozoic Era.

On an Expansion Tectonic Earth the Mesozoic Era commenced as a result of rupture and break-up of the Pangaea supercontinent to form the modern continents, along with draining of the continental Tethys, Panthalassa, and Iapetus Seas to initiate formation of the modern oceans (Figure 19.7).

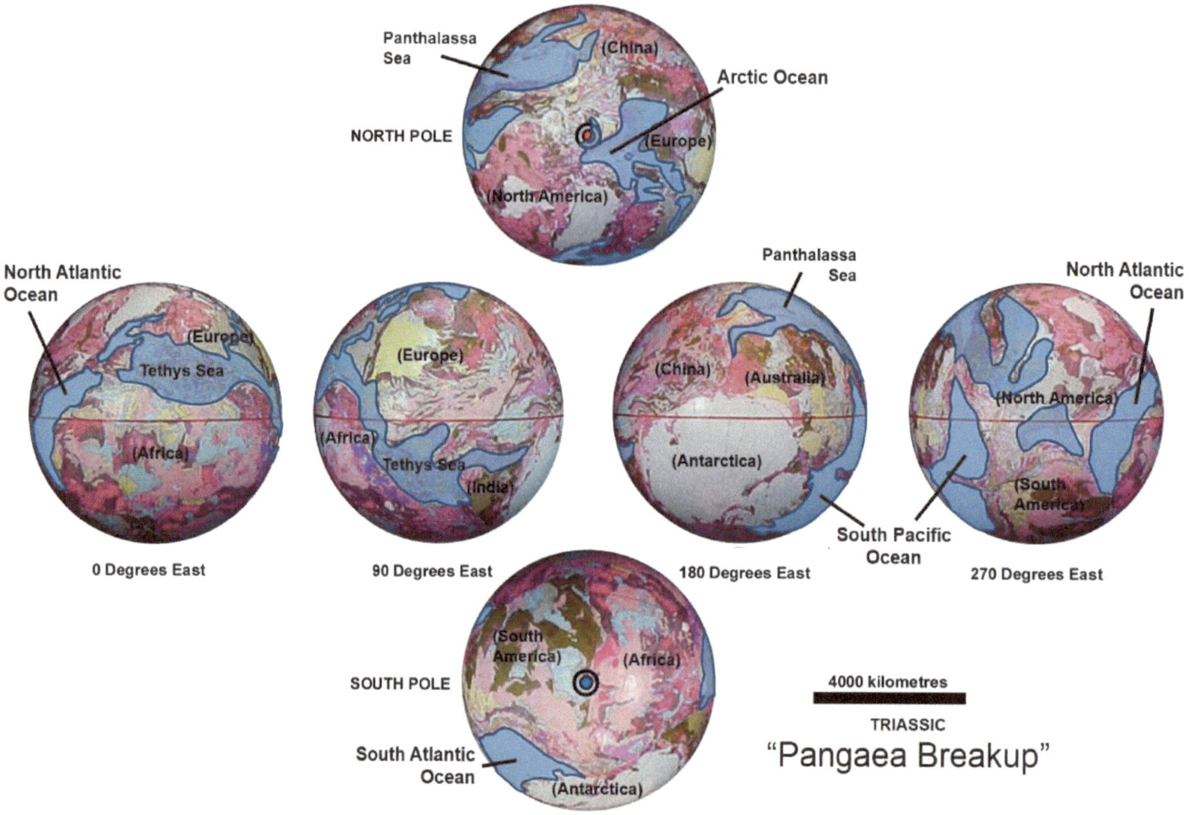

Figure 19.7 Triassic post-Pangaea break-up Expansion Tectonic small Earth crustal assemblages. The model shows the ancient coastline distribution (blue lines) as well as remnants of the ancient Tethys, Iapetus, and Panthalassa Seas (blue shaded areas) forming part of a global network of separate continental seas. The figure also shows the locations of the modern north and south Pacific, Atlantic, and Arctic Ocean regions.

This draining was accompanied by a shift in sedimentary deposition, changing from continental to shallow marine and marginal continental basin settings. Draining of the continental seas into the newly opening marine basins gave rise to a number of separate and discrete seas and embryo oceans, each with their own separate sea-levels, salinities, ocean currents, temperatures, and marine species. The location of these newly opening marine basins coincided with the precursors to the modern mid-ocean-rift zones. The influx of sediments within these zones, and mixing with intruded magma and volcanism, also initiated what are now referred to as *island-arcs*, which were first initiated and located between the emerging modern continents.

Continental break-up and opening of the modern oceans also initiated seafloor spreading which was accompanied by the establishment of well-defined mid-ocean-rift zones and expulsion of new water and atmospheric gases. During this early continental break-up phase the seafloor spreading process was initially masked by sedimentary input into the small marine sedimentary basins. Deposition of sediments during that time was confined to isolated basins and around the continental shelves. As opening of the modern oceans progressed further, beyond the reach of sedimentary deposition, the intruded seafloor volcanic lava was eventually exposed on the seafloor as now shown preserved in the *Geological Map of the World*. Modern seafloor mapping shows that these mid-ocean-rift zones also coincide with zones of high mantle-derived heat flow.

During the Mesozoic, this break-up of the supercontinental crusts initiated migration of the continents and opening of new oceans. Migration of the continents, in turn, affected the geographical location of the magnetic poles and severely disrupted established climate zones and global climates. Break-up and disruption, in turn, affected the existing habitats and migration routes of both marine and terrestrial life forms. This disruption forced species to imperceptibly migrate over time, along with the migrating continents and climate zones, to evolve or better adapt to changing climate, or simply perish if unable to keep pace with environmental change.

Tectonism during the Mesozoic was essentially confined to the newly emerging mid-ocean-rift zones, along with crustal movement during stabilisation of the continental crusts as surface curvature continued to adjust and compensate for crustal break-up. Draining of the continental seas and formation of new waters from the mid-ocean-rift zones was well advanced by that time (Figure 19.7) allowing erosion of the newly exposed land surfaces to continue along with colonisation by rapidly evolving terrestrial plant and animal species.

On an Expansion Tectonic Earth the Mesozoic Era is also noted for its progressive breaching and merging of previously separate seas and oceans, resulting in a number of sea-level related extinction events. Breaching and merging of the North Pacific and South Pacific Oceans gave rise to the end-Triassic extinction event, and similarly breaching between Australia and Antarctica during the Paleocene Epoch initiated opening of the South Pacific Ocean and gave rise to the end-Cretaceous extinction event.

19.7 Cenozoic Earth

The Cenozoic Era began in the wake of the end-Cretaceous extinction event some 66 million years ago and continues through to the present-day. The Cenozoic is also known as the Age of Mammals because the extinction of many previous groups allowed mammals to successfully take over. Early in the Cenozoic the Earth was dominated by relatively small fauna, including small mammals, birds, reptiles, and amphibians. From a geological perspective, it did not take long for the mammals and birds to greatly diversify in the absence of the large reptiles that had previously dominated during Mesozoic times. Mammals increased in size and came to occupy almost every available niche—including in the oceans, on the lands, and in the air. Climate-wise, the Earth had moved into a drying and cooling trend, culminating in the glaciations of the Quaternary Epoch.

On an Expansion Tectonic Earth the Cenozoic is marked by establishment of symmetric seafloor spreading in all of the modern oceans. This seafloor spreading represents a preservation of the seafloor spreading and growth history of the Earth and is testament to opening and enlarging of all of the modern oceans. The Cenozoic was also a time when all of the oceans had merged into a single global ocean, thus removing the threat of further sea-level-related mass extinction and allowing for relative stability of marine and terrestrial life forms.

Crustal extension as a result of Earth expansion is now predominantly confined to the mid-ocean-rift zones, with minor basin extension and rifting within the continents during adjustment for change in surface curvature. Crustal break-up of both seafloor and continental crust continues to be closely related to lengthening and propagation of the mid-ocean-ridges, and includes focused seismic activity, heat-flow, and magmatism along well-defined zones of crustal weakness.

20 A Paradigm Shift in Understanding

"Don't confuse me with any more facts; I have already made up my mind."

By far the single most important contribution to modern scientific understanding of the concept of Expansion Tectonics has been the completion of bedrock geological mapping and age dating of all the continental and seafloor crusts. This bedrock mapping has enabled the assemblage of crustal plates to be accurately constrained on models of an Expansion Tectonic Earth and for the first time has enabled small Earth modelling studies to be extended back to earliest Archaean times. It has also provided a means to mathematically calculate an increase in Earth radius over time, as well as a means to mathematically investigate additional related physical processes throughout Earth history.

What this seafloor bedrock mapping has shown is that the mapping can be used to accurately constrain the location and assemblage of seafloor crustal plates on smaller radius Earth models. Assemblage of these seafloor crustal plates on models of an Expansion Tectonic Earth consistently show that each plate assembles with a single unique fit, with all plates assembling with a very high degree of accuracy along each of the mid-ocean-ridge spreading zones.

If these crustal plate assemblages were mere coincidence it should be expected that the seafloor mapping, as well as geological and geographical evidence from adjoining continents, would not match across plate boundaries on any of these small Earth models. The evidence instead shows that seafloor bedrock mapping does indeed match across these plate boundaries, all continental sedimentary basins do merge to form a global network of ancient continental seas, orogenic and fold mountain belts coincide, and ancient crustal regions assemble together precisely.

This assemblage of seafloor crustal plates is shown to extend for 200 million years back in time to the Triassic Period and demonstrates the viability and uniqueness of a post-Triassic Expansion Tectonic process. This contrasts strongly with Plate Tectonic reconstructions for the same time interval, where assemblage of crustal plates is based on palaeomagnetic apparent-polar-wander evidence with multiple plate-fit options to contend with. The uniqueness of Expansion Tectonic small Earth assemblages also contrasts strongly with the Plate Tectonic requirement to arbitrarily fragment continents in order to comply with the seafloor mapping data. It also contrasts with the requirement to dispose of large areas of inferred pre-existing crust beneath subduction zones in order to maintain a constant surface area.

The utilisation of continental crustal bedrock geology to constrain a pre-Triassic Expansion Tectonic Earth assemblage has never been done before. Most previous Expansion Tectonic Earth researchers were limited to simply removing the seafloor crusts and visually fitting the remaining continents together on a small Earth at a radius of about 50 to 55 percent of the present radius. At that time the continental crusts were considered by most researchers to be rigid, which had essentially stabilised by about 200 million years ago. Beyond closing off the oceans and assembling the continental crusts, at no point were these early researchers able to extend crustal reconstructions back in time to recreate a primordial Archaean Earth.

What has been seen from the Expansion Tectonic small Earth model studies is that all continental crusts unite precisely to form a complete global Pangaean supercontinental crust during the Permian Period. At that time the bulk of the seafloor volcanic lava, along with the atmosphere and hydrosphere, was retained within the mantle. From this supercontinental assemblage, continental sedimentary basins were then shown to merge to form a global network coinciding with continental seas and where ancient supercontinents and seas were, in turn, defined by the variation in coastal outlines and sea-levels.

When sediments from within the network of sedimentary basins were progressively returned to the exposed land surfaces, from where they were first eroded, and basins were progressively closed the remaining ancient crustal fragments were again shown to assemble and retain a unique spatial assemblage throughout the pre-Triassic Earth history. This unique spatial assemblage of ancient crusts was maintained throughout a long history of Precambrian and Palaeozoic crustal extension, prior to crustal rupture during the late-Permian followed by continental break-up and dispersal of the modern continents during the Mesozoic and Cenozoic Eras.

This unique assemblage of all continental and seafloor crustal fragments on models of an Expansion Tectonic Earth demonstrates that Expansion Tectonics, extending back 4,000 million years to the beginning of Earth's geological past, is indeed viable. What the full range of Archaean to present-day small Earth models demonstrate is that, rather than being a random, arbitrary, amalgamation-dispersal-amalgamation crustal forming process as we are currently lead to believe, crustal development on an

Expansion Tectonic Earth is instead shown to be a simple, evolving, and predictable crustal forming process.

It is significant to reiterate that on Expansion Tectonic small Earth models each of the established Precambrian and Palaeozoic sedimentary basin settings merge together to form a global network, with a single common pan-global continental crust. It was shown that the sedimentary basins from each of the continents coincide with a global network of crustal weakness, operating throughout Precambrian and Palaeozoic times. It was within this network of global crustal weakness that crustal extension, generated during on-going Earth expansion, was focused, as well as on-going crustal mobility, mantle-derived heat flow, magmatic activity, crustal rupture, continental break-up, and eventual opening of each of the modern oceans. The merging of each of these crustal settings shows that global crustal forming processes correlate precisely with the overall development of the continental crusts.

On small Earth models of an Expansion Tectonic Earth, the eventual break-up of the ancient continental crust during late-Permian times resulted in disruption of the established polar ice-sheets, disruption of the ancient continental seas, changes to sea-levels, and a disruption of established climatic zones. These disruptive changes, in turn, affected plant and animal species habitats and either drove evolution of these species, affected their long-term decline, or caused their periodic extinction.

It has also been shown that when the imposed constant Earth surface area and constant Earth radius premises are removed from geophysical observations, these same geophysical observations, when applied to models of an Expansion Tectonic Earth, demonstrate that the data is consistent with an Expansion Tectonic Earth. Similarly, geographical and biogeographical information, when applied to Expansion Tectonic Earth models, aptly quantify crustal development on an Expansion Tectonic Earth and quantify the location of the ancient magnetic poles and equators as determined from unconstrained geophysical pole data.

The application of palaeomagnetism to models of an Expansion Tectonic Earth has shown that all ancient magnetic poles cluster as diametrically opposed north and south poles on each small Earth model constructed. This diametrically opposed clustering of poles is impossible on Plate Tectonic reconstructions, where pole data is instead used to generate complex apparent-polar-wander-paths. When used to determine an ancient Earth radius, this same palaeomagnetic evidence, traditionally used to negate Earth expansion, is instead shown to confirm Earth expansion.

When published coastal geography is plotted on Expansion Tectonic small Earth models it has been shown that the large, Plate Tectonic

Panthalassa, Iapetus, and Tethys Oceans were not present on a smaller radius Earth. Instead, this same coastal geography defines the presence of more restricted continental Panthalassa, Iapetus, and Tethys Seas, which represent precursors to the modern Pacific and Atlantic Oceans and emergent Eurasian continent respectively. From this coastal geography the emergent land surfaces and coastal outlines on models of an Expansion Tectonic Earth then defined the ancient Rodinia, Gondwana, and Pangaea supercontinents and smaller sub-continents. This coastal geography was shown to amply demonstrate an evolutionary progression and development of each of these ancient supercontinents throughout Earth history.

On each Expansion Tectonic small Earth model the evolutionary development of the ancient supercontinents was shown to be intimately related to changes in sea-level, with no requirement for random continental assemblage or break-up. The coastal geography shows that the outlines of emergent supercontinents were intimately related to changes to the outlines of continental sedimentary basins, to changes incurred during crustal mobility, and changes to sea-levels as the modern oceans rapidly opened to the present-day.

When dinosaur distributions were plotted on Expansion Tectonic small Earth models, their global distributions illustrated the ease and simplification of migration and biogeographical development. The identified cosmopolitan and provincial distributions and inter-relationships of the dinosaurs were shown to have been maintained without the need for complex continental assemblage-dispersal requirements. This contrasts strongly with Plate Tectonic reconstructions, where assemblage and movements of the continents do not correspond to the known, or necessary migration routes required by their established biogeographical boundaries.

During continental break-up and opening of the modern oceans, the traditional migration routes of the various dinosaur species were shown to be disrupted, enabling species endemic to the various regions to interact and extend their boundaries with time. The timing of species development was shown to be intimately related to changes in sea-level, disruptions to the ancient continental seas, and opening of the modern oceans. This timing either facilitated species migration by extending and expanding on existing migration routes, or caused species extinction because of a failure to adapt to changing environmental conditions.

From the distribution of climate-dependent rocks and animal species it was shown that their distributions coincide precisely with climatic zones anticipated on an Expansion Tectonic Earth. It was also shown that these climatic indicators display a distinct latitudinal zonation paralleling the ancient equator, suggesting that an inclined Earth rotational axis inclined to the pole of the ecliptic was well established during at least the Palaeozoic Era and has persisted to the present-day.

To round off this investigation into the concept of an Expansion Tectonic Earth we were compelled to at least suggest a causal model for Earth expansion. The proposed causal model for Earth expansion involves an on-going input of ionised electron and proton particulate matter originating from the Sun. Particulate matter enters the Earth and recombines as new matter most likely within the 200 to 300 kilometres thick D" region, located at the base of the mantle directly above the core-mantle boundary. This particulate matter generation process then represents the basis for formation of all elements and mineral species present on Earth. The increase in volume of new matter then results in a swelling of the mantle. Mantle swell is then transferred to the outer crust as continental crustal extension which is currently seen and preserved as extension along the mid-ocean-rift zones, and expulsion of new basaltic lava, sea water, and atmospheric gases.

So, what does Earth really have to say? The extensive evidence presented here tells us that an Expansion Tectonic Earth is indeed a viable and demonstrable global tectonic process. At no stage has any fundamental physical law been violated. We simply removed the commonly held assumption that Earth radius has remained constant throughout time and allowed the Earth to tell its own story. In order to create small Earth models we simply removed from the Earth what was not previously there–intruded seafloor volcanic lava and eroded sediments–to end up with a Pangaean Earth comprising an assemblage of continental crustal components, and ultimately to a primitive Earth comprising an assemblage of equally primitive crustal components. We then simply displayed published physical data on the Expansion Tectonic small Earth models created and each of these model studies was shown to indeed complement each other and fully substantiate an Expansion Tectonic Earth process.

21 Glossary of Terms

Accretion: a process by which material is added to a tectonic plate or landmass. This material may be sediment, volcanic material, or other igneous rocks.
Age dating: see radiometric age dating.
Albedo: a measure of the reflecting power of the Earth surface, the chemical composition of the atmosphere, and atmospheric humidity.
Angular momentum: a vector quantity that represents the product of a body's rotational inertia and rotational velocity about a particular axis.
Anoxic event: a total depletion in the level of oxygen, said to occur when the Earth's oceans become completely depleted of oxygen below the surface levels.
Apparent-polar-wander: a Plate Tectonic concept referring to the apparent movement of the ancient magnetic pole locations as determined for each continent.
Applied force: a force is any influence that causes an object to undergo a certain change, either concerning its movement, direction, or geometrical construction. In other words, a force can cause an object with mass to change its velocity, to accelerate, or for a flexible object to deform.
Applied load: or actions are forces, deformations, or accelerations applied to a structure or its components. Loads cause stress, deformation, and displacement in structures.
Applied stress: is a physical quantity that expresses the internal forces that neighbouring particles of a continuous material exert on each other. For example, when a solid vertical bar is supporting a weight, each particle in the bar pulls on the particles immediately above and below it.
Archosaurs: a group of species that includes the extinct dinosaurs and whose living representatives consist of birds and crocodiles.
Asthenosphere: the highly viscous, mechanically weak and ductile region of the upper mantle of the Earth lying at depths between 100 and 200 kilometres below the surface. It is considered the source region of mid-ocean-ridge basalt.
Asymmetric spreading ridge: a one-sided, near-continental-margin-spreading zone.
Atmosphere: a layer of gases surrounding the planet Earth that is retained by Earth's gravity.
Aurora: a natural light display in the sky, particularly in the high latitude Arctic and Antarctic regions, caused by the collision of energetic charged particles with atoms in the high altitude atmosphere.
Auroral zone: a zone of auroral activity typically extending from 10° to 20° of latitude around the Earth's magnetic poles.
Back-arc basin: a Plate Tectonic geologic marine basin feature, typically very long (several hundreds to thousands of kilometres) and relatively narrow (a few hundred kilometres) found at some convergent plate boundaries, presently concentrated in the Western Pacific ocean and associated with island arcs and subduction zones.
Banded Iron Formation: a distinctive, mainly Precambrian sedimentary rock consisting of repeated thin layers of iron oxides alternating with bands of iron-poor shale and chert, often red in colour.
Basalt: an extrusive volcanic lava rich in iron and magnesium. It forms the bedrock of the ocean floors and also occurs on land as lava flows.

Basin: an area that is underlain by a substantial thickness of sedimentary rock that possess unifying characteristics of both sediment type and deformation history.
Basin inversion: the uplifting and exposure of basin sediments as distinct from a slowly sinking sedimentary basin.
Bathymetry: is the study of underwater depths and topography of lakes or ocean floors.
Bedrock: the common term for consolidated rock underlying the surface of the Earth. Above the bedrock is usually an area of broken and weathered unconsolidated rock often referred to as the regolith. Bedrock may also experience subsurface weathering at its upper boundary.
Bedrock geologic map: a geologic map of an area showing the distribution of differing rock types that would be exposed at the surface if all soil or other superficial deposits were removed.
Bilaterian animals: animals having a bilateral symmetry comprising a front and a back end, as well as an upside and downside.
Biogeography: the study of the distribution of plant and animal species and ecosystems in geographic space and through geological time.
Biosphere: the global sum of all ecosystems integrating all living beings and their relationships, including their interaction with the elements of the lithosphere, hydrosphere, and atmosphere.
Black smoker: a type of hydrothermal vent found on the seabed, typically appearing as chimney-like structures that emit a cloud of black sulphur-bearing minerals. Black smokers are formed in fields hundreds of metres wide when superheated water from below the Earth's crust comes through the seafloor.
Brittle fracture: occurs in brittle crystalline materials by cleavage as the result of tensile stress acting normal to crystallographic planes with low bonding.
Carbonate limestone: a sedimentary rock composed largely of the minerals calcite and aragonite, which are different crystal forms of calcium carbonate. Many carbonate limestones are composed from skeletal fragments of marine organisms such as coral or foraminifera.
Cartography: the study and practice of making maps.
Catastrophism: a theory where the Earth is considered to have been affected in the past by a number of sudden, short-lived, violent events, possibly worldwide in scope.
Chalk: a soft, white, porous sedimentary rock; a form of limestone composed of the mineral calcite, formed under reasonably deep marine conditions from the gradual accumulation of minute calcite plates shed from marine micro-organisms.
Chemical weathering: a gradual and on-going process of change or decomposition of rocks, often transforming them when water interacts with minerals to create various chemical reactions. New or secondary minerals develop from the original minerals of the rock.
Chord: a line joining two points on a curve, such as but not limited to a circle or ellipse.
Circles of latitude: an imaginary east-west circle connecting all locations on the surface with the same north or south latitude.
Climate indicator: various rocks or fossil organisms specific to certain climate zones, e.g. coral reefs indicative of equatorial climate zones.
Climatic zone: Earth's surface comprises five main geographical and climatic zones, which are defined by major circles of latitude. The main differences between each of these zones relate to climate distribution, as well as the attitude of the Sun.
Cnidarians: a species of animal found mostly in marine environments from about 580 million years ago.

Coal swamp: a general term referring to dense forests in low-lying wetland areas and swamps prevailing during the Carboniferous and Permian Periods where, after being covered and deeply buried by sediments, it forms coal.
Colatitude: the angular distance from a point location on the Earth surface to the North or South Pole, where the point is located in the northern or southern hemisphere respectively.
Collisional boundary: a Plate Tectonic term also known as a convergent boundary. An actively deforming region where two (or more) tectonic plates move toward one another and collide.
Continental collision: a Plate Tectonic phenomenon that occurs at convergent boundaries whereby a subduction zone may be destroyed, mountains are produced, and two continents may be sutured together.
Continental Crust: the layer of igneous, sedimentary, and metamorphic rocks that forms the continents and the areas of shallow seabed close to their shores.
Continental Drift: movement of the Earth's continents relative to each other by appearing to drift across the ocean bed.
Conservative plate boundary: also called conservative plate margin or transform boundary, are faults that occur when two plates slide against one another.
Continental margin: the zone of the seafloor comprising the continental shelf, continental slope, and continental rise that separates the thin oceanic crust from thick continental crust.
Continental sea: a body of seawater that covers low-lying parts of the continental crusts.
Continental shelf: the extended perimeter of each continent and associated coastal plain that slopes gradually to the ocean bottom.
Convection currents: a Plate Tectonic mechanism for carrying heat from the interior of the Earth to the surface.
Convergent boundary: a Plate Tectonic term referring to an actively deforming region where two or more tectonic plates move toward one another and collide.
Conveyor belt principle: a Plate tectonic explanation for seafloor spreading and the motion of the plates making up the Earth's crust.
Coral reef: underwater structures made from calcium carbonate secreted by corals, plus accumulated shelly fragments from deceased marine animals.
Core–Mantle boundary: lies between the Earth's mantle and its liquid iron-nickel outer core, located at approximately 2900 kilometres depth beneath the Earth's surface. The boundary is observed via the discontinuity in seismic wave velocities at that depth.
Coriolis Effect: a deflection of moving objects when they are viewed in a rotating reference frame such as the Earth. This force causes moving objects, such as the atmosphere, to be deflected in a clockwise sense in the northern hemisphere, and in a counter-clockwise sense in the southern hemisphere.
Coronal mass ejection: a massive burst of solar wind and magnetic fields originating from active regions on the Sun's surface, such as groupings of sunspots associated with solar flares.
Cosmopolitan distribution: a species is said to have a cosmopolitan distribution if its range extends across all or most of the world in appropriate habitats. For instance, the whale has a cosmopolitan distribution, extending over most of the world's oceans.
Craton: a part of the Earth's crust that has attained relative crustal stability by about 2,400 million years ago, and since then has undergone little deformation compared to adjacent parts of the crust.
Crustal interaction: an Expansion Tectonic term for geologic processes acting at the interface between the various crustal plates.

Crustal jostling: an Expansion Tectonic term for a global compensating process of adjustment of each of the crustal fragments relative to adjoining crusts.
Crustal rifting: a linear zone where the Earth's crust is being pulled apart. Where rifts remain above sea-level they form a rift valley, which may be filled by water forming a rift lake. Major rifts occur along the central axis of mid-ocean-ridges, where new seafloor crust is created between two tectonic plates.
Crustal rebound: the relaxation and rise of a continent after glacial melting, estimated to be less than 10 millimetres per year.
Crustal rupture: an Expansion Tectonic term for the moment in time when the continental crusts first begin to fail and break-up.
Crustal shredding: an Expansion Tectonic term referring to the break-up and fragmentation of flexible crusts, such as the sedimentary basins, during opening of the modern oceans.
Crustal stretching: the ability of crustal rocks to stretch and thin under an applied extensional force.
Degassing: the removal or expulsion of dissolved gases from liquids, including molten magma.
Density: of a material is defined as its mass per unit volume.
Desertification: a type of land degradation in which a relatively dry land region becomes increasingly arid, typically losing its bodies of water as well as vegetation and wildlife.
Devolatilization: a process of removing volatile elements or compounds from a material.
Diapiric orogen: upward rising crust.
Differentiation: a chemical processes whereby molten magmas undergo bulk chemical change during crystallisation.
Displaced terrain: a Plate Tectonic term referring to a piece, or fragment of continent that has broken away from its parent continent and become accreted to another continent.
Divergent boundary: a linear feature that exists between two tectonic plates which are moving away from each other. The space that this creates is filled with new crustal material sourced from molten magma below.
D" layer: a thermal and chemical boundary layer located at the bottom of the lower mantle where an important change in the speed of seismic P waves appears.
Docking-point: an Expansion Tectonic term for the re-uniting of continental crustal fragments along a section of coastline or continental landmass.
Down warping: downward flexing of the crust, particularly within a sedimentary basin.
Dyke: a discordant tabular or sheet-like body of intruded magma that cuts vertically or almost vertically through and across older rocks.
Earth core: is divided into two parts, a liquid outer core that is much less viscous than the mantle, and a solid inner core.
Earth crust: the outermost solid shell of the Earth, which is physically distinct from the underlying mantle, ranging from 5–70 km in depth. The thin seafloor crusts underlie the ocean basins and are composed of dense iron-magnesium-rich basalt rocks. The thicker continental crusts are less dense and mainly composed of sodium-potassium-aluminium-silicate-rich rocks such as granite.
Earth magnetic field: a magnetic field that extends from the Earth's inner core to where it meets the solar wind, a stream of energetic particles emanating from the Sun.
Earth mantle: extends to a depth of 2,890 km and is composed of silicate rocks that are rich in iron and magnesium relative to the overlying crust.

Earthquake: a violent movement of the rocks in the Earth's crust. Earthquakes are usually quite brief, but may repeat over a long period of time. They are the result of a sudden release of energy in the Earth's crust.
Earth Sciences: an all-embracing term for the sciences related to the planet Earth, which includes the study of the atmosphere, hydrosphere, oceans and biosphere, as well as the solid earth.
Electromagnetic radiation: a form of energy emitted and absorbed by charged particles which exhibits wave-like behaviour as it travels through space.
Empirical evidence: a source of knowledge acquired by means of observation or experimentation.
Empirical research: a way of gaining knowledge by means of either direct or indirect observation.
Eon: (sometimes spelt Aeon) a measure of time in the geologic time scale.
Epeiric Sea: a continental sea comprising a body of seawater that covers low-lying parts of the ancient or modern continental crusts.
Epoch: a subdivision of the geologic timescale that is longer than an age and shorter than a period.
Era: a subdivision of geologic time that divides an eon into smaller units of time.
Erosion: the gradual wearing down of land surface materials, especially rocks, sediments, and soils by the action of water, wind, or glaciers.
Eukaryotes: animal cells with complex internal structures.
Eustatic sea-level change: global-scale changes in sea-level, where sea levels rise or fall the same amount in all oceans.
Evaporate rocks: water-soluble sedimentary rocks, such as salt deposits, that result from concentration and crystallization by evaporation from an aqueous solution.
Exotic continental terrane: a Plate Tectonic term referring to a fragment of crustal material with its own distinctive geologic history formed on, or broken off from, one tectonic plate and accreted to crust lying on another plate.
Expanding Earth: [Earth Expansion] a scientific theory that allows Earth radius to increase over time in order to describe the large-scale motion and assemblage of Earth's crustal plates.
Expansion Tectonics: a study of the evolution and behaviour of the Earth's crust as a result of an exponential increase in Earth's surface area.
Exponential rate of increase: the growth rate of the value of a mathematical function is proportional to the function's current value and is used to model a relationship in which a constant change in the independent variable gives the same proportional change in the dependent variable.
Extinction: a widespread and geologically rapid decrease in the amount of life on Earth.
Extrusive volcanic rock: an igneous volcanic rock formation in which hot magma from inside the Earth flows out (extrudes) onto the surface as lava or explodes violently into the atmosphere to fall back as ash or tuff.
Fault: a planar or gently curved fracture in the rocks of the Earth's crust, where compressional or tensional forces cause relative displacement of the rocks.
Flat Earth theory: a largely redundant belief that the Earth's shape is a flat plane or disk.
Flexible crust: the ability of crusts to readily flex, fold, fault, shear, stretch or distort, as highlighted in many sedimentary rock exposures throughout the world today.
Fluid dynamics: a sub-discipline of fluid mechanics that deals with fluid flow, the natural science of fluids, liquids, and gases in motion.

Fold: a fold occurs when one or a stack of originally flat and planar surfaces, such as sedimentary strata, are bent or curved as a result of permanent deformation. Folds in rocks vary in size from microscopic crinkles to mountain-sized folds.
Fossil: the preserved remains or traces of animals, plants, and other organisms from the remote past.
Fossil record: the presence of fossils preserved in rock formations or sedimentary layers.
Fracture: any local separation or discontinuity plane in a geologic formation, such as a joint or a crack that divides the rock into two or more pieces.
Geocentric axial dipole model: a first order approximation of the Earth's magnetic field.
Geocentric magnetic poles: antipodal north and south points where the axis of a theoretical magnetic dipole intersects the Earth's surface.
Geochronology: the science of age dating of rocks.
Geodetics: the science of determining the size and shape of the earth.
Geodynamic mechanism: the Plate Tectonic mechanism for explaining how mantle convection leads to subduction and other geologic phenomena such as seafloor spreading, mountain building, volcanoes, earthquakes, faulting and so on.
Geography: the science that studies the lands, the features, the inhabitants, and the natural phenomena of the Earth.
Geographic coordinate system: a coordinate system that enables every location on the Earth to be specified by a set of numbers or letters. A common choice of coordinates is latitude, longitude, and elevation.
Geographical latitude and longitude: latitude of a point on the Earth's surface is the angle north or south of the equator. The 0° parallel of latitude is designated the equator, which divides the globe into Northern and Southern Hemispheres. The longitude of a point on the Earth's surface is the angle east or west from a reference meridian, such as Greenwich, to another meridian that passes through that point. All meridians of longitude converge at the north and south poles.
Geology: the science comprising the study of solid Earth, the rocks of which it is composed, and the processes by which they evolve.
Geological time scale: a system of chronological measurement that encompasses the history of the Earth and relates stratigraphy—the succession of rock strata—to time.
Geomagnetic field: the magnetic field that extends from the Earth's inner core to where it meets the solar wind; a stream of energetic particles emanating from the Sun.
Geomagnetic storm: a temporary disturbance of the Earth's magnetosphere caused by a solar wind shock wave interacting with the Earth's magnetic field.
Geophysics: the science comprising a study of the Earth using quantitative physical methods.
Geosyncline: a subsiding low-lying linear trough or down-warp in the Earth's crust where sediments accumulate.
Geothermal gradient: the rate of increasing temperature with respect to increasing depth in the Earth's interior.
Glacial period: an interval of time (thousands of years) within an ice age that is marked by colder temperatures and glacier advances, followed by interglacial periods, comprising periods of warmer climate between glacial periods.
Glacigenic: a generic term for all glacial-related rocks, formations, or rock debris that has a glacial origin and includes debris that was deposited both on the lands and on the seafloors.
Global geodetic network: a three dimensional geodetic framework of the Earth.

Global Positioning System (GPS): a space-based satellite navigation system that provides location and time information anywhere on or near the Earth.
Global Tectonics: an all-embracing term relating to Earth movements and interactions on a global scale, especially as they relate to the causes and results of the dynamics of the crustal plates and sea-floor spreading, often mistakenly considered to be synonymous to Plate Tectonics.
Gneiss: a common and widely distributed type of metamorphic rock formed by high-grade regional metamorphic processes from pre-existing formations that were originally either igneous or sedimentary rocks.
Gondwana: the name given to the more southerly of two supercontinents (the other being Laurasia) which were part of the Pangaea supercontinent that existed from approximately 510 to 180 million years ago.
Granite: a common type of generally light coloured intrusive, igneous rock that is granular in texture consisting mainly of quartz, mica, and feldspar minerals.
Granite-greenstone belts: zones of variably metamorphosed volcanic rocks that occur within Archaean and Proterozoic cratons associated with granite and gneiss bodies.
Granitization: the alteration and melting of rocks by extremes in heat or pressure to form granite.
Greenhouse gas: a gas in an atmosphere that absorbs and emits radiation within the thermal infrared range. The primary greenhouse gases are water vapour, carbon dioxide, methane, nitrous oxide, and ozone.
Greenstone: altered volcanic rocks. The name greenstone comes from the green hue imparted by the colour of the metamorphic minerals within the volcanic rocks.
Guyot: also known as a tablemount, is an isolated underwater volcanic mountain with a flat top over 200 metres below the surface of the sea.
Gymnosperms: a group of seed-producing plants represented mainly by the conifers.
Hadean Eon: the first geologic eon on Earth occurring before the Archaean.
Heavy minerals: a rock mineral with a density that is greater than 2.9 grams per cubic centimetre, most commonly referring to dense components of siliciclastic sediments such as zircon, ilmenite, monazite, rutile, and so on.
Heliopause: the theoretical boundary where the Sun's solar wind is stopped or balanced by the interstellar medium; where the solar wind's strength is no longer great enough to push back the stellar winds of the surrounding stars.
Heliosphere: a region of space dominated by the Sun where a "bubble" of charged particles in the space surrounding the Solar System is blown into the interstellar medium by the solar wind.
Homeostasis: a process by which the body's internal environment is kept stable.
Horst and graben: terms referring to regions that lie between faults and are either above or lower than the area beyond the faults. A horst represents rocks pushed upward by the faulting, and a graben is a rock that has dropped due to the faulting.
Hot-spot: volcanic regions thought to be fed by areas of anomalously hot underlying mantle, which may provide a great deal of molten magma.
Hydrocarbon: an organic compound consisting entirely of hydrogen and carbon. The majority of hydrocarbons found on Earth naturally occur in crude oil, where decomposed organic matter provides an abundance of carbon and hydrogen in the form of complex chemical chains.

Hydrosphere: in physical geography describes the combined mass of water found on, under, and over the surface of the Earth.
Hydrostatic equilibrium: a fluid is said to be in hydrostatic equilibrium or hydrostatic balance when it is at rest, or when the flow velocity at each point is constant over time. This occurs when external forces such as gravity are balanced by a pressure gradient force.
Icehouse state: refers to the state of the Earth as it experiences an ice-age where ice-sheets are present, and these sheets wax and wane throughout times known as glacial periods and interglacial periods. During an icehouse state, greenhouse gases tend to be less abundant, and global temperatures tend to be cooler.
Ice-sheet: a mass of glacier ice greater than 50,000 square kilometres that covers surrounding terrane, also known as continental glaciers.
Impact event: the collision of an asteroid, comet, meteoroid, or other celestial object with another celestial object such as Earth.
Ion: an atom or molecule in which the total number of electrons is not equal to the total number of protons.
Ionization: the process of converting an atom or molecule into an ion by adding or removing charged particles such as electrons or ions. In the case of ionization of a gas, ion pairs are created consisting of a free electron and a positive ion.
Ionosphere: a shell of electrons and electrically charged atoms and molecules that surrounds the Earth closer than the Van Allen belts, stretching out from a height of about 50 kilometres to more than 1000 kilometres above the Earth.
Insolation: a measure of solar radiation energy received on a given surface area and recorded during a given time.
Interglacial period: an interval of warmer global average temperature lasting thousands of years that separates consecutive glacial periods within an ice age.
Intergalactic plasma: low density plasma existing between galaxies.
Interplanetary medium: the material that fills the solar system, and through which all the larger solar system bodies, such as planets, asteroids, and comets, move.
Intrusive volcanic rock: a rock formed from magma that cools and solidifies within cracks or faults in the crust of a planet.
Invertebrates: animal species that do not develop a vertebral column, including the insects, worms, clams, crabs, octopus, snails, and starfish.
Impact event: a collision between celestial objects causing measurable effects.
Island-arc: a Plate Tectonic term for a curved chain of volcanic islands bounded on the convex side by a deep oceanic trench.
Isostatic equilibrium: a term used in geology to refer to the state of gravitational equilibrium between the Earth's lithosphere and asthenosphere such that the tectonic plates "*float*" at an elevation which depends on their thickness and density.
Isotopic signature: a ratio of stable or unstable isotopes of particular elements found in an investigated material where the atomic mass of different isotopes affect their chemical kinetic behaviour, leading to natural isotope separation processes.
K-T event: The Cretaceous–Palaeogene extinction event, formerly known as the Cretaceous–Tertiary (K-T) extinction, was a mass extinction of some three-quarters of plant and animal species on Earth—including all non-avian dinosaurs—that occurred over a geologically short period of time 66 million years ago. It marked the end of the Mesozoic Era and beginning of the Cenozoic Era which continues today.

Joint: a fracture in rock where the displacement associated with the opening of the fracture is greater than the displacement due to lateral movement in the plane of the fracture (up, down or sideways) of one side relative to the other.
Juvenile water: is water that exists within, and in equilibrium with, a magma or water-rich volatile fluids that are derived from a magma. This water is released to the atmosphere during a volcanic eruption. Juvenile magmatic water may also be released as hydrothermal fluids during the late stages of magmatic crystallization or solidification within the Earth's crust.
Land-bridge: an isthmus or wider land connection between otherwise separate areas, over which animals and plants are able to cross and colonise new lands.
Laterite: soil types rich in iron and aluminium, formed in hot and wet tropical areas, developed by intensive and long-lasting tropical weathering of the underlying parent rock.
Laterization: a prolonged process of chemical weathering which produces a wide variety in the thickness, grade, chemistry and ore mineralogy of the resulting soils. The majority of the land area containing laterites is between the tropics of Cancer and Capricorn.
Latitude: the angular distance north or south of the equator.
Lava: refers both to molten rock expelled by a volcano during an eruption and the resulting rock after solidification and cooling.
Law of superposition: a given rock layer is probably older than those above it and younger than those below it.
Limestone: a sedimentary rock composed largely of the minerals calcite and aragonite—which are different crystalline forms of calcium carbonate.
Listric fault: a type of fault in which the fault plane is curved where the dip of the fault plane becomes shallower with increased depth.
Lithosphere: the rigid outermost shell the Earth, comprising the crust and the portion of the upper mantle that behaves elastically on time scales of thousands of years or greater.
Magma: a mixture of molten or semi-molten rock, volatiles, and solids, including suspended crystals and dissolved gas. Magma often collects in magma chambers that may feed a volcano or are intruded into adjacent rocks, extruded onto the surface as lava, or ejected to form pyroclastic rock.
Magnetic compass: a navigational instrument that measures directions relative to the Earth's magnetic North Pole.
Magnetic pole: Either of the two locations on a planet's surface at which the planet's magnetic field lines are vertical, referred to as the North Magnetic Pole and South Magnetic Pole.
Magnetic polarity: A pair of electric charges or magnetic poles of equal magnitude but of opposite sign or polarity, separated by a small distance.
Magnetic striping: stripes of alternating normal and reversed magnetic polarities preserved in seafloor crust, caused by Earth's changing magnetic poles.
Magnetite: one of the two common naturally occurring iron oxide minerals (chemical formula Fe_3O_4). Naturally magnetized pieces of magnetite, called lodestone, will attract small pieces of iron, and this was how ancient people first noticed the property of magnetism.
Magnetometer: a geophysical instrument used to measure the Earth's magnetic field and in geophysical surveys to detect magnetic anomalies of various types.
Magnetopause: the abrupt boundary between a magnetosphere and the surrounding plasma, and in planetary science it is also defined as the boundary between the planet's magnetic field and the solar wind.

Magnetosphere: the area of space around the Earth where charged particles are controlled by the magnetic field.
Magnetotail: the opposite end of the compressed magnetic field of a magnetosphere, where the magnetosphere extends far beyond the astronomical object. It contains two lobes, referred to as the northern and southern tail lobes.
Mantle: a layer between the crust and the outer core of the Earth comprising a silicate shell about 2,900 kilometres thick that constitutes about 84 percent of Earth's volume. Two main zones are distinguished in the upper mantle: the inner asthenosphere composed of plastic flowing rock about 200 kilometres thick, and the lowermost part of the lithosphere composed of rigid rock about 50 to 120 kilometres thick.
Mantle convection cell: a Plate Tectonic term for the slow creeping motion of Earth's mantle caused by convection currents carrying heat from the interior of the Earth to the surface.
Mantle diaper: upward rising magmatic rocks such as granite.
Mantle plume: upwelling of hot magmatic rock within the Earth's mantle.
Marginal basin: a sedimentary basin located around the margins of a continent.
Marine basin: or oceanic basin is a large geologic basin that is below sea-level and covered by seawater, including the flanking continental shelves and shallow, epeiric seas.
Marine transgression: a geologic event during which sea-level rises relative to the land and the shoreline moves toward higher ground, resulting in flooding of the land.
Mass extinction: a change in the diversity and variety of life forms, occurring when the rate of extinction increases with respect to the rate of speciation.
Meme: a notion or idea that has remained both untested and uncontested for so long that it is unconsciously considered as a truth.
Metamorphism: the alteration of rocks by high heat (temperatures greater than 150 to 200°C) and pressure (1500 bars), causing profound physical and/or chemical changes to the rocks.
Metamorphic rock: a rock that has undergone metamorphism. The original rock may have been sedimentary rock, igneous rock, or another older metamorphic rock. Some examples of metamorphic rocks are gneiss, slate, marble, schist, and quartzite.
Metasomatism: the alteration of rocks by hot fluids.
Meridian: an imaginary great circle on the Earth's surface connecting points of equal longitude, terminated by the North Pole and the South Pole.
Microbial mats: multi-layered colonies of different types of bacteria.
Mid-ocean-ridge: a general term for an underwater mountain system that demarcates the boundary between two tectonic plates, often termed a divergent plate boundary. The mid-ocean ridges of the world are inter-connected and form a single global mid-oceanic ridge system about 65,000 kilometres long.
Milankovitch cycle: the collective effects of changes in the Earth's movements upon its climate, named after Serbian geophysicist and astronomer Milutin Milanković, based on the variations in eccentricity, axial tilt, and precession of the Earth's orbit.
Mohorovičić discontinuity: usually referred to as the Moho, is the boundary between the Earth's crust and the mantle. Named after the Croatian seismologist Andrija Mohorovičić, the Moho is located 5 to 10 kilometres below the ocean floor, and 20 to 90 kilometres, with an average of 35 kilometres, beneath typical continents.
Motion velocities: vector motions of the Earth's space-based geodetic observational stations in three dimensional coordinates.
Mountain building: the geological processes that underlie the formation of mountains.

Obliquity of the ecliptic: the angle between an object's rotational axis and its orbital axis. The Earth currently has an obliquity, or axial tilt, of about 23.4°.
Ocean basin: large geologic basins that are below sea-level and covered by seawater, including the flanking continental shelves and shallow, epeiric seas.
Ocean circulation: a continuous, directed movement of ocean water generated by the forces acting upon the mean flow, such as breaking waves, wind, Coriolis effect, cabbeling, temperature and salinity differences and tides caused by the gravitational pull of the Moon and the Sun.
Oceanic crust: part of the Earth's lithosphere that surfaces in the ocean basins, primarily composed of volcanic rocks rich in iron and magnesium.
Oceanic plateau: large, relatively flat submarine regions of the oceans that rise well above the level of the ambient seabed. While many oceanic plateaux are composed of continental crust, some plateaux are undersea remnants of large igneous provinces.
Oceanised continental rocks: a former belief that the oceanic crust simply comprised of continental rocks that have been covered with ocean waters.
Ophiolite: a part of the Earth's ancient seafloor crust and the underlying upper mantle that has been uplifted and exposed above sea-level and often emplaced onto continental crustal rocks.
Organelles: a specialized subunit within a cell, such as mitochondria.
Organism: carbon and water-based cellular living systems with complex organization and heritable genetic information, such as an animal, fungus, micro-organism, or plant, which are capable of responses to stimuli, reproduction, growth, and development, as well as maintenance of homeostasis.
Orocline: a bend or curvature of an orogenic belt imposed after it was formed, first introduced by S. Warren Carey in 1955.
Orogen: a belt of rocks characterised by regional folding, metamorphism, and intrusion of magmatic rocks.
Orogenic belt: long parallel zones of deformed and folded rock exhibiting similar characteristics along the length of the belt.
Orogeny: refers to forces and events leading to a large structural deformation of the Earth's crust and uppermost mantle due to the engagement of tectonic plates.
Outgassing: the release of a gas that was dissolved, trapped, frozen or absorbed in a material.
Oxygenic photosynthesis: the process of generating molecular oxygen through chemical reaction.
Palaeo: ancient.
Palaeobiogeography: a study of the Earth's past geography, in particular that of the ancient land surfaces.
Palaeoclimatology: the study of the Earth's past changes in climate taken on the scale of the history of Earth.
Palaeogeography: the study of the Earth's past historical geography. The term generally refers to the study of physical landscapes, but it can also refer to the study of human or cultural environments.
Palaeomagnetics: the study of Earth's remnant magnetism preserved in rocks that contain magnetic minerals, in particular the mineral called magnetite.
Palaeontology: a study of fossils across geological time, how they were formed, and the evolutionary relationships between the various organisms.

Pan-African tectonic event: also known as the Pan-African orogeny, was a series of major orogenic mountain building events which were related to the formation of the Gondwana supercontinent about 600 million years ago.

Pangaea: a supercontinent that formed about 300 million years ago, existed during the late-Paleozoic and early-Mesozoic Eras and began to break apart around 200 million years ago.

Pan-global: covering the entire Earth surface.

Paradigm: universally recognized scientific achievements that, for a time, provide model problems and solutions for a community of researchers.

Paradigm paralysis: a stage where the inability or refusal to see beyond the current models of thinking has set in.

Paradigm shift: a successive transition from one paradigm to another via revolution.

Passive margin basin: a simple ocean basin opening process accompanied by extensive deposition of shallow marine sediments.

Peninsula: a piece of land that is bordered by water on three sides but connected to a mainland.

Petrology: the study and observations of rocks under the microscope.

Period: a time unit subdivision of geologic time defined as a span of years into which the larger era time units are divided into smaller timeframes.

Photosynthesis: a process used by plants and other organisms to convert light energy, normally from the sun, into chemical energy that can be used to fuel the organisms' activities.

Plasma: a state of matter similar to a gas in which portions of the particles are both ionized and moving in a stream analogous to an electric current.

Plasma cosmology: a cosmological model, whose central postulate is that the dynamics of ionized gases and plasmas play important roles in the physics of the universe beyond the Solar system.

Plasmapause: the outer boundary of the plasmasphere, which is defined by an order of magnitude drop in plasma density.

Plasmasphere: or inner magnetosphere, is a region of the Earth's magnetosphere consisting of cool, low-energy plasma located above the Earth's ionosphere.

Plastic deformation: a permanent deformation or change in shape of a solid body under the action of a sustained force without fracture.

Plastic flow: an irreversible deformation process whereby a change in the shape or size of an object due to an applied force is characterized by strain hardening, followed by necking, and finally, fracture (also called rupture).

Plateau: a high plain or tableland, an area of highland usually consisting of relatively flat terrain.

Plate Tectonics: a scientific theory that assumes a status radius Earth premise to describe the large-scale motions of Earth's lithosphere, originally built on the concepts of Continental Drift.

Polar Region: the areas of the globe surrounding the North Pole and South Pole, also known as frigid zones. These regions are dominated by the polar ice-caps, resting respectively on the Arctic Ocean and the continent of Antarctica.

Precambrian Eon: the span of time in Earth's history before the current Phanerozoic Eon, extending from the formation of Earth to the beginning of the Cambrian Period, about 541 million years ago, when macroscopic hard-shelled animals first appeared in abundance. The Precambrian Eon accounts for 88 percent of geologic time.

Radiometric age dating: a technique used to date materials such as rocks, usually based on a comparison between the observed abundance of a naturally occurring radioactive isotope elements and their decay products, using known decay rates.
Refraction: a phenomenon that often occurs when waves travel from a medium with a given refractive index to a medium with another refractive index at an oblique angle.
Remnant magnetism: or natural remnant magnetization, is the permanent magnetism preserved in a rock or sediment and its usage forms the basis of palaeomagnetism and magnetostratigraphy.
Rheology: the study of the flow of matter, primarily in the liquid state, but also as solids under conditions in which they respond during plastic flow rather than deforming elastically in response to an applied force.
Rift valley: or rift basin is a linear zone where the Earth's crust and lithosphere are being pulled apart during continental crustal extension.
Rift zone: zones of continental crustal extension.
Seafloor spreading: a process that occurs at mid-ocean ridges, where basaltic magma rises up along fractures in the crust and cools on the ocean floor to form new seafloor crust.
Secular variation: refers to changes in the Earth's magnetic field over periods of a year or more, reflecting changes in the Earth's core. Phenomena associated with this secular variation include geomagnetic jerk, westward drift and geomagnetic reversals.
Sedimentary basin: a regional topographical down-warp of the Earth's surface generally filled with eroded sediments and water.
Sedimentary rocks: types of rock formed by weathering and erosion of pre-existing rocks, and then transported to a place of deposition by water, wind, ice, mass movement or glaciers.
Seismic activity: the frequency, type, and size of earthquakes experienced over a period of time.
Shear: a form of fault in rocks spread over a broad zone.
Shield volcano: a type of volcano usually built almost entirely of fluid lava flows, named for their large size and low profile, resembling a warrior's shield.
Sill: a tabular igneous rock that has been intruded between older layers of sedimentary rock, beds of volcanic lava or tuff, or into metamorphic rock.
Sinusoidal projection: a pseudo-cylindrical equal-area map projection, sometimes called the Sanson–Flamsteed or the Mercator equal-area projection.
Snowball Earth hypothesis: a hypothesis that considers that the Earth's surface became entirely or nearly entirely frozen at least once during Precambrian times.
Solar flare: a sudden outburst of energy from the Sun's atmosphere.
Solar radiation: the electromagnetic radiation given off by the Sun, particularly infrared, visible, and ultraviolet light.
Solar System: comprises the Sun and its planetary system of eight planets, as well as a number of dwarf planets, moons, and other objects that orbit the Sun.
Solar wind: a stream of charged particles released from the upper atmosphere of the Sun consisting mostly of electrons and protons.
Space geodetics: a relatively new physical science, which basically uses a network of radio telescopes, satellites, and ground-based receiver and transmitter stations to routinely measure the precise dimensions and continental plate motions of the Earth.
Speciation: the evolutionary process by which new biological species arise.

Spherical trigonometry: a branch of spherical geometry that deals with the relationships between trigonometric functions of the sides and angles of the spherical polygons (especially spherical triangles) defined by a number of intersecting great circles on the sphere.
Stadion: ancient Greek measure. The length of a Hellenic stadium is about 185 metres.
Stereographic projection: a particular mapping function that projects a sphere onto a plane.
Stratigraphy: a branch of geology that studies rock layers and layering (stratification). It is primarily used in the study of sedimentary and layered volcanic rocks.
Stratum: (plural strata) a layer of sedimentary rock or soil with internally consistent characteristics that distinguishes it from other layers. The "stratum" is the fundamental unit in a stratigraphic column and forms the basis of the study of stratigraphy.
Stromatolites: microorganisms, commonly known as blue-green algae, which form layered structures in shallow water by the trapping, binding, and cementation of sedimentary grains.
Subduction: a Plate Tectonic term referring to processes that take place at convergent boundaries whereby one tectonic plate moves under another tectonic plate and sinks into the mantle as the plates converge.
Supercontinent: a single landmass consisting of fragments of all or several of the modern continents.
Super-elevation: the difference in elevation (height) between two outer edges of an object.
Suturing: a Plate Tectonic term for the joining together of fragments of various continents.
Symmetric spreading ridge: a parallel, two-sided spreading zone that straddles a mid-ocean-ridge.
Synapsids: terrestrial vertebrates, including the ancestors of mammals.
Synkinematic metamorphism: the alteration of rocks by heat and pressure at the same time as a folding or deformation event.
Taxa: a group of one or more populations of organisms.
Tectonic plates: a feature of the Earth's crust and mantle where the Earth's lithosphere is divided into a number of large rigid blocks (plates) that are floating on semi-fluid rock and are thus able to interact with each other at their boundaries.
Tectonics: the processes which control the structure and properties of the Earth's crust, and its evolution through time.
Tectonism: the processes of mountain building, the growth and behaviour of the plates that comprise the Earth's outer shell, and how the plates interact with each other.
Terrane: an extensive fault-bounded body of rock whose geological history differs from those of abutting terranes.
Terrella: Latin for "little" Earth but used here in the context of small Earth.
Therapsids: a group of the most advanced terrestrial vertebrates, including the ancestors of mammals.
Thermal expansion: the tendency of matter to change in volume in response to a change in temperature.
Thermal plume: or mantle plume, is a thermal abnormality in the Earth's crust where hot rock nucleates at the core-mantle boundary and rises through the Earth's mantle and crust as a diapir.
Time-dependent deposition: where sediments or seafloor volcanic rocks were deposited or intruded over different periods of time at the prevailing Earth curvature existing at the time of deposition or intrusion.
Transform-fault boundary: also known as a conservative plate boundary is a type of fault seen on the seafloor whose relative motion is predominantly horizontal.

Transgressive-regressive sea-level changes: relatively small, local, or regional variations in sea-level where the term transgressive refers to when the sea encroaches on the land for a period of time, and regressive refers to when the sea recedes from the land.
Trench: or oceanic trenches are hemispheric-scale long but narrow topographic depressions of the seafloor. They are also the deepest parts of the ocean floor and are a distinctive morphological feature of convergent plate boundaries.
Triple-junction: the junction of three separate plates along spreading ridges.
Upper mantle: the upper part of the mantle located between the crust and the outer core.
Upwelling: an oceanographic phenomenon that involves wind-driven motion of dense, cooler, and usually nutrient-rich water towards the ocean surface, replacing the warmer, usually nutrient-depleted surface water.
Van Allen radiation belt: one of at least two layers of energetic charged particles (plasma) that is held in place around the planet Earth by the planet's magnetic field.
Vascular plants: a large group of land plants that have lignified tissues for conducting water and minerals throughout the plant, including the club mosses, horsetails, ferns, gymnosperms (conifers), and angiosperms (flowering plants).
Vector: a geometric entity endowed with magnitude and direction as well as a positive-definite inner product, used to represent physical quantities that have both magnitude and direction, such as force, in contrast to scalar quantities, which have no direction.
Vertebrates: animals with backbones and spinal columns, including the jawless fishes, bony fishes, sharks and rays, amphibians, reptiles, mammals, and birds.
Viscosity: a measure of a fluid's resistance to gradual deformation by shear stress or tensile stress.
Volcanism: all phenomena resulting from and causing magma within the crust or mantle of the Earth to rise through the crust and form volcanic rocks on the surface.
Volatile-bearing minerals: minerals capable of retaining various elements, such as carbon, oxygen, hydrogen, and sulphur in their crystal lattice.
Wadati–Benioff zone: a Plate Tectonic term referring to a deep active seismic area in a subduction zone, producing deep-seated earthquakes, the foci of which may be as deep as about 670 kilometres.
Weathering: the breaking down of rocks, soils, and minerals as well as artificial materials through contact with the Earth's atmosphere, biota, and waters.
Wilson cycle: a Plate Tectonic supercontinental cycle describing the quasi-periodic aggregation and dispersal of Earth's continental crust.
Zoophytes: a type of animal that visually resembles a plant but is now obsolete in modern usage.

22 References

BAILEY D. K. & STEWART A. D. 1983. Problems of Ocean Water Accumulation on a Rapidly Expanding Earth. 67-69 in Carey S. W. (ed): Expanding Earth Symposium, Sydney 1981, University of Tasmania.

BARNETT C. H. 1962. A suggested reconstruction of the land masses of the Earth as a complete crust. Nature 195, 447-448.

BARRON E. J. 1983. A warm, equable Cretaceous: the nature of the problem. Earth Science Reviews 19, 305-338.

BLATT H. & JONES R. L. 1975. Proportions of exposed igneous, metamorphic and sedimentary rocks. Geological Society of America Bulletin 86, 1085-1088.

BLINOV V. F. 1983. Spreading rate and rate of expansion of the Earth. In: Carey S.W. ed. Expanding Earth Symposium, Sydney, 1981. University of Tasmania, 297-304

BOSS A. P. & SACKS I. S. 1985. Formation and Growth of Deep Mantle Plumes. Geophysical Journal of the Royal Astronomical Society 80, 241-255.

BRUNNSCHWEILER R. O. 1983. Evolution of Geotectonic Concepts in the Past Century. 9-15, in Carey S. W. (ed): Expanding Earth Symposium, Sydney 1981, University of Tasmania.

BURG J-P. & FORD M. 1997. Orogeny through time: an overview. In: Burg J-P. & Ford M. eds. Orogeny through time. Geological Society Special Publication 121, 1-17.

BUTLER R. F. 1992. Paleomagnetism: Magnetic Domains to Geologic Terranes. Blackwell Scientific Publications, USA.

CAREY S. W. 1958. The tectonic approach to continental drift. In: Continental Drift, a Symposium. University of Tasmania, Hobart, 177-355.

CAREY S. W. 1963. The Asymmetry of the Earth. Australian Journal of Science, 25, 369-383 & 479-488.

CAREY S. W. 1975. The expanding Earth - an essay review. Earth Science Reviews 11, 105-143.

CAREY S. W. 1976. The Expanding Earth. Elsevier, Amsterdam.

CAREY S. W. 1983a. Earth expansion and the null Universe. In: Carey S.W. ed. Expanding Earth Symposium, Sydney, 1981. University of Tasmania, 365-372.

CAREY S. W. 1983b. The necessity for Earth expansion. In: Carey S.W. ed. Expanding Earth Symposium, Sydney, 1981. University of Tasmania, 375-393.

CAREY S. W. 1988. Theories of the Earth and Universe: A History of Dogma in the Earth Sciences. Stanford University press, Stanford, California.

CAREY S. W. 1996. Earth, Universe, Cosmos. University of Tasmania, Hobart.

CEBULL S. E. & SHURBET D. H. 1992. Conventional plate tectonics and orogenic models. In: Chatterjee S. & Hotton N (III) eds. New Concepts in Global Tectonics. Texas Technical Press, USA, 111-117.

CHATTERJEE S. 1997. The Rise of Birds. Baltimore: Johns Hopkins University Press.

CGMW & UNESCO 1990. Geological Map of the World. Commission for the Geological Map of the World, Paris.

CHUMAKOV N. M. & ELSTON D. P. 1989. The paradox of late-Proterozoic glaciations at low latitudes. Episodes 12, 115-1120.

COX A. & DOELL R. R. 1961. Palaeomagnetic evidence relevant to a change in the Earth's radius. Nature 189, 45-47.
COX C. B. 1990. New geological theories and old biogeographical problems. Journal of Biogeography 17, 117-130.
CREER K. M 1965. An Expanding Earth. Nature 205, 539-544.
CREER K. M., IRVING E. & RUNCORN S. K. 1954. The direction of the geomagnetic field in remote epochs in Great Britain. Journal of Geomagnetics and Geoelectricity 6, 163-168.
CROWLEY T. J. 1994. Pangean climates. In: Klein G. D. ed. Pangea: paleoclimate, tectonics, and sedimentation during accretion, zenith, and break-up of a supercontinent. Geological Society of America Special Paper 288, 25-39.
CWOJDZINSKI S. 2003. The Tectonic Structure of the Continental Lithosphere Considered in the Light of the Expanding Earth Theory – A Proposal of a New Interpretation of Deep Seismic Data. Ed: Majewska A. Polish Geological Institute, Special Papers.
DENNIS J. G. 1962. Fitting the Continents. Nature 196, 364
DODSON P. & TATARINOV L. P. 1990. Dinosaur extinction. In: Weishampel D. B., Dodson P. & Osmólska H. eds. 1990. The Dinosauria. University of California Press, Los Angeles, 55-62.
EGGLER D. H. 1987. Solubility of major and trace elements in mantle metasomatic fluids: experimental constraints. In: Menzies M. A. & Hawkesworth C. J. eds, Mantle Metasomatism. Academic Press, London. 21-41.
EGYED L. 1960. Some remarks on continental drift. Geofisica Pura Applica 45, 115-116.
EGYED L. 1963. The expanding Earth? Nature 197, 1059-1060.
EICHLER J. B. 2011. A New Mechanism for Matter Increase within the Earth. Ettore Majorana Foundation and Centre for Scientific Culture, 37th Interdisciplinary Workshop of the International School of Geophysics, Erice, Sicily, 4-9 October 2011, 159-162.
ERWIN D. H. 1993. The great dying: life and death in the Permian. Columbia University Press, New York.
EYLES N. & YOUNG G. M. 1994. Geodynamic controls on glaciation in Earth history. In: Deynoux M., Miller J. M. G., Domack E. W., Eyles N., Fairchild I. J. & Young G. M. eds. International Geological Correlation Program Project 260: Earth's Glacial Record. Cambridge University Press, 1-28.
FLÜGEL E. 1994. Pangean shelf carbonates: Controls and paleoclimatic significance of Permian and Triassic reefs. In: Klein, G. D. ed. Pangea: paleoclimate, tectonics, and sedimentation during accretion, zenith, and break-up of a supercontinent. Geological Society of America Special Paper 288, 247-266.
FRANCIS J. E. & FRAKES L. A. 1993. Cretaceous climates. Sedimentology Review 1993. 17-28.
GARFUNKEL Z. 1975. Growth, shrinking, and long-term evolution of plates and their implications for the flow pattern in the mantle. Journal of Geophysical Research 80, 4425-4432.
HEEZEN B.C. 1960. The rift in the ocean floor. Scientific America, 203, 98-110.
HILGENBERG O. C. 1933. Vom wachsenden Erdball. Selbstverlag, Berlin.
HILGENBERG O. C. 1962. Paläopollagen der Erde. Neues Jahrb. Geol. Und Paläontol, Abhandl 116, Struttgart.
HOFFMAN P. F., KAUFFMAN A. J., HALVERSON G. P. & SCHRAG D. P. 1998. A Neoproterozoic snowball earth. Science 281, 1342-1346.

HOTTON N. (III) 1992. Global distribution of terrestrial and aquatic tetrapods, and its relevance to the position of continental masses. In: Chatterjee S. & Hotton N. (III) eds. New Concepts in Global Tectonics. Texas Technical Press, USA, 267-285.
HURRELL S. W. 1994. Dinosaurs and the Expanding Earth. OneoffPublishing.com.
JEFFREYS H. 1962. A Suggested Reconstruction of the Land Masses of the Earth as a Complete Crust: Comment. Nature 195, 148.
KLEIN G. D. 1994. Pangea: paleoclimate, tectonics, and sedimentation during accretion, zenith, and break-up of a supercontinent. Geological Society of America Special Paper 288.
KOZIAR J. 1980. Ekspansja den oceanicznych I jej zwiazek z hipotaza ekspansji Ziemi. Sprawozdania Wroclawskiego Towarzystwa Naukowego, 35, 13-19.
KREMP G. O. W. 1992. Earth Expansion Theory Versus Statical Earth Assumption. In: Chatterjee S. & Hotton N (III) eds. New Concepts in Global Tectonics. Texas Technical Press, USA, 297-307.
KRÖNER A. 1982. Archaean to early-Proterozoic tectonics and crustal evolution: A review. Revista Brasileira de Geociências 12, 15-31.
KUHN T. S. 1970. The Structure of Scientific Revolutions. University of Chicago Press.
LAMBERT I. B. 1982. Early geobiochemical evolution of the Earth. Revista Brasileira de Geociências 12, 32-38.
LARSON K. M., FREYMUELLER J. T. & PHILIPSEN S. 1997. Global plate velocities from the Global Positioning System. Journal of Geophysical Research 102, 9961-9981.
LARSON R. L., PITMAN W. C. (III)., GOLOVCHENKO X., CANDE S. C., DEWEY J. F., HAXBY W. F. & LaBRECQUE (mapcompilers) 1985. The Bedrock Geology of the World. Freeman & Co., New York.
Le PICHON X. & HEIRTZLER J. R. 1968. Magnetic anomalies in the Indian Ocean and sea floor spreading. Journal of Geophysical Research, 73:2, 119-2,136.
LINDERMAN B. 1927. Kettengebirge, Kontinentale Zerspaltung und Erdexpansion. Gustav Fischer Publishers, Jena. 186p.
LOWMAN P.D. Jr. 1992. Plate tectonics and continental drift in geologic education. In: Chatterjee S. & Hotton N. (III) eds. New Concepts in Global Tectonics. Texas Technical Press, USA, 3-9.
MAXLOW J. 1995. Global Expansion Tectonics: the geological implications of an expanding Earth. Unpublished Master of Science thesis, Curtin University of Technology, Perth, Western Australia.
MAXLOW J. 2001. Quantification of an Archaean to recent Earth Expansion Process using Global Geological and Geophysical Data Sets. Unpublished PhD thesis, Curtin University of Technology, Perth, Western Australia. http://adt.curtin.edu.au/theses/available/adt-WCU20020117.145715
MAXLOW J. 2005. Terra non Firma Earth: Plate Tectonics is a Myth. TerrellaPress@bigpond.com.au
McCARTHY D. 2003. Biogeography and Scientific Revolutions. The Systematist 2005 No. 25, 3-12.
McELHINNY M. W. & BROCK A. 1975. A new paleomagnetic result from East Africa and estimates of the Mesozoic paleoradius. Earth and Planetary Science Letters 27, 321-328.
McELHINNY M. W. & LOCK J. 1996. IAGA paleomagnetic databases with Access. Surveys in Geophysics, 17, 575-591.
MEYERHOFF A. A., BOUCOT A. J., MEYERHOFF-HULL D. & DICKINS J. M. 1996. Phanerozoic faunal and floral realms of the Earth: The intercalary relations of the

Milvinokaffric and Gondwana faunal realms with the Tethyan faunal realm. Geological Society of America, Memoir 189.
MILANOVSKY E. E. 1980. Problems of the tectonic development of the Earth in the light of concept on its pulsations and expansion. Revue de Geologique et de Geographie Physique 22, 15-27.
MOORBATH S. 1982. Crustal evolution in the Early Precambrian. Revista Brasileira de Geociências 12, 39-44.
NOËL D. 1989. NUTEERIAT: Nut Trees, the Expanding Earth, Rottnest Island and All That. Published for the Planetary Development Group TREE CROPS CENTRE. Cornucopia Press, Australia.
OLLIER C. D. 1985. Morphotectonics of passive continental margins: Introduction. Z. Geomorphology Suppl. Bd. N. F. 54, 1-9.
OLLIER C. D. & PAIN C. F. 2000. The origin of mountains. Routledge, London.
OWEN H. G. 1976. Continental displacement and expansion of the Earth during the Mesozoic and Cenozoic. Philosophical Transactions of the Royal Society, London 281. 223-291.
ORESKES N. 1999. The Rejection of Continental Drift: Theory and Method in American Earth Science, Oxford University Press.
OWEN H. G. 1983. Atlas of Continental Displacement, 200 Million Years to the Present. Cambridge Earth Series, Cambridge University Press, Cambridge.
PARSONS B. 1982. Causes and consequences of the relation between area and age of the ocean floor. Journal of Geophysical Research 87, 289-302.
RICKARD M. J. 1969. Relief of curvature on expansion - a possible mechanism of geosynclinal formation and orogenesis. Tectonophysics 8, 129-144.
ROBAUDO S. & HARRISON C. G. A. 1993. Plate tectonics from SLR and VLBI global data. In: Smith D. E., & Turcotte D. L. eds. Contributions of Space Geodesy to Geodynamics: Crustal Dynamics. Geodynamics Series, Volume 23. American Geophysical Union.
RUBEY W. W. 1975. Geologic history of sea water; an attempt to state the problem. In: Kitano Y. ed. Geochemistry of Water. Dowden, Hutchinson & Ross Inc., Stroudsburg, Pennsylvania.
SAULL V. A. 1986. Wanted: Alternatives to plate tectonics. Opinion. Geology 14, 1986
SCALERA G. 1990. General Clues Favouring Expanding Earth Theory. In: Critical Aspects of the Plate Tectonics Theory: Volume II, Alternative Theories. Theophrastus Publishers, Athens, Greece, 65-93.
SCOTESE C. R. 2001. PALEOMAP Project http://www.scotese.com/
SHEN W., SUN R., CHEN W., ZHANG Z., LI J., HAN J., & DING H. 2011. Evidences of Earth Expansion from Space-Geodetic and Gravimetric Observations. Ettore Majorana Foundation and Centre for Scientific Culture, 37th Interdisciplinary Workshop of the International School of Geophysics, Erice, Sicily, 4-9 October 2011, 131-134.
SHIELDS O. 1996. Geologic significance of land organisms that crossed over the Eastern Tethys "Barrier" during the Permo-Triassic. Palaeobotanist 43, 85-95.
SHIELDS O. 1997. Is plate tectonics withstanding the test of time? Annali di Geofisica, Vol XL, 1-8.
SHULDINER V. I. 1982. The oldest high grade terranes: Possible relicts of primeval Earth crust. Revista Brasileira de Geociências 12, 45-52.

SMILEY C. J. 1992. Paleofloras, faunas, and continental drift: some problem areas. In: Chatterjee S. & Hotton N. (III) eds. New Concepts in Global Tectonics. Texas Technical Press, USA, 241-257.

SOUDARIN L., CRÉTAUX J-F. & CAZENAVE A. 1999. Vertical crustal motions from the DORIS space-geodesy system. Geophysical Research Letters 26, 1207-1210.

STEINER J. 1977. An expanding Earth on the basis of sea-floor spreading and subduction rates. Geology 5, 313-318.

STEVENS G. R. 1990. The influences of paleogeography, tectonism and eustacy on faunal development in the Jurassic of New Zealand. In: Pallini G., Cecca F., Cresta S. & Santantonio M. eds. Atti del secondo convegno internazionale. Fossili, evoluzione, ambiente. Pergola, 1987, 441-457.

STORETVEDT K. M. 1992. Rotating Plates: New Concepts in Global Tectonics. In: Chatterjee S. & Hotton N. (III) (eds), New Concepts in Global Tectonics. Texas Technical Press, USA, 203-220.

VAN ANDEL S. I. & HOSPERS J. 1968. Palaeomagnetic calculations of the radius of the ancient Earth by means of the palaeomeridian method. Physics of the Earth and Planetary Interiors 1, 155-163.

VAN HILTEN D. 1963. Palaeomagnetic indications of an increase in the Earth's radius. Nature 200, 1277-1279.

VINE F. J. & MATHEWS D. H. 1963. Magnetic anomalies over oceanic ridges. Nature 199, 947-949.

VOGEL K. 1983. Global models and Earth expansion. In: Carey S.W. ed. Expanding Earth Symposium, Sydney, 1981. University of Tasmania, 17-27.

VOGEL K. 1984. Beitrage zur frage der expansion der Erde auf der grundlage von globenmodellen. Z. geol. Wiss. Berlin 12. 563-573.

VOGEL K. 1990. The expansion of the Earth - an alternative model to the plate tectonics theory. In: Critical Aspects of the Plate Tectonics Theory; Volume II, Alternative Theories. Theophrastus Publishers, Athens, Greece, 14-34.

WEGENER A. 1915. *Die entstehung der kontinente und ozeane* [The Origin of Continents and Oceans.]

WEIJERMARS R. 1986. Slow but not fast Global Expansion may Explain the Surface Dichotomy of the Earth. Physics of the Earth and Planetary Interiors, 43. 67-89.

WEISHAMPEL D. B. 1990. Dinosaurian distribution. In: Weishampel D. B., Dodson P. & Osmólska H. eds. 1990. The Dinosauria. University of California Press, Los Angeles.

WESSON P. S. 1973. The implications for geophysics of modern cosmologies in which G is variable. Quarterly Journal of the Royal Astrological Society 14, 9-64.

YUEN D. A. & PELTIER W. R. 1980. Mantle Plumes and the Thermal Stability of the "D" Layer. Geophysical Research Letters, 7, 625-628.

ZAREBSKI T. 2009. Toulmin's model of argument and the "logic" of scientific discovery. Studies in logic, Grammar and Rhetoric 16(29) 2009.

Appendix A

Kinematics of Earth Expansion

A mathematical equation for an exponential increase in Earth radius extending from the Archaean to present-day is derived by considering the mathematical equation for linear regression:

$$y = Ax + B \qquad \text{Equation A1}$$

Where: y = the Y axis variable, x = the X axis variable, A = the gradient of a line, and B = the y-intercept of the line.

For a linear increase in ancient radius this equation can be written as:

$$R = At + B \qquad \text{Equation A2}$$

Where: R = Earth radius, t = time before the present (negative), B = the y-intercept of the line.

To determine an exponential increase in ancient radius, Equation A2 can be written as:

$$\ln R_a/R_0 = At + B \qquad \text{Equation A3}$$

Where: ln = natural logarithm, R_0 = present Earth radius at time t_0, R_a = ancient Earth radius at time t_a.

Equation A3 simply expresses the exponential curve as a straight line suitable for analysis using the linear regression Equation A2. Rearranging Equation A3 for ancient Earth radius:

$$Ra = R_0 e^{(At + B)} \qquad \text{Equation A4}$$

Where: e = base of natural logarithm.

Mathematical modelling of the post-Triassic increase in Earth radius in Maxlow (1995) demonstrated that the y-intercept B is negligible and can be disregarded. The gradient of the line A, representing the ancient Earth radius, is a constant k. Equation A4 can then be written as:

$$R_a = R_0 e^{kt} \qquad \text{Equation A5}$$

The Archaean-Mesoproterozoic primordial Earth radius p, determined from empirical modelling, is approximately 1,700 kilometres. This represents the radius of an Archaean to early-Proterozoic proto-Earth at formation or crustal stabilisation. It is inferred that Earth radius increases exponentially from this primordial proto-Earth to the present radius. An equation for exponential increase in Earth radius from the Archaean to Recent is expressed as:

$$\mathbf{R_a = (R_0\text{-}R_p)e^{kt} + 1700} \qquad \text{Equation A6}$$

Where: $k = 4{\cdot}5366 \times 10^{-9}$/year and the constant k is determined from Equation A3 by using post-Triassic Earth radii derived from empirical seafloor data and 1,700 kilometres as the limiting primordial proto-Earth radius to solve for gradient A.

Equation A6 represents a mathematical expression for exponential change in the ancient radius of the Earth from the Archaean to the Recent, and assumes that oceanic and continental crust is fully fixed in the rock record with little or no requirement for removal of crusts by subduction. This equation therefore represents the fundamental Expansion Tectonic equation for determining ancient Earth radius at any time back or forward in time from the present-day.

Radius, Circumference, Surface Area, Volume

An investigation into the kinematics of an Earth expansion with time uses Equation A6 for ancient radius:

Known parameters of the present-day Earth at time t_0 are:

Radius	$R_0 = 6.3708 \times 10^6$ m
Surface Area	$S_0 = 5.1000 \times 10^{14}$ m^2
Volume	$V_0 = 1.0830 \times 10^{21}$ m^3
Mass	$M_0 = 5.9730 \times 10^{24}$ kg
Mean Density	$D_0 = 5.5150 \times 10^3$ kg/m^3
Universal Gravitation	$G_0 = 6.6732 \times 10^{-11}$ m^3 kg^{-1} sec^{-2}
Surface Gravity	$g_0 = 9.780317$ m sec^{-2}

Derived quantities for the variables radius, circumference, surface area and volume can then be calculated for the time interval t_0 (present-day) to t_{4000} (early-Archaean) using the following equations:

Ancient Radius	$R_a = (R_0 - R_p)e^{kt} + R_p$
Ancient Circumference	$C_a = 2\pi\, R_a = 2\pi((R_0 - R_p)e^{kt} + R_p)$
Ancient Surface Area	$S_a = 4\pi\, R_a^2 = 4\pi((R_0 - R_p)e^{kt} + R_p)^2$
Ancient Volume	$V_a = (4/3)\pi(R_a)^3 = (4/3)\pi((R_0 - R_p)e^{kt} + R_p)^3$

Where: R_a = past Earth radius, R_0 = present mean Earth radius = 6370.8 km, R_p = primordial Earth radius = 1,700 km, e = an exponent, k = $4.5366\text{x}10^{-9}$/year, t = time before present (negative).

Derived quantities for radius, circumference, surface area and volume for the various time intervals used for each of the Expansion Tectonic small Earth models are listed in Table A1.

Geologic Time Scale		Kinematics			
Event	Beginning Age	Radius	Circumference	Area	Volume
	(m.y.)	(km)	(kmx 10^4)	(km^2x10^5)	(km^3x10^8)
Future	300	19916.4	12513.81	49845.91	330916.26
Future	200	13272.8	8339.54	22137.79	97943.42
Future	100	9052.2	5687.64	10297.07	31070.24
Future	5	6478	4070.22	5273.35	11386.84
Present-day	0	6370.8	4002.89	5100.32	10831.05
Quaternary	-1.9	6330.7	3977.7	5036.34	10627.87
Pliocene	-5.9	6247.4	3925.38	4904.72	10213.97
Miocene	-23	5908	3712.11	4386.23	8637.94
Oligocene	-37.7	5636.5	3541.54	3992.39	7501.00
Eocene	-59.2	5270.7	3311.68	3490.97	6133.29
Paleocene	-66.2	5159.1	3241.55	3344.69	5751.85
Late-Cretaceous	-84	4890.7	3072.94	3005.79	4900.19
Mid-Cretaceous	-118.7	4426	2780.93	2461.68	3631.79
Early-Cretaceous	-143.8	4132.6	2596.59	2146.14	2956.38
Late-Jurassic	-160	3960.2	2488.29	1970.84	2601.67
Early-Jurassic	-205	3542.9	2226.05	1577.32	1862.74
Triassic	-245	3237	2033.89	1316.75	1420.79
Permian	-286	2976.2	1869.98	1113.07	1104.22
Carboniferous	-360	2612.2	1641.32	857.5	746.66
Devonian	-408	2433.7	1529.16	744.31	603.82
Silurian	-438	2340.4	1470.5	688.3	536.96
Ordovician	-505	2172.5	1365.04	593.11	429.52
Cambrian	-570	2051.9	1289.22	529.06	361.85
Late-Proterozoic	-900	1778.7	1117.61	397.59	235.74
Mid-Proterozoic	-1,600.00	1703.3	1070.21	364.57	206.99
Early-Proterozoic	-2,500.00	1700.1	1068.18	363.19	205.82
Archaean	-4,000.00	1700	1068.14	363.17	205.8

Table A1 Derived quantities for the variables radius, circumference, surface area and volume. Calculated from radius Equation A6.

Mass Density and Surface Gravity

The kinematics of mass, density and surface gravity, assuming a simplistic spherical Earth model with a homogeneous mass and density distribution is governed by the standard equations:

Density	$D = \text{Mass} / \text{Volume}$
Surface gravity	$g = GM/R^2$

In the density equation, volume is calculated from Equation A6 and the variables mass M and density D are unknown. In the surface gravity equation, universal gravitation G is assumed to be constant. Three possible scenarios exist for an expression representing the kinematics of mass, density and surface gravity with time:

- Mass remains constant requiring density to decrease exponentially as volume increases throughout geological time.
- Density remains constant requiring mass to increase exponentially as volume increases throughout geological time.
- Both mass and density are variable as volume increases throughout geological time.

Derived quantities for the variables mass, density and surface gravity for conditions of a) constant mass and b) constant density are calculated by incorporating Equation A6 for the time interval t_0 to t_{4000} using the following equations:

a) Constant Mass: $M = M_0$

$$D = M/V = M_0/(4/3)\pi((R_0\text{-}R_p)e^{kt}+R_p)^3$$
$$g = GM/R^2 = GM_0/((R_0\text{-}R_p)e^{kt}+R_p)^2$$

b) Constant Density: $D = D_0$

$$M = DV = D_0(4/3)\pi((R_0\text{-}R_p)e^{kt}+R_p)^3$$
$$g = GM/R^2$$
$$GD_0(4/3)\pi((R_0\text{-}R_p)e^{kt}+R_p)^3/((R_0\text{-}R_p)e^{kt}+R_p)^2$$
$$= GD_0(4/3)\pi((R_0\text{-}R_p)e^{kt}+R_p)$$

Each of these variables is shown graphically in Figure A1 and are discussed further in Chapter 6.

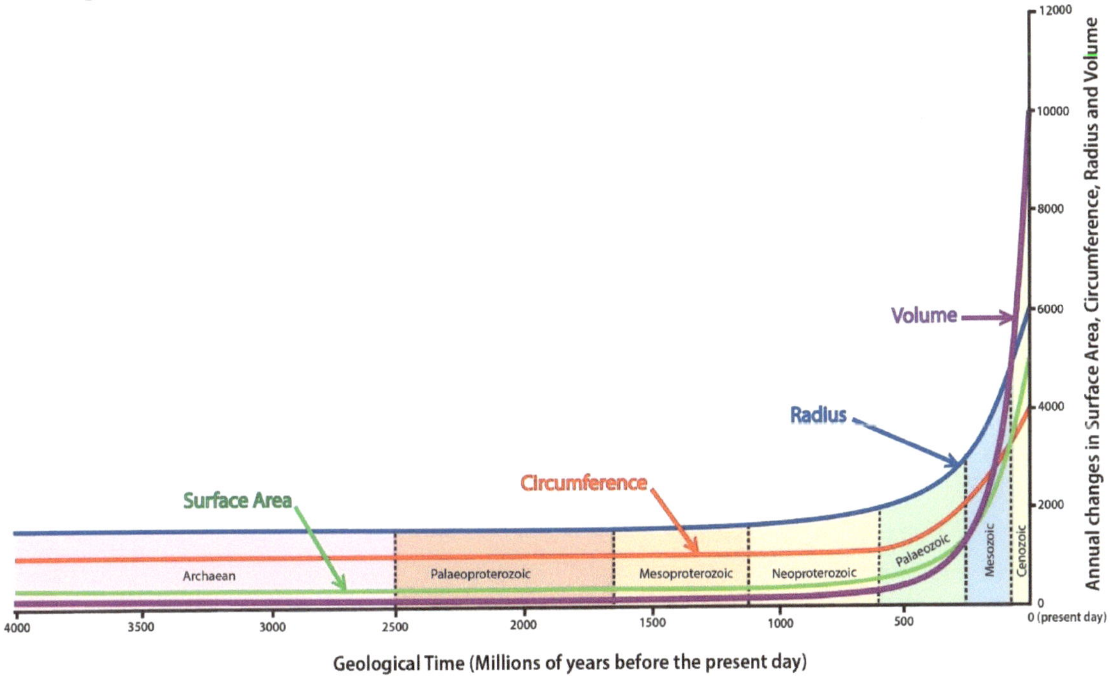

Figure A1 The annual changes in Earth radius, circumference, surface area and volume over time. The graphs range from the Archaean to the present-day and are derived using the established Earth radius formula. Annual change in Radius is in kilometres, Circumference is times 10 kilometres, Surface Area is times 100 million square kilometres, and Volume is times 100 million cubic kilometres.

Secular Rate of Earth Expansion

The secular rate of Earth expansion is defined as an incremental increase in the physical dimensions of the Earth throughout geological time and may be quantified by considering the variables radius, circumference, surface area and volume. Mass, density and surface gravity are considered speculative and mass is assumed to increase with time (Chapter 6). The application of Equation A6 enables the secular rate of Earth expansion to be determined for the following variables between time t_1 and t_2:

$$dR/dt = (R_1-R_2)/(t_1-t_2) = (((R-Rp)e^{kt1}+Rp)-((R_0-Rp)e^{kt2}+Rp))/(t_1-t_2)$$

$$dC/dt = (C_1-C_2)/(t_1-t_2)=((2\pi((R_0-Rp)e^{kt1}+Rp)-(2\pi((R_0- Rp)e^{kt2}+ Rp))/(t_1-t_2)$$

$$dS/dt = (S_1-S_2)/(t_1-t_2) = ((4\pi((R_0- Rp)e^{kt1}+ Rp)^2-(4\pi((R_0- Rp)e^{kt2}+ Rp)^2)/(t_1-t_2)$$

$dV/dt = (V_1\text{-}V_2)/(t_1\text{-}t_2) = ((4\pi/3)((R_0\text{-}Rp)e^{kt1} + Rp)3\text{-}(4\pi/3)((R_0\text{-} Rp)e^{kt2} + Rp)^3/(t_1\text{-}t_2)$

Assuming a constant Earth density with variable mass and surface gravity:

$dD/dt = 0$

$dM/dt = (M_1\text{-}M_2)/(t_1\text{-}t_2)$
$= (D_0(4/3)\pi((R_0\text{-}R_p)e^{kt1}+R_p)^3\text{-}D_0(4/3)\pi((R_0\text{-}R_p)e^{kt2}+R_p)^3)/(t_1\text{-}t_2)$

$dg/dt = (g_1\text{-}g_2)/(t_1\text{-}t_2)$
$= (GD_0(4/3)\pi((R_0\text{-} R_p)e^{kt1}+ R_p)\text{-}GD_0(4/3)\pi((R_0\text{-}R_p)e^{kt2}+R_p))/(t_1\text{-}t_2)$

The present secular rates of Earth expansion derived from the above formulae at time t_0 are:

Radius	dR/dt_0 = 22 mm/year
Circumference	dC/dt_0 = 140 mm/year
Surface Area	dS/dt_0 = 3.50 km^2/year
Volume	dV/dt_0 = 11,000 km^3/year
Mass	dM/dt_0= 60 x 10^{12}tonnes/year
Density	$dD/dt_0 = 0$
Surface Gravity	dg/dt_0 = 3.4 x 10^{-8} msec^{-2}/year

Appendix B

Palaeomagnetic Formulae

The mathematical equation for an exponential increase in Earth radius extending from the Archaean to present-day (Equation A6) is used to modify conventional palaeomagnetic formulae.

Magnetic Dipole Equations

The geocentric axial dipole model is central to the principles of palaeomagnetism (Figure B1). In this conventional Earth model a magnetic field is produced by a single magnetic dipole M, located at the centre of the Earth and aligned with the rotation axis.

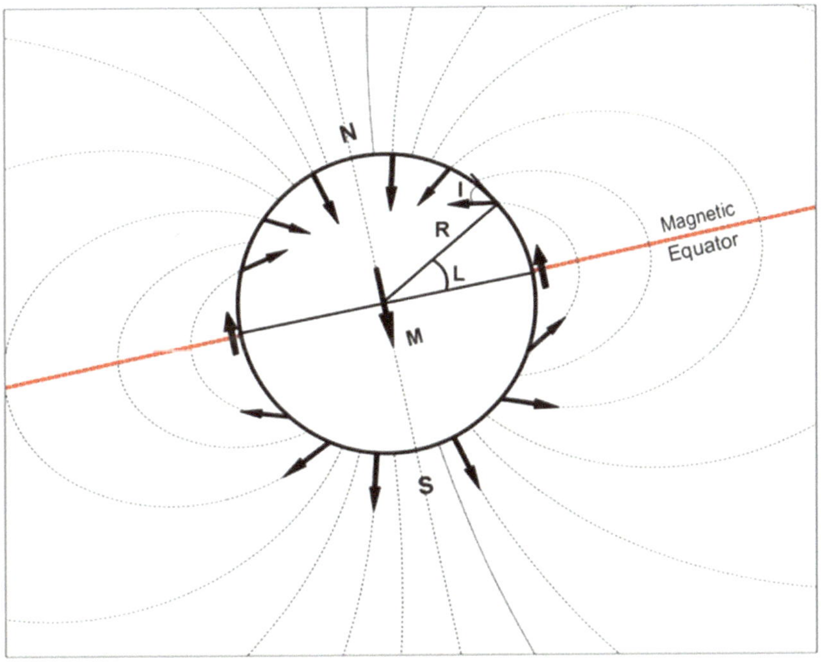

Figure B1. Geocentric axial dipole model. A magnetic dipole M is located at the centre of the Earth and aligned with the rotation axis. The geographic latitude is L, the mean Earth radius is R, the magnetic field directions at the Earth's surface, produced by the geocentric axial dipole, are schematically shown, inclination I is shown for one location and N is the North geographic pole.

Derivation of the conventional geomagnetic dipole equation is given in Butler (1992) and defined as:

$$\tan I = 2 \tan L = 2 \cot P \qquad \text{Equation B1}$$

Where: I is the mean inclination of the magnetic field at a particular site location, increasing from -90° at the geographic South Pole to +90° at the geographic North Pole, L is the geographic latitude determined from I, and P is the geographic colatitude determined from I.

Rearranging the dipole equation gives colatitude P as:

$$P = \cot^{-1} (\tan I/2) = \tan^{-1} (2/\tan I) \qquad \text{Equation B2}$$

Equation B2 represents a means of calculating a great-circle angular distance from the sample site to the magnetic pole and, like latitude, is independent of any Earth radius or time constraints imposed by the sample. In fact, this dipole equation is a general equation applicable to any sized magnetic sphere that obeys the geocentric axial dipole model (Figure B1). On the sphere shown in Figure B1 the magnetic lines of force behave in exactly the same manner, irrespective of the scale or size of the object, and, for a given site inclination, the colatitude (and latitude) calculated using the dipole equation will always remain the same—simply because it is an angular measurement.

Figure B2 demonstrates this very important characteristic of the dipole equation on a magnetized spherical object of different sizes. For a given site inclination value I, the dipole equation remains true for an infinite number of sample sites along a radius vector R, passing from the centre of the Earth through the site location and beyond. Colatitude (Equation B2) calculated from the inclination I at sites S_1 to S_n, located along the radius vector R, is, by definition, equal to a constant angular measurement P. As can be seen in Figure B2, at site S_1 the colatitude P is also defined by a surface distance labelled Distance D_1 extending from the sample site to the ancient magnetic Pole P_1. This distance D_1, of course, is not equal to distances D_2 or D_n for sites S_2 and S_n using the identical values of inclination I and colatitude P.

To simulate a simplistic radial expansion of the Earth from Radius R_1 to R_2 to R_n the ancient magnetic pole positions Pole P_1, Pole P_2 and Pole P_n, calculated using the conventional dipole Equation B2, are shown in Figure B2. Because the conventional dipole equation uses angular measurements and has no provision for either a radial or time component to compensate for the shift in actual ancient pole position with increase in radius, the calculated pole positions at any moment in time will always coincide with the same geomagnetic pole. This is why the conventional dipole equation cannot be used on an Expansion Tectonic Earth, because it insists on using angular measurements.

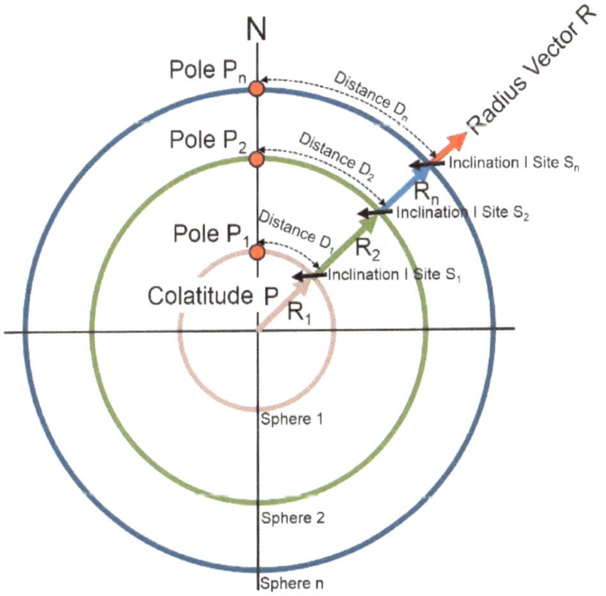

Figure B2. Cross sections of the Earth showing a number geocentric axial dipolar spheres. The ancient magnetic pole locations determined using the conventional dipole equation are shown at Poles P_1 to P_n.

Determining the ancient pole position Pole P_a on the present-day Earth where radius varies exponentially with time, is shown schematically in Figure B3. At Site S_n located on the present-day Earth of Radius R_n, using the conventional colatitude equation (Equation B2) an inclination I gives a colatitude P which equates to the ancient colatitude locked in the rock record from its original location at site S_a at radius R_a.

At site S_a the surface distance D_a from the sample site to the ancient pole is equal to:

$$D_a = R_a P$$

Where P is in radians. Rearranging:

$$P = (D_a/R_a)$$

To determine the ancient pole position P_a located on the present radius Earth, which equates to the ancient palaeopole position P_a determined from site S_a, colatitude:

$$P = (D_a/R_0) \times 180/\pi = (R_a P_a \times (\pi/180)/R_0) \times 180/\pi = R_a P_a /R_0$$

Where P is in degrees

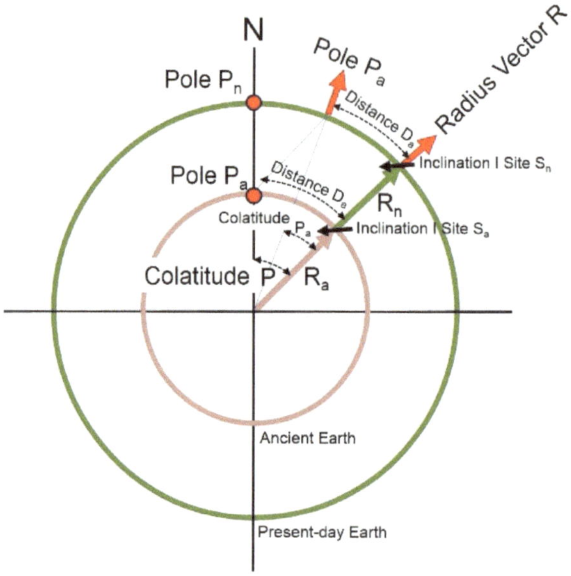

Figure B3. Determining the actual ancient pole position on the present-day Earth. The cross section shows an ancient Earth at radius R_a and a present Earth at present radius R_n.

The mathematical relationship for an exponential increase in the Earth's radius, derived from empirical measurements of seafloor and continental surface area data was previously found to be (Equation A6):

$$R_a = (R_0 - R_p)e^{kt} + R_p \qquad \text{Equation A6}$$

Incorporating Equation A6 for R_a gives:

$$P = ((R_0 - R_p)e^{kt} + R_p)P_a/R_0$$

Incorporating Equation B2 for p_a gives:

$$P = ((R_0 - R_p)e^{kt} + R_p)) (\tan^{-1} (2/\tan I))/R_0)$$

For any site sample, constrained by the age of the rock sequence containing the site data, the ancient latitude from site to the ancient pole position on an Earth of present radius is equal to:

$$P = ((R_0 - R_p)e^{kt} + R_p)) (\tan^{-1} (2/\tan I))/R_0 \qquad \text{Equation B3}$$

The application of this modified dipole equation to palaeomagnetic site data enables the ancient colatitude to be converted to the present geographical grid system. This then correctly locates the ancient pole position on the present Earth. The conventional dipole equation (Equation B2) can only be used to determine the ancient latitude and ancient latitudinal zonation.

The modified dipole Equation B3 can now be used to develop formulae to convert the ancient pole geographical coordinates to present geographical coordinates from ancient site mean data located on the present Earth. This enables the ancient poles to be correctly located on the present Earth, constrained by both time and ancient Earth radius.

Modified Dipole Formulae

Site mean data determined from a set of site samples are assumed by palaeomagneticians to represent a time-averaged field which compensates for any secular variation caused by non-dipole components. For a geocentric axial dipole field (Figure B1) the time-averaged and structurally corrected inclination I, determined from site data establishes the ancient colatitude existing at the site when the site data were locked into the rock-record and a time-average declination D determines the direction along an ancient meridian to the ancient pole. Calculation of the ancient magnetic pole position on the present Earth surface uses spherical trigonometry, based on the dipole Equation B1, to determine the distance travelled from the observing locality to the pole position (Figure B4).

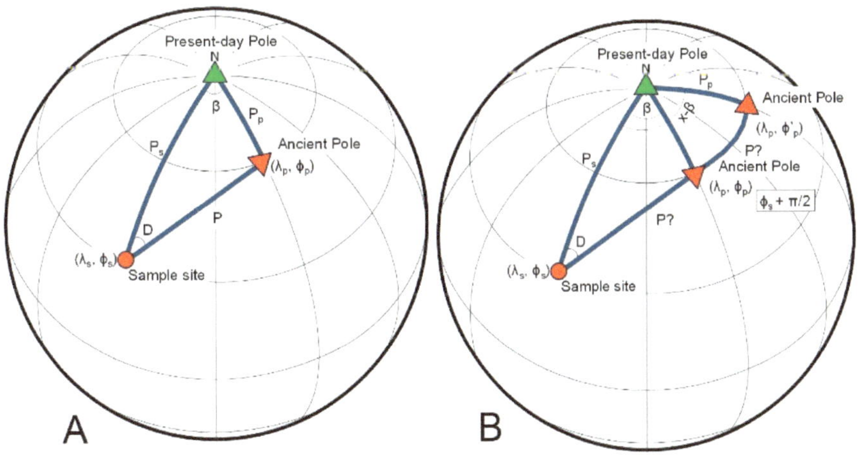

Figure B4 Determination of a magnetic pole from a magnetic field direction, using the palaeomagnetic dipole equation. Orthorhombic projection with latitude and longitude grid in 30° increments. Figure B4a: the site latitude and longitude is (λs, ϕs); the ancient magnetic pole is located at (λp, ϕp); site colatitude is P_s; colatitude of the magnetic pole is P_p; and the longitudinal difference between the magnetic pole and site is β. Figure B4b: illustrates the ambiguity in magnetic pole longitude. The pole may be at either (λp, ϕp) or (λp, ϕ'p; the longitude at ϕs + π/2 is shown by the heavy meridian line. (From Butler, 1992).

Details and derivation of the following conventional palaeomagnetic equations, sign conventions, and symbols for geographic locations, are adopted from Butler (1992):

- Latitude increases from -90° at the south geographic pole to 0° at the equator and +90° at the north geographic pole;
- Longitudes east of the Greenwich meridian are positive, while westerly longitudes are negative and;
- (λp, ϕp) is the pole position calculated from a site-mean direction (I_m, D_m) measured at site location (λs, ϕs).

The pole latitude derived from spherical trigonometry (Figure B4) is:
$\lambda p = \sin^{-1} (\sin\lambda s \cos p + \cos\lambda s \sin p \cos D_m)$ Equation B4

The longitudinal difference between pole and site, denoted by β, is positive towards the east, negative towards the west and is:

$\beta = \sin^{-1} (\sin p \sin D_m / \cos\lambda p)$ Equation B5

Where if:

$\cos p \geq \sin\lambda s \sin\lambda p$ Equation B6

Then the pole longitude:

$\phi p = \phi s + \beta$ Equation B7

But if:

$\cos p \leq \sin\lambda s \sin\lambda p$ Equation B8

Then the pole longitude:

$\phi p = \phi s + 180° - \beta$ Equation B9

For any site-mean direction (I_m, D_m) the associated circular confidence limit ($\alpha 95$) is transformed into an ellipse of confidence about the calculated pole position with semi-axes of angular length given by (Figure B5):

$d_p = \alpha 95 ((1 + 3\cos 2p)/2) = 2\, \alpha 95 (1/ (1 + 3\cos 2\, I_m))$ Equation B10

And:

$d_m = \alpha 95 (\sin p/\cos I_m)$ Equation B11

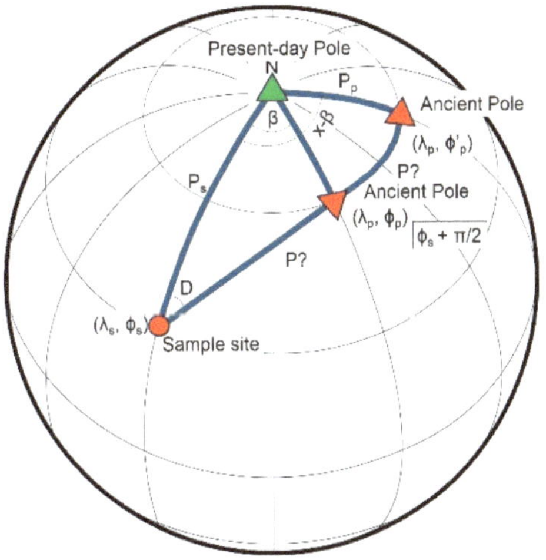

Figure B5 Ellipse of confidence about a magnetic pole position, determined using the palaeomagnetic dipole equation. For a magnetic colatitude p, d_p is the semi-axis of a confidence ellipse along the great-circle path from site to pole; d_m is the semi-axis perpendicular to that great-circle path. Orthorhombic projection, with latitude and longitude grid in 30° increments. (From Butler, 1992)

While Equations B4 to B11 represent the basis for conventional Plate Tectonic determination of pole positions and derived apparent polar wander paths, they do not and cannot acknowledge any variation in angular dimension of ancient geographical grids as a result of a variable ancient radius or time constraint. What is assumed by these equations is that the angular dimension of the ancient geographic co-ordinate system, indicated by the site-data, equals the angular dimension of the geographical co-ordinate system represented by the present site location. By using these conventional ancient magnetic equations to determine the ancient pole position the two systems are simply added, using spherical trigonometry to give an ancient pole latitude and longitude.

To determine actual ancient pole locations on the present Earth each site sample must be qualified by a time constraint to convert the ancient geographic co-ordinate system to the present system. For an Earth undergoing an exponential increase in radius from the Archaean to the Recent, derivation of ancient colatitude from site data is (Equation B3):

$p = ((R_0 - R_p)e^{kt} + R_p)) (\tan^{-1} (2/\tan I))/R_0$ Equation B3

By incorporating the ancient colatitude Equation B3 into the pole coordinate Equations B4 to B11 of Butler (1992), equations that constrain the ancient colatitude arcuate distances to the present geographical co-ordinates are established. These are used to determine the actual ancient pole location on the present Earth surface.

Ancient pole latitude becomes:

$\lambda p = \sin^{-1} (\sin\lambda s \cos[((R_0 - R_p)e^{kt} + R_p)) (\tan^{-1} (2/\tan I))/R0] + \cos\lambda s \sin[((R_0 - R_p)e^{kt} + R_p)) (\tan^{-1} (2/\tan I))/R_0] \cos D_m)$ Equation B12

Longitudinal difference becomes:

$\beta = \sin^{-1} (\sin[((R_0 - R_p)ekt + R_p)) (\tan^{-1} (2/\tan I))/R_0] \sin D_m/\cos\lambda p)$
Equation B13

Where if:

$\cos[((R_0 - R_p)e^{kt} + R_p)) (\tan^{-1} (2/\tan I))/R_0] \geq \sin\lambda s \sin\lambda p$ Equation B14

Then the ancient pole longitude:

$\phi p = \phi s + \beta$ Equation B15

But if:

$\cos[((R_0 - R_p)e^{kt} + R_p)) (\tan^{-1} (2/\tan I))/R_0] \leq \sin\lambda s \sin\lambda p$ Equation B16

Then the ancient pole longitude:

$\phi p = \phi s + 180° - \beta$ Equation B17

And the ellipse of confidence:

$dp = \alpha 95 ((1 + 3\cos 2[((R_0 - R_p)e^{kt} + R_p)) (\tan^{-1} (2/\tan I))/R_0])/2)$
$= 2 \alpha 95 (1/ (1 + 3\cos 2 I_m))$ Equation B18

And:

$dm = \alpha 95 (\sin[((R_0 - R_p)e^{kt} + R_p)) (\tan^{-1} (2/\tan I))/R_0]/\cos I_m)$ Equation B19

Index

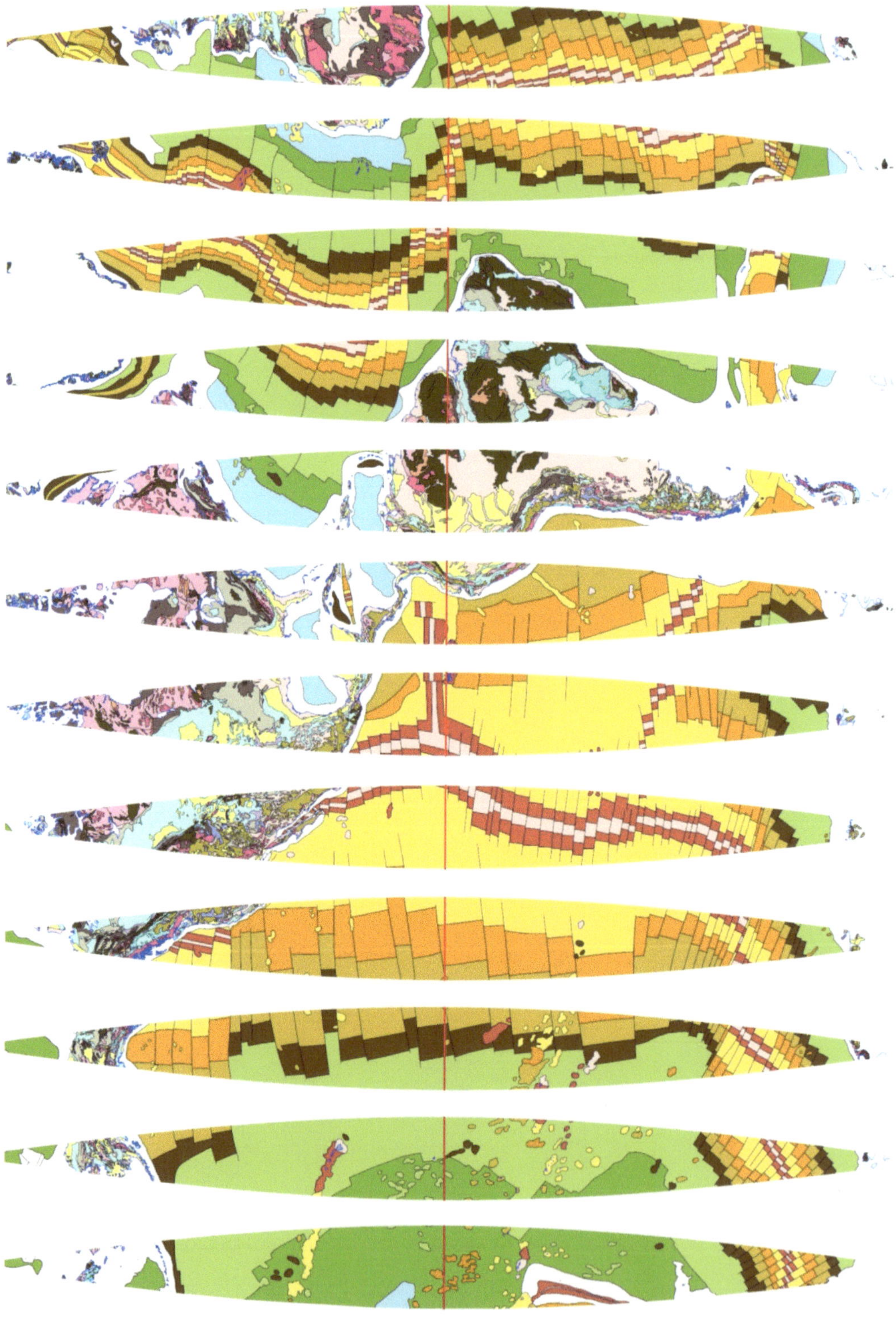

www.ingramcontent.com/pod-product-compliance
Ingram Content Group UK Ltd.
Pitfield, Milton Keynes, MK11 3LW, UK
UKHW061544051026
15625UKWH00012B/613
9780992565206